Recent Trends in
Fuel Cell Science and Technology

Recent Trends in
Fuel Cell Science and Technology

Edited by

Suddhasatwa Basu

Department of Chemical Engineering
Indian Institute of Technology Delhi
New Delhi-110 016, India

A C.I.P. catalogue record for the book is available from the Library of Congress

ISBN 0-387-35537-5 (HB)

Copublished by Springer
233 Spring Street, New York 10013, USA
with Anamaya Publishers, New Delhi, India

Sold and distributed in North, Central and South America by Springer,
233 Spring Street, New York, USA

In all other countries, except India, sold and distributed by Springer,
P.O. Box 322, 3300 AH Dordrecht, The Netherlands

In India, sold and distributed by Anamaya Publishers
F-154/2, Lado Sarai, New Delhi-110 030, India

springeronline.com

Printed in India.

In loving memory of my parents

Professors N.M. Bose and N. Bose

who nurtured and inspired me

Preface

Fuel cell science and technology is evolving fast for the past two decades as it is thought to be an efficient way of transforming chemical energy of hydrogen rich compounds to electrical energy. Although this idea of direct conversion of chemical energy to electrical energy was first demonstrated by Sir William Grove in 1839 using a fuel cell, it was only in the middle of the twentieth century when Bacon's pioneering work led to the use of fuel cell in space missions. The interest in commercialization of fuel cell for civilian use has caught up with government organizations and private corporations for the past decade on account of fluctuating oil prices and environmental concerns. It is well known that the conventional fossil fuel, which is a primary source of gasoline, is not going to last more than a hundred years in the face of ever-increasing demand in the developed and developing countries. Although the reserves of natural gas, coal and tar sands may last another two to three hundred years with the current rate of production, their conversion is not efficient and pollution-free. Thus, scientists all over the world have taken up fuel cell development work in their quest of solution to the energy crises looming largely on global population. This book aims to script the present status of the rapidly developing field of fuel cell science and technology. Since fuel cell (FC) science and technology is multidisciplinary in nature, contributions from the world experts in different areas of fuel cell technology are brought under one umbrella. There are different types of fuel cells, which work on different principles on the basis of different electro-reactions, temperatures, electrodes, electrolytes and fuels. Thus, instead of a single authored book, it is more appropriate to present the work carried out by various experts in the abovementioned areas. The reader should note that FC technology is fast developing towards commercialization and it is not possible to provide crucial details of the patented technology. However, this book provides sufficient information on FC technology so that new researchers from similar areas and those readers who are working in FC technology would be able to take up problems in the area of industry needs.

Introduction to fuel cells and the common topics related to fuel cells, e.g. electro-analytical techniques and power conditioning, are included in the beginning and end of the book in their logical sequence. In-between, chapters describe the state-of-the-art of different types of fuel cell systems. Chapter 1 describes introduction to fuel cell technology and different types of fuel cells. Chapter 2 on electro-analytical techniques in fuel cell research and development deals with evaluation of kinetics of electro-catalytic reaction in half-cell, electro-analytical tools for single cell and stack design. Polymer electrolyte membrane fuel cell (PEMFC) has a promise of providing higher efficiency of the drive system and is regarded as the future motive power source for the transport sector. Thus, special emphasis is given on PEMFC. Chapters 3 to 5 discuss, elaborately, the latest developments in PEMFC, gas diffusion layer and water management in PEMFC. There are certain disadvantages associated with hydrogen such as large-scale and economical production of hydrogen from fossil fuel or other renewal route, emission of polluting gases as by-products during production of hydrogen from fossil fuel, dispensing and storage problems of hydrogen and safety issues. On the other hand, alcohols are produced from renewal sources, easy to handle, store and dispense. Use of alcohols directly into the fuel cell as fuel has been investigated for the past ten years. In Chapters 6 and 7, micro fuel cells, which are based on direct alcohol PEM fuel cell and direct alcohol alkaline fuel cell are presented. High temperature fuel cells are covered in Chapters 8 to 11 in the order of operating temperature. Phosphoric acid, molten carbonate fuel cells and direct conversion of coal in fuel cell are elaborately discussed in Chapters 8 to 10. Chapter 11 describes the principles, designs and state-of-the-art of solid oxide fuel cells (SOFC). SOFCs, operated in the range of 700-1000°C with an efficiency of 60-80%, have a tremendous potential in the future as stationary power source in the order of kilowatt to megawatt range. Since it is operated at high temperature, material issues related to SOFC are discussed in Chapter 12. Chapter 13 covers the power conditioner system for fuel cell. Finally, future directions and challenges of fuel cell science and technology are presented in Chapter 14.

SUDDHASATWA BASU

Acknowledgements

In writing this book, I was inspired by memories of working with my teachers in the area of Interfacial and Electrochemical Engineering and Fuel Cells. Generous funding and the platform provided by Indian Institute of Technology (IIT) Delhi and Ministry of Non-conventional Energy Sources, Government of India, has drawn me into the research and development of fuel cell technology. While covering the subject of fuel cell technology during teaching of electrokinetic transport course to the graduate students of IIT Delhi and in course of discussion with my research students, I was motivated to write a book on fuel cells. I have mentioned in the Preface why I have chosen to bring out an edited book whereby contributions from world experts in different areas of fuel cell technology are sought. I owe my thanks to all the contributors for sharing valuable state-of-the-art knowledge and experience on different types of fuel cells and associated topics. Reviewing chapters was not an easy task as they dealt with interdisciplinary fields of sciences and technologies. The objective of the book is fulfilled through patient and careful reading by myself and further revision carried out by the respective authors. Several occasions and informal discussions held with Dr. T.K. Roy, CMDC Ltd. and Dr. V.V. Krishnan, IIT Delhi were helpful in writing the last chapter on future directions of fuel cell science and technology. Encouraging discussions with my research students, Anil Verma, Amit K. Jha, Krishna V. Singh and Hiralal Pramanik brought confidence in believing in future potential of fuel cell technology.

Finally, without the support of my wife and son, from whom I took away important family time, this book would not have been published in its present form.

SUDDHASATWA BASU

Contents

Recent Trends in Fuel Cell Science and Technology
Edited by S. Basu
Anamaya Publishers, New Delhi, India

1. Introduction to Fuel Cells

R.K. Shah

Subros Ltd., Noida-201304, India
E-mail: rkshah@gmail-com[†]

1. Introduction

A fuel cell is an electrochemical device (a galvanic cell) which converts free energy of a chemical reaction into electrical energy (electricity); byproducts are heat and water/steam if hydrogen and air are the reactants; in some fuel cell types, the additional byproducts may be carbon dioxide and leftover lower forms of hydrocarbons depending on the fossil fuels used. There is no combustion in this process and hence no NO_x are generated. Sulfur is poison to all fuel cells so it must be removed from any fuel before feeding to any fuel cell type; hence, no SO_x are generated. A fuel cell produces electricity on demand continuously as long as the fuel and oxidant are supplied. For reference, primary cell or battery is also an electrochemical energy producing device (one-way chemical reaction producing electricity) and needs to throw away once the battery is discharged. A rechargeable or secondary battery is an electrochemical energy storage device having reversible chemical reaction producing or using electricity, but it also has a limited life.

The components of a fuel cell are anode, anodic catalyst layer, electrolyte, cathodic catalyst layer, cathode, bipolar plates/interconnects and sometimes gaskets for sealing/preventing leakage of gases between anode and cathode. The stack of such fuel cells (a repeated stack of such components) is connected in series/parallel connections to yield the desired voltage and current. The anode and cathode consist of porous gas diffusion layers, usually made of highly electron conductivity materials (and having zero proton conductivity theoretically) such as porous graphite thin layers. One of the most common catalysts is platinum for low temperature fuel cells and nickel for high temperature fuel cells, and other materials depending on the fuel cell type. The electrolyte is made of such material that it provides high proton conductivity and *theoretically* zero electron conductivity. The charge carriers (from the anode to the cathode or vice versa) are different depending on the type of the fuel cells. Some details are presented in Table 1. The bipolar plates (or interconnects) collect the electrical current as well as distribute and separate reactive gases in the fuel cell stack.

The anode reaction in fuel cells is either direct oxidation of hydrogen, or methanol or indirect oxidation via a reforming step for hydrocarbon fuels. The cathode reaction is oxygen reduction from air in most fuel cells. For hydrogen/oxygen (air) fuel cells, the overall reaction is

$$H_2 + \tfrac{1}{2} O_2 \rightarrow H_2O \quad \text{with} \quad \Delta G = -237 \text{ kJ/mol} \tag{1}$$

where ΔG is the change in Gibbs free energy of formation. The product of this reaction is water released at cathode or anode depending on the type of the fuel cell. The theoretical voltage E^0 for an ideal H_2/O_2 fuel

---

[†]Formerly at Department of Mechanical Engineering, Rochester Institute of Technology, Rochester, NY 14623, USA.

cell at standard conditions of 25°C and 1 atmosphere pressure is 1.23 V. The typical operating voltage is about 0.6–0.7 V for high performance fuel cells.

Some general information for fuel cells: Cell voltage is 0.6–0.7 volts in high performance fuel cells and may be 0.2–0.4 V for DMFC. Stack voltage depends on the number of cells in a stack and their series/parallel connection. Cell current depends on the area (the size) of a cell. Energy density is the amount of energy stored in a fuel cell per unit volume. Power is voltage times current (VI) and also energy per unit time. Specific power is defined as power per unit mass. Cell power density is power per unit volume of the cell. A fuel cell for portable devices (laptop) needs low power density (few W) but needs high energy density so that one can run the laptop for a week for example. A car during high acceleration requires high power density. A car also needs high energy density so that one can drive the car with full fuel charge for about 400–500 km.

Table 1.   Some characteristics of important fuel cells

	PEMFC	DMFC	AFC	PAFC	MCFC	SOFC
Primary applications	Automotive and stationary power	Portable power	Space vehicles and drinking water	Stationary power	Stationary power	Vehicle auxiliary power
Electrolyte	Polymer (plastic) membrane	Polymer (plastic) membrane	Concentrated (30–50%) KOH in H_2O	Concentrated 100% phosphoric acid	Molten Carbonate retained in a ceramic matrix of $LiAlO_2$	Yttrium-stabilized Zirkondioxide
Operating temperature range	50–100°C	0–60°C	50–200°C	150–220°C	600–700°C	700–1000°C
Charge carrier	H^+	H^+	OH^-	H^+	$CO_3^=$	$O^=$
Prime cell components	Carbon-based	Carbon-based	Carbon-based	Graphite-based	Stainless steel	Ceramic
Catalyst	Platinum	Pt-Pt/Ru	Platinum	Platinum	Nickel	Perovskites
Primary fuel	H_2	Methanol	H_2	H_2	H_2, CO, CH_4	H_2, CO
Start-up time	Sec-min	Sec-min		Hours	Hours	Hours
Power density (kW/m^3)	3.8–6.5	~0.6	~1	0.8–1.9	1.5–2.6	0.1–1.5
Combined cycle fuel cell efficiency	50–60%	30–40% (no combined cycle)	50–60%	55%	55–65%	55–65%

The major types of fuel cells being developed/used are: proton exchange membrane fuel cells (PEMFC) for transportation power generation, direct methanol fuel cells (DMFC) for portable power generation, alkaline fuel cells (AFC) for space program for producing electricity and drinking water for astronaughts; phosphoric acid fuel cells (PAFC), molten carbonate fuel cells (MCFC) and solid oxide fuel cells (SOFC) for stationary power generation applications. Many other fuel cells are also being developed as well as some of the foregoing fuel cells are also used for other applications than those mentioned.

Characteristic of fuel cell systems is generally high efficiency since it is not limited by Carnot efficiency. Efficiency can be very high (up to 55–65%) for the fuel cells with a combined cycle and/or cogeneration compared to the system efficiency of current power generation of up to about 40–45%. Fuel cell power can reduce costly transmission lines and transmission losses for a distributed system. No moving parts in fuel cells and a very few moving parts in the fuel cell system so that it has higher reliability compared to an internal combustion or gas turbine power plant.

Fuel cell power plant emissions are at least 10 times lower than the most stringent early 2000 California emissions standards. The fuel cell power plant has water as a byproduct so that it requires very low water usage, if any, compared to the steam power plants. Also the water discharged from fuel cell is clean and does not require any pretreatment. The fuel cell power plant produces very little noise compared to conventional steam or gas turbine power plant. Noise is generated only from the fan/compressor used for pumping/pressurizing the cathode air. No ash or large volume wastes are produced from fuel cell operation. However, the fuel cell power plant produces CO_2 emissions if fossil fuels are used for generating hydrogen. As mentioned earlier there are no NO_x or SO_x emissions from fuel ells.

Unique operating characteristics of fuel cell power plants are as follows. Beneficial operating characteristics of fuel cells saves cost and other benefits include load following, power factor correction, quick response to generating unit outages, control of distribution line voltage and quality control; can control real and reactive power independently; control of power factor, line voltage and frequency can minimize transmission losses, reduce requirement for reserve capacity and auxiliary electric equipment; fuel cells have an excellent part load heat rate and can respond to transmission loads.

Fuel cell power plant provides good planning flexibility. Fuel cell performance is independent of the power plant size. In general, the efficiency does not deteriorate going from MW to kW to W size power plants. Fuel cells generally can meet the electric demand as needed without the cost of overcapacity or undercapacity. The unit can slow down or accelerate its response to the growth. Distributed power supply to consumers in any capacity can benefit in many ways for cost, reliability, high efficiency, meeting load demand, etc.

For automotive and stationary power generation applications, the fuel cells have very low emissions, high efficiency, quiet if no bottoming cycle is used, and gradual shift from fossil fuels to other fuels. However, the most challenges at present are: high cost and packaging, low performance and durability, hydrogen infrastructure, and availability of suitable fuels. References [1–4] provide more detailed information on the areas covered in this chapter.

2. Electrochemistry and Thermodynamics

The electrical work done in any system is represented by the Gibbs free energy—energy available to do external work, neglecting any work done by changes in pressure and/or volume. In a fuel cell, the external work involves moving electrons around an external circuit.

When we analyze any electrochemical reaction, there will be a change in the Gibbs energy of formation due to energy release. We evaluate it with the difference in the free energy of the products minus that of the reactants or input. The theoretical/ideal electrical potential E^0 for hydrogen/oxygen fuel cell is 1.23 V at 25°C and 1 atm pressure. The theoretical electrical potential E for a fuel cell with operating pressure and temperature different from 25°C and 1 atm pressure for hydrogen and oxygen is given by Nernst equation as follows:

$$E = E^0 + \frac{RT}{2F} \ln\left(\frac{P_{H_2} P_{O_2}^{0.5}}{P_{H_2O}} \right) \qquad (2)$$

where P's are the operating partial pressures of appropriate reactants and products in atmospheric units, T

the temperature in K, R the gas constant for a particular gas in J/kg K and F is the Faraday constant in C/g mole. The theoretical operating voltage E increases with increasing partial pressures of hydrogen and oxygen, and increasing concentration of oxygen or the system pressure, and decreases with increasing fuel and oxygen utilization (reducing P) and increasing operating temperature.

The actual useful voltage V obtained from a fuel cell with the load is different from the theoretical/ideal voltage E from thermodynamics. This is due to losses associated with the operation, fuel cell materials used, and the design. These losses are: ohmic ir, activation $A \ln(i/i_0)$, fuel crossover and internal current leakage $A \ln(i_n/i_0)$, and mass transport or concentration losses $m \exp(ni)$ (Larminie and Dicks, 2003):

$$V = E - ir - A \ln\left(\frac{i + i_n}{i_0}\right) + m \exp(ni) \tag{3}$$

where i is the current density, A/cm^2; r the electric resistance per unit area, Ω/cm^2, A the coefficient in natural logarithm form of Tafel equation, V; i_n the fuel crossover current density, Ω/cm^2; i_0 the exchange current density at an electrode/electrolyte interface, Ω/cm^2; m is a constant, V; and n the constant, cm^2/A. Usually the desired actual/operating cell voltage is about 0.6 to 0.7 V, although the methanol fuel cell may have one half or lower of that voltage. The typical voltage V versus current density i for a PEMFC and SOFC are shown in Fig. 1 (a, b). We can see that for the low temperature fuel cell (PEMFC), all losses are important while for the high temperature fuel cell, only the ohmic and concentration losses are important.

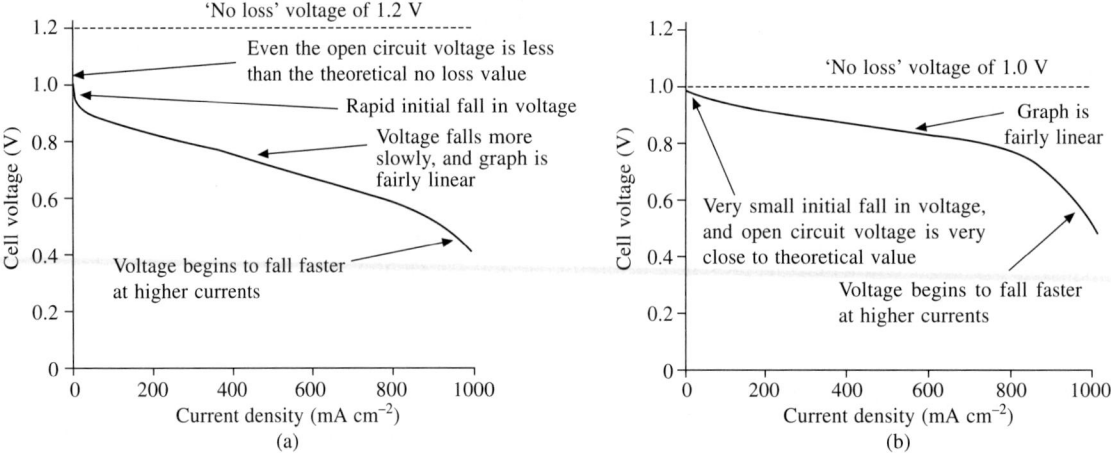

Fig. 1 Voltage versus current density curve for a typical: (a) low temperature and (b) high temperature fuel cells (from Larminie and Dicks, 2003).

There are many fuel cell efficiencies depending on the fuel cell type, fuel utilization and the fuel type. The overall fuel cell efficiency for a specified fuel cell stack is then

$$\varepsilon_{fc} = \varepsilon_r^{cell}\varepsilon_v\varepsilon_F\mu_F\varepsilon_H \tag{4}$$

where ε_r^{cell} is the thermodynamic fuel cell efficiency, defined as the ratio of the electric energy produced divided by the enthalpy change of electrochemical reaction or energy released in burning the fuel; ε_v the electrochemical efficiency, defined as the ratio of the actual cell voltage V to theoretical maximum voltage

E; ε_F the Faradaic efficiency, defined as the ratio of experimental (actual) current I_{exp} to the maximum possible current I_{max}; μ_F is the fuel utilization coefficient, defined as the actual fuel reacted to the fuel supplied to the fuel cell on a mass basis; and ε_H the hydrocarbon efficiency, defined as the ratio of the heating value of all fuel components that are converted electrochemically (e.g., H_2) to the heating value of the fuel supplied (e.g., natural gas). Note that not all of the above fuel cell efficiencies are important for every fuel cell type to be discussed in this chapter. Note also that the total fuel cell efficiency is still 40–50% with all these factors for the fuel cells considered in this chapter except for DMFC. It is higher than any fossil fuel power generation by steam or gas turbine including microturbines.

3. Proton Exchange Membrane Fuel Cells (PEMFC)

The PEMFC, also referred to as the solid polymer fuel cell, is derived from the special "plastic" membrane used as an electrolyte. The components of a single cell are: an electron conducting anode (a porous gas diffusion layer as an electrode and an anodic catalyst layer), a proton conducting electrolyte (hydrated solid membrane), an electron conducting cathode (a cathodic catalyst layer and a porous gas diffusion layer as an electrode), and current collectors with the reactant (gas) flow fields. Current collectors are bipolar plates in a stack; they contain over 90% of the volume and 80% of the mass of a fuel cell stack. The bipolar plate is the most expensive part of the fuel cell. Platinum or platinum alloys in nanometer size particles are the electrocatalysts used with Nafion membranes. The anode-electrolyte-cathode assembly is referred to as Membrane Electrode Assemblies (MEAs), only a few hundred micron thickness; it is the heart of PEMFC. If heat generated in the fuel cell due to exothermic reaction is large, cooling passages are provided by a central channel in each bipolar plate.

A stack of cells are connected in series, and cell stacks (modules) connected in series and parallel to obtain the desired current and voltage. Sources of pressurized air and CO free (very essential to have CO less than 10 ppm) hydrogen gas are required for generating the desired electric power. The cell voltage at the design point is around 0.7 V and power densities of up to 1 W/cm^2 of electrode area when supplied with hydrogen and air. Oxygen reduction is more complex and results in significant overpotential at the cathode. The PEMFC relies on the presence of liquid water to conduct protons effectively through the membrane, and moisturization of the membrane limits the operating temperature of the PEMFC. Systems for thermal management in the cells and water management in the MEAs are essential for efficient operation of the PEMFC. Use of pure oxygen instead of air results in about 30% performance improvement. The PEMFCs are used in applications that require small, medium and large electric power generation.

4. Direct Methanol Fuel Cells (DMFC)

The DMFC is a special form of low-temperature fuel cell based on the PEM technology. It produces power by direct conversion of liquid methanol (no need of hydrogen) to hydrogen ions on the anode side of the fuel cell. The DMFC has all fuel cell components (anode, cathode, membrane and catalysts) the same as those of a PEMFC. The system is simpler to use and easy to refill due to methanol in liquid form. The main difference between the DMFC and PEMFC is the sluggish reaction (significant activation overvoltage) at the anode and hence requires: (1) about the same amount of catalyst on the anode as on the cathode of the DMFC; (2) high surface area 50:50% Pt/Ru (more expensive bimetal) alloy as the anode catalyst to overcome the sluggish reaction; and (3) the catalyst loading 10 times higher than that for the hydrogen/air PEMFC. And still the DMFC results in the useful voltages only 0.2–0.4 V. Hence, DMFC can compete in the marketplace where higher costs are sustainable. There are a number of advantages for the methanol fuel cell: liquid fuel, high energy density, no reforming needed, no problem of membrane humidification, quick start-up, etc. There are several major drawbacks of this fuel cell: low power density due to poor kinetics of the anode reaction, significant fuel crossover, safety concerns, etc. Flooding by methanol/water mixture is required in DMFC, and CO_2 must be expelled quickly. In hydrogen PEMFC, flooding is highly undesirable and gas has to be drawn in. Currently, the immediate applications of DMFC are where low power but higher energy density is required, particularly for the applications where lithium-ion batteries are used for portable power applications.

5. Alkaline Electrolyte Fuel Cells (AFC)

As the name indicates, the electrolyte is an alkaline solution for AFC (OH^- ion moving across the electrolyte). There are three types of AFCs: mobile electrolyte, static electrolyte and dissolved fuel. The first two are being used in the space program.

The AFCs have the major advantages of lower activation overpotential at cathode, typical high operating voltage (0.875 V), inexpensive electrolyte material; electrodes can be made from non-precious metals and no need for bipolar plates. Also, the water management problem is simpler than the PEMFC. The major disadvantages of the AFCs are: low power density, and CO_2 is poison to the fuel cell. The first real application of fuel cells started with the space program in late 1950s with the AFC and they continue till today for power generation in the space program. The mobile electrolyte system was used in the first AFCs in 1940s and is used in terrestrial systems. The Shuttle Orbiter uses a static electrolyte system.

6. Phosphoric Acid Fuel Cells (PAFC)

This fuel cell operates at 200°C, is well developed, and is commercially available. The PAFC, like PEMFC, uses gas diffusion electrodes. Platinum or platinum alloys are used as the catalyst at both electrodes. The carbon is bonded with PTFE (about 30–50% weight) to form an electrode support structure. Electrolyte is a matrix 0.1–0.2 mm thick made up of silicon carbide particles held together with a small amount of PTFE. The pores of the matrix contain 100% phosphoric acid by a capillary action. Bipolar plates are multi-component in which a thin impervious carbon plate serves to separate the reactant gases in the adjacent cells in the stack; separate porous plates with ribbed channels are used for directing gas flow. The stack consists of a repeating arrangement of a ribbed bipolar plate, the anode, electrolyte matrix and cathode. Cooling channels are provided in the bipolar plates to cool the stack. Water cooling is used with 100 kW and larger power generation systems. Since the freezing point of phosphoric acid H_3PO_4 is 42°C, the PAFC must be kept above this temperature once commissioned to avoid the thermal stresses due to freezing and re-thawing. There will be some loss of H_3PO_4 over long periods depending upon the operating conditions. Hence, generally sufficient acid reserve is kept in the matrix at the start. The operating current densities are 150–400 mA/cm^2. The operating cell voltages are 600–800 mV. The ohmic loss in PAFCs is small. Although 300–400 units have been sold for the clean uninterruptive power applications, the market for the PAFC has not picked up due to its high cost.

7. Molten Carbonate Fuel Cells (MCFC)

The MCFC is a high temperature fuel cell operating at 600–650°C. The anode is made of a porous chromium-doped sintered Ni-Cr/Ni-Al alloy. Because of the high temperatures resulting in a fast anode reaction, high surface area is not required on the anode compared with the cathode. Partial flooding of the anode with molten carbonate is acceptable and desirable to act as a reservoir for carbonate. This partial flooding of anode also replenishes carbonate in a stack during prolonged use. The cathode is made up of porous lithiated nickel oxide. Because of the high operating temperatures, no noble catalysts are needed in this fuel cell. Nickel is used on the anode and nickel oxide on the cathode as catalysts. Like in the PAFC, liquid electrolyte is immobilized in a porous matrix in the MCFC. Capillary pressure is responsible for establishing the electrolyte interfacial boundaries in the porous electrodes. Currently, the electrolyte matrix is made using tape-casting methods commonly used in the ceramic and electronics industry. The ceramic matrix has a large effect on the ohmic resistance of the electrolyte. It accounts for 70% of the ohmic losses. Electrode areas of 1 m^2 are now achieved. Bipolar plates or interconnects are made from thin stainless steel sheets with corrugated gas diffusion channels. The anode side of the plate is coated with pure nickel to protect against corrosion. Because of the high temperature operation, CO is no more poison, and acts as a fuel. Note that unlike all other common fuel cells, CO_2 must be supplied at cathode in addition to oxygen for generation of carbonate ions and these ions convert back to CO_2 at the anode reaction. Hence, there is a net transfer of two ions with one molecule of CO_2. Uniqueness of the high temperature fuel cells are: CO is not a poison, internal reforming of natural gas and partially cracked hydrocarbons is possible in the inlet chamber of the fuel cell thus eliminating the separate fuel processing of natural gas or some other fuels. The need for CO_2 makes the digester gas (sewage, animal waste, food processing waste, etc.) an ideal fuel. Other fuels used in MCFC are: natural gas, landfill gas, propane, coal gas and liquid fuels (diesel, methanol, ethanol, LPG, etc.). High operating temperatures results in high temperature exhaust gas which can be utilized for heat recovery for secondary power generation or co-generation, thus improving overall efficiency. MCFC is a well developed fuel cell and is commercially viable technology for stationary power plant compared to other fuel cell types. A good number of MCFC prototype units are operating around the world in power range of 200 kW to 1 MW and higher. However, the cost and useful life issues must be comparable to existing (thermal or other) electric power generation before MCFC use can become widespread.

8. Solid Oxide Fuel Cells (SOFC)

The solid oxide fuel cells operate at temperatures where certain oxidic electrolytes become highly conducting oxygen ions O^{2-}. The oxides used are mixtures of yttria and zirconia first demonstrated by Nernst in 1899. Thus the charge carrier is an oxygen ion and not a proton. Overall cell reaction is the formation of water and standard reversible potentials are the same as for other hydrogen/oxygen fuel cells. The typical operating temperature of the SOFC is 700–1000°C. Very unique tubular SOFC design concept has been developed to avoid the sealing problem for preventing anode/cathode gas leaks at the operating temperature of about 950–1000°C (no seals are available at these temperatures). This fuel cell is very expensive and has low power density. It has been developed and proven the durability requirements; however, due to high cost, it has not been marketable. Planar fuel cells in about 5 kW power generation capacities are being developed now. SOFC is of considerable interest since it has considerably high system efficiency compared to other fuel cell systems with cogeneration as a result of high operating temperatures, and negligible deterioration in performance over several years.

9. Fuel Processing/Reforming

As mentioned earlier, the fuel cell electrochemically oxidizes hydrogen to generate electricity, heat and water/steam. Hydrogen as a gas or liquid is not available in nature. It must be produced from the following sources: (1) fossil fuels (near term source of H_2): natural gas, petroleum (gasoline, diesel, JP-8), and coal and coal gases; (2) bio-fuels (such as produced from biomass, landfill gas, biogas from anaerobic digesters, syngas from gasification of biomass and wastes, and pyrolysis gas; generally they contain mixtures of CH_4, CO_2 and N_2, together with various organic materials); (3) chemical intermediates (methanol, ethanol, NH_3, etc,); and (4) renewable energy sources, such as solar, wind, hydro, geothermal, etc. from which electricity is generated (which is a non-continuous supply) and is used to electrolyze water to generate hydrogen. From the cost consideration, at present, fossil fuels (natural gas) are considered for hydrogen source. Bio-fuels and chemical intermediates have some specific applications. Renewable energy sources are not cost competitive today.

Fuel processing is defined as conversion of the primary fuel (gaseous or liquid hydrocarbons) supplied to a fuel cell system to the fuel cell gas (H_2 and may be CO) supplied to the stack. Each fuel cell type has some particular fuel requirements. The fuel needs to be hydrogen rich and should contain less than 5-10 ppm CO for PEMFC, 0.5% CO for PAFC, and CO acts as a fuel (through water shift gas reaction) for MCFC and SOFC. Natural gas (methane) is also acceptable for SOFC and internal reforming MCFC, but is not acceptable for PAFC and PEMFC. Sulfur must be removed from all fuels to less than 0.5-2 ppm since it is poison to all fuel cells. Major fuel processing techniques are steam reforming, partial oxidation (catalytic and non-catalytic), autothermal reforming, and other techniques such as dry reforming, direct hydrocarbon oxidation and pyrolysis. CO clean up techniques for low temperature fuel cells are preferential oxidation, methanation, separation by membrane. Special reforming techniques for high temperature fuel cells are Direct Internal Reforming (DIR), Indirect Internal Reforming (IIR), and a combination of both DIR and IIR. Considerable advancement has taken place to reform fossil fuels to get hydrogen.

10. Hydrogen Infrastructure, Production, Safety and Storage

Hydrogen infrastructure needs to be developed to produce hydrogen from some of the techniques mentioned in the previous paragraph. Hydrogen molecule is the smallest molecule and hence all joints in a fuel stack system must be very tight or a good number of joints should be eliminated from the system. Hydrogen is a highly volatile and flammable gas, but being very light, it immediately goes up in the atmosphere. Hydrogen safety issue is a very important consideration, although it is not as dangerous as the gasoline.

Hydrogen can be stored as a compressed gas or liquid, or in reversible or alkali metal and chemical hydrides. The advantages of storing hydrogen as a compressed gas are simplicity, indefinite storage time and no purity limits on the hydrogen. Metal hydride compounds must have the following characteristics:

(1) The storage device must have important characteristic to release the hydrogen easily when desired.
(2) The manufacturing process to store hydrogen in metal hydrides must be simple and inexpensive both from the cost and energy usage.
(3) They must be safe to handle. Considerable R&D efforts and developments are going on for hydrogen infrastructure, production, safety and storage.

11. Fuel Cell Balance of Power Plant

In addition to the fuel cell stack system, the rest major systems of fuel cell power plants are:

(1) Fuel reforming system that requires a fuel reformer, chemical reactors, heat exchangers, fans/blowers, burner, etc. Fuel flow rate to the fuel cell could use an ejector system to eliminate the fan for the fuel flow to the stack.
(2) Air management system that requires a compressor, turbine, heat exchangers, fan, motor, water tank, etc.
(3) Power conditioning system that has inverter, converter, batteries, motor, etc.

We need to regulate the electric power output from the fuel cell stack since the voltage generated will not be constant, and will depend upon the operating current density (Fig. 1) and power. Increasing the current will cause the voltage to fall in all electrical power generators, but this drop is much greater in fuel cells. Voltage regulators, DC/DC converters and chopper circuits are used to control the fuel cell voltage to a fixed value, higher or lower than the fuel cell operating voltage. Fuel cell generates the DC power. If we need to connect to the grid lines or use all appliances and other business and household needs, the DC power needs to be converted to AC power using the inverters. Electric motors are used to drive any mechanical system by converting electrical power to the mechanical power. The fuel cell system requirements are as follows: The motors are used to circulate the reactant gases or cooling fluids, operate all the time when the fuel cell is in use, should have the highest efficiency and the longest life, and must not have sparks generated (a common occurrence in a brushed motor) which could be a disaster in the presence of hydrogen. Hence, the fuel cell requires brushless motors such as induction motors, brushless DC motors, and switched reluctance motors. For optimum cost of the system, the fuel cells are designed for maximum steady state power. The operating peaks are then taken care by rechargeable batteries or capacitors. When the fuel cell is not using the maximum power, the excess power is used to charge the batteries or the capacitors. This is the case in many systems, but if the power requirement is quite constant, there is no need for a hybrid system.

12. Concluding Remarks

A brief overview is presented in this chapter for introduction to various fuel cell types and associated issues. For further studies, there are many references available (Hoogers, 2003; Larminie and Dicks, 2003; Vielstich et al, 2003). Fuel cell technology is in the infant stage today, it has to compete with the current technology of power generation at the same or lower cost and the same or better reliability, durability and all other functional requirements. Methanol and other fuel cells will first be introduced commercially for portable power applications by about 2007 since the cost of these fuel cells are comparable with lithium ion and lithium potassium batteries. Fuel cell power plant for stationary power generation application will become acceptable and widespread as the cost of such a power plant becomes competitive with the current thermal/hydro/nuclear power plants. For transportation applications, fuel cells will utilize hydrogen as the fuel for fuel cells, generated at the fueling stations from fossil fuels or transported from the hydrogen generating station utilizing the renewable sources. In addition, infrastructure for hydrogen fueling stations needs to be developed before fuel cell operated vehicles become common. It may become reality by 2015 or later in developed countries. Within next 25–50 years, the fuel cell power plant use will become a reality in portable, stationary and mobile applications using a variety of fuels/methods for hydrogen generation. Eventually, with the renewable energy sources becoming cost effective for hydrogen generation, dependency on the fossil fuels will be diminished for electric power generation and also will eliminate the global warming.

References

G. Hoogers, Editor, Fuel Cell Technology Handbook, CRC Press, Boca Raton, FL, 2003.

J. Larminie and A. Dicks, Fuel Cell Systems Explained, Second Edition, John Wiley, New York, 2003.

R.H. Thring, Editor, Fuel Cells for Automotive Applications, Professional Engineering Publishing, UK, 2004.

W. Vielstich, A. Lamn and H.A. Gasteiger, Editors, Handbook of Fuel Cells: Fundamentals, Technology and Applications, Four Volumes, John Wiley, New York, 2003.

Recent Trends in Fuel Cell Science and Technology
Edited by S. Basu
Anamaya Publishers, New Delhi, India

2. Electro-Analytical Techniques in Fuel Cell Research and Development

Manikandan Ramani

Plug Power, Latham, NY 12110, USA

1. Introduction

Fuel cells are electrochemical devices, which convert chemical energy of the fuel to electrical energy and heat energy. In a combustion engine, fuel is mixed with air in appropriate stoichiometric ratios to initiate a combustion reaction that further creates work. In a combustion reaction, the gas species are not spatially separated, but in proton exchange membrane fuel cell (PEMFC), these combustion reactions are split into two half-cell reactions, namely, fuel oxidation and oxygen reduction, that occurs in two separate chambers called anode and cathode, respectively. If all of the chemical energy can be converted into electrical energy, then we start with the following transfer function:

$$\Delta G = \text{Electrical work} \tag{1}$$

where ΔG (Gibbs free energy) determines the maximum useful work that can be extracted from the system of interest. For a fuel cell reaction with hydrogen as fuel at temperature 25°C, the overall exothermic chemical reaction is given as

$$H_2 + \frac{1}{2} O_2 \leftrightarrow H_2O_{liq} \ (\Delta G = -237.2 \text{ kJ/mole}) \tag{2}$$

The half-cell reactions, i.e. hydrogen oxidation reactions (HOR) and oxygen reduction reactions (ORR) are represented as follows.

Anodic electrochemical pathway can be explained as follows (acidic electrolytes):

$$\frac{1}{2} H_2 + H_2O \rightarrow H_3O^+ + e^- \text{ (Overall) } E \text{ [H}_2/\text{H}^+] = 0 \text{ V (Standard Hydrogen Electrode)} \tag{3}$$

$$H_{2 \text{ (solvated)}} \rightarrow H_{2,\text{adsorbed}} \rightarrow 2H_{\text{adsorbed}} \text{ (Tafe-Volmer mechanism)} \tag{4}$$

$$H_{\text{adsorbed}} + H_2O \rightarrow H_3O^+ + e^- \text{ (Volmer-Heyrovsky mechanism)} \tag{5}$$

While the cathodic pathway of oxygen reduction to water proceeds through a peroxide pathway, the rate quantification of several parallel steps has been investigated and mechanisms have been explained by multi-parallel pathways (Damjanovic et al, 1966). Following is the overall reaction proposed in an acidic medium

$$O_2 + 4H^+ + 4e^- \rightarrow 2H_2O \text{ (Overall) } E \text{ [O}_2/\text{H}_2\text{O]} = 1.23 \text{ V} \tag{6}$$

Electrical work is defined as the work done by the charge Q moving in the electric field across a voltage E. Hence, electrical work done is

$$\text{Electrical work} = Q \times E \tag{7a}$$

$$Q = n \times F \tag{7b}$$

Using (1), we can obtain the relation

$$\Delta G = nFE \tag{8}$$

where n is the number of moles of electrons consumed or generated and F is the Faraday's constant (96487 Coulombs/mole of species).

A unit fuel cell can be constructed using a membrane electrode assembly (MEA) and two bipolar plates as shown in Fig. 1(a). Multiple cells can be stacked up to obtain more power and such a configuration is shown in Fig. 1(b). The charge carriers and their flow direction are identified in Fig. 1(a, b).

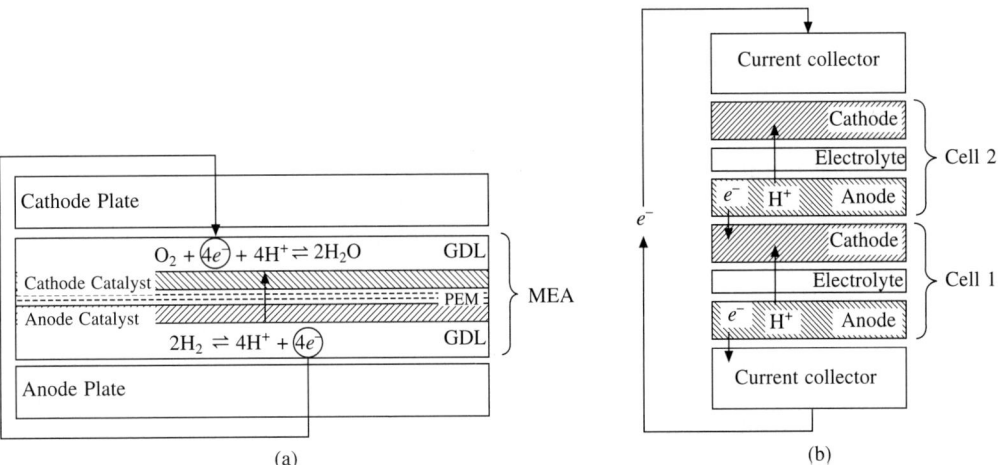

(a)   (b)

Fig. 1 (a) Single unit cell: once electrons are rejected through external circuit by hydrogen oxidation, the reaction shifts in forward direction due to Le Chatlier's principle. GDL represents gas diffusion layer and PEM is proton exchange membrane. (b) Multiple-cell configuration (2-cell): the electrons from cell 2 anode flows into the cell 1 cathode plate enabling series configuration.

2. Fuel Cell Electrodes

Electrodes primarily include electro-catalysts to catalyze a specific reaction, as in PEMFC, hydrogen oxidation reaction (HOR) at the anode or oxygen reduction reaction (ORR) at the cathode surface. These catalysts need to have lower overpotential for the corresponding half-cell reactions in an energy-producing device. An ideal electrode should possess a structure capable of transporting gases to the electrode-electrolyte interface. Moreover, the need for high current densities for practical fuel cells stresses the importance of a porous structure that is capable of handling high turnovers of a reaction per site, with surface areas as high as 100 m^2/g. High surface area supports enable effective dispersion of the electro-catalysts providing more active reaction sites per weight of the catalyst and the unit geometric surface. The primary functions of these electrodes are:

(a) Ability to transport reactants and products through the porous structure.
(b) Capacity to adsorb the reactants and enable charge transfer through electrolytic and electronic continuity throughout the matrix (translates to low overpotentials for a particular electrochemical reaction).
(c) Exhibit low energies of adsorption for products.
(d) Ability to selectively oxidize and reduce reactants.

Table 1 shows the variety of electro-catalysts that are being used in various types of fuel cells.

Table 1. Fuel cell types and electro-catalysts

Fuel cell type	Temperature (°C)	Electro-catalysts
PEMFC/DMFC	50-90	Pt/C, Pt-Ru/C, Pt-Mo/C, Pt/Cr, Pt/Ni
Alkaline fuel cells	50-240	Pt alloys, Ni/NiO, Au/Ag alloys
Phosphoric acid fuel cells	140-220	Pt, Pt alloys
Molten carbonate	600-700	Ni/Lithiated Ni oxides
Solid oxide	700-950	Ni alloys

As shown in Table 1, the electro-catalysts are selective to the type of fuel cell for a variety of applications and the types of electro-catalysts depend on electrolytes and operating temperature. This article focusses on polymer electrolyte membrane (PEM) based applications due to the extensive research in this area over the past decade. These electrodes are designed with the intent of catalyzing the reactions (3) to (6) through multiple steps, multiple pathway reactions.

2.1 Issues/Limitations of the Conventional Approach

Catalyst selection and their application, namely, coating and fabrication techniques have heavily influenced the performance and durability of the catalyst layers. Controlling the amount of metal loading and utilization of these precious metal electrodes were primarily governed by the exploitation of three-phase boundary (reactant-electrode-electrolyte). The variability in process conditions seriously impacts this three-phase boundary. The conventional method (Fuel Cells and Fuel Batteries, 1968) included bonding platinum (Pt) black to the polymer membrane. This active layer comprised a film of Pt and polytetra flouro-ethylene (PTFE) dispersed from an emulsion and this paste was applied to the membrane at an appropriate temperature and pressure. A thick film and high loading (sometimes as high as 3-4 mg/cm^2) was typical in these cases, very much limiting the gas transport. This approach presented a variety of problems that include: (a) low catalyst utilization, hence high Pt loading and (b) inefficient gas permeability.

2.2 Evolutionary Approach: Pt Supported on Carbon

Many researchers further investigated this disadvantage and a Pt on carbon architecture took over and eliminated most of the disadvantages of the conventional architecture. Ticianelli et al (1991) demonstrated and studied low Pt loading using this support structure and also showed good performance for these electrodes. Researchers at Los Alamos National Lab in the 1990's have significantly contributed to the optimization of such a structure, enabling reproducible electrodes and membrane electrode assemblies (MEA) (Wilson and Gottesfeld, 1992). A simple calculation indicating the dispersion and surface area of Pt on a carbon matrix can shed light on the importance of a high surface area support. Given a support surface area of around 1000 m^2/g of carbon, roughly 0.5×10^{15} sites/cm^2 of carbon are available to disperse Pt and obtain as high as 80 m^2/g. A pure Pt black with a similar particle size can yield between 10 and 40 m^2/g. This rationalizes the advantage of a Pt supported on a carbon matrix. Such a structure is elucidated through Fig. 2.

To enable a structure with high performance catalysts, the opted route in development included the following strategies:

1. Selection of catalysts with low overpotential for both anode and cathode. For example, Pt supported on carbon or Pt alloy supported on carbon.
2. Catalyst layer preparation: Catalyst utilization depends on the Pt dispersion, electrolyte availability and fabrication techniques.

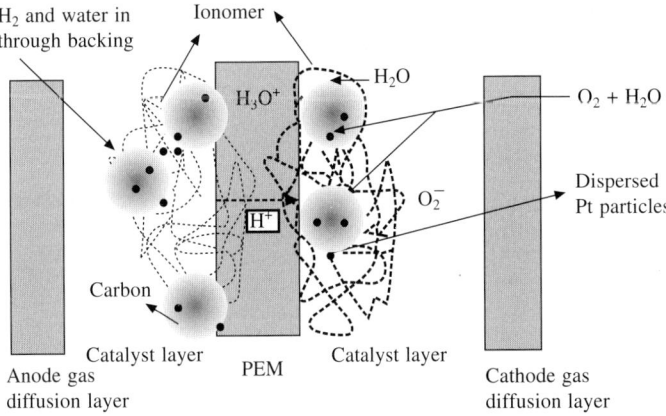

Fig. 2 Membrane and electrode interface in PEM systems (not to scale). Catalyst layer is around 4-20 μm thick, the gas diffusion layer is 150-250 μm thick and PEM is around 25-125 μm thick.

3. Mode of application in the MEA: Bonded on to the membrane or to the backing layer.

Current mental model of catalyst layer evaluation is shown in Fig. 3. Density functional theory (DFT) calculation enables the candidate selection followed by a catalyst layer synthesis and fabrication. Kinetic evaluations are further conducted for a particular half-cell reaction either HOR or ORR.

3. Electrode/Electrolyte Interface Characteristics: PEMFC

3.1 Materials Selection for Catalyst
Pt black or Pt supported on carbon was primarily chosen as the catalysts because of the following reasons:

(a) High exchange current densities of Pt for both anodic and cathodic reactions.
(b) Reasonable tafel slopes on Pt at all potentials.
(c) Ability to oxidize carbon-monoxide (CO) impurity on the anode side (bi-functional electrodes).

Overpotential is a term often used in fuel cells and can be defined as the electrochemical potential above the equilibrium potential where the current flows.

A simplified and naive model of activation overpotential versus current density expression from a Butler-Volmer equation can be written as:

$$\eta_{\text{activation}} = \frac{RT}{\alpha\, nF}\, \ln\left(\frac{i}{i_0}\right) \tag{9}$$

where $\eta_{\text{activation}}$ is the activation overpotential, i_0 the exchange current density, T the temperature, n the number of moles of electrons taking part in the reaction, α the transfer coefficient and $\dfrac{2.303 \times RT}{\alpha nF}$ is the tafel slope for either anodic or cathodic reaction.

Along with the exchange current density and tafel slope, the concentration of the reactant plays a major role when the reaction is not first order. For a Pt site, at a given concentration of available reactants and for a given reaction turnover per site, this overpotential is low compared to other elements from transition

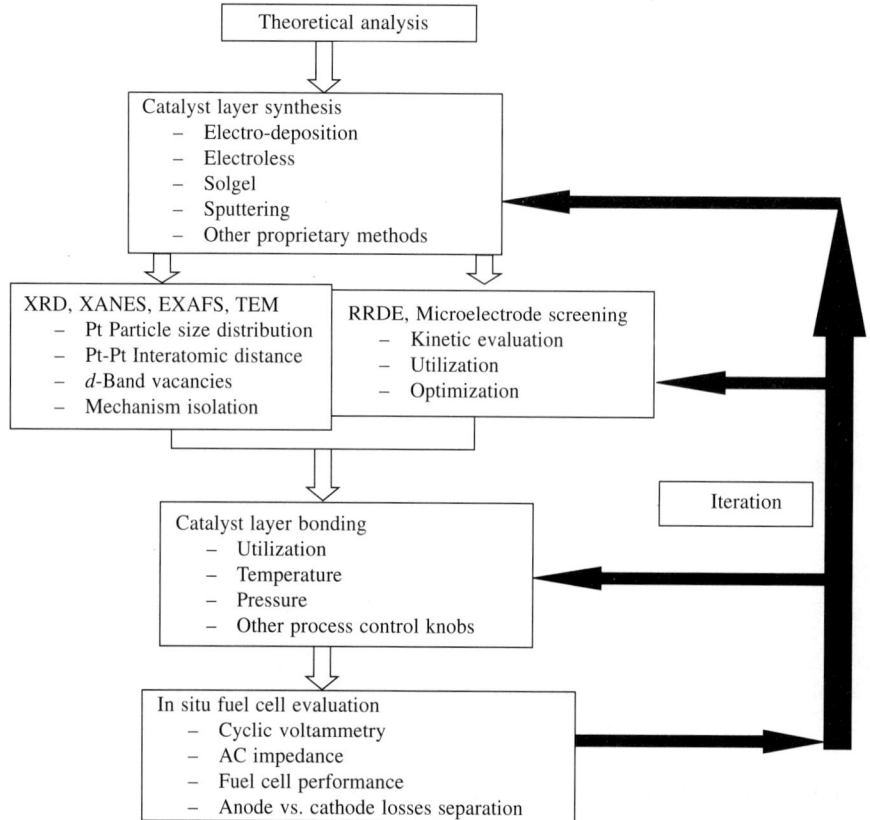

Fig. 3 Catalyst layer evaluation flow-chart. Material selection, process variables significantly influence fuel cell performance.

metal series. The reactant adsorbs onto the catalyst and then the charge transfer initiates and once complete, the product should leave the surface for multiple turnovers of the reaction on a particular site. An activity versus metal-reactant binding energy plot is often called volcano plot and demonstrates the critical parameters for catalyst selection. Fig. 4 shows such a plot for the hydrogen evolution reaction (HER) on different metal surfaces. Norskov et al (2005) calculated hydrogen chemisorption energies for various metals at two different monolayer coverages.

For the cathodic oxygen reduction reaction, the overpotential is high even in the case of Pt, because of the extremely low exchange current densities as low as 1.7×10^{-9} A/cm^2 Pt (Gasteiger et al, 2001). Similar to the hydrogen evolution reaction, the activity of the catalyst is represented as exchange current densities and such values collected from DFT calculations from Norskov et al (2004, 2005) shows the Pt and Pd (Palladium) as optimum candidates for catalysts. The metal-oxygen bond calculations as well as the exchange current densities for ORR are the critical parameters. Pd and Pt seem to exhibit high exchange current densities with appropriate bond energy and are likely to be the natural choice. One would prefer a catalyst with the optimum binding energy that would facilitate charge transfer and at the same time enables dissociation once the charge transfer is complete.

If the fuel is reformed, it might contain impurities like CO and the selected anode catalyst should have the ability to selectively oxidize hydrogen. A careful look at the adsorption isotherms for CO and hydrogen

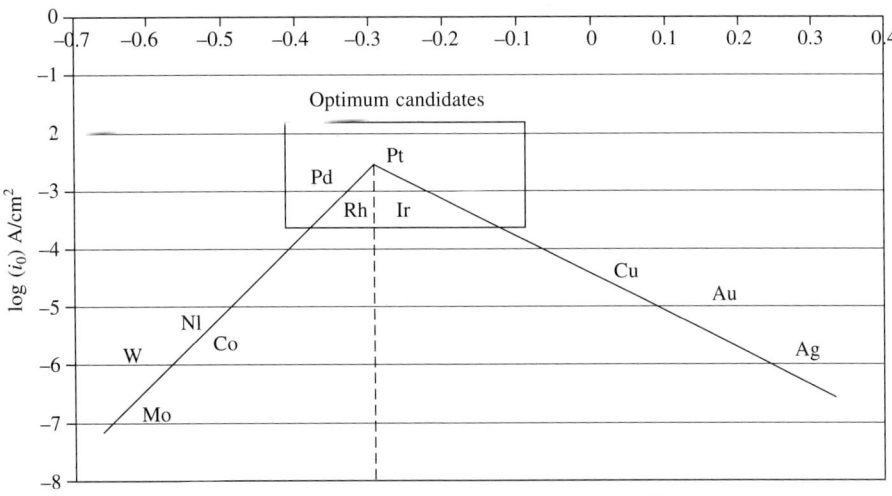

Fig. 4 Experimentally estimated exchange current densities as a function of chemisorption energies for hydrogen evolution reaction by Norskov et al (2005). Optimum candidates are identified based on optimum exchange current density and adsorption energies (reproduced by permission of The Electrochemical Society, Inc.).

at low temperatures indicates a competitive scenario with preferred adsorption of CO over hydrogen onto the Pt sites. This has been dealt with in two ways:

1. Chemical oxidation by bleeding air into the anode chamber to oxidize CO (Uribe et.al, 2002).
2. Electrochemical oxidation (Ru adsorbs OH species at lower overpotential and oxidizes CO).

Ru adsorbs $OH_{adsorbed}$ species around 0.35-0.45 V (vs. SHE) whereas water activation in Pt begins beyond 0.6 V vs. standard hydrogen electrode (SHE). The lower overpotential for water activation in Ru resulted in choosing Pt-Ru alloy with a suitable atomic weight % for each metal. Tungsten (W), Molybdenum (Mo) alloys offered similar bi-functionality with respect to electrochemical oxidation of CO.

3.2 Selection of Support Material (Carbon)
Carbon has been a favorite contender for support because of its ready availability, ease of preparation and low cost. To be a candidate as support, the material should be or have

(a) High surface areas to disperse Pt
(b) High electronic conductivity
(c) Inert
(d) Less expensive and readily available

High surface area carbon supports have been demonstrated and Brunaer Emmett Teller method (BET) determined areas as high as 1500 m^2/g capability and Pt can be dispersed over this very high surface area. Several other approaches including nanotechnology have created avenues to control the dispersion thereby controlling carbon corrosion and the Pt size, which in turn affects the activity of the Pt for the ORR reaction. Pourbaix potential-pH relationships can clearly illustrate the thermodynamic stability of a support material at a specific pH and a potential that the metal experiences due to M/M^{n+} equilibrium across the metal-electrolyte interface.

The overpotential on carbon for both fuel cell reactions at anode and cathode is very high and generally the 4-electron transfer process does not include carbon in its pathway. Having stated that this charge transfer depends on the state of Pt and its d orbital vacancy. The objective is to attach Pt on the carbon and hence retain the relatively similar activity as Pt black for the reactions through Pt participation. This is highly influenced by the surface coverage of Pt with other species namely $OH_{adsorbed}$ and other organic species (Gottesfeld et al, 1988). This implies that if a contaminant or a film covers Pt, the reduction process can prefer a 2-electron pathway instead of a 4-electron pathway through carbon as catalyst. On a cathodic surface, any anodic reaction will generate a galvanic couple, leading to a mixed potential. This will result in reduction in potential and cell voltage.

For CO tolerance when using reformed anode streams, bi-functional alloy catalysts are synthesized using chemical vapor deposition (CVD) and sputtering techniques. Growing monolayers of RuO_x on the Pt improves bi-functionality on the anode side when the fuel is contaminated with CO (Haug et al, 2002; Brankovic et al, 2001). With this architecture of high surface areas and smaller metal loadings, matrix conductivity is a key parameter and cannot be compromised due to its impact on performance and utilization. Typically for a carbon, electrical specific resistivity values ranging from 4×10^{-5}-1×10^{-2} ohm-cm were reported in Kinsohita's work (Carbon, 1992). Addition of ionomer to the catalyst enhances the three-phase interface utilization and appropriate addition is required balancing electrical conductivity and the hydrophobicity of the catalyst layer.

Presence of higher voltage at the cathode electrode surface at low overpotentials presents a mixed voltage scenario with carbon oxidation pathway on the same electrode. At potentials, greater than 0.8 V, the carbon goes through the gas phase oxidation, converting to oxides of carbon, and at potentials less than that in acidic medium, the carbon initially forms a surface layer and this oxide layer grows with time. Though the gas phase oxidation and the formation of surface oxide can occur separately, the potential dependant mechanism proposed, involves the surface oxide formation initially and then at elevated potentials, prefers the gas phase oxidation

$$CO + 2H^+ + 2e^- \leftrightarrow C + H_2O \quad E_0 = 0.518 \text{ V} \tag{10}$$

$$C + O_2 \leftrightarrow CO_2 \quad E_0 = 0.8 \text{ V} \tag{11}$$

Though these corrosive currents are extremely small and indicate very slow kinetics, for a 40,000 hr life requirement for residential systems and frequent start and stopping in automotive systems, corrosion rates as low as 1.5×10^{-8} A/cm^2 pose a significant challenge in the durability of the electro-catalysts. This polarizes the electrode below the Nernst potential for the cathode electrode.

This mixed potential is explained in Fig. 5 through an Evans diagram. In an operating fuel cell, along with this polarization close to open circuit voltage (OCV), there are losses due to hydrogen permeation into cathode electrode from anode chambers in PEMFC and methanol crossover in direct methanol fuel cell (DMFC). In a half-cell system, the crossover losses do not exist, but the polarization due to the carbon oxidation or any other contaminant participating in a side-reaction depresses the OCV.

3.3 MEA: Functional Unit (PEMFC)

A typical catalyst layer preparation for a functional PEMFC cell is illustrated in Fig. 6. There are several other techniques for catalyst synthesis which includes electro- and electroless deposition, sol-gel and sputtering.

Figure 6 explicitly conveys a variety of process control parameters that need to be controlled to attain reproducible results every time. The control parameters that are of particular interest are:

1. Temperature
2. Compaction load and time

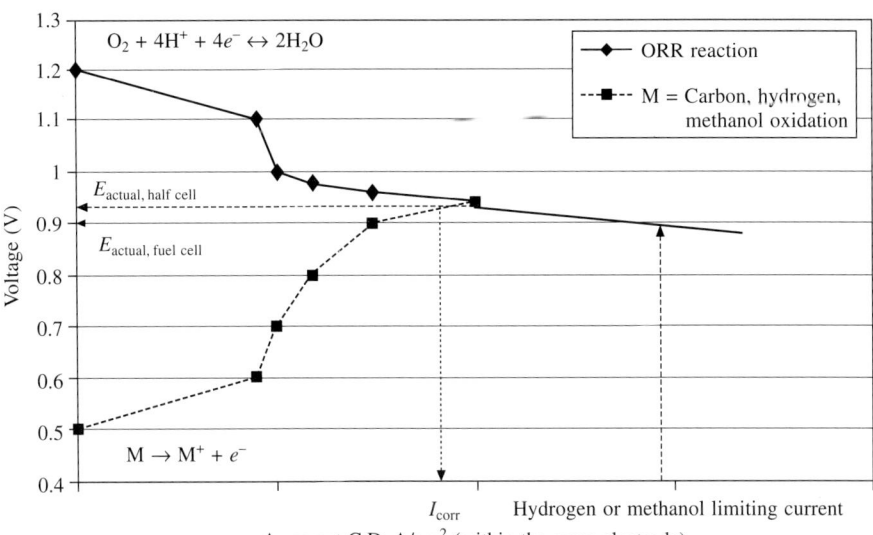

Fig. 5 Evans diagram showing mixed potential at the Pt/C electrode interface on cathode. Several situations identified. An internal current gets generated due to anodic reaction on a cathodic surface and polarizes the cathode electrode (carbon oxidation, hydrogen and methanol oxidation). M is the species participating in the anodic reaction.

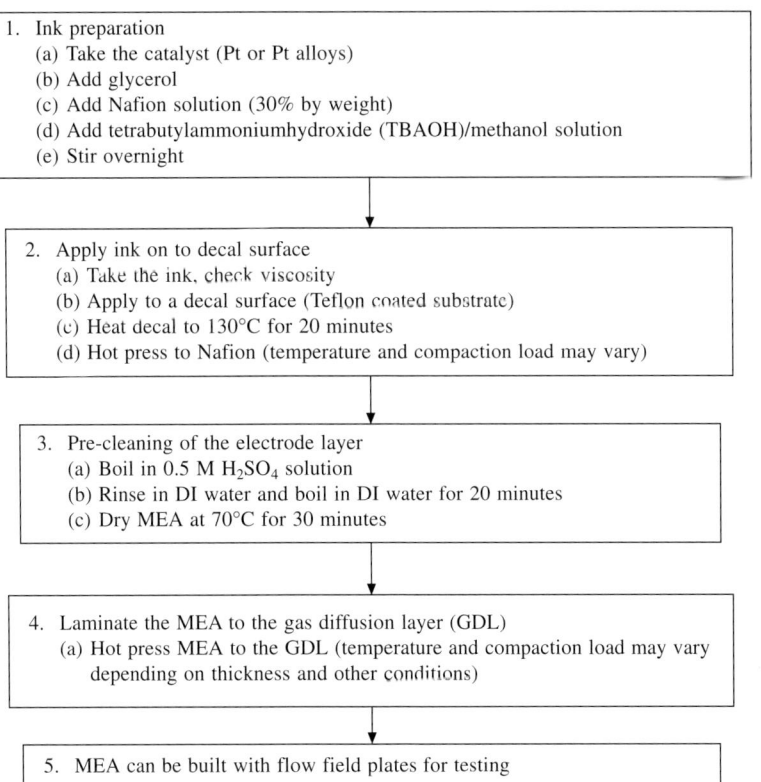

Fig. 6 Typical decal process for electrode preparation.

3. Coating uniformity
4. Nafion content
5. Cleaning requirements
6. Other process design parameters

All these can affect a variety of critical parameters of the membrane electrode assembly (MEA) and include:

1. Protonic conductivity
2. Electronic conductivity
3. Catalyst utilization
4. Porosity and gas permeability

The components of an MEA include a proton exchange membrane (PEM), a catalyst layer with electrode and electrolyte impregnated and a gas diffusion layer (GDL) (see Fig. 7 for MEA input/output diagram).

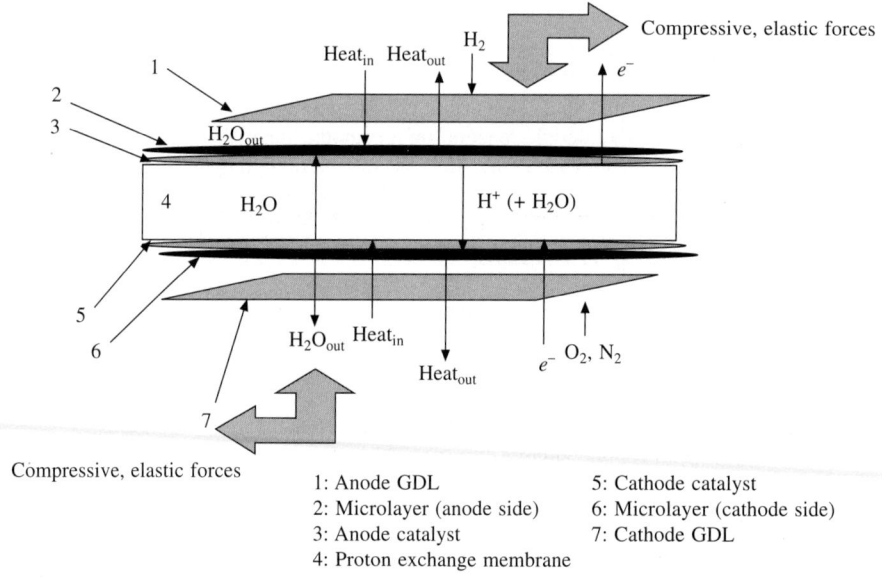

1: Anode GDL   5: Cathode catalyst
2: Microlayer (anode side)   6: Microlayer (cathode side)
3: Anode catalyst   7: Cathode GDL
4: Proton exchange membrane

Fig. 7 Membrane electrode assembly: Components and physical processes occurring at the interfaces.

Functions of the MEA can be classified as follows:
1. Separate reactants (H_2 and O_2)
2. Provide electrical insulation between anode and cathode electrode
3. Transport reactants to electrode
4. Oxidize hydrogen
5. Reduce oxygen
6. Conduct electrons from electrode to external circuit
7. Conduct protons intrinsic between electrode layer and membrane
8. Conduct heat to the flow-fields
9. Remove products

10. Oxidize CO (in reformer systems with low amounts of CO)

All the above functions may or may not be weighted equally, but failure to perform any one of the stated functions, can lead to suboptimal performance. The functions alone are insufficient, since we need to determine appropriate function and its physical model in order to translate these functional requirements into component specifications. The ability to relate performance to the component specifications is critical and appropriate diagnostic tools needs to be used to characterize the functionality of the subcomponents of an MEA.

4. Approaches in Kinetics Evaluation of Electro-Catalysts in Half-Cell Reactions

4.1 Half-Cell Approaches in HOR and ORR

Several experimental techniques have been used in the recent years to understand the half-cell reactions of a fuel cell system. These set-ups have been used to:

1. Understand the ability of an electrode in terms of its utilization
2. Characterize the electrode/electrolyte interface at various temperatures
3. Estimate reactant solubility and diffusion rates across a diffusion length
4. Elucidate the electron-transfer mechanism (quantify side-reactions-current efficiency)
5. Estimate wettability of electrode in desired electrolyte
6. Determine electrode polarization at various current densities
7. Screen various catalysts for variety of fuel cell applications
8. Enables selection of binary, ternary, quaternary alloys and their optimum atomic wt% composition specifically with respect to each of the following:

   (a) CO oxidation
   (b) Methanol oxidation
   (c) ORR gain due to alloying (surface restructuring in $OH_{adsorbed}$ species)

4.2 Rotating Disc Electrode (RDE) Approach

The RDE system is a hydrodynamic approach and came after pitfalls with extensive polarographic studies with dropping mercury electrode (DME). The changing area during DME measurements complicated the analysis with the change in pseudo-steady state current. Using RDE with forced convection, the repeatability in the specific activity measurements and limiting current density measurements improved dramatically and it is quite understandable that natural convection will vary easily due to a variety of noises during experimentation. Fig. 8(a) shows how convection when applied to the solution in the vicinity of the electrode, enables easier mass transport. A typical RDE set-up is shown in Fig. 8(b). This pine instrument set-up is accompanied with a motor controller that can control the angular rotation speed of the disc. The electrode is coated on the disc and is pre-treated before kinetic studies. The electrolyte carrying the reactant species flow near the electrode and then spiral out from the middle to the periphery of the electrode. The electroactive species move from the disc electrode to the ring electrode in a rotating ring disc electrode (RRDE). At varying speeds and different overpotential, one can understand the reaction rate of an electrode for a controlled half-cell. As in most of the electrochemical systems, this is a three-electrode configuration, with reference/non polarizable electrode of choice (SHE, saturated calomel electrode (SCE), Ag/AgCl etc.), a working electrode of intent and a counter electrode to complete the circuit. This technique is very sensitive and pre-cleaning of the electrode and a clean electrolyte is a key step in obtaining reproducible results. We can independently control the electrochemical potential of the disc and ring electrode thereby studying the intermediate species produced at the disc electrode.

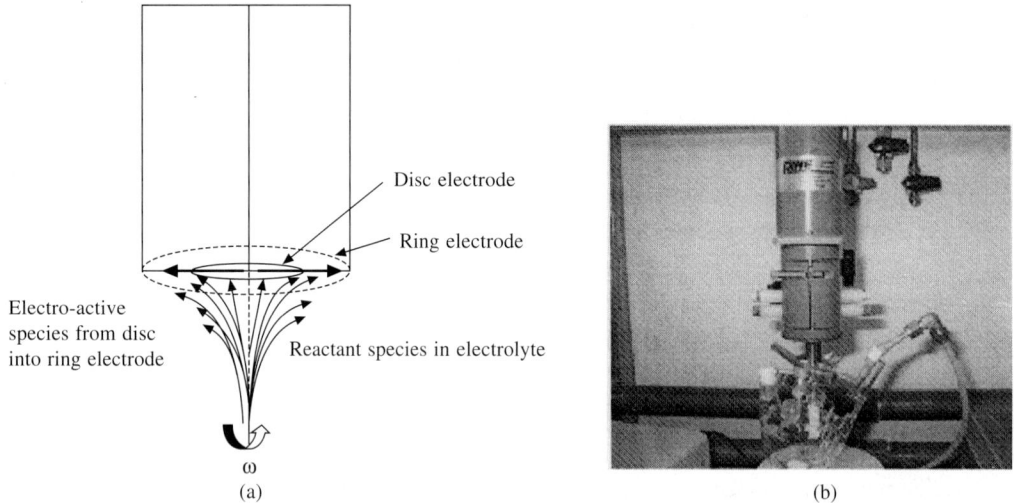

Fig. 8 (a) Rotating ring disc electrode: electrode/solution interface at an angular frequency of ω and (b) RDE set-up (Pine Instruments): the RDE is connected to a motor controller that controls the angular velocity of the disc.

Linear sweep voltammetry (LSV) is often used to study the intrinsic kinetics of the catalyst provided the total current is actually a fraction of the mass transport limited current density. The concentration of the electrolyte can also impact the measurement unless the reaction kinetics is zeroth order in either the soluble species or the electrolyte concentration.

Levich solved the family of equations and provided an empirical relationship between limiting current and rotation speeds. This equation can be used to ascertain solubility and diffusion of oxygen or hydrogen at different temperatures and with different electrolytes. One such outcome is shown in Fig. 9 where the limiting current changes as a function of rotation speed.

$$I_1 = 0.62 \ nFACD^{2/3}v^{-1/6}\omega^{1/2} \tag{12}$$

where n is the number of moles of electrons, F is Faradays constant, A the geometric area (cm^2), C the bulk concentration, D the diffusivity of species, and v and ω represent the scan rate and rotation speed of the electrode, respectively.

Both transmission electron microscopy (TEM) and X-ray diffraction (XRD) have complemented the kinetics evaluation with RDE, through particle size distribution information, to determine Pt particle size effect on the specific activity of Pt and Pt alloys. Fig. 10 shows the change in Pt particle size off-cell aging. The aged catalyst now can be applied in RDE and evaluated for kinetic constants. From XRD in particular, one can use the Scherrer's equation to determine the Pt crystalline size through the following equation:

$$L_{hkl} = \frac{k\lambda}{(B - b)\cos\theta} \tag{13}$$

where k is a constant ($= 0.9$), λ the X-ray wavelength, θ the Brag angle, B the observed half-line width of the peak and b the line correction due to instrumental factor.

A slight modification using rotating ring disc electrodes creates the additional ability of studying the

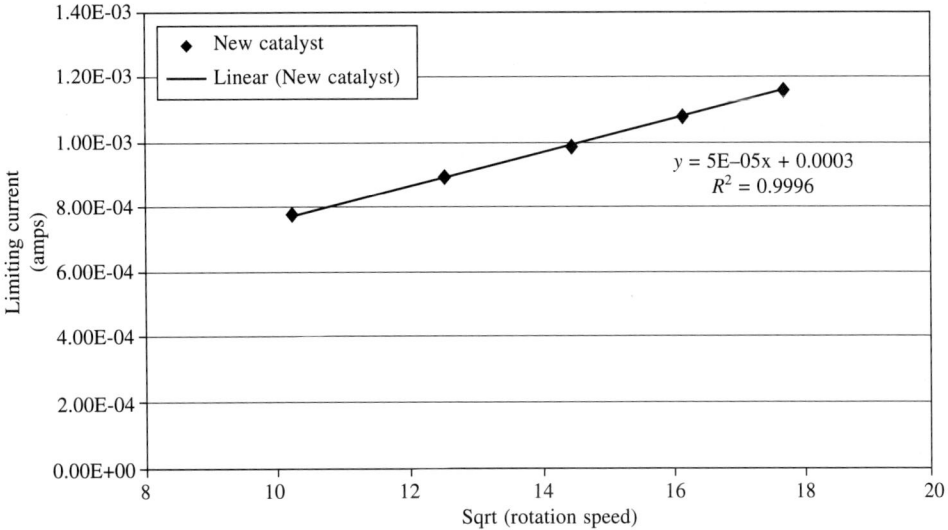

Fig. 9 RDE output during a 2 mV/sec scan at different rotation speeds. Temperature = 25°C, pH = 2, Reactant: Oxygen-Pt/C catalysts were coated with a syringe on a RDE tip and evaluated under T = 25°C in 0.5 M H_2SO_4. Karuppaiah and Tang (Plug Power) extensively used these techniques to characterize performances on various catalysts.

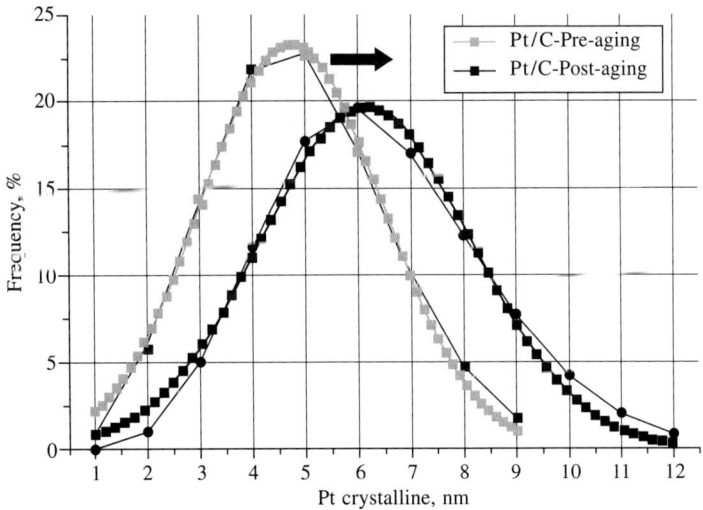

Fig. 10 XRD Pt crystalline size distribution analysis of a catalyst pre and post aging. Study helped characterize the impact of Pt agglomeration on catalyst active area and activity. Experiments performed by Tang (Plug Power) and Wang (GE Corporate Research and Developement), Wei (GE Corporate Research and Developement).

side reactions like the 2-electron pathway through peroxide generation during the 4-electron pathway in oxygen reduction reactions. By controlling the potential of the ring as high as 1.2 V, the peroxide decomposition current can be calculated by the following equations:

$$I_D = I_{water} + I_{peroxide} \tag{14}$$

$$I_{\text{peroxide}} = \frac{I_R}{N} \tag{15}$$

where I_D is the disc current, I_R the ring current, N the collection efficiency and I_{water} and I_{peroxide} are the currents generated while generating the respective end species. From the results of linear voltammetric studies at RDE (Paulus et al, 2001), we may extrapolate that the anode chamber can produce as high as 5-6% peroxide at potentials less than 0.1 V, which will result in membrane decomposition.

Several researchers have used micro-electrode studies through RDE set-up to investigate the exchange current densities of the ORR at humidity-controlled environment (Mitsushima et al, 2002; Uribe et al, 1992).

Other techniques that are commonly applied while using RDE system are electrochemical impedance spectroscopy (EIS) and cyclic voltammetry (CV). Using EIS, one can extrapolate the exchange current density of an electrode. This is achieved by completing the EIS spectrum close to the rest potential of the electrode. Fig. 11 shows the EIS curves developed at two overpotentials and the low frequency intercept at close to rest potential or equilibrium potential helps quantify the exchange current density. Let us apply simple mathematical treatment to attain the i_0 (exchange current density). Butler-Volmer equation can be simplified and written as

$$i = i_0 \left(\exp\{\alpha nF\eta/RT\} - \exp\{(1-\alpha)nF\eta/RT\} \right) \tag{16}$$

Using (16) and linearizing the Butler-Volmer for low overpotential approximation, one can obtain the following expression:

$$i_0 = RT/(nF\alpha R_{\text{CT}}) \tag{17}$$

where i is the electrode current per geometric area, η the overpotential of the electrode for the ORR, α the transfer coefficient (0.5-1.0 depending on the potential), R is molar gas constant, J/mol-K, T is temperature in degrees Kelvin and R_{CT} is the charge transfer resistance from the EIS spectrum close to rest potential.

Such an estimation of exchange current density may not be possible in-situ in a fuel cell, since the cathode electrode is polarized due to other side reactions. While the output is exchange current density per geometric area, a cyclic voltammetric experiment can yield the ratio of Pt real surface area to geometric

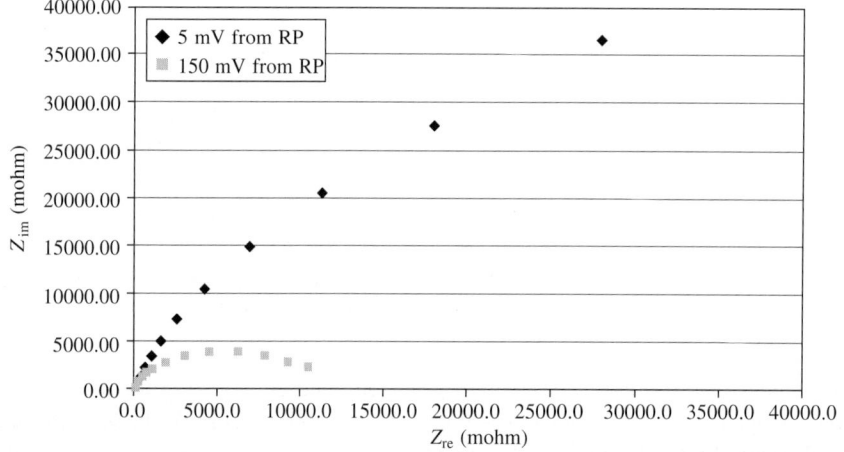

Fig. 11   EIS Spectrum at 500 rpm close to rest potential (RP) and at an overpotential of 150 mV.

area and this value can be used to obtain exchange current density normalized for real Pt active surface area.

4.3 Micro-Electrodes and Other Applications

Over the years several microprobes like ultra-microelectrodes have been developed and acquire signal from the substrate generating current or due to the faradaic current sensed by ion-sensitive electrodes. Mini arrays have been developed (Liu and Smotkin, 2002; Chan et al, 2005) using state of the art micro-electronics fabrication and these approaches have the capability of screening various catalysts in a short period of time. In a typical differential electrochemical mass spectroscopy (DEMS) experiment, the ionization currents of gas species of interest is recorded in addition to faradaic current during the potential scan. Recently, modifications of these systems have been used to study methanol oxidation reactions for direct methanol fuel cells. Jambunathan and Hillier (2003) modified this technique and used scanning differential electrochemical mass spectroscopy (SDEMS) to study CO oxidation and methanol oxidation on Pt based substrates with in situ mass spectroscopy. Recent SDEMS evaluations have indicated that the spatial resolution for current collection and compositional analysis are adequate. The advantage of this scheme over conventional RDE systems is that along with the I-V characteristics, one could also obtain the reaction mechanism by allocating electron transfer steps to converted species tracked through mass spectroscopy. This tool could also be used for analyzing variety of catalysts in situ at the same time. However the scaling implications have to be understood to effectively apply a successful catalyst selected through RDE and such kinetic evaluations through triple phase boundary investigations were carried out by O'Hyare and Prinz (2004).

Along with the reaction rate information, one could quantitatively assess the current efficiency using a mass spectrometer in line with the substrate. The scanning differential electrochemical mass spectroscopy technique employed a simple electrochemical cell with a mass spectrometer and a 3-D positioning system connected to a potentiostat and all controlled through a lab-controlled program. A schematic of the set up is shown in Fig. 12 (a, b). A capillary probe collects the reaction products and is held 100 µm over the current collecting substrate and this substrate-capillary distance is the critical distance between the substrate and the collection capillary across an electrolyte and controls the time for the evolved gases to reach the pump and eventually to the ionization chamber in the mass spectrometer. The rotary vane pump was

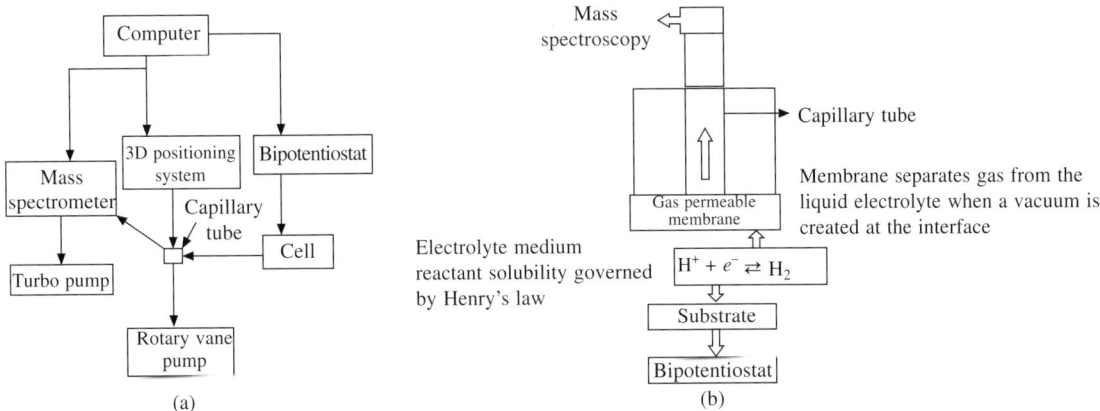

Fig. 12 (a) SDEMS set-up used by Jambunathan and Hillier (2003) and (b) cell interface dynamics with potentiostat and mass spectrometer (reproduced by permission of The Electrochemical Society, Inc.).

connected through the T-connector to provide differential pressure. Ultimate pressure of 10^{-7} Torr was achieved using a turbo-molecular pump in line with the capillary tubes. It was clearly demonstrated that SDEMS has the resolution to detect partial and complete oxidation products of methanol oxidation by detecting CO_2 and methyl formate on an array of Pt-Ru electrodes.

Jambunathan and Hillier (2003) discussed the capillary diameter as an important parameter in trading off imaging resolution and ion signal generation due to the fact that smaller diameter provides higher spatial resolution while limiting the flux of products entering mass spectrometer. Baltruschat (2004) compiled a detailed review on the differential electrochemical mass spectrometry and SDEMS and their ability in detection of products in nano-mole levels and compared the technique over its evolution.

4.4 Potential Step Methods of Catalyst Evaluation

Voltammetry has been a popular technique that measures current as a function of applied voltage across a working electrode with respect to a non-polarizable electrode. This potentiodynamic technique used in electrochemical systems captures the processes occurring at the surface of the electrode. The results and analysis in some cases may be affected by the geometry of the electrode. Several mathematical treatments of various geometries have been studied in the past and the constructions of microelectrodes have been congruent to the development of the technique. The instrumentation involves a potentiostat/galvanostat set-up with an external electrometer as shown in Fig. 13 that interfaces with an electrochemical cell. The recorder instrumentation is required to possess response times significantly faster than the experimental time scale during transient techniques. The equipment may include useful features such as IR correction, background correction and high scan rates. In a linear sweep voltammetry, the voltage imposed is of the nature below

$$E_t = E_i + (R) \times t \tag{18}$$

where E_t is instantaneous voltage, E_i is initial voltage, R the sweep rate (mV/sec) and t time in sec.

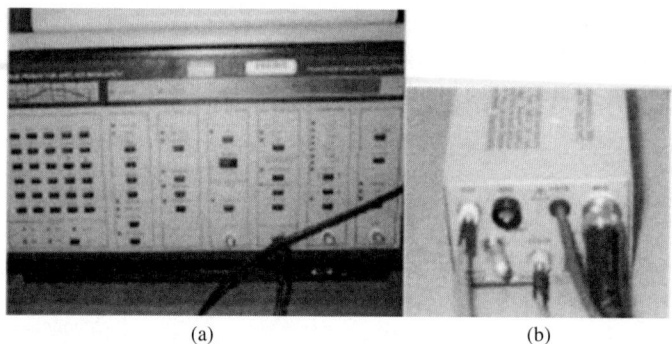

(a)   (b)

Fig. 13 EG&G 273A model: (a) Potentiostat/galvanostat and (b) electrometer.

The electrometer has very high input impedances with respect to voltage measurement and zero input impedance with respect to current sensing along with high precision and resolution. Here we focus on the aspects of this potentiodynamic technique. Aspects of instrumentation are beyond the scope of this chapter. There are several other pseudo-steady state methods that are often used in other electrochemical systems. Some I-V tests include:

(1) Steady state voltammetry

(2) Potential step chronoamperometry
(3) Linear sweep voltammetry
(4) Cyclic voltammetry (CV)

While investigating electrodes using voltammetry, one must carefully consider faradaic current separation from the double layer charging and discharging at the electrode/electrolyte interface and other competing side reactions. During half-cell cyclic voltammogram, the cell should be free of any contaminants that may contribute to additional faradaic current or may depress the current generated by the reaction of interest. In the case of a fuel cell, these measurements can be made in-situ using hydrogen on the anode side and nitrogen on the cathode side. In this case, the background current will include hydrogen permeation or crossover current apart from the double layer charging and discharging currents. The capacitive currents are proportional to scan rates and can be calculated between 0.4 and 0.7 V with respect to standard hydrogen electrode. In fact, the major focus on developing super capacitor materials has been on utilizing the double layer capacitance formed at the interface of the electrode and the electrolyte. Although this capacitance per unit area is very low (10-30 μFarads/cm^2), it can be enhanced significantly by the use of materials with high specific surface areas. The capacitive currents can be calculated per the following equation:

$$C_{DL} = I_{peak} \times R \tag{19}$$

where C_{DL} is the double layer capacitance, I_{peak} is the peak current in the capacitance region and R is the scan rate in mV/sec.

This double layer capacitance is mostly from carbon in a Pt/C system though Pt also contributes to some double layer capacitance. This value may be monitored to understand the state of Pt and carbon in an electrode, as reduced capacitance might indicate (1) less wetted area, (2) carbon oxidation and (3) organic contamination.

A cyclic voltammogram of polycrystalline Pt in 0.5 M H$_2$SO$_4$ is shown in Fig. 14. The electrode is swept between -0.3 and 1.2 V with respect to saturated calomel electrode. Q_a is the integrated anodic charge in Coulombs representing desorption of hydrogen whereas Q_c is the cathodic charge representing the adsorption

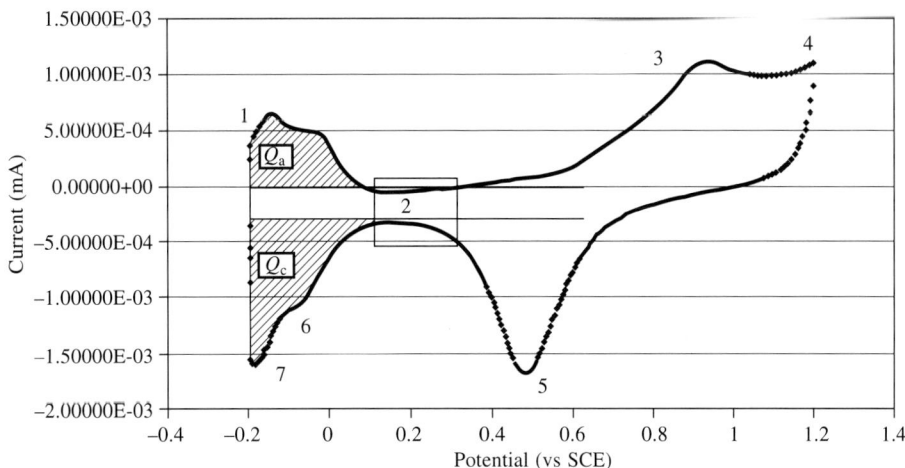

Fig. 14 Cyclic voltammogram on Pt electrode in 0.5 M H$_2$SO$_4$ at temperature = 25°C. Reference electrode = Saturated calomel electrode.

of hydrogen. The broken lines indicate the background current subtraction to accurately integrate the faradaic contribution. The ratio of Q_A/Q_C gives the reversibility of the hydrogen adsorption and desorption reactions. A charge of 210 $\mu C/cm^2$ is contributed from three basal planes and this value can be used to calculate the roughness factor (RF) of the electrode, which is given by

$$RF_{experimental} = \text{Active area (C/cm}^2 \text{ geometric)}/(210 \, (\mu C/cm^2 \, Pt) \qquad (20)$$

$$RF_{theoretical} = \text{Total Pt from surface area} \times \text{Pt Loading} \qquad (21)$$

$$\% \text{ Pt utilization} = RF_{experimental}/RF_{theoretical} \qquad (22)$$

In Fig. 14, different regions (1 to 7) are named to describe different processes occurring at the electrode surface and they can be summarized as follows:

$$Pt\text{-}H_{ads} \rightarrow Pt + H^+ + e^- \{1\text{-Hydrogen desorption}\} \qquad (23)$$

$$DL_{charge} \leftrightarrow DL_{discharge} \{2\text{-Double layer charging and discharging}\} \qquad (24)$$

While in a half cell, the double layer charging and discharging during the reverse sweep are the dominant contributors to the current, in a fuel cell during insitu CV measurement across a fuel cell, region 2 might also contain hydrogen oxidation limiting current due to hydrogen crossover from the anode chamber while evaluating the cathode catalyst.

Above 0.6 V vs SCE, the Pt catalyzes water activation through the following reaction and close to 1.2 V, the Pt is covered with an oxide layer. Between 1.0 and 1.2 V, Pt/Pt^{2+} equilibrium sets up

$$Pt + H\text{-}O\text{-}H \rightarrow Pt\text{-}O\text{-}H + H^+ + e^- \text{ (region 3)} \qquad (25)$$

$$Pt\text{-}O\text{-}H + H\text{-}O\text{-}H \rightarrow Pt\text{-}(O\text{-}H)_2 + H^+ + e^- \qquad (26)$$

$$Pt\text{-}(O\text{-}H)_2 \rightarrow Pt\text{-}O + H\text{-}O\text{-}H \text{ (region 4)} \qquad (27)$$

Beyond 1.2 V with respect to SCE, the oxygen gas evolution starts to dominate so a reverse sweep would immediately follow.

At potentials close to 0.45 V (w.r.t. SCE)

$$\begin{array}{cc} O & O \\ \| & \| \\ \end{array}$$
$$Pt - Pt + 4\,H^+ + 4e^- \rightarrow Pt - Pt + 2\,H_2O \text{ (region 5)} \qquad (28)$$

During reverse sweep back to hydrogen evolution potentials (region 7), the following surface reactions take place:

$$Pt + H^+ + e^- \rightarrow Pt\text{-}H_{ads} \{6\text{-Hydrogen adsorption}\} \qquad (29)$$

5. Electro-analytical Tools in Single Cell and Stack Design

5.1 In Situ Voltammetric Measurements in an Assembled MEA

In an ideal cyclic voltammetric (CV) set-up, one would choose the appropriate scan rate and a potential window for scanning. In an aqueous electrolyte, one can scan the 1.5 V potential window to get mechanistic information, but in a real fuel cell, potentials above 0.75 V (vs. SHE) is not advised if the electro-active Pt is supported on carbon. At these elevated potentials, the carbon corrosion current density increases significantly and such a useful technique could impact the cell performance post diagnostics. Scan rates are typically fixed by the electrode geometry and the maximum current the equipment can tolerate before overload. Some potentiodynamic methods like anode-stripping schemes may be very useful in quantifying CO coverage in monolayers (Brett et al, 2004) and investigating other contamination like H_2S in the anode reactant streams (Mohtadi et al, 2003).

To calculate the active area of the cathode electrode, N_2 is purged over the cathode while the anode serves as counter and reference electrode as shown in Fig. 15. In some cases, a sensing wire is used to detect instantaneous IR drop and the potentiostat compensates for the imposed voltage. One of the differences between the half-cell and the in situ measurement in a fuel cell is that the reference electrode is a dynamic hydrogen electrode, since it also acts as the counter electrode. The current is small enough that the polarization at the electrode is negligible. However, proper selection of humidity (RH), temperature and background current subtraction (due to crossover of the hydrogen and double layer charging/discharging) is required to produce the repeatable and meaningful measurements. The data

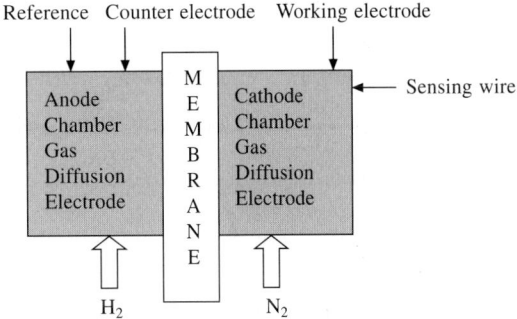

Fig. 15 CV set-up in a fuel cell (Example: Working electrode-cathode, reference and counter electrodes-anode).

shown in Fig. 16 indicates that up to 40% error may be induced during measurement if the RH is not properly controlled.

Several other interesting analyses apart from the Pt surface area determination that can be performed using a CV, includes the following:

1. Crossover Current and Double Layer Capacitance Determination
Molecular hydrogen permeates through the membrane and when the cathode voltage is between 0.4 V and

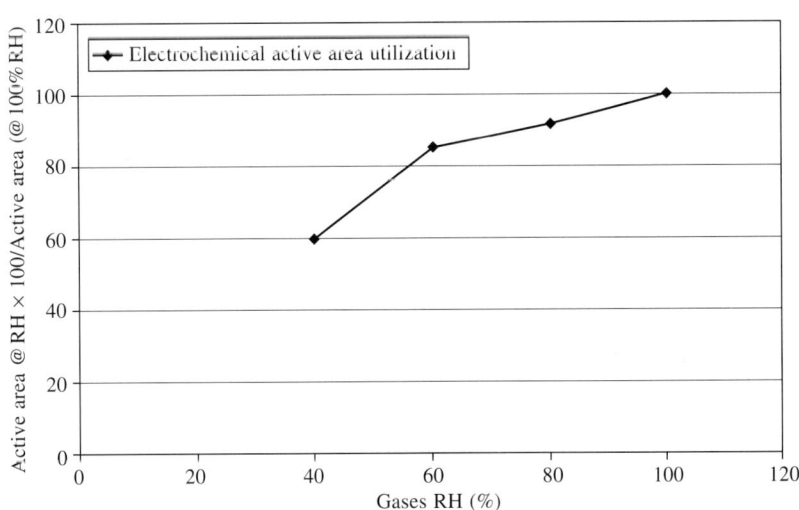

Fig. 16 Active area measured at different RH of the gases (H_2 and N_2). It is clearly seen that access to active sites is a clear function of RH and if the RH is not controlled, the error induced can be as high as 40%.

0.5 V, during the anodic or the forward sweep towards higher potentials, the total current is due to summation of the limiting current due to hydrogen crossover to cathode side and the double layer charging of the electrode/electrolyte interface. During reverse sweep, the total current is the hydrogen crossover and the double layer discharging current as shown in Fig. 17. The double layer contribution can also be separated by performing a N_2/N_2 capacitance cyclic voltammogram. Such a cyclic voltammogram is showed in Fig. 18 and the capacitance calculated from Eq. (19) is 2.07 milli Farads.

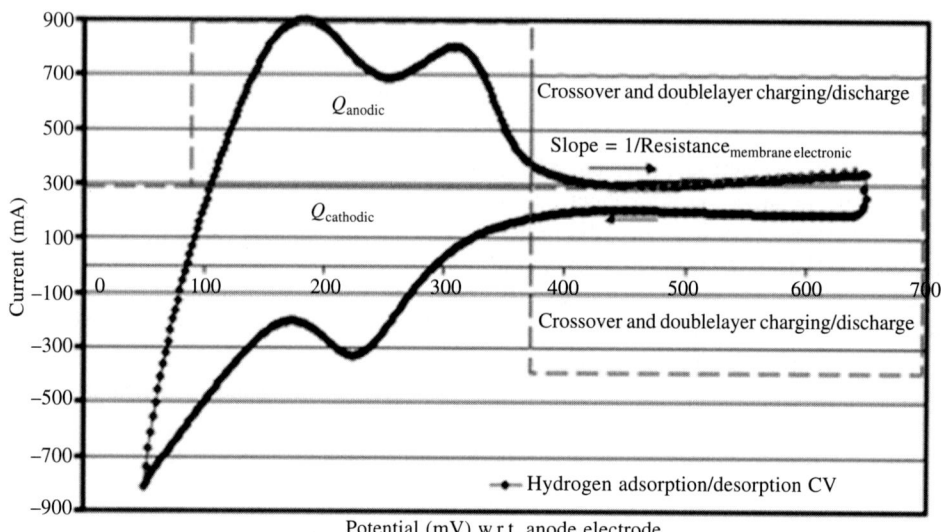

Fig. 17 In situ CV measurement in an assembled MEA. Area under the peak can be integrated to estimate the roughness of the electro-active area.

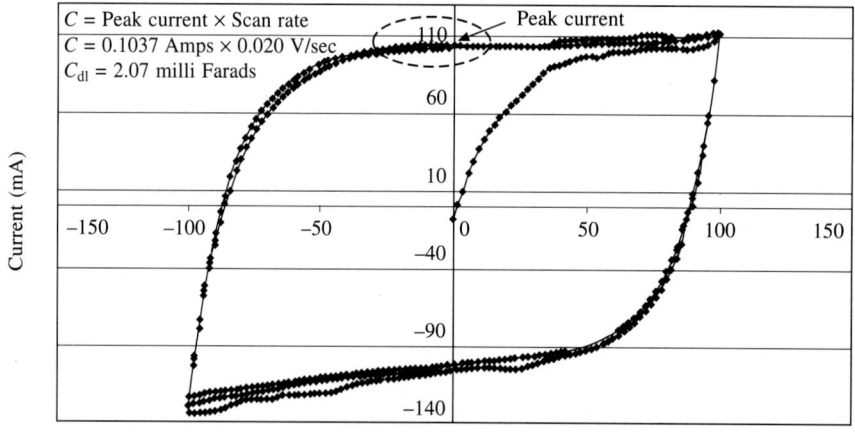

Fig. 18 A typical capacitance CV measurement. Using Eq. (19), the double layer capacitance can be calculated.

It is not necessary to complete a CV, when one needs to measure just crossover or capacitance. For example, crossover measurements can be made by chrono-amperometry technique, where the potential of the cathode side is elevated to 0.4 V from a rest potential of 0.08 V. During the chrono-amperometry experiment, inducing backpressure on the anode chamber elevates the partial pressure of the hydrogen. If there are pinholes, more hydrogen permeates increasing the background current substantially beyond the step changes as indicated in Fig. 19.

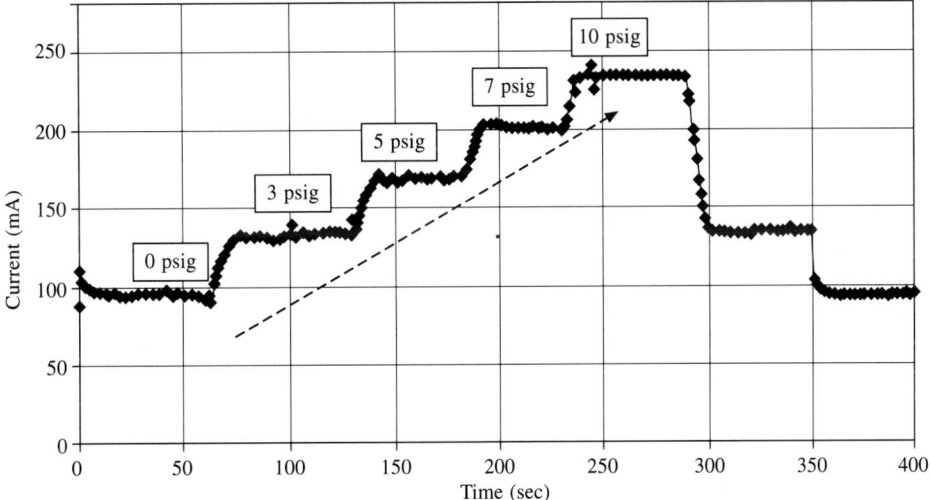

Fig. 19 Chrono-amperometric experiment at different elevated backpressures. Experiments performed by Guo (Plug Power) demonstrated increase in crossover current with increased anode backpressures from 0-10 psig.

Once the crossover current is determined, the losses expected at lower current densities due to mixed potential scenario can be evaluated (see Fig. 5). A galvanic cell sets up as below:

$$2H_2 \underset{K_{r,H_2}}{\overset{K_{f,H_2}}{\rightleftharpoons}} 2H^+ + 4e^- \quad \Big| \text{ Depletion of electrons}$$

$$4H^+ + 4e^- + O_2 \rightleftharpoons 2H_2O \tag{30}$$

2. Membrane Intrinsic Electronic Resistance

One of the functions of the membranes is to provide an electronic insulation barrier between the anode and the cathode electrodes in order to spatially separate the electrons to drive a load in the external circuit. A certain amount of current will be split between internal and external circuit depending on the electronic resistance ratio of the intrinsic membrane and the external pathway to the current collector. A clear description is shown in Fig. 20 (a, b) where a relay switch keeps the fuel cell at open circuit, but in reality the fuel cell is actually dissipating energy across the membrane resistor.

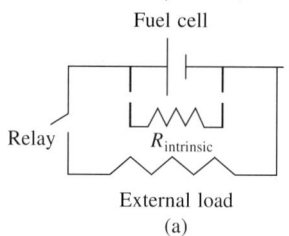

Fuel cell

Relay $R_{intrinsic}$

External load

(a)

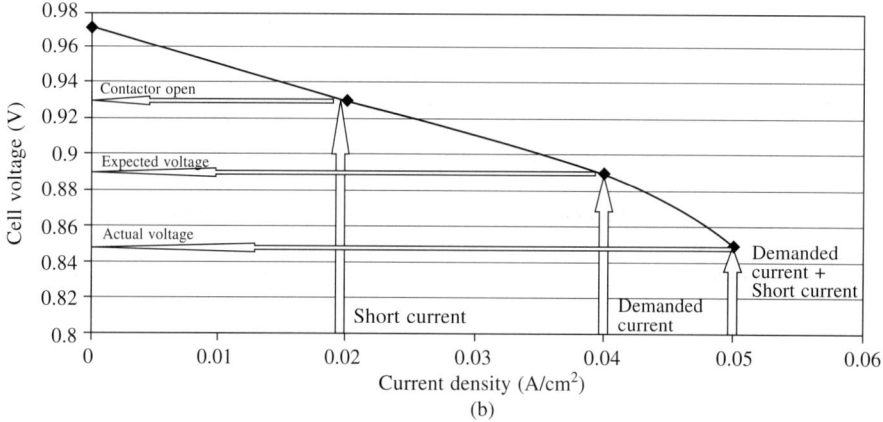

(b)

Fig. 20 (a) Electrical circuit showing an intrinsic electrical resistive pathway through membrane and (b) Electrical short through membrane-mechanistic explanation. When the relay is open, $E_{measured} = E_{OCV} - iAR_{intrinsic}$, where iA is the current through membrane and $R_{intrinsic}$ is the electronic resistance of the membrane, as shown in (b).

$$2H_2 \underset{K_{r,H_2}}{\overset{K_{f,H_2}}{\rightleftarrows}} 4H^+ + 4e^-$$

Discharges
through membrane

$$4H^+ + 4e^- + O_2 \rightleftarrows 2H_2O$$

(31)

This small parasitic load is an issue at low current densities, where the corrective action will be to increase the fuel demand to sustain the increase in current demand, which has a negative impact on the efficiency of the fuel cells. This electrical resistance, which is a function of how an MEA is assembled and compressed, can be measured through a variety of direct current (DC) techniques, but CV provides this measurement as the inverse of the slope from 0.5 to 0.6 V during the forward sweep as shown in the in situ measurement in Fig. 17.

5.2 Polarization Losses in an Operating Fuel Cell

In a fuel cell, there are several types of polarization losses that result in lower performance and can be summed up as follows:

$$V_i = V_{OCV} - \Delta V_{Nernst} - \Delta V_{activation} - \Delta V_{mass\ transport} - \Delta V_{ohmic} \tag{32}$$

In brief, there are multiple reasons for losses namely nernst, activation, ohmic and mass transport losses. Voltage of a fuel cell is a lumped response and instantaneously one may not be able to provide information if the loss is at anode or cathode chamber. It lumps all the losses shown in Eq. (32) and provides first level information about the fuel cell performance. In terms of relating material property to performance, one needs a powerful tool to separate these losses and assign these loss contributions to specific components. In a typical fuel cell, the losses are shown in Fig. 21 (a) and these losses are quantified and separated as through vigorous experimentation that involves measuring high frequency resistance, fuel pumping and developing a constant stoich polarization curve. ΔV_{Nernst} is covered earlier and the biggest contributors are electronic shorts through the membrane and fuel crossover to the cathode side. A typical polar curve can be modified as an overpotential versus log i plot and tafel slope can be determined after taking into account IR and mass transport losses. Such a plot is shown in Fig. 21 (b) for a theoretical polar curve of a tafel slope of 85 mV/decade and an additional catalyst performance parameter-mass specific activity (40 mA/mg) was calculated from the current density at 0.9 V.

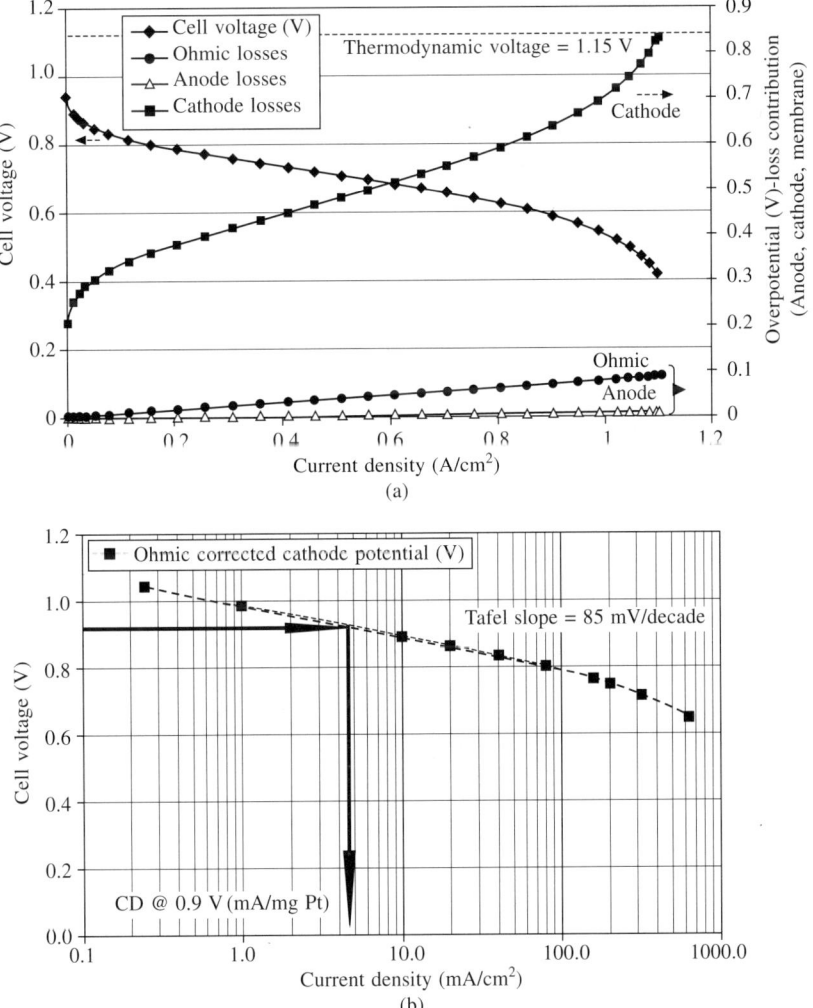

Fig. 21 (a) Polarization losses in a fuel cell. Cathode losses are significant and (b) Catalyst performance parameter extraction from IR and mass transport free polar curve. Tafel slope = 85 mV/decade at 70°C.

Ohmic and mass transport losses can be mathematically calculated based on the following equation:

$$\eta_{\text{ohmic}} = i \times R, \quad \eta_{\text{mass transport}} = \frac{RT}{\alpha n F} \ln\left(1 - \frac{i}{i_{\text{limiting}}}\right) \tag{33}$$

where i_{limiting} is the limiting current density.

$\Delta V_{\text{activation}}$ can be due to several reasons as follows:

(a) Intrinsic Pt activity and surface area (function of Pt size, three-phase interface)
(b) Contamination
(c) Gas water content

$\Delta V_{\text{mass transport}}$ may be due to:

(a) Insufficient fuel and air stoichiometry
(b) GDL/electrode flooding
(c) GDL/flow field flooding

ΔV_{ohmic} may be due to the following reasons:

(a) Loss of contact between MEA and flow field or between MEA subcomponents
(b) Inorganic contamination, etc.

Performance of a PEM fuel cell changes with time due to change in material and electrochemical properties and can be mathematically expressed as

$$\left(\frac{\partial V}{\partial t}\right)_i = \left(\frac{\partial V^{\text{OCV}}}{\partial t}\right)_i + \left(\frac{\partial V^{\text{act}}}{\partial t}\right)_i + \left(\frac{\partial V^{\text{mass}}}{\partial t}\right)_i + \left(\frac{\partial V^{\text{ohm}}}{\partial t}\right)_i \tag{34}$$

In order to eliminate failure modes, it is important to understand if failure mechanism involves the membrane, the GDL, the electrode or external influences. In small-scale fuel cells, some cell architectures have a Pt black reference electrode incorporated on the membrane, which will help assign the cell voltage contribution between the anode and cathode chamber. Even with these dynamic reference electrodes, it is still not possible to understand the physics causing the lower cell performance. Techniques like current interrupt, high frequency impedance and full electrochemical impedance spectrum can give us additional information. Now losses can be assigned due to ohmic, mass transport, activation or nernst because the physical processes contributing to the losses have different time constants and can be separated by varying frequency during such measurements especially when using the electrochemical impedance spectroscopy (Fig. 22). The flow rates (litres/min) and diffusion coefficient (cm^2/sec) determine the mass transport time constants whereas the reactions turnover per second (reaction rates) determine the frequency at which activation resistance can be separated from the EIS Nyquist spectrum.

5.3 Electrochemical Impedance Spectroscopy (EIS)

Impedance spectroscopy is a powerful tool that enables clear understanding of activation losses, mass transport and ohmic losses in an operating fuel cell. Most electrochemical systems are non-linear and require careful treatment of data to understand and extract properties of materials. The induced perturbations could be in the form of voltage, monitoring the current response or current drawn from the fuel cell and monitoring the voltage response. The perturbation amount is fixed and this is superimposed on top of the existing DC voltage or current as illustrated in Fig. 23. Different regions of the polar curve may be opted to be perturbed and impedance at different current densities will provide separation between activation and

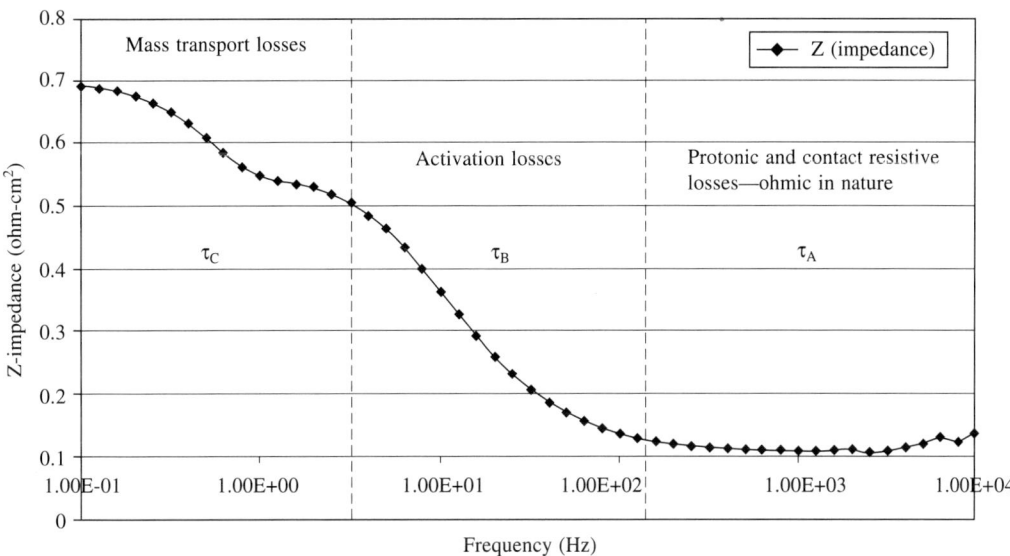

Fig. 22 Bode Plot: magnitude vs frequency (EIS) showing separation of losses through resistances, *t* is the time constant of the physical process, τ_C the time constant of the mass transport processes, τ_B the time constant of the reaction controlled processes and τ_A the time constant for the ohmic processes.

mass transport losses in the system. In both cases, the departure from linearity should be minimal to make sure the system is steady, implying careful selection of peak-to-peak amplitude during perturbations. Significant reduction in amplitude will increase the noise in the data collection and hence optimum is the key.

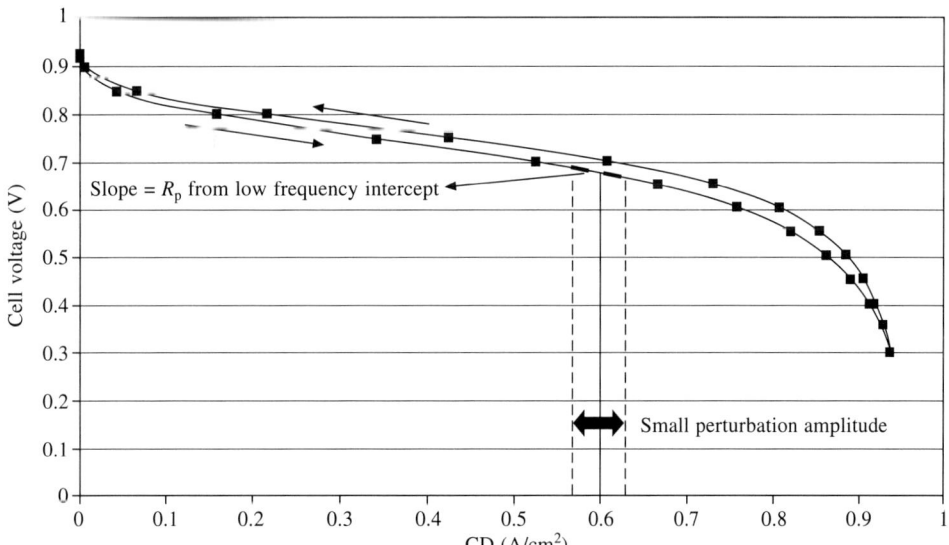

Fig. 23 Perturbation amplitude and current density selection from polar curve for linearity.

This tool can be used with carefully designed experiments to separate:

(a) Ohmic losses from catalyst layer and membrane.
(b) Catalyst activity under certain conditions (CO effect on performance etc.).
(c) Mass transport losses in catalyst layer versus GDL and versus channels of the plate and their interfaces.

EIS can be used to characterize the properties of an MEA. Following are the two ways to extract the cell parameters from the Nyquist response by using a:

(a) system of differential equations describing the physics and electrochemistry.
(b) finite transmission line of resistors and capacitors (Fig. 24).

The parameter extraction through equivalent circuit has been applied by several researchers and can help in understanding the material and electrochemical properties of the subcomponents in the MEA. A typical impedance set-up is shown in Fig. 25 that includes a fast Fourier transform (FFT) based frequency response analyzer with a signal generator, an oscilloscope to check the generated waveform, a current measuring device

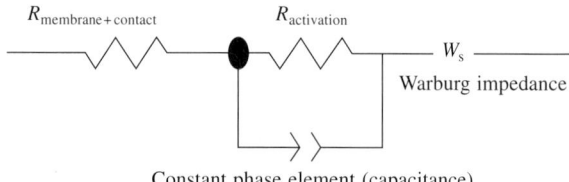

Fig. 24 A simple equivalent circuit-transmission line model.

and the fuel cell with a voltage monitor. This is interfaced with software like Z-plot that can plot and help analyze data through transmission line models (Springer and Raistrick, 1989; Ciureanu and Roberge, 2001). Typically once these measurements are performed, the data analysis includes error structure analysis through Kramer-Kronig transformation (Agarwal et al, 1995). Measurement artifacts need to be understood

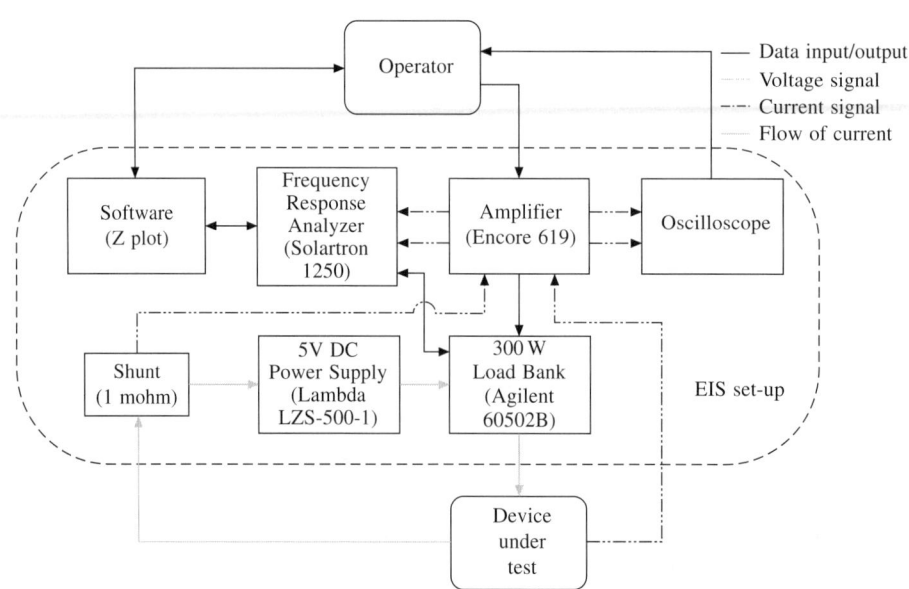

Fig. 25 AC impedance set-up structure and interfaces.

prior to getting signatures from an active fuel cell so that inductances and resistances from various sources like wires can be minimized before experimentation. Several researchers have used this tool effectively to understand the performance limiters and were able to provide mathematical treatment to correlate losses to material properties. Springer et al (1996) showed the effect of RH on the ac impedance Nyquist and Bode responses of a fuel cell and provided interesting interpretations on their data. Lefebvre et al (1999) tried to separate the electrolyte layer properties from the membrane using AC impedance spectroscopy and Makharia et al (2005) soon followed this scheme. Studies performed by Brett et al (2003) effectively demonstrated the cathode stoichiometry requirements and its effect on the mass transport properties along the flow field channels using AC impedance spectroscopy. Several other studies have used AC impedance tool to study other transient characteristics and steady state responses. On the transient characteristics, Wagner and Schulze (2003) mapped impedance for low temperature CO poisoned system and showed that a low frequency inductance like behavior is seen in such systems. A spectrum of high frequencies ($\sim 10^2 - 10^4$ Hz) is necessary for separating the membrane impedance $R_{membrane}$. According to this model, at high frequencies, the capacitors act like short circuits, bypassing the polarization and charge transfer resistances.

The double layer capacitance values calculated from impedance technique and cyclic voltammogram should be the same for a particular cell configuration. On a Nyquist plot, 1 kHz is near the high-frequency intercept of the real axis. It is known qualitatively that the high-frequency impedance is more sensitive to a drying condition than to a supersaturated condition. At intermediate frequencies around 1-4 Hz, the first semicircle is seen and at low current densities, whereas in pure kinetic control, the diameter of the semicircle is proportional to tafel slope. At higher current densities, diameter of the loop is equal to $\beta RT/\alpha nF$, where β is a multiplication factor. However, if liquid water condenses in the flow fields or in the MEA interface, then the impedance spectrum at low frequencies ($\sim 10^{-1} - 10^0$ Hz) is affected. One of the significant losses in low temperature PEM is due to transport of reactants to the electrodes through the relatively thick GDL. A PEM fuel cell is especially vulnerable to transport limitations on the cathode side, where the GDL must simultaneously repel product water while allowing fresh oxygen into the catalyst. Condensation of water on the cathode can block oxygen transport pathways and increase polarization impedance. The low-frequency intercept of the real impedance on Fig. 26 corresponds to the total polarization resistance and clearly shows the difference in R_P between a flooded and an un-flooded MEA.

Current interrupt technique is also commonly used to determine ohmic losses across the membrane electrode assembly (Buchi et al, 1995). A load contactor/relay switch is in series with the load bank and fuel cell. This relay is latched off while operating at a particular current. This along with an oscilloscope in line can help separate ohmic losses from the total losses in a fuel cell. The switching speeds of the relay switch are to be of the order of milli-seconds and a gigahertz oscilloscope is required to record the waveform when the contactor is open.

In the current interrupt technique, the fuel cell is the probe, and the current drop and voltage change measurements are continuously collected and stored for further processing. A current-interrupt response is shown in Fig. 27.

6. Additional Diagnostics

In the recent years, computational fluid dynamic (CFD) modeling has gained popularity among electrochemists, with user-friendly electrochemical subroutines and has shed light on the intrinsic variations within a fuel cell with respect to current density and membrane water content distribution (Eldrid et al, 2003 and Dutta et al, 2000). In an operating fuel cell, this information is not available since the current density and membrane water content (λ) are averaged over the active area of the fuel cell. This information is particularly useful in designing flow-fields and gas diffusion electrodes so that an optimum distribution is obtained since local stresses in a fuel cell may result in higher performance degradation reaction. In order to validate

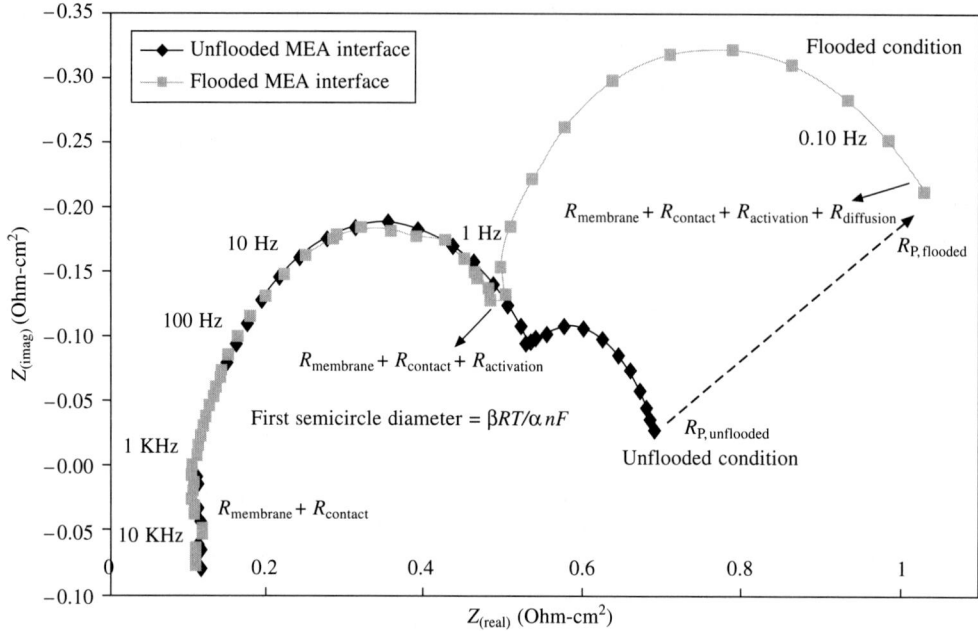

Fig. 26 EIS comparison of a flooded and an unflooded MEA.

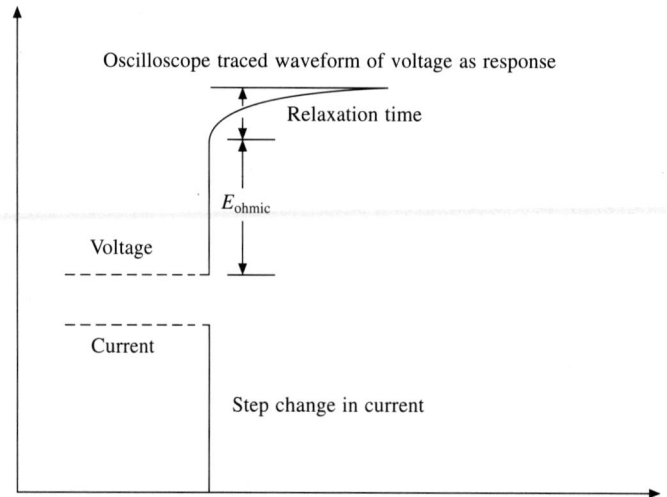

Fig. 27 IR loss estimation through current interrupt (CI) method.

the model and gain better understanding, several researchers began to segment their fuel cell to obtain distributions of current density and species concentration as a function of flow field length from inlet to outlet. This type of study explains performance losses with respect to local defects and mass flow rates for a particular flow-field geometry. Multiple load banks with precise current and voltage control are needed along with the appropriate technique to segment the fuel cells with good spatial resolution. Lateral communication through the electrodes and gas diffusion layer has been avoided by segmenting the electrodes along with segmenting the current collectors (Natarajan and Nguyen, 2004). Hicks (2005) used a printed circuit board technology dividing a serpentine 50 cm^2 into 121 segments and reported 30% higher current at the inlet compared to the outlet. The segmented cell has been used with in situ electrochemical impedance spectroscopy to explain the performance limitations and Brett et al (2003) mapped the spectrum from inlet to outlet as a function of air utilization and used the low frequency real axis intercept to explain the effect of channel distance and stoich on mass transport limitations. Along with the intrinsic information, one needs to understand the extrinsic impacts on the fuel cell namely fuel as a source of contamination and some of them being CO, H$_2$S in ppm level. Though in small quantities, these impurities have serious effect on performance as studied by Mohtadi et al (2003) from University of South Carolina. Zhang and Dutta (2002) studied the potential oscillations induced due to CO poisoning and re-activation of the catalysts and provided a kinetic model that can be used to assess the state of health of the catalyst. The resulting oscillation in cell voltage is due to self-cleaning due to increased anode over potentials greater than 0.35 V for Pt/Ru alloy and greater than 0.6 V for a Pt bare metal and then poisoning again due to a steady source of feed. In Plug Power, Karuppaiah (2000) and Du (2004) utilized these tests to understand the impact on the anode electrodes by fuel starvation. Fig. 28 shows the voltage oscillation of a 50 cm^2 fuel cell at a constant current density of 0.6 A/cm^2 at elevated CO feed in the anode stream. The response was traced through an oscilloscope for high-speed data acquisition to acquire amplitude and frequency information that can be gathered for further kinetic evaluation.

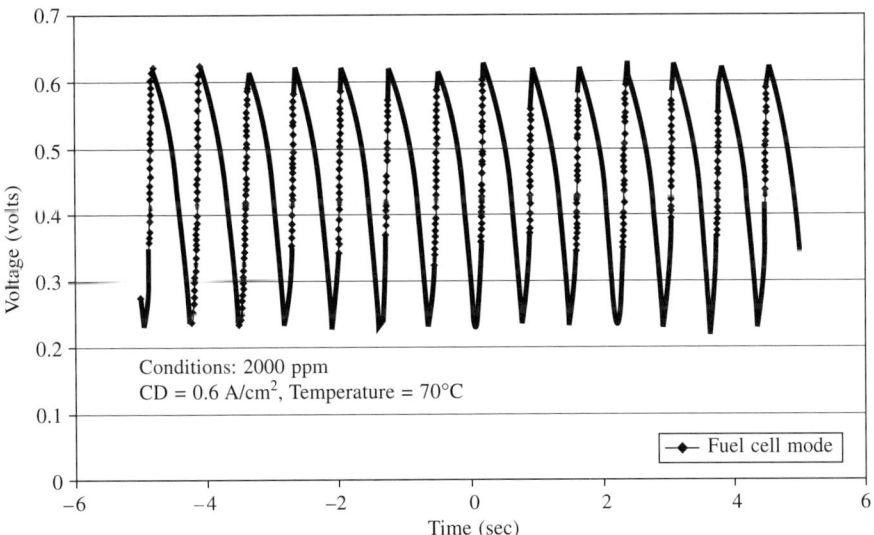

Fig. 28 Cell voltage oscillations due to CO (2000 ppm) deposition and self-cleaning (temperature = 70°C, current density = 0.6 A/cm^2).

7. Conclusions

In conclusion, several conventional electro-analytical techniques have been modified to choose appropriate electro-catalysts, processing conditions, understand mechanisms of losses during scale up and fuel cell operation. These techniques have been instrumental in developing relationships between performance loss and material properties. Microelectrode development through SDEMS, RRDE and other approaches provided screening techniques that catalyzed fresh investigations of binary, ternary alloys that resulted in higher performance.

8. Future Direction

We believe that the performance in the fuel cell electrodes have improved significantly but utilization of the three-phase interface has to be improved and metal loading can be further reduced. Alloyed catalysts show significant promise but long-term durability tests need to be performed on the corrosion properties of these alloys. Nano-scale structural investigation and nano-scale architecture can improve the utilization and the material properties of these electrode interfaces. Integration of the electrodes into a fuel cell stresses the importance of processing and fabrication conditions. Perfection of in-situ tools for non-intrusive performance evaluation is of utmost importance, since the performance parameters can be studied as a response while investigating the optimum conditions for MEA fabrication. In situ detection of performance losses can further provide insights on the latitude in operating conditions along with material specifications. Additional understanding of reaction kinetics as a function of time with degrading interfaces may provide direction in creating a robust MEA architecture.

Acknowledgements

I thank Lakshmi Krishnan, Qunhui Guo, Bin Du, Dhirendra Danji, Dustin Blodgett and Krishna Ohlsson for their contribution. I also thank John F. Elter (Chief Technology Officer, Plug Power Inc.) and Daniel Beaty for reviewing the details of this article.

References

Agarwal, P., Orazem, M.E., Garcia-Rubio, L.H., *Journal of the Electrochemical Society*, **142** (1995) 4159.

Baltruschat, H., *Journal of American Society for Mass Spectrometry*, **15** (2004) 1693.

Brankovic, S.R., Wang, J.X., Adzic, R.R., *Electrochemical and Solid State Letters*, **4** (2001) A217.

Brett, D.J.L., Atkins, S., Brandon, N.P., Vesovic, V., Vasileiadis, N., Kucernak, A.R., *Electrochemical and Solid State Letters*, **6** (2003) A63.

Brett, D.J.L., Atkins, S., Brandon, N.P., Vesovic, V., Vasileiadis, N., Kucernak, A.R., *Journal of Power Sources*, **133** (2004) 205.

Buchi, F.N., Marek, A., Scherer, G.S., *Journal of the Electrochemical Society*, **142** (1995) 1895.

Carbon, Electrochemical and Physico-chemical properties, John Wiley & Sons, 1992.

Chan, B.C., Liu, R., Jambunathan, K., Zhang, H., Chen, G., Mallouk, T.E., Smotkin, E.S, *Journal of the Electrochemical Society*, **152** (2005) A594.

Ciureanu, M., Roberge, R., *Journal of Physical Chemistry B*, **105** (2001) 3531.

Damjanovic, A., Genshaw, M.A., and Bockris, J.O'M., *Journal of Chemical Physics*, **45**, 1966, 4057.

Du, B., Plug Power Internal Communication, 2004.

Dutta, S., Shimpalee S., Vanzee J.W., *Journal of Applied Electrochemistry*, **30** (2000) 135.

Eldrid, S., Shahnam, M., Prinkey, M.T., Dong, Z., First International Conference on Fuel Cell Science, Engineering and Technology, Rochester, NY, April 2003.

Fuel Cells and Fuel Batteries—A Guide to the Research and Development, Liebhafsky & Cairns, John Wiley & Sons, New York, 1968.

Gasteiger, H., Mathias, M., Yan, S., Catalyst Development Needs, Presented at the NSF Workshop, Washington DC, Nov. 2001.

Gottesfeld, S., Pafford, J., *Journal of the Electrochemical Society*, **135**, 2651, (1988).

Haug, A.T., White, R.E., Weidner, J.W., Huang, W., *Journal of the Electrochemical Society*, **149** (2002) A862.

Hicks, M., DoE Hydrogen Program Review, Washington DC, May 2005.

Jambunathan, K., Hillier, A.C., *Journal of the Electrochemical Society*, **150** (2003) E312.

Karuppaiah, C., Plug Power Internal Communication, 2000.

Lefebvre, M.C., Martin, R.B., Pickup, P.G., *Electrochemical and Solid State Letters*, **2** (1999) 259.

Liu, R., Smotkin, E.S., *Journal of Electroanalytical Chemistry*, **535** (2002) 49.

Makharia, R., Mathias, M.F., Baker, D.R., *Journal of the Electrochemical Society*, **152** (2005) A970.

Mitsushima, S., Araki, N., Kamiya, N., Ota, K., *Journal of the Electrochemical Society*, **149** (2002) A1370.

Mohtadi, R., Lee, W-K., Cowan, S., Van Zee, J.W., Murthy, M., *Electrochemical and Solid State Letters*, **6** (2003) A272.

Natarajan, D., Nguyen, T.V., *Journal of Power Sources*, **135** (2004) 95.

Norskov, J.K., Bligaard, T., Logadottir, A., Kitchin, J.R., Chen, J.G., Pandelov, S., Stimming, U., *Journal of the Electrochemial Society*, **152** (2005) J23.

Norskov, J.K., Rossmeisl, J., Logadottir, A., Lindqvist, L., Kitchin, J.R., Bligaard, T., Jonsson, H., *Journal of Physical Chemistry B*, **108** (2004) 17886.

O'Hyare, R., Prinz, F.B., *Journal of the Electrochemical Society*, **151** (2004) A756.

Paulus, U.A., Schmidt, T.J., Gasteiger, H.A., Behm, R.J., *Journal of Electroanalytical Chemistry*, **495** (2001) 134.

Springer, T.E., Raistrick, I.D., *Journal of the Electrochemical Society*, **136** (1989) 1594.

Springer, T.E., Zawodzinski, T.A., Wilson, M.S., Gottesfeld, S., *Journal of the Electrochemical Society*, **143** (1996) 587.

Ticianelli, E.A., Beery, J.G., Srinivasan, S., *Journal of Applied Electrochemistry*, **21** (1991) 597.

Uribe, F.A., Springer, T.E., Gottesfeld, S., *Journal of the Electrochemical Society*, **139** (1992) 765.

Uribe, F.A., Zawodzinski, T.A., *Electrochimica Acta*, **47** (2002) 3799.

Wagner, N., Schulze, M., *Electrochimica Acta*, 48 (2003) 3899.

Wilson, M.S., Gottesfeld, S., *Journal of Electrochemical Society*, **139** (1992) L28.

Zhang, J., Dutta, R., *Journal of the Electrochemical Society*, **149** (2002) A1423.

Recent Trends in Fuel Cell Science and Technology
Edited by S. Basu
Anamaya Publishers, New Delhi, India

3. Polymer Electrolyte Membrane Fuel Cell

K.S. Dhathathreyan and N. Rajalakshmi

Centre for Fuel Cell Technology, International Advanced Research Centre for Powder Metallurgy and
New Materials (ARCI), Medavakkam, Chennai-601 302, India

1. Introduction

The polymer electrolyte membrane fuel cell (PEMFC) also known as proton exchange membrane fuel cell, polymer electrolyte fuel cell (PEFC) and solid polymer fuel cell (SPFC) was first developed by General Electric in the USA in the 1960's for use by NASA in their initial space applications. The electrolyte is an ion conducting polymer membrane, described in more details in Section 2.2. Anode and cathode are bonded to either side of the membrane. This assembly is normally called membrane electrode assembly (MEA) or EMA which is placed between the two flow field plates (bipolar plates) (Section 2.5) to form what is known as "stack". The basic operation of the PEMFC is the same as that of an acid electrolyte cell as the mobile ions in the polymer are H^+ or proton.

The first PEMFC such as the one used in the NASA Gemini flight had a life time of about 500 h which was sufficient for those limited early missions (Warshay, 1990). However, in the subsequent space flights (Apollo) NASA used the alkaline fuel cells (AFC) as the cost of PEMFC was very high. Nearly 28 mg of platinum was needed for each sq. cm area of the electrode. The development of PEMFC went more or less into abeyance in the 1970's and early 1980's. In the later half of 1980's and early 1990's there was a renaissance of interest in PEMFC (Prater, 1990). From 1967, Nafion membrane, a trade mark of DuPont, became available as electrolyte for use in PEMFC which revolutionized the technology development, variants of which are continually being used till today. The developments in recent years have brought current densities upto 1 A /sq. cm or more, while at the same time reducing the use of platinum by a factor of over 100. These improvements have led to huge reduction in cost per kW of power and much improved power density.

PEMFCs are being developed for use in transportation applications as well as in a variety of portable and stationary applications. There have been some demonstrations of combined heat and power generation systems albeit the low grade heat generated by PEMFC. A sign of dominance of PEMFC in recent times is reflected in the number of companies that have sprouted "manufacturing" these units, various demonstration programs, and increase in patents that have appeared. A growing number of global corporations are becoming involved in fuel cells, both as developers and strategic partners. Large established manufacturers, such as DuPont, Gore, SGL, 3M and Johnson Matthey, are positioning themselves to become world suppliers of PEMFC components. The drive for zero emission vehicles has led to great technological strides in the development of PEMFC. Several demonstrations in cars, buses as well as highly publicized investment by leading car manufacturers have given the technology a high media profile. Most of the world's largest automotive manufacturers including GM, Daimler Chrysler, Ford, Toyota, Nissan, Hyundai and Honda have also recognized the importance of early fuel cell commercialization and are also involved in the development of stationary fuel cells as a means of building their overall capacity in automotive fuel cell applications for the longer term.

Initially PEMFC was considered less suitable for stationary applications than the other fuel cell types. However, the opinion has changed by the rapid technical progress made and by the considerable reductions in projected manufacturing costs.

Among the various applications of PEMFC two aspects remain similar, viz. (i) electrolyte used and (ii) electrode structure and the catalyst.

However, depending on the applications the following options vary:

(a) Water and thermal management.
(b) Interconnection of the cells.
(c) Reactants to be used.
(d) Pressure of operation.

The lower operating temperature of a PEMFC results in both advantages and disadvantages. Low temperature operation is advantageous because the cell can start from ambient conditions quickly, especially when pure hydrogen fuel is available. Present cells operate at 80°C, nominally, 0.285 MPa (30 psig) and a range of 0.10 to 1.0 MPa (10 to 100 psig) have been reported. Using appropriate bipolar plates and supporting structure, PEMFC should be capable of operating at pressures up to 3000 psi and differential pressures up to 500 psi which is a great advantage. PEMFC stacks are modular, simple to construct and hence they find wide variety of applications ranging from 0.1 watt to 100 kW. The solid electrolyte in PEMFC exhibit excellent resistance to gas crossover. No moving liquid electrolyte and hence no replenishment of electrolyte as in the case of AFC. PEMFC can operate at very high current densities compared to the other fuel cells. These attributes lead to a fast start capability and the ability to make a compact and lightweight cell. Other beneficial attributes of the PEMFC include lower sensitivity to orientation. As a result, the PEMFC is particularly suited for vehicular power application. The disadvantage is that the low quality thermal output cannot be used effectively in all places. Although the low grade heat produced in PEMFC cannot be used in co-generation, Mitsubishi (2004) has recently shown a total efficiency with a PEMFC system (can be as high as 83%) by utilizing this low grade heat for heating the house using their proprietary technology. Another disadvantage associated with PEMFC is that platinum catalysts are required to promote the electrochemical reaction If a reformate is used as fuel, carbon monoxide (CO) binds strongly to platinum sites at temperatures below 150°C, which reduces the sites available for hydrogen chemisorption and electro-oxidation. Because of CO poisoning of the anode, only a few ppm of CO can be tolerated with the platinum catalysis at 80°C. The reformed hydrocarbons contain about 1% of CO, a mechanism to reduce the level of CO in the fuel gas is needed. The low temperature of operation also means that little, if any, heat is available from the fuel cell for any endothermic reforming process (Krumpelt et al., 1992, 1993).

2. PEMFC Description and Construction Methods

2.1 Components of a PEMFC Stack

The primary components of a PEMFC are an ion conducting electrolyte, a cathode and an anode. The use of organic cation exchange membrane polymers in fuel cells was originally conceived by Grubb (1957, 1959). The desired function of the ion membrane was to provide an ion conductive gas barrier. Strong acids were used to provide a contact between the adjacent membrane and catalytic surfaces. After further development, it was recognized that the cell functioned well without adding acid. As a result, present PEMFCs do not use any electrolyte other than the hydrated membrane itself (Grune, 1992). The basic cell (Fig. 1) consists of a proton conducting membrane, such as a per fluorinated sulfonic acid polymer, sandwiched between two platinum impregnated porous carbon electrodes. The other side of the electrodes is made hydrophobic by coating with an appropriate compound, such as Teflon which provides a path for gas diffusion to the catalyst layer. Together, these three are often referred to as membrane electrode assembly (MEA), or simply a single fuel cell. In the simplest example, a fuel such as hydrogen is brought into the anode compartment and an oxidant, typically oxygen, into the cathode compartment. The other components are gas flow distribution plates for reactants, and mechanical components like end plates, current collectors, gaskets, bolts and nuts. The voltage of a fuel cell is small, about 0.7 volts when drawing a useful current, that is, to produce a useful voltage many cells have to be connected in series. Such a collection of fuel cells in series is known as a stack. The most obvious way to do this is to connect the anode plate with the adjacent cathode plate of the next cell with a wire all along the stack. However a better method to do this is to use a "bipolar plate" where the entire face of the anode plate (opposite the gas distribution side) is in contact with the obverse

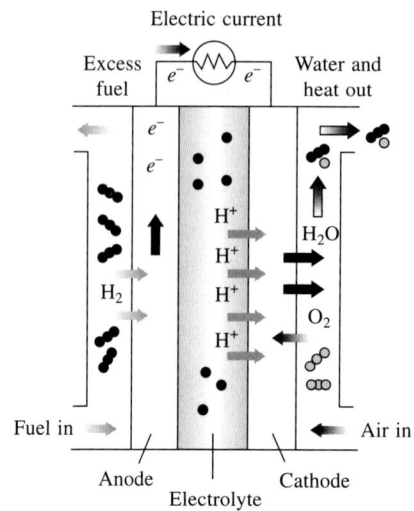

Fig. 1 PEM fuel cell (from U.S Department of Energy, Energy Efficiency and Renewable Energy).

of the cathode plate. So a major component of the fuel cell stack is the bipolar plate which provides the gas feeds to the cell and also transfers the current produced in each cell. Besides the gas bipolar plates, the fuel cell stack is interspaced with gas/water plates wherein on one side, one of the reactant gases is fed and the other side a coolant is supplied which helps in maintaining the temperature of the stack (Fig. 2).

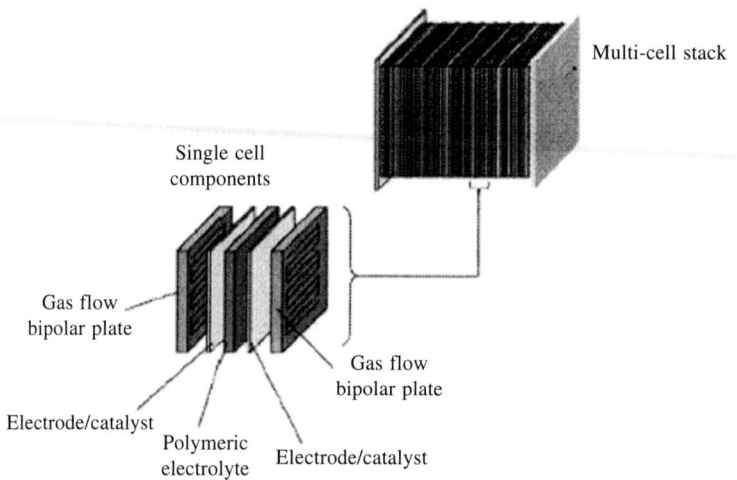

Fig. 2 Fuel cell stack assembly (from Costamagna 2002).

2.2 Membrane Electrolytes

PEMFC requires an ion exchange polymer in the form of a continuous pore free sheet. The properties which characterize the ideal ion exchange membrane fuel cell electrolyte will include the following: high ionic conductivity, zero electrical conductivity, low gas permeability, dimensional stability, high mechanical strength, low transference of water by conducting ions, high resistance to degradation, chemical stability to oxidation and hydrolysis. A number of factors affect the conductivity of the membrane, e.g. ionic size, charge and the solvation. The most highly conducting membrane electrolytes for fuel cell application are those in which the mobile ion is the hydrogen ion and the solvate is water at saturation. Water transport is of importance among the various properties required by the ion exchange membranes, both with regard to electro-osmotic transport of water and back diffusion of water once a gradient is established.

2.2.1 Perfluorosulfonic Acid (PFSA) Membranes

DuPont's Nafion® is the most advanced commercially available proton conducting polymer material, which is produced in membrane form with thickness between 25 and 250 μm. Nafion® is the electrolyte against which other membranes are judged and is in a sense an 'Industrial Standard'. It is a copolymer of tetrafluoroethylene (TFE), and perfluoro (4-methyl-3,6-dioxa-7-octene-1-sulfonyl fluoride) or "vinyl ether", as shown in Fig. 3. The methods of creating and adding the side chains are highly complicated and the process involving many steps is proprietary. One of the modern methods is described by Kiefer et al. (1999). The Teflon-like molecular backbone of the copolymer imparts chemical and thermal stability rarely available with non-fluorinated polymers. A lifetime of over 60,000 hours under fuel cell conditions has been achieved with commercial Nafion membranes. The ionic functionality is introduced when the pendant sulfonyl fluoride groups (SO_2F) are chemically converted to sulfonic acid (SO_3H).

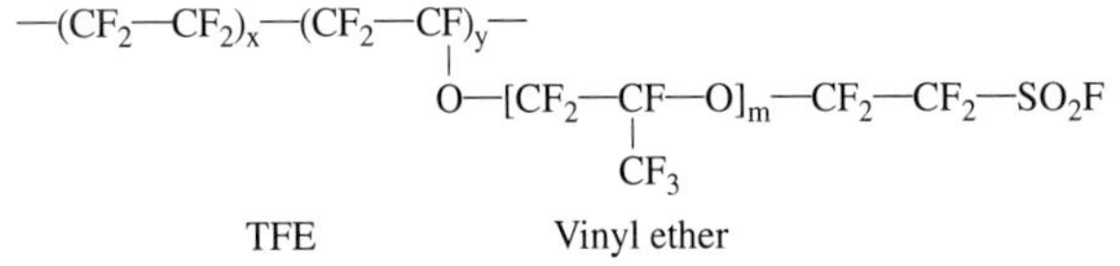

$$-(CF_2-CF_2)_x-(CF_2-CF)_y-$$
$$|$$
$$O-[CF_2-CF-O]_m-CF_2-CF_2-SO_2F$$
$$|$$
$$CF_3$$

TFE   Vinyl ether

Perfluoro (4-methyl-3,6-dioxa-7-octene-1-sulfonyl fluoride)

Fig. 3 Nafion® polymer structure before conversion to the sulfonic acid form.

The result is the presence of SO_3^- and H^+ ions in close proximity leads to clustering within the overall structure of the polymer. The copolymer's acid capacity is related to the relative amounts of co-monomers specified during polymerization, and can range from 0.67 to 1.25 meq. g^{-1} (1500-800 EW, respectively). The traditional extrusion-cast membrane manufacturing process was developed for "thick" films, typically greater than 125 μm. The extruded polymer film must be converted from the SO_2F to the SO_3K form using an aqueous solution of potassium hydroxide and dimethyl sulfoxide, followed by an acid exchange with nitric acid to the final SO_3H form (Smith, 1984). The PTFE backbone imparts hydrophobicity and the sulfonic acid group imparts hydrophilicity. The hydrophilic regions around the clusters of sulfonated side chains can lead to the absorption of large quantities of water (as much as 50% from the dry weight). Within these hydrated regions, the H^+ ions are weakly attracted to the SO_3^- group and are able to move. The polymer membrane thus has different phases—dilute acid regions within a tough and strong hydrophobic structure. In this micro-phase separated morphology, although the hydrated regions are somewhat separate, it is still possible for the H^+ ions to move through the supporting long molecule structure. However, for this to happen the hydrated regions must be as large as possible. Nafion membranes exhibit a protonic conductivity as high as 0.10 S cm^{-1} under fully hydrated conditions. For a membrane thickness of, say, 175 μm (Nafion 117), this conductivity corresponds to a real resistance of 0.2 ohm cm^2, i.e., a voltage loss of about 150 mV at a practical current density of 750 mA cm^{-2}. As an electrolyte, the polymer membrane provides an environment for electrode reactions at the electrolyte electrode interfaces. Compared with phosphoric acid, for example, the catalytic activity of carbon-supported noble metal catalysts for oxygen reduction is high in the PFSA electrolyte, due to the non-adsorbing nature of the sulfonic acid anions on the Pt catalyst surface among other factors. Solubility of hydrogen and oxygen are also found to be 20-30 times higher than that in phosphoric acid. As a result of the fast electrode reaction kinetics, the performance of PEMFC is high, especially at low noble metal loadings. The membrane also serves as a catalyst support and an effective gas separator. At 23°C and 50% relative humidity (RH), for example, the tensile strength of Nafion® membranes is about 40 MPa and the elongation is larger than 200%. The permeability of both oxygen and hydrogen through the membrane is of the order of 10^{-11} to 10^{-10} mol cm^{-1}s^{-1}atm^{-1}, corresponding to an equivalent current loss of 1–10 mA cm^{-2}, about 1% of the performance.

However, Nafion® has some limitations. The polymer only functions as a proton conductor when in a highly hydrated state, hence, a secondary hardware system must be used in PEMFC to humidify the gases before they pass into the stack. A further limitation, related to Nafion's® high dependence on humidification, is that it does not function well above 80°C in PEMFC (under normal operating conditions). But there is a necessity to operate PEMFC at higher temperatures as this would allow the fuel cell to tolerate much higher levels of CO (produced as a by-product in the fuel reformer), if the fuel is reformate hydrogen. When operating above 100°C, PEMFC will also benefit from enhanced gas transport in the electrode layers as liquid water would not be present in the structure. If the membrane is not strongly dependent on water to maintain its proton conductivity, it will not be necessary to humidify the gas feeds before entering the stack, simplifying the system further.

Dow Chemical Company and Asahi Chemical Company have made advanced perfluoro sulfonic acid membranes with shorter side chains and a higher ratio of SO_3H to CF_2 groups (Wakizoe 1995). The lower equivalent weights of these membranes compared to Nafion account for their higher specific conductivities, which enabled significant improvements in PEMFC performance, i.e. about 50-100 mV increase in cell potential at 1 A cm^{-2} over that on the control Nafion® 115, with about the same thickness (~100 μm). Since the ohmic overpotential is predominant in PEMFC in the intermediate to high current density range (0.3-1 A cm^{-2}), a logical approach to enhance power densities is to use membranes thinner than Nafion® 115. Experiments have shown that H_2/O_2 PEMFCs with Nafion® 112 membranes (50 μm thick) exhibit a cell potential of about 0.75 V at a current density of 1 A cm^{-2}. However, there have been problems of: (i) small amount of cross-over of the reactant gases, which reduced the open circuit potential of the cell by about 0.1 V and (ii) mechanical stability of the thin membranes, which have created hot-spots and cell failure. The first problem could partially be overcome and significantly the second problem too by employing supported Nafion® membranes (Bahar et al., 1997). In these membranes, solubilized Nafion is incorporated in a fine-mesh Teflon support. These membranes have a high mechanical strength, even when the thickness is as low as 10 μm. Furthermore, though the active proton conductor (Nafion) occupies only a fraction of the overall volume of the supported membrane, there is a compensation because the recast Nafion® in the membrane probably has a lower equivalent weight and hence a higher proton conductivity than the conventional Nafion film. Till date only M/s GORE associates seemed to have perfected this technology and GORE's MEAs (Kato, 2000) exhibit the best performance in PEMFCs as shown in Fig. 4 and several fuel cell developers (GM, IFC, energy partners, Plug Power) are using these MEAs in their stacks. A number of research papers on the development and performance of such composite membranes are appearing frequently after the easy availability of the expanded Teflon (Fig. 5) (Yu et al., 2004). In recent times DuPont also supplies solution cast membranes (Preishel et al., 2001, Kohler et al., 2002) with improved properties. The process used in the solution casting is shown in Fig. 6. A base film (1) is unwound and measured for thickness (2). Polymer dispersion is applied (3) to the base film, and both materials enter a dryer section (4). The composite membrane/backing film is measured for total thickness (5), with the membrane thickness the difference from the initial backing film measurement. The membrane is inspected for defects (6), protected with a coversheet (7), and wound on a master roll (8). The membrane is produced in a clean room environment (9). Master rolls are slit into product rolls, which are individually sealed and packaged for shipment. This process has several key advantages: (a) pre-qualification of large dispersion batches for

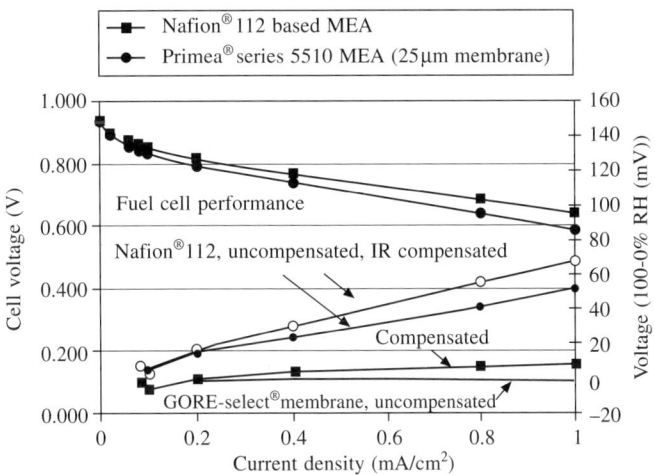

Fig. 4 Comparison of Nafion-112 and GORE-select membrane 25 mm based MEAs: Polarization performance and difference in polarization performance in cells operated with either dry or 100% RH cathode humidification (100% RH anode in both experiments). General Motors (GM) Global Alternative Propulsion Center (GAPC) 500 cm^2 fuel cell stack at cell temperature 80°C, hydrogen and air reactants at 2.0× stoichiometric flow and 270 kPa operating pressure (from Cleghorn et al., 2003).

Fig. 5 SEM micrograph (×5000) of the surface of porous PTFE membrane. SEM micrograph (×2500) of PN-50 composite membrane: (a) surface; (b) cross section (from Yu et al., 2004).

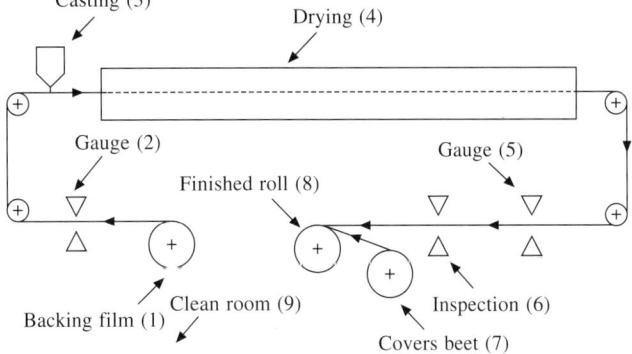

Fig. 6 Solution-casting process for Nafion® membranes (from Curtin et al., 2004).

quality (e.g., free of contamination) and expected performance (e.g., acid capacity); (b) increased overall production rates for H⁺ membrane from solution-casting as compared to polymer extrusion followed by chemical treatment; and (c) improved thickness control and uniformity, including the production capability of very thin membranes (e.g., 12.7 μm) (Curtin et al., 2004).

One of the major problems with the perfluoro sulfonic acid membranes has been and still is their high cost (~US\$ 700 m^{-2}) (Hogarth and Glipa, 2001). Thus, for a PEMFC operating at the desired power density of about 0.6 W cm^{-2}, the cost of membrane alone will be about US\$ 120 kW^{-1}. According to DuPont and Asahi Chemical, increasing the production of perfluoro sulfonic acid membranes to that required for at least a million vehicles per year could make it possible to reduce the cost of the membrane by a factor of 10. The high cost of perfluoro sulfonic acid membranes is due to the expensive fluorination step. Thus, partially-fluorinated and non-fluorinated ionomer membranes are currently under study. The other aspect that needs careful consideration is the safety concerns when the polymers are made (Hogarth and Glipa, 2001). Decomposition products could be a concern during manufacturing emergencies or the vehicle accidents and could limit fuel cell recycling options. However, recently DuPont has developed new perfluorinated polymers where the emission of fluoride ions has been brought down considerably (Fig. 7) (Curtin et al., 2004).

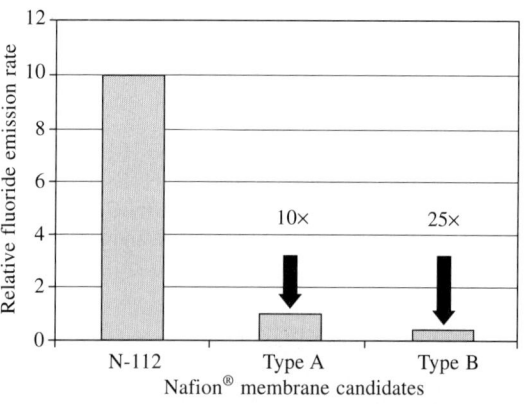

Fig. 7 Reduction in fluoride emissions for developmental Nafion® membranes made using DuPont's proprietary protection strategies (from Curtin et al., 2004).

2.2.2 Alternate Membranes

To overcome the deficiencies with PFSA membranes most notably with Nafion mentioned earlier, a number of research groups are carrying out extensive research on developing alternative PEM materials, some relying on other species than water for proton conduction (Kerres, 2001, Kreuer, 2001, Steele et al., 2001). Iojoiu et al. (2005) and Smitha et al. (2005) have reviewed all the polymer electrolytes that have been developed for fuel cell applications. There have also been significant developments in producing fluoropolymers of controlled architecture. An excellent review is provided by Souzy and Ameduri (2005) who have identified three families of such fluoropolymers. They can be separated into three main families of alternatives. The first concerns the direct radical copolymerization of fluoroalkenes with fluorinated functional monomers. The latter are either fluorinated vinyl ethers, α, β, β-trifluorostyrenes or trifluorovinyl oxy aromatic monomers bearing sulfonic or phosphonic acids. The resulting membranes are well-known: Nafion, Flemion, Hyflon, Dow, Aciplex or BAM3G (Table 1). The second route deals with the chemical modification of hydrogenated polymers (e.g. polyparaphenylenes) with fluorinated sulfonic acid synthons. The third alternative concerns the synthesis of FP-γ-poly(M) graft copolymers where FP and M stand for fluoropolymer and monomer, respectively, obtained by activation (e.g. irradiation arising from electrons, gamma-rays, or ozone) of FP polymers followed by grafting of M monomers. The most used M monomer is styrene, and a further step of sulfonation on FP-γ-PS leads to FP-γ-PS sulfonic acid graft copolymers.

The polymer membranes (other than the perfluorinated membranes) are classified into three groups, viz. (a) modified PFSA membranes, (b) alternate sulfonated hydrocarbon polymers and their inorganic composite membranes and (c) acid-base complex membranes.

Table 1. **Fuel cell membranes prepared from poly (Perfluorosulfonic acid) arising from copolymers of tetrafluoroethylene (TFE) and perfluorovinyl ether alkyl sulfonyl fluoride (from Souzy, 2005)**

Structural parameters and monomer contents	Supplier and trademark	Eq.wt (IEC: mequi.g^{-1})	Thickness (μm)
$n = 1; x = 5\text{-}13.5; p = 2$	DuPont		
	Nafion-120	1200 (0.83)	250
	Nafion-117	1100 (0.91)	175
	Nafion-115	1100 (0.91)	125
	Nafion-112	1100 (0.91)	50
$n = 0\text{-}1, p = 1\text{-}5$	Asahi glass		
	Flemion-T	1000 (1.00)	120
	Flemion-S	1000 (1.00)	80
	Flemion-R	1000 (1.00)	50
$n = 0, p = 2\text{-}5, x = 1.5\text{-}14$	Asahi Chemicals		
	Aciplex-S	1000-1200 (0.83-1.00)	25-100
$n = 0, p = 2, x = 3.6\text{-}10$	Dow Chemical		
	Dow	800 (1.25)	1.25
	Solvay		
	Hyflon Ion	900 (1.11)	

Modifications to PFSA Membranes

Considerable efforts are being made to modify the PFSA membranes to achieve high temperature operation. In one approach the water is replaced with non-aqueous and low volatile media. This approach has met with limited success. The other approach is to develop methods to improve water management. The water balance in a PEMFC involves the following mechanisms:

(1) Water supply along with the fuel and oxidant (humidification).
(2) Water produced at the cathode (current density).
(3) Water drag from the anode to the cathode (current density, humidity, temperature).
(4) Back-diffusion of water from the cathode to the anode (concentration gradient and capillary forces).

Accordingly, approaches have been developed for low humidification operation at both low (80°C) and high (above 100°C) temperatures. These approaches include reducing the thickness of membranes, impregnating the membranes with hygroscopic oxide nanoparticles, and solid inorganic proton conductors.

Nafion-silica composite has been studied by many groups (Antonucci et al., 1999, Dhathathreyan et al., 2001, Miyake et al., 2001, Yang 2001, Adjemian 2002, Costamagna 2002). All these groups, except Miyake (2001), have relied on solution casting method to make this composite membrane who used a sol-gel process. However, this membrane, although contained more water, showed decreasing conductivity with increased silica content, and was lower than that in the unmodified membrane under all conditions investigated. While Antonucci (1999) claimed a high membrane conductivity (1×10^{-1} Scm^{-1}) when operating in a liquid feed DMFC at 140°C, the data reported by this group was significantly poorer than that of Ren et al. (1996), whose measurements were carried out with Nafion-112 membranes of similar thickness. In addition, the material showed poor durability losing 100 mV cell voltage over 1000 h. Both these examples are typical of the available published data on this type of membrane.

Yang (2001), Adjemian et al. (2002) and Costamagna (2002) have also prepared Nafion containing SiO$_2$

and zirconium phosphate particles. They found, for example, that the silicon oxide modified PEMs showed improved robustness and water retention, which resulted in high conductivities at 130°C for 50 h.

Savadogo (2000) have prepared membranes based on Nafion, silicotungstic acid and thiophene. The modified membranes are reported to have a higher water uptake and conductivity than the unmodified membrane, resulting in improved fuel cell characteristics. Staiti et al. (2001) have also doped phosphotungstic acid and silicotungstic acid in the Nafion-silica membranes, which were found to show suitable properties for operation at 145°C in a direct methanol fuel cell. Ramani et al. (2004) have reported a Nafion-HPA composite which can operate at low humidity and high temperature. Choi et al. (2005) have prepared composite polymers containing Nafion and sulfated ZrO_2 and studied their thermodynamic and proton transport properties.

The most logical approach to developing a high temperature membrane would be to adopt materials which do not require water to maintain their proton conductivity. Other possible approaches include materials which have sufficient proton conductivity at reduced water contents and materials which have resistance to dehydration. DuPont have carried out some studies with doped Nafion membranes containing molten acidic salts (Doyle et al., 2000) such as 1-butyl, 3-methyl imidazolium trifluoromethane sulfonate (BMITf). This non-volatile proton conducting material gave membranes with high proton conductivities of around 0.1 Scm^{-1} at 180°C, even under anhydrous conditions. Savinell et al. (1994) doped Nafion-117 membrane with 11 M phosphoric acid to increase its conductivity at higher temperatures. They demonstrated reasonable proton conductivities around 5×10^{-2} Scm^{-1} at 175°C. It remains to be proven whether the dopants (e.g. H_3PO_4 or BMITf) are retained in the membrane over extended periods of operation. However, it should be noted that all of these materials employ modified perfluorinated sulfonic acid membranes, which are themselves costly to produce. The preparation of the composite materials would contribute further to an already expensive commodity making these materials economically unfriendly.

Hydrocarbon Membranes
Development of sulfonated aromatic polymer membranes (Table 2) as alternatives to PFSA has been an active area, principally motivated to lower the material cost and facilitate high temperature operation. The most widely investigated systems include sulfonation of polysulfones (PSF) or polyethersulfone (PES), polyetheretherketone (PEEK), poly(benzimidazoles) (PBI), poly(imides) (PI), polyphenylenes (PP), poly(4-phenoxybenzoyl-1,4-phenylene) (PPBP), rigid rod poly(p-phenylenes) (PP), and other polymers such as poly(phenylenesulfide) (PPS), poly(phenyleneoxide) (PPO), poly(thiophenylene), poly(phenylquinoxaline), and poly(phosphazene). Besides improved water retention at high temperatures, the membrane morphology of these ionomer-based membranes have been investigated. The membrane morphology is important for the performance, and is linked to the nature of the ionomer and the membrane formation process. It typically depends strongly on the water content, and on the concentration and distribution of the acidic moieties (Kreuer, 2001, Tang et al., 2001, Ding et al., 2002). For example, it has been shown that hydrated membranes based on sulfonated poly(etherketone) have a less pronounced separation into hydrophilic and hydrophobic domains, as well as a larger distance between the acidic moieties, as compared to the Nafion membrane (Kreuer 2001). High temperature polymers are reviewed by Jannash (2003). A variety of sulfonated polymers containing diarylsulfone units are being developed by different groups. Wang et al. (2002) have prepared high molecular weight polysulfones containing randomly distributed disulfonated diarylsulfone units. Analysis of the membrane morphology by atomic force microscopy revealed hydrophilic phase domains that increased in size, from 10 to 25 nm, with increasing degree of sulfonation. The membranes were stable up to 220°C in air, and highly sulfonated ones showed conductivities of 0.17 S cm^{-1} at 30°C in water. Poppe et al. (2002) have produced flexible PEMs based on carboxylated and sulfonated poly (arylene-co-arylene sulfone)s. The carboxylated materials as one would expect showed lower water uptake

Table 2. Non-fluorinated polymers used in PEMFC (from Souzy, 2005)

Polymers	Structure
Sulfonated polystyrenes	
Sulfonated polyimides	
Sulfonated poly(aryl ether sulfones)	
Sulfonated poly(aryl ether ketones) (S-PEEK)	
Sulfonated phenol formol resins	
Sulfonated poly(phenylene oxide)	
Sulfonated poly(phenoxybenzoyl-1,4-phenylene)	
Phosphonic poly(phenylene oxide)	R : CH$_3$ or CH$_2$P(O)(OH)$_2$
Sulfonated poly(benzimidazole)	
Sulfonated silicates	
Polyphosphazenes	(-PNR$_2$)$_n$

and lower conductivity in comparison with the sulfonated ones. Sulfonated polysulfones have also been blended with basic polymers such as polybenzimidazole (PBI) and poly(4-vinyl pyridine) in order to improve the performance in direct methanol fuel cells (Jonissen et al., 2002).

Aromatic polyimides (Genies et al., 2001, 2001a, Guo et al., 2002, Besse et al., 2002) show high levels of conductivity, but the hydrolytic stability is reported to be very sensitive to the chemical structure of the polyimide main-chain. A membrane-electrode assembly based on a sulfonated polyimide evaluated in a fuel cell at 70–80°C, was found to have a performance similar to Nafion.

Sulfonated PBI has also been investigated by various research groups as PEM for fuel cell applications. At low water contents, PEMs of PBI grafted with sulfopropyl units showed a proton conductivity in the order of 10^{-3} S cm^{-1} in the temperature range from 20 to 140°C, which is superior to Nafion under the same conditions (Kawahara et al. 2000). As the operation temperature of the PEMs is increased to temperatures above 100°C, loss of the sulfonic acid unit though hydrolysis, was observed (Bae et al., 2002). Poly (aryloxyphosphazenes) functionalized with phenyl phosphonic acid units have been developed for use in direct methanol fuel cells (Allcock et al., 2002). Poly(aryloxyphosphazenes) having sulfonimide units (Hoffmann et al., 2002) are also known. Blending and radiation cross linking have been investigated as means to reduce water swelling and methanol permeation of poly(aryloxyphosphazene) ionomers (Carter et al., 2002).

In sulfonated hydrocarbon polymers, the hydrocarbon backbones are less hydrophobic and the sulfonic acid functional groups are less acidic and polar. As a result, the water molecules of hydration may be completely dispersed in the nanostructure of the sulfonated hydrocarbon polymers. Both PFSA and sulfonated hydrocarbon membranes have similar water uptakes at low water activities, whereas at high relative humidity (100%) PFSA membranes have a much higher water uptake due to the more polar character of the sulfonic acid functional groups. The sulfonated aromatic polymers have different microstructures from those of PFSA membranes (Fig. 8) (Li et al., 2003).

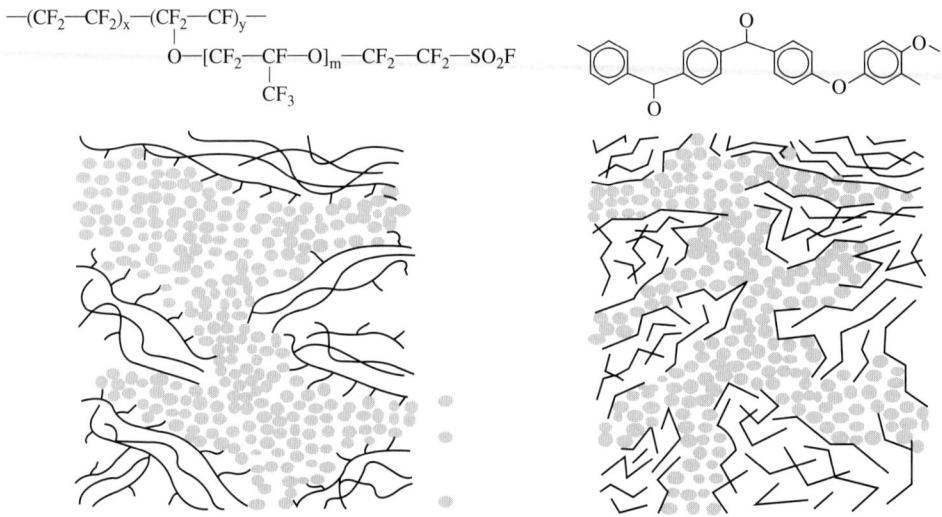

Fig. 8 Schematic illustration of the microstructures of Nafion-117 and SPEEK (from Li et al., 2003).

Another group of polymers that are being investigated are the organic and inorganic composites. These composite polymers based on sulfonated membranes, including partially fluorinated, silicone, and hydrocarbon polymers. It is interesting to note that some of these materials, especially the sulfonated hydrocarbons, exhibit improved performance at relatively high temperatures.

Hybrid membranes based on different arylene main chain polymers have also been investigated. For example, Bonnet et al. (2000) have studied the properties of hybrid membranes based on sulfonated poly (etheretherketone) and particles of amorphous silica, zirconium phosphate and sulfophenyl phosphate as a function of temperature and humidity. In all cases the presence of the particles led to increased conductivities at 100°C. Nanocomposite PEMs based on phosphotungstic acid in sulfonated polysulfones have been prepared by Hickner et al. (2001). Interestingly, the presence of the nanoparticles was found to increase the proton conductivity, while at the same time decreasing the water absorption. In addition, the mechanical modulus of the material was improved after addition of the particles. Genova-Dimitrova et al. (2001) incorporated phosphatoantimonic acid particles into sulfonated polysulfone and obtained PEMs with improved mechanical properties and conductivities close to Nafion, while avoiding excessive water swelling at 80°C. Staiti (2001a), has attached silicotungstic acid on SiO-support particles, and then used PBI as a binder to prepare membrane films. The materials are reported to be thermally stable with a conductivity of 10^{-2} S cm^{-1} at 160°C and 100% relative humidity. The use of a phosphonated PBI gave membranes with twice the conductivity at the same operating conditions. Ma et al. (2004), Staudt (2005) studied the conductivity of phosphoric acid doped PBI membranes for high temperature fuel cells and found that the upper limit of the conductivity is limited by the conductivity of liquid state H$_3$PO$_4$. Asensio et al. (2004) prepared phosphoric acid Impregnated poly 2,5-benzimidazole membranes and found that the conductivity is of the order of 6×10^{-2} S cm^{-1} at 150°C at 30% RH. Kim et al. (2005) evaluated the proton conductivity of benzimidazole/monododecyl iron phosphate hybrids for their suitability in fuel cell applications.

A somewhat different approach has been pursued by Honma et al. (2001) and Nakajima et al. (2002) who prepared different organic–inorganic hybrid materials by forming networks containing nanoparticles covalently linked by oligo ether segments. After doping the networks with various hetero poly acids, they reported proton conductivities of 10^{-3} Scm^{-1} in the temperature range 20-140°C under fully humidified conditions. Also, the thermal stability of the oligo ethers was greatly improved after formation of the hybrids. Stangar et al. (2001) have shown that a similar material based on silica functionalized by poly(propylene glycol) and doped with a heteropolyacid showed better results than Nafion in a methanol fuel cell, mostly due to a lower methanol cross over. Although the incorporation of various nanoparticles seems to be very encouraging, a great deal remains to be understood in these rather complex hybrid materials.

Acid-Base Polymer Membranes
Acid-base complexation is another approach to develop proton-conducting membranes. Basic polymers can be doped with an amphoteric acid, which acts both as a donor and an acceptor in proton transfer and therefore allows for the proton migration. Polymers bearing basic sites such as ether, alcohol, imine, amide, or imide groups generally react with strong acids such as phosphoric acid or sulfuric acid. The basicity of polymers enables the establishment of hydrogen bonds with the acid. In other words, the basic polymers act as a solvent in which the acid undergoes to some extent dissociation. Because of their unique proton conduction mechanism by self-ionization and self-dehydration, H$_3$PO$_4$ and H$_2$SO$_4$ exhibit effective proton conductivity even in their anhydrous form. When a basic polymer is present, the interaction between these acids and the polymer through hydrogen bonding or protonation would increase the acid dissociation, compared to that of anhydrous acids.

A number of basic polymers have been investigated: PEO, PVA, poly(acrylamide) (PAAM), and poly(ethylenimine)(PEI). Most of these polymers blended with acids exhibit proton conductivity less than

10^{-3} S cm^{-1} at room temperature. High acid contents result in high conductivity but the mechanical stability is poor, especially at temperatures above 100°C. Another concern is the oxidative stability of the tertiary C-H bonds in applications for fuel cells.

To improve the mechanical strength, several attempts have been made, viz. (1) Cross-linking of polymers (e.g., PEI), (2) Using high T_g polymers such as PBI and polyoxadiazole (POD) and (3) Adding inorganic filler or/and plasticizer

The combination of the acid and polymer forms a solid poly cation at low acid contents. When the acid content is higher, the plastifying effect of the excessive acid sometimes leads to the formation of a soft paste, which is unable to be processed into membranes. Addition of an inorganic filler such as high surface-area SiO_2 would make the materials stiffer, as demonstrated in systems of PEI-H_3PO_4-SiO_2, SiO_2-PVDF-acid, and Nylon-H_3PO_4/H_2SO_4-SiO_2. The latter was reported to exhibit a room-temperature conductivity as high as 10^{-1}Scm^{-1}. Shin et al. (2004) evaluated the transport properties of PEO complexes with SiO_2 and Al_2O_3 fillers and found that the fillers have no effect on transport properties under dry condition.

Most of the studied acid/polymer systems are not entirely anhydrous, as water is present as a necessary plasticizer for improving conductivity and mechanical properties. Gel electrolytes, as often termed, are obtained by introduction of organic plasticizers such as propylene carbonate (PC), dimethylformamide (DMF), and glycols. DMF and PC/DMF have also been used as plasticizers in H_3PO_4-PVDF and acid-PMMA systems.

New ionomer blend membranes have been synthesized by combining polymeric nitrogen-containing bases (N bases) with polymeric sulfonic acids. The sulfonic acid groups interact with the N-base either to form hydrogen bonds or by protonation of the basic N-sites (Kerres et al. 1999). The most advanced acid-base polymer blends are those based on sulfonated poly(etheretherketone) (S-PEEK) or *ortho*-sulfone-sulfonated poly(ethersulfone) (SPSU) as the acidic component, and poly(benzimidazole) (PBI) as the basic component. These membranes show excellent thermal stabilities (decomposition temperatures ranging between 270 and 350°C) and good proton conductivities. Their performance in direct hydrogen fuel cells at 70°C is similar to that of Nafion 112 membrane, however, only limited durability of around 300 h has been demonstrated. In addition to direct hydrogen testing, preliminary studies in DMFCs has shown their suitability for this application and it is reported that their methanol permeability is significantly lower than that of Nafion (Walker et al. 1999). Quantitatively, the methanol crossover rate is reduced by a factor of about 8 and 15, respectively, for S-PSU/PBI and S-PEEK/PBI membranes.

Walker et al. (1999) have also developed another polymer blend approach which is based on the mixing of sulfonated poly(phenylene oxide) and poly(vinylidene fluoride) (PVDF). These blend membranes show a particularly strong composite effect in that the non-conducting component enhances the intrinsic conductivity of the S-PPO. Some combinations of this material have higher ionic conductivity than pure sulfonated PPO. The performance of these membranes in direct hydrogen fuel cells at 45°C are reported to be much higher than that of Nafion 112 membrane. However, only 200 hours of durability has been demonstrated so far. Other advantages of these particular polymer blend membranes are that they have higher flexibility and mechanical strength, and lower water uptake than pure S-PPO. However, they are only thermally stable up to about 160°C making them unsuitable for fuel cell applications above 120–130°C. This is only a small disadvantage which is easily offset by the materials numerous advantages. In practice, S-PPO/PVDF is found to require less water than S-PPO while attaining similar conductivities. Quantitatively, this corresponds to a specific hydration number (ratio of H_2O:SO_3H) of 9.9 for the best S-PPO/PVDF blends, which compares favorably with S-PPO, Nafion 117 and Dow which have values of 18.5, 21 and 25 (Ren et al., 1996), respectively.

Other Polymer Systems

Kreuer et al. (1998) have outlined a very interesting approach to obtain proton conducting polymeric systems based on nitrogen-containing heterocycles, such as imidazole, benzimidazole and pyrazole . These heterocycles form hydrogen bonded networks similar to that found in water, and also their transport properties are similar to that of water with proton transfer occurring via structure diffusion. An important advantage of the heterocycles over water is that they can be covalently incorporated into polymer structures to obtain all-polymeric proton conductors, thus avoiding any volatile low molecular weight species. It is, however, important that the incorporation is accomplished in such a way that the heterocyclic groups retain a high mobility. Schuster et al. (2001) showed that imidazole terminated ethylene oxide oligomers can reach conductivities of up to 10^{-2} S cm^{-1} at 120°C. The conductivity was further enhanced after acid-doping. In another study, Yoon et al. (2001) prepared a polyurethane with imidazole units in the main-chain which reached conductivities of 0.1 S cm^{-1} at 140°C. Notably, these levels of conductivity were obtained in the complete absence of water. Uda and Haile (2005) have reported a solid acid fuel cells which utilizes an anhydrous non polymeric proton conducting electrolyte that can operate at 240°C with peak power densities as high as 415 mW/cm^2.

Several new strategies for developing polymer membranes which can operate at high temperature and low humidity are being worked out at Case Western University, USA in a collaborative program involving several partners (Zawodzinski, 2005). These efforts include development of multi block polymers, network structures, new materials based on polymer version of ionic liquids and multi-site bases, C$_{60}$ doped polymers (Fig. 9), sulfonimide proton conducting polymers, new polymer architecture with imidazole, etc., In another approach, Niyogi et al. (2005) are developing dendritic macromolecules and inorganic/organic hybrids. In one such effort they have attached dendrimers with polyepichlorohydrin to form water insoluble polymers. Dimensional

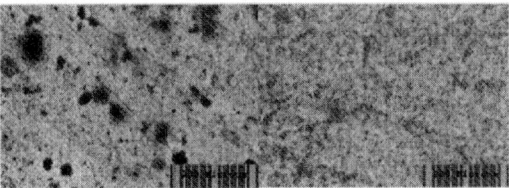

1 wt% C$_{60}$-Nafion composite 1 wt% C$_{60}$(OH)$_a$-Nafion composite

Fig. 9 Morphologies of fullerene-Nafion composite membranes (from Zawodzinski, 2005).

stability and conductivity have been improved by cross linking. Reichman et al. (2005) have claimed a superior low cost polymer composite based on PTFE/PVDF with ceramic nano powder and sulphuric acid which performs better than Nafion with methanol and DMG as fuel.

Teranishi et al. (2005) have used magnetic resonance imaging (MRI) to investigate the effect of thickness of the membrane on the fuel cell performance. They also used SECM technique to compare the results.

Theoretical Studies

To understand the mechanism of proton transport in ionomer membranes a number of theoretical studies have been carried out. Eikerling et al. (2001, 2001a) were able to predict values of PEM conductivities by using a model based on a heterogeneous membrane structure, and addressing relevant experimental parameters such as the concentration of acidic moieties and the level of hydration. In another study, the same authors (2002) carried out computations to evaluate the proton dissociation of various acidic moieties at different levels of hydration. They found that the sulfonimide moieties have higher degrees of proton dissociation at low water contents as compared to triflic acid, CF$_3$SO$_3$H, which has a higher tendency towards ion pair formation. Paddison et al. (2002) have taken morphological parameters obtained from SAXS data into account when calculating proton diffusion coefficients, which were obtained at different hydration levels and distances from the pore walls. Also in this study, the computed values were close to experimentally measured values. Li et al. (2000) have studied interactions of the hydronium ion with water and model Nafion structures using ab initio, density functional theory, and molecular dynamics simulations. The results indicated, that the flexible sulfonated perfluorinated side chain is stretched in the aqueous phase, and that the sulfonate-hydronium contact ion pair is very stable. Ab initio molecular dynamics simulations have also been performed to investigate the diffusion process of an excess proton in hydrogen bonded imidazole chains (Munch et al., 2001). The diffusion mechanism was described by a Grotthus mechanism involving a proton transfer step and a rate-determining molecular reorientation step.

2.3 Electrodes, Electrode Structures and Electro Catalysis

2.3.1 Gas Diffusion Electrodes

Gas diffusion electrodes have assumed a fundamentally important role in the technology of PEMFC. The treatment of the gas diffusion electrode is very complex, involving the physics of the porous matrix, the analysis of its equilibrium with the electrolyte and the gas as well as the thermodynamics and kinetics of the approximate transport processes. The surface of a gas electrode depends on the successful maintenance of a three phase equilibrium involving the electrode, porous matrix, the reactant gas and the electrolyte.

The porous gas diffusion layer in PEM fuel cells ensures that reactants effectively diffuse to the catalyst layer. In addition, the gas diffusion layer is the electrical conductor that transports electrons to and from the catalyst layer. Typically, gas diffusion layers are constructed from porous carbon paper, or carbon cloth, with a thickness in the range of 100–300 μm. The gas diffusion layer also assists in water management by allowing an appropriate amount of water to reach, and be held at, the membrane for hydration. In addition, gas diffusion layers are typically wet-proofed with a PTFE (Teflon) coating to ensure that the pores of the gas diffusion layer do not become congested with liquid water. Despite its many functions, Gas diffusion media has received very little development attention, as evidenced by the scarcity of publications on PEMFC diffusion media in the literature. Currently available diffusion media do not meet long-term requirements for cost, and development of less expensive materials is needed. Additionally, issues of flooding under steady-state and transient (e.g., start-up) conditions as well as low-current stability issues demand careful diffusion media design. Moreover, it is likely that durability is significantly impacted by diffusion-media substrate and treatment in ways that are not yet understood. There are many candidate materials and process variables that can be adjusted to develop optimum materials for a given application. In support of this, much remains to be done in terms of establishing characterization methods and property-performance relationships. The diffusion media will need to receive much more focused attention to development before the widespread commercialization of PEMFC fuel cells becomes a reality (Mathias et al., 2003).

2.3.2 Electrodes, Electro Catalysis and Kinetics

There are two basic criteria which must be satisfied before a chemical reaction can be considered as a source of energy in fuel cell. The first criteria is that at least one of the reactants must be ionisable at the operating conditions and the ionisable reaction system is the source of energetic electron flow that does work in the external circuit. During this process, the formed ions establish oxidation-reduction which when traversed by ions ultimately provides the energy that will be used at the terminals. For practical purposes, high flow rate of electron flow is required, hence the rapid rates of electron supplying and consuming reactions is a second criterion. The first criterion is the static system and is the object of thermodynamic inquiry. The second criterion is dynamic in nature and is the subject of chemical kinetics. However, the equal rates of the backward and forward electrode reactions that occur at static conditions establish the equilibrium. Hence first criterion may also be considered as dynamic and is also the subject of kinetics. Successful development of PEMFC depends on how well these kinetic criteria are met.

The PEM fuel cell consists of a gas diffusion electrode in which the reaction is considered to proceed through all of the following elementary steps:

1. Bulk flow and diffusion of reactant molecules through large electrode pores.
2. Adsorption of molecules on reaction site viz. platinum or platinum alloy catalysts.
3. Discharge of ionic species, proton to electrolyte.
4. Surface reactions between adsorbed molecules, discharged ions or radicals.
5. Desorption of products and transport into the electrolyte or pores.

Activation energy must be supplied before these processes and the magnitudes depend on the properties of the reactants, products and intermediates for most physical and chemical reactions. If any one of the reactions listed is slower than the activation, thermal and electrical output will control the cell's current voltage characteristics. The slow step may occur before or after the electron transfer step and may not involve charged particles. If the products or reactants of the non-electrical reaction participate in electron transfer, then potential will affect the kinetics by its effect on reactant activity. This potential dependent for the non-current producing reactions is due to chemisorption prior to electron transfer and a surface reaction following the ion discharge.

In a fuel cell electrode process, the reactant is assumed to be adsorbed at an active site on the electrode prior to electron transfer or surface reaction. The activation energy may often be low for gases such as hydrogen on a typical fuel cell catalyst like platinum, the adsorption of a hydrocarbon may require as much as 10 kcal, thereby presenting a formidable barrier to higher currents. Further limitations occur, if the surface is covered with products that are difficult to desorb or poisons like CO, that reduce the number of active sites. For these reasons, it is possible that the rate of chemisorption may be the slowest reaction step and therefore the sole source of polarization. At high current densities, depletion of reactants at the electrode may occur when they are removed by the electrode process faster than that can be supplied through diffusion. Product water may also accumulate faster than that can diffuse away from the reaction sites, which results in electrode potential change. Generally these potential losses are very large in comparison to those attributed to activation polarization. When the reactant activity approaches zero, the current density is limited by the rate of mass transfer and the electrode is being described as concentration polarized. In PEMFC, slow transport of gas through porous electrodes or ions through the electrolyte may result in concentration polarization. An improvement in diffusion layer characteristics, minimizing the mass transport problems, is particularly useful for the cathode, mainly working at elevated pressures in H_2/air operation.

The electrochemical oxidation of hydrogen is extremely facile on a Pt group catalyst. When operating with pure hydrogen at practical current densities, the anode potential is less than 0.1 V (vs. reversible hydrogen electrode). Under such operating conditions, the cell potential is only slightly lower than the cathode potential and the fuel cell performance effectively reflects the cathode operation.

In fuels like reformate or methanol, the methanol oxidation potential is not zero, and also leads to CO poisoning of the catalyst. Under these circumstances, the anode potential reaches 0.45 V. Hence the fuel cell performance not only depends on the cathode potential but also on the anode potential. In order to reduce the anode potential and to avoid CO poisoning, development of alternative Pt alloy electro catalysts like Pt-Ru and Pt-Sn are in progress.

The over potential for the hydrogen oxidation reaction is considerably lower than that for the oxygen reduction e.g., in a PEMFC operating at current densities of 1 A cm^{-2}, the over potential at the hydrogen electrode is about 20 mV and at the oxygen electrode it is about 400 mV. About one half of the over potential at the oxygen electrode is due to its loss at open circuit. The departure of the potential of the PEMFC from the reversible value is due to the extremely low exchange current density (i_0) for oxygen reduction (about 10^{-9} A cm^{-2}, very low if compared to that for the electro oxidation of hydrogen, 10^{-3} A cm^{-2}) on smooth platinum electrodes. Even after over 50 years of research, a conclusive mechanism for the intermediate and the rate determining steps for this reaction on different types of electrocatalysts have not been arrived.

This is unlike in the case of the two electron transfer hydrogen oxidation reaction, where there is definitive evidence for the reaction pathway.

One of the major problems with the Pt electro catalysis for hydrogen electrode is its low tolerance to CO in H_2 from reformed fuels. Furthermore, according to the US Department of Energy, an increase of the cell potential to about 0.75-0.8 V is necessary for PEMFCs to compete with compression injection direct ignition (CIDI) engines in order to meet the goal efficiency of 45% for fuel consumption in the PNGV program. The improvement can only be possible by reduction of oxygen (ORR) over potential by 50-100 mV. Such an improvement is possible by using inter metallic electro catalysts of platinum with a transition metal; Many investigations (Mukerjee et al., 1995; Fernandez et al., 2005; Gasteiger et al., 2005; Gonzalez-Huerta, 2005; Ismagilov, 2005; Rao and Trivedi, 2005; Reiner et al., 2005; Travitsky et al., 2005; Wells et al., 2005; Xie et al., 2005; Yu et al., 2005) have shown that some Pt-based alloy catalysts, such as Pt-M (where M = Co, Ni, Fe, V, Mn and Cr), exhibited an enhanced electro catalytic activity for the ORR compared to Pt alone. The improvement in the ORR electro catalysis on Pt alloy catalysts has been due to several factors such as electronic and structural effects. Usually, such carbon-supported Pt alloy catalysts were prepared by the impregnation of the second metal on Pt/C and then by alloying at temperatures above 700°C under inert gas or hydrogen. This heat treatment at high temperatures gives rise to an undesired alloy particle growth, which may result in the decrease in Pt mass activity (MA) for the ORR. Also, the control of the particle size distribution with this preparation method is quite limited. Pt alloy catalysts could also be prepared by the co-reduction of the metallic salt precursors at low temperatures and that the obtained Pt alloy particle sizes are relatively small. Another alternative way to tailor the nanosized Pt-based alloys for the different purposes is the use of organo metallic compounds as precursors. By thermal decomposition or reduction treatment of precursors, small nanoparticles of metal or alloy with narrow size distribution could be obtained. Among the various precursors used, metal-carbonyl complexes are often employed for preparing carbon-supported metal or alloy catalysts. It is known that, among the various Pt-based alloy catalysts used for the ORR, the Pt-Cr alloy is stable in acidic and oxidizing media at high temperature, whereas the Pt-Cu and Pt-Fe alloys are unstable under fuel cell operating conditions. Pharkya et al. (2005) have used a novel method (high energy ball milling) to prepare Pt-Co catalyst for oxygen reduction and find that the performance of this catalyst is superior to the conventional Pt-Co catalyst. A study on platinum supported on hydrous metal oxides (Swider-Lyons, 2003) has shown that these catalysts are at least six times more active than 20% Pt/C. These materials have open framework structures. However the open circuit potential is lower.

Cavaliere et al. (2004) studied the electrocatalytic activity of capped platinum nanoparticles towards oxygen reduction and established that the organic crown particles did not decrease the activity. Guo et al. (2004) analyzed the cathode performance from the perspective of five parameters and established that ionic conductivity play a major role in the performance over a wide range of current densities, while gas diffusion influences at high current densities. Bouwmann et al. (2004) and Rajalakshmi et al. (2005) studied the oxygen reduction reaction using Pt catalyst supported on iron phosphate and carbon nanotubes instead of Vulcan XC, respectively, which opens up several possibilities beyond the present capabilities of Vulcan XC support. Fe based ORR catalysts with various carbon supports have been reported by Villers et al. (2004) and they showed the increased catalytic activity by enhancing with N atoms and surface treatment. Williams et al. (2004) characterized the gas diffusion electrodes by porosity measurements, which gives the correlation between the permeability and the limiting current with respect to temperatures and humidity. Neverlin et al. (2005) have found that relative humidity plays a major role in the oxygen reduction kinetics. PEMFC face an efficiency loss , so called "oxygen gain", when cathode gas is changed from oxygen to air. Prasanna et al. (2004) found that oxygen gain can be reduced by optimizing the surface area and porosity of the carbon support and by coating the catalyst on membrane than on the GDL. Recent study by

Ismagilov (2005) showed the development of active catalysts by surface tailoring of nano carbon materials. The ionomer content, solvent composition and evaporation rate during the preparation of catalyst ink also affects the catalyst layer microstructure which in turn affects the mass transport and kinetics of the reaction as reported by Fernandez et al. (2005), Ahn et al. (2005), Siroma et al. (2003), Chen et al. (2004). The kinetics of the oxygen reduction without double layer and ohmic contribution has been investigated by Jayaraman et al. (2001), Liu et al. (2002), Fernandez (2003, 2005a) using scanning electrochemical microscope technique.

One of the challenges in PEMFC (as well as other acid fuel cells and AFC) research has been to find non-platinum containing electro catalysts for the fuel cell reaction. Platinum and/or platinum alloys are still the best electro catalysts and are used in the state-of-the-art fuel cells. Significant development took place in the late 1970s and early 1980s with a heat-treated metal-organic macro cyclic (e.g. cobalt tetra phenyl porphyrin) as the electro catalyst for the oxygen electrode reaction in alkaline media. The electro catalytic activity of these non-noble metals was close to that on Pt or Pt alloys. Due to the low corrosion rate of this transition metal, as well as of other transition metals such as iron or nickel which were tested as similar type metal-organic macro cyclics (Holze et al., 1986). However, in the perfluoro sulfonic acid polymer electrolyte, there was a considerable degradation in performance. In the 1990s, studies on ruthenium-oxide pyrochlore ($Pb_2Ru_{2-x}Pb_xO_{7-x}$) (Zen et al., 1994) and on pyrolyzed Fe(II) acetate adsorbed on 3, 4, 9, 10-perylenetetracarboxylic dianhydride (Faubert et al., 1999) were reported to show reasonable activities for the oxygen reduction reaction in acidic media, but these electro catalytic activities were less than that of platinum. Recently Brosha et al. (2005) evaluated the transition metal macrocycles like pyrolyzed TPP and TMPP chelates of Co and Co/Fe, metal chalcogenides (Ru-based and Ru-free catalysts) and metal oxides namely NiO, Co_2O_3, $NiCoO_2$, perovskitic LaSrCo oxides, CuMn oxides as oxygen reduction catalyst and found that COTPP shows a higher activity compared to all other chalcogenides and oxides. A novel cathode for PEMFC has been reported by Ishihara et al. (2005) who have used tantalum oxynitride as the catalyst.

Popov et al. (2005) have carried out durability studies and reported a molecular modeling method for selection of novel non-precious metals for PEMFC Catalyst for oxygen reduction. Lundbad et al. (2005) have identified a porphyrin based catalysts for oxygen reduction reactions. A non platinum electrocatalyst (cobalt hexacyano ferrate precursor dispersed in carbon support) showing higher activity than Pt/C catalysts has also been reported (Sawai and Suzuki, 2004). Self-assembled Pt nanoparticle electrode and membrane electrode-assembly have been successfully prepared by using charged Pt nanoparticles. The performance of the self-assembled MEA was 2.3 mW cm^{-2}, corresponding to a Pt utilization of 821 W per 1 g Pt. The results show that the self-assembled Pt nanoparticles were able to form a Pt monolayer and such a mono layered structure could potentially offer a powerful tool in the fundamental studies in the PEFC systems (Pan et al., 2005).

Hydrogen is the ideal fuel for PEMFCs, generating the highest level of electrochemical performance. But since hydrogen is not available readily, generation of hydrogen from hydrocarbons is a common method employed to generate hydrogen rich gases. But the level of CO impurities in the hydrogen produced via the steam-reforming or partial oxidation route is too high for PEMFC applications. The performance behavior of a PEMFC in the presence of CO in the fuel stream has been analyzed since the 1980s and it has been observed that concentrations as low as 10 ppm lead to a decrease of performance by about 0.2-0.3 V at 0.8 A cm^{-2} and when the concentration of CO is between 25 and 250 ppm in the hydrogen fuel stream the loss in performance is very severe. Significant developments towards the goal of realizing a CO tolerant PEMFC have been made in the 1980s and 1990s, by: (i) the use of binary Pt-Ru catalysts and (ii) the technique of oxygen bleeding in the fuel. The use of Pt-Ru alloy catalysts for PEMFCs was first proposed in the 1980s (Eisman, 1986), and results by Schmidt et al. (1999) show that the cell potential is 0.4 V at 1 A cm^{-2} with an electro catalyst loading of 1 mg cm^{-2} of $Pt_{0.5}Ru_{0.5}$ when 250 ppm of CO are present in the hydrogen fuel: the same cell exhibits a potential of 0.68 V at 1 A cm^{-2} when operated with pure hydrogen. One explanation for the enhanced electro catalytic activity of Pt–Ru is related to the changes in the lattice structure and in the surface properties due to alloying, which decrease the strength of the CO adsorption without increasing the over potential for electro reduction of hydrogen; another is that the ruthenium in the electro catalyst is in a partially oxidized state and provides the radical for the oxidative removal of CO adsorbed on neighboring platinum sites. Oxygen bleeding, the second technique proposed in the 1990s to solve the CO poisoning problem, involves injection of 0.4–2% of O_2 into a CO contaminated hydrogen stream to rapidly oxidize CO adsorbed on the platinum electro catalyst. The performance of a PEMFC with only the Pt electro catalyst operating with up to 100 ppm of CO in hydrogen was found to be identical with that using pure hydrogen (Gottesfeld et al., 1993; Wilson et al., 1993). However, a drawback of this method is that it cannot be used for higher concentrations of CO in the fuel stream, because higher concentrations of oxygen will be needed for its removal, and the limit in O_2 concentration is about 4–5%, which is close to the threshold value for causing an explosion. Further, oxygen bleeding could create hot spots and the subsequent development of pin-holes in the membrane. Another drawback is that the oxygen remaining after the reaction with CO (e.g. most of the oxygen added) reacts with the hydrogen in the fuel, causing a loss in columbic efficiency of the fuel cells. A related technique which was tested was to add small amounts (1–5%) of hydrogen peroxide to the humidification system; this led to the oxidative removal of CO by nascent oxygen (Divisek et al., 1998) the PEMFC tolerance level was about 100 ppm of CO on a pure Pt electro catalyst. A high percentage of CO_2 in the fuel stream can lead to a higher anodic over potential than that expected from hydrogen dilution effects (Vickers, 1996). The reason is that the water gas shift reaction causes the electro reduction of CO_2 to CO in the PEMFC, and hence electrode poisoning. Another method for increasing the CO tolerance of PEMFCs could be made by operating the cell at high temperature (Gottesfeld et al., 1998). This is because the strength of the CO adsorption on the Pt electro catalyst at above 150°C is considerably decreased as compared with that of hydrogen adsorption; as a consequence the CO tolerance of the PEMFC increases. However such membranes are not available commercially as yet. Zeolites as support for catalyst has been tried by Rosso et al. (2004) for CO preferential oxidation. Yamada et al. (2004) developed a new method namely IR thermography to screen the anode catalysts for high throughput screening. The CO tolerant sulfided catalysts developed by Venkataraman (2004) showed the CO tolerance of 100 ppm but the sulphur present in the catalysts resulted in redox activity, at lower potentials. The Pt-Mo based catalysts developed by Santiago (2004) and Mukherjee et al. (2004) showed a CO concentration of 100 ppm and 1500 hrs lifetime with superior performance compared to Pt/Ru catalysts (Fig. 10). Recently Lebedeva et al. (2005) have reported a Pt Mo/C bimetallic catalyst for anode reaction.

2.3.3 Gold as an Electrocatalyst

The application of gold as an electrocatalytic component within the actual fuel cell has to date been limited primarily to the historical use of an Au-Pt bi-metallic electrocatalyst for oxygen reduction in the Space Shuttle/Orbiter alkaline fuel cells (AFC) (Bockris and Appleby, 1986) and the recently claimed use of gold for borohydride oxidation in the direct borohydride alkaline fuel cell (DBFC) (Reeve et al., 2003, Lakeman and Scott, 2003). In order to make PEMFC commercially viable alternate catalysts are continuously being investigated. With gold presently approximately half the cost of platinum on a weight for weight basis, research programmes are evaluating gold as a potential electrocatalyst component, particularly as part of a bi-metallic system with platinum group metals. Recent results on gold and gold-platinum alloy nanoparticles as potential fuel cell electrocatalysts are encouraging (Zhong and Maye, 2001, Matsushita Electric Ind. Co., 2002, Zhong et al., 2003, Maye et al., 2003, Zhong et al., 2003a, Matsuoka, 2004). This work has focused on refining the synthesis, assembly and thermal treatment of shell capped Au and AuPt nanoparticles in the 2-5 nm size range and comparing the electrocatalytic oxygen reduction reaction (ORR) and methanol

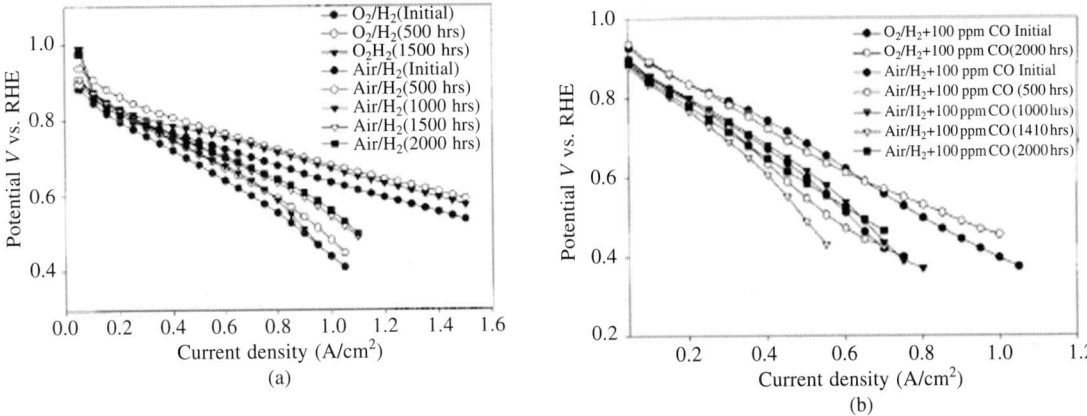

Fig. 10 Comparison of single-cell polarization profiles in a PEMFC measured at different intervals during the long-term steady-state test under an operating condition of 85°C under fully humidified conditions with anode cathode backpressure at 50/60 psig, respectively. Polarization curves with (a) pure H_2 as anode feed and cathode feed varied between air and O_2 and (b) anode feed kept at H_2 (100 ppm CO) and cathode feed varied between air and O_2 (from Mukherjee 2004).

oxidation reaction (MOR) activities of the Au and AuPt nanoparticle catalysts with commercially available Pt/C and PtRu/C catalysts. The AuPt catalysts with >70% Au and 10–25% metal loading exhibited at least comparable, and in some cases much higher catalytic activities than Pt (ORR) and PtRu catalysts (MOR) in alkaline electrolytes. World Gold Council commissioned a range of gold reference catalysts from Sud Chemie in Japan in 2002, under the guidance of AIST, Japan's National Institute of Advanced Science and Technology (Corti et al., 2005). The availability of the reference catalysts enables researchers to compare their own experimental results with those of other groups and it is believed that this supports the development and application of gold catalysts.

2.4 Catalyst Deposition and Membrane Electrode Assembly

2.4.1 Catalyst Deposition

Electrochemical energy conversion for technical applications relies on a high catalytic reactivity. The electro catalytic reactivity is strongly influenced by the structure and composition of the surface of the catalysts. In PEMFC the catalysts consist of Pt or Pt alloys, which are of nanometer size and often supported on carbon in order to optimize the surface area and the costs of the noble metals. However, the important relation between the reactivity and the structure is obscured in technical electro catalysts by a variety of parameters, like, e.g. the properties of the carbon support, the preconditioning of the catalyst, and the structure of the interface between the electrolyte and the active layer (e.g. the Nafion® content). An excellent review on the PEM fuel cell electrodes is given by Litster and McLean (2004)

The catalyst layer is in direct contact with the membrane and the gas diffusion layer, referred to as the active layer. In both the anode and cathode, the catalyst layer is the location of the half-cell reaction.

As can be seen in the flow chart of the preparation of PEMFC electrodes shown in Scheme 1, the development of PEMFC electrodes in recent years went from a two layer (A) to a three layer (B) electrode, using the same chemical composition for both anode and cathode. The next step was to design anodes and cathodes with different chemical composition of both the diffusion layer and the catalyst layer (C). Better water management in the cell was achieved using anode and cathode with different content of the hydrophobic agent. The evolution of the active layer of both anodes and cathode is shown in Scheme 2.

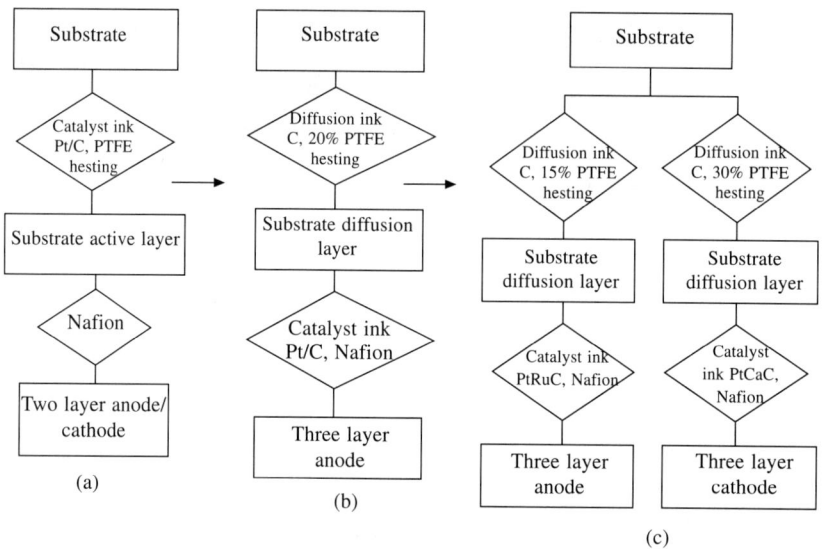

Scheme 1 Evolution of the preparation method of PEFC electrodes from two-layer (a) to three-layer (b) structure, and of anode and cathodes from the same (b) to different (c) composition of both diffusion and catalyst layer (from Antolini, 2004).

Essentially there are three types of PEMFC electrodes:

(a) Carbon paper + Catalyst layer, used particularly in the early 1990s. The standard dual layer structure is composed of a porous catalyst layer and a hydrophobic support layer.

(b) Carbon paper + Diffusion layer + Catalyst layer. Three layer electrodes consist of a porous backing layer, a diffusion layer formed by carbon particles and polytetrafluoroethylene (PTFE) and a catalyst layer formed by carbon supported platinum (Pt/C) and ionomer.

(c) Carbon cloth + Two diffusion sublayers (one on the catalyst side and the other on the gas side of the support) + Catalyst layer.

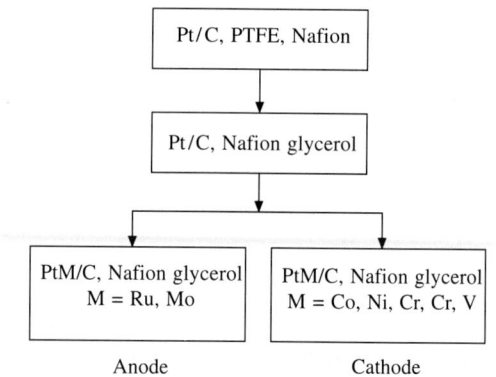

Scheme 2 Electrode configuration and preparation (from Antolini 2004).

Each of these electrodes have some advantages and disadvantages. The carbon paper electrode are more expensive than the carbon cloth electrodes. Efforts to reduce the cost of this component has been reported by Rajalakshmi et al. (2005). Evolution of the preparation method of PEMFC electrodes is described by Antolini (2004).

The two widely employed electrode designs are the PTFE-bound and thin-film electrodes. Other methods include those featuring catalyst layers formed with electrodeposition and vacuum deposition (sputtering). In general, electrode designs are differentiated by the structure and fabrication of the catalyst layer.

In PTFE-bound catalyst layers, the catalyst particles were bound by a hydrophobic PTFE structure commonly cast to the diffusion layer. In order to provide ionic transport to the catalyst site, the PTFE-bound catalyst layers are typically impregnated with Nafion by brushing or spraying. However, platinum utilization in PTFE-bound catalyst layers remains approximately 20% (Murphy et al., 1994, Cheng et al., 1999) The present convention in fabricating catalyst layers for PEM fuel cells is to employ thin-film methods. In this method the hydrophobic PTFE traditionally employed to bind the catalyst layer is replaced with hydrophilic perfluorosulfonate ionomer (Nafion). Thus, the binding material in the catalyst layer is composed of the same material as the membrane. Thin-film catalyst layers have been found to operate at almost twice the power density of PTFE-bound catalyst layers. This correlates with an active area increase from 22 to 45.4% when a Nafion-impregnated and PTFE-bound catalyst layer is replaced with a thin-film catalyst layer (Cheng et al., 1999).

The catalyst layer is either applied to the gas diffusion media (Mode 1) or to the membrane (Mode 2). In either case, the objective is to place the catalyst particles, platinum or platinum alloys, in close proximity of the membrane.

For Mode 1, there exist five methods for catalyst preparation and application to fabricate a GDL/catalyst assembly.

Spreading: The spreading method described by Srinivasan et al. (1994) consists of preparing a catalyzed carbon and PTFE dough by mechanical mixing and spreading it on a wet-proofed carbon cloth using a heavy stainless steel cylinder on a flat surface. This operation leads to a thin and uniform active layer on the GDL/catalyst assembly for which the Pt loading is directly related to the thickness.

Spraying: In the spraying method (Srinivasan et al., 1994, Subramaniam et al., 2000, Baturina et al., 2005) the electrolyte is suspended in a mixture of water, alcohol, and colloidal PTFE. This mixture is then repeatedly sprayed onto wet-proofed carbon cloth. Between each spraying, the electrode is sintered in order to prevent the components from re-dissolving in the next layer. The last step is rolling of the electrode. This operation has been found to produce a thin layer of uniform thickness and of low porosity on the GDL/catalyst assembly. For commercialization the spray technique is the most promising because a production line can be fully automated, therefore the electrode fabrication can be readily scaled up. In a study by Shin et al. (2002) who followed spraying method to coat the catalyst on to GDL, it was shown that Nafion in colloidal form gives better fuel cell performance than when it is in solution phase.

Ionomer impregnation: In the ionomer impregnation method described by Gottesfeld and Zawodzinski (1997) the catalytically active side of GDL is painted with solubilized PFSA in a mixture of lower aliphatic alcohols and water. To improve reproducibility of the GDL/catalyst assembly, the catalyst and ionomer are premixed before the catalyst layer is deposited, rather then ionomer impregnation of Pt/C//PTFE layer.

Electro-deposition: Electro-deposition as described by Taylor et al. (1992) and Gottesfeld and Zawodzinski (1997) involves impregnation of the porous carbon structure with ionomer, exchange of the cations in the ionomer by a cationic complex of platinum and electrodeposition of platinum from this complex onto the carbon support. This results in deposition of platinum only at sites that are accessed effectively by both carbon and ionomer.

Catalyst powder deposition: In catalyst powder deposition described by Bevers et al. (1998) the components of the catalytic layer (Vulcan XC-72, PTFE powder, and a variety of Pt/C loadings) are mixed in a fast running knife mill under forced cooling. This mixture is then applied onto a wet-proofed carbon cloth. Also applying a layer of carbon/PTFE mixture flattens out the roughness of the paper and improves the gas and water transport properties of the MEA.

In Mode 2, there are six methods for catalyst application to prepare a membrane/catalyst assembly.

Impregnation Reduction: In impregnation reduction (electroless deposition) as described by Fedkiw and

Her (1989), Foster et al. (1994) and Rajalakshmi et al. (2005), the membrane ion exchanged to the Na^+ form is equilibrated with an aqueous solution of platinum salt and a co-solvent of H_2O/CH_3OH. Following impregnation, vacuum dried PFSA in the H^+ form is exposed on one face to air and the other to an aqueous reductant $NaBH_4$.

Evaporative deposition: In evaporative deposition as described by Fedkiw and Her (1989) and Foster et al. (1994), $(NH_3)_4PtCl_2$ is evaporatively deposited onto a membrane from an aqueous solution. After deposition of the salt, metallic platinum is produced by immersion of the entire membrane in a solution of $NaBH_4$. The method has been found to produce metal loadings of the order of ≤ 0.1 mg Pt/cm^2 on the membrane/catalyst assembly.

Catalyst decaling: In the catalyst decaling method described by Gottesfeld and Wilson (1992, 1992a) and Chun et al. (1998)

Pt ink is prepared by thoroughly mixing the catalyst and solubilized PFSA. The protonated form of PFSA in the ink is next converted to the TBA + (tetrabutylammonium) form by the addition of TBAOH in methanol to the catalyst and PFSA solution. The paintability of the ink and the stability of the suspension can be improved by the addition of glycerol. Membranes are catalyzed using a "decal" process in which the ink is cast onto PTFE blanks for transfer to the membrane by hot pressing. When the PTFE blank is peeled away, a thin casting layer of catalyst is left on the membrane. In the last step, the catalyzed membranes are rehydrated and ion-exchanged to the H^+ form by immersing them in lightly boiling sulfuric acid followed by rinsing in deionized water. Bender et al. (2005) have established an improved method for decaling the catalyst layer onto the membrane by adjusting the properties of the catalyst ink and selecting a suitable substrate.

Casting Method: In a novel method suggested by Matsubayashi et al. (1994) PFSA solution is mixed with the catalyst and dried in a vacuum. Then, the PFSA coated catalyst is mixed with a PTFE dispersion, calcium carbonate used to form pores, and water. The mixture is passed through a filter and the filtrate is formed into a sheet. The sheet is then dipped in nitric acid to remove any calcium carbonate. The sheet is then dried and PFSA solution is applied to one side of the electrode catalyst layer. Finally catalyst layer is applied to the membrane.

Painting: In the painting method described by Gottesfeld and Wilson [1992] Pt ink is prepared as described for the decaling method. A layer of ink is painted directly onto a dry membrane in the Na^+ form and baked to dry the ink. When using thinner membranes or heavy ink applications, there will be considerable amount of distortion of the painted area. The distortion is managed through drying on a specially heated and fixtured vacuum table. Also, the bulk of the solvent is removed at a lower temperature to alleviate cracking and the final traces of solvent are rapidly removed at higher temperatures. In the last step, the catalyzed membranes are rehydrated and ion-exchanged to the H^+ form by immersing them in lightly boiling sulfuric acid followed by rinsing in deionized water.

Dry Spraying: In the dry spraying method described by Gulzow et al. (2000, 2002), reactive materials (Pt/C, PTFE, PFSA powder and/or filler materials) are mixed in a knife mill. The mixture is then atomized and sprayed in a nitrogen stream through a slit nozzle directly onto the membrane or GDL. Although adhesion of the catalytic material on the surface is strong, in order to improve the electric and ionic contact, the layer is fixed by hot rolling or pressing. Depending upon the degree of atomization, a completely, uniformly covered reactive layer with thickness down to 5 µm can be prepared with this technique. Some of the benefits of the dry layer technique are its simplicity because of the lack of evaporation steps, and its ability to create graded layers with multiple mixture streams. In addition, the platinum loading in the electrode fabricated is reported to be as low as 0.08 mg/cm^2. The cell performance results presented by this group depict a preparation method with good future potential for use in MEA mass production. Dry deposition technique has also been reported by Yu et al. (2005).

2.4.2 Other Methods

Colloidal Method

An alternative method to conventional thin-film techniques is the colloidal method. Typically, the catalyst layers are applied as a solution. It is well known that Nafion forms a solution in solvents with dielectric constants greater than 10. When a solvent which has a dielectric constant of 5.01 is employed as the solvent, a colloid forms in lieu of a solution. Shin et al. (2002) suggested that in the conventional solution method the catalyst particles could be excessively covered with ionomer, which leads to under-utilization of platinum. In addition, it was proposed that in the colloidal method the ionomer colloid absorbs the catalyst particles and larger Pt/C agglomerates are formed. The colloidal method is known to cast a continuous network of ionomer that enhances proton transport. The thickness of a catalyst layer that Shin et al. (2002) formed by the colloidal ink was twice that of the 0.020 mm thick layer formed with solution ink. In addition, the size of Pt/C agglomerates increased from 550 to 736 nm with the introduction of the colloidal method. The colloidal method dramatically outperformed the solution method at high current densities in single cell experiments.

Controlled Self Assembly

Middelman (2002) reported on the development of a catalyst layer that features a controlled morphology to enhance performance. A fabrication method to create a highly oriented catalyst morphology has been developed as an alternative to conventional methods that typically create a random morphology. To create highly oriented structures, Middelman increased the mobility of the catalyst layer with high temperatures and chemical additives. Then an electric field was employed as the driving force to orient the strands. Middelman suggests that this method could increase Pt utilization to almost 100%, and states that increases in voltages of 20% are obtained with this process. However details of this process are not available. Self-assembled Pt nano particle electrode and membrane-electrode-assembly of PEMFC have been successfully prepared by using charged Pt nanoparticles. Performance of the self-assembled MEA was 2.3 mW cm^{-2}, corresponding to a Pt utilization of 821 W per 1 g Pt. The results show that the self-assembled Pt nanoparticles were able to form a Pt monolayer and such a mono layered structure could potentially offer a powerful tool in the fundamental studies in the PEFC systems (Pan et al., 2005).

In Modes 1 and 2, sputtering among the various vacuum deposition methods, can also be used as a single step option to catalyst preparation and application as it is known for providing denser layers than the alternative evaporation methods (Cavalca et al., 2001).

In Mode 1, Srinivasan et al. (1994) describe a method in which a ~5 μm layer is sputter deposited on the wet-proofed GDL. In Mode 2, Dhar (1994, 1996) describe a method in which the catalyst is sputtered onto both sides of the membrane. To enhance the performance, a mixture of PFSA solution, carbon powder, and isopropyl alcohol is brushed on the catalyzed surfaces of membrane/catalyst assembly. The assembly is then dried in a vacuum chamber to remove any residential solvent. Sputtering and application of the ink is repeated to form a second layer of catalyst.

O'Hayre et al. (2002) have reported on their development of a catalyst layer with ultra-low platinum loading. Their paper suggests that they are developing these electrodes for use in micro-fuel cells since it was stated that the sputtering process is compatible with many other integrated circuit fabrication techniques.

Graded Catalyst Deposition

A graded or composite catalyst layer refers to a variety of catalyst layers that are produced with multiple deposition methods. A typical form is a supported catalyst layer, PTFE-bound or thin-film electrode, with an additional sputtering of platinum on the surface of the membrane or electrode. The objective of this method is to reduce the thickness of the supported catalyst layer and increase the catalyst concentration at the interface between the electrode and polymer electrolyte membrane. Cavalca et al. (2001) has described this procedure. The inventors combined thin-film methods and vacuum deposition techniques, such as electron beam-physical vapor deposition (EB-PVD) and dc magnetron sputtering, to fabricate a catalyst layer with progressive loading. The preparation of the catalyst layer began by mixing a common thin-film ink that contained carbon supported platinum, Nafion solution, and solvents, which was then brushed onto a PTFE blank for transfer-printing. Subsequently, a layer of catalyst, single metal or bimetallic, was deposited via EB-PVD or sputtering onto either the thin-film catalyst layer or the polymer electrolyte membrane. The inventors preferred method of vacuum deposition was EB-PVD because it exhibited greater surface texture, which aids the reaction kinetics. Thus, this method produces a dense pure catalyst layer directly adjacent to the membrane and places dispersed platinum further from the membrane with ionic transport provided by the impregnated Nafion. This technique has also been followed by Xie et al. (2005) to study the effect of nafion distribution on the PEMFC performance.

Multiple Layer Sputtering

Cha and Lee (1999) presented a novel strategy for depositing the catalyst layer onto the membrane (Nafion 115) of a PEM fuel cell. The process consisted of multiple short sputterings separated by an application of carbon-Nafion ink. The process was carried out on both sides of the membrane. After each sputtering, the newly formed film was brushed with a Nafion solution and then again with a Nafion–XC-72 carbon powder-isopropyl mixture. The addition of the carbon powder increases the electrical conductivity in the intermediate Nafion layer. CL found that after enough catalyst had accumulated on the surface, additional sputtering of platinum does not contribute to the amount of active area. A single sputtering thickness of 5 nm was found to be ideal. However, when the Nafion-carbon powder-alcohol mixture was applied between additional 5 nm thick sputterings the performance increased considerably. But, the marginal increase in performance was negligible after five sputterings.

Electro Spraying

Electro spraying of Pt/C-Nafion-alcohol dispersions was employed as a new method to deposit catalyst layers on Nafion membranes for hydrogen/oxygen air fuel cells, suggesting that control of electrospray processing parameters can lead to tailored electrode structures where such mass transport losses are mitigated (Baturina et al., 2005). Wei et al. (2005) have also used this technique.

The effect of pore formers and Nafion content in the catalyst layer have been investigated by Yoon et al. (2003) and Benitez et al. (2005). The dependence of performance on the catalyst loading was reinvestigated and studied by Gasteiger (2004), and development of new catalysts by incorporating a transition metal layer in between two platinum layers by Ross (2005).

The other new techniques for catalyst layer preparation are impregnation of polypyrole with Nafion (Park et al., 2004), grafting of polymer into the catalyst (Mizuhata, 2004), incorporation of organic solvents (Yang et al., 2004).

Membrane Electrode Assembly

The membrane and electrode assembly (MEA) is the 'heart' of the PEMFC. Its structure and composition are of vital importance:

  (i) to minimize all forms of over potential and maximize the power density.

(ii) to minimize the noble metal loading (and thus, the cost per kW of the PEMFC) in the gas diffusion electrodes by high utilization of the surface areas of nano-sized particles of the electro catalyst.

(iii) for effective thermal and water management (the latter including operation at the PEMFC without external humidification).

(iv) to attain lifetimes of PEMFCs.

The design of MEA could vary depending on the application of the fuel cell. Major breakthrough in this development was achieved during late 1980s and early 1990s. There was 10-fold reduction in platinum loading from about 4 mg cm^{-2} (as used in the Gemini space flights) to 0.4 mg cm^{-2} or less. The main reasons for making it possible to reduce the platinum loading from more than 4–0.4 mg cm^{-2} are: (i) considerably higher BET surface area of the carbon supported electro catalysts (particle size about 30 Å) than that of the unsupported previously developed PEMFCs electro catalyst (particle size about 100-200 Å); and (ii) extension of the three dimensional zone in the electrode by the impregnation of the proton conductor so that the utilization of the electro catalyst is similar to that in a fuel cell with a liquid electrolyte (e.g. phosphoric acid, potassium hydroxide). This led to the demonstration of high power density PEMFCs with electrodes containing a platinum loading of 0.4 mg cm^{-2} or less by an optimization of not only the structure of the electrode, but also by that of the MEA. In the late 1990s, other significant increases in power densities, with even further reduction in platinum loading (to a level of about 0.05 mg cm^{-2} for the hydrogen electrode and 0.1 mg cm^{-2} for the oxygen electrode) were achieved by deposition of thin active layers of the supported electro catalyst and proton conductor on an uncatalyzed electrode or on the proton conducting membrane. These active layers are only about 10-20 nm and contain no Teflon as in conventional electrodes. Because the active layers are considerably thinner than the conventional electrodes (10 nm vs. 50 nm), the ohmic and mass transport over potentials in the electrodes, (generally predominant at intermediate and high current densities) are greatly minimized. An equally important advantage of such types of electrodes is the increase in platinum utilization from about 20-25% to 50-60%. It is worthwhile stressing at this point that minimizing ohmic over potentials is vital for attaining high power densities; and this was made possible by using supported membranes (prepared by impregnation of Nafion into micro porous Teflon mesh, invented by W.L. Gore et al. (S. Cleghorn, 2000) and by deposition of very thin active layers (about 10 nm), containing only the carbon supported platinum nano crystallite and Nafion, directly on the supported membrane. This process reduced the contact resistance. Thus, this MEA has shown the best PEMFC performance to date and is being widely used by fuel cell developers in the USA and Japan (Costamagna et al., 2001). MEA's based on hybrid membranes prepared by a sol-gel process have been evaluated for fuel cell studies by Thangamuthu and Lin (2005). The hybrid membranes are made from alkoxysilane end capped poly(ethylene glycol) and 4-dodecylbenzene sulfonic acid. Frey et al. (2005) studied the fuel cell performance with respect to the preparation of MEAs. They found that changing electrode compositions or the use of different GDLs have a larger effect on the fuel cell performance than changing preparation parameters like hot pressing or spray conditions and fabrication of high performance MEA's has been evaluated by Bender et al. (2005). The degradation of MEA's with respect to performance over a period of time has been studied and reported by Cheng et al. (2004) that the formation of metal oxides at the anode catalyst layer leads to large particle size and hence decrease in catalytic activity. Xie et al. (2005) recently characterized the ionomer segregation in MEA's by AFM and identified the formation of ionomer skin due to decal process at the interface which affects the mass transport and hence performance at higher current densities. They also suggested that the skin formation can be prevented by alternate decal substrate with less hydrophobic, multiple ink application and elimination of hydrophobic lubricants. Fabrication of MEA's by electrophoretic deposition was carried out by Morikawa et al. (2004) and they exhibited a higher platinum utilization of 56% compared to normal hot press method. Lindermeir et al. (2004) have reported a tuned layer preparation and coating technology for making MEAs which is suitable for large scale production. The production sequence used is: wet ball milling, wet spray coating and calendaring. Tek de Nora has developed a dual beam assisted platinum multiplayer deposition in high temperature membrane V for improved mass transfer with less interfacial contact resistance (De Castro, 2005). Similarly, 3M has developed a technique for MEA fabrication in high volume with enhanced operating conditions (Debe, 2005).

2.5 Bipolar Plates

In a fuel cell stack, bipolar (also known as flow field or separator plates) typically have four functions:

(1) Distribution of fuel and oxidant within the cell
(2) Separation of the individual cells in the stack
(3) Facilitation of water and thermal management within the cell
(4) Current collection

Mehta and Cooper (2003) note that plate topologies and materials facilitate these functions. Topologies can include straight, serpentine, or inter-digitated flow fields, rigid or flexible plates, internal or external manifolding, internal or external humidification, and integrated cooling. In fact, Mehta and Cooper reviewed over 100 topology-material combinations and related fabrication options for PEMFC bipolar plates. For many design options, bipolar plate design requirements have been proposed by many researchers and are summarized in Table 3. Here, requirements have been grouped into four categories: stack performance related design criteria, system performance (for the vehicle, building, or other product or process needing power) related design criteria, manufacturing related design criteria, and environmental impact related design criteria. Although most researchers who discuss bipolar plate design discuss the former two categories, fewer investigate manufacturing and environmental design requirements. Cooper (2004) analyzed bipolar

Table 3. Summary of PEMFC bipolar plate design requirements (from Cooper, 2004)

Category	Requirement
Stack performance related design criteria	Electrical resistance is minimized
	Thermal resistance is minimized
	Allows distribution of the fuel, oxidant, residual gases and water without leaks
	Withstand mechanical loads during operation
	Resistant to corrosion in contact with electrolyte, oxygen, heat and humidity
	Minimize coefficient of thermal expansion between metal plates and coatings
System performance related design criteria	Mass/kW is minimized (plates should be lightweight)
	Volume/kW is minimized plates should be slim)
	Stacks must operate in freeze and cold conditions
	The design life is maximized
Manufacturing related design criteria	The stack is inexpensive to manufacture
	Plates should call for manufacturing processes with high yields relative to mass production
	Length/width should be system defined (flexible cross section)
	The plate surface finish requirements are minimized to increase manufacturing options
	Plate tolerance should be maximized to increase manufacturing options
Environmental impact related design criteria	Plate materials are made of recycled materials

plate design focusing on requirements for stack and automotive performance, Design for Manufacturing (DFM) and Life Cycle Design (LCD) and has arrived at lists of 51 requirements and 69 engineering characteristics which can be used for quantitative analysis or as a qualitative guide.

There are several types of materials that are being used in bipolar plates. The selection of a bipolar plate is based on the flow fields, system requirements and materials. Flow fields are generally machined, stamped on the bipolar plates. Bipolar plate materials include graphite, metallic plates with or without coating, and a number of composite structures. Sheet metal, graphite foil and graphite polymer composites are potentially low cost materials, and in principle suitable for mass production. High purity electro graphite is an excellent material for machining prototype plates, but material costs and process costs are generally considered high for mass production. Carbon–carbon composite is not expected to achieve cost price targets, and needs expensive post processing. Flexible graphite is a thin, low density, inexpensive material made from expanded natural graphite. Being based on natural graphite, purity and consistency of quality are real concerns for this material. Another drawback of graphite foil is the very limited formability and poor dimensional stability. Roser et al. (2005) have recently reported a low cost graphite foil sheet for portable applications by integrating the gas diffusion media. Graphite filled polymer composite can offer a combination of inexpensive material and economical processing. Thin sheet metal, for example, 125 mm stainless steel can be stamped establishing a mass production process, but has a drawback of increased contact resistance and ionic contamination of membrane and catalyst, thus limiting the life of the stack. The corrosion behaviour of sheet metals in fuel cell environment has been studied by Li et al. (2004). Lee et al. (2004) have shown that the corrosion resistance of SS can be increased by surface modification by electrochemical methods. TiN coated SS bipolar plates have been investigated as bipolar plates by Wang et al. (2004, 2004a and 2004b) and by Cho et al. (2005). The heterogeneous bipolar plate with low contact resistance under very low compression force, less stack weight and volume, full electrode utilization, has been studied by Lee et al. (2005a) for high performance and low cost. Joseph et al. (2005) have studied the effect of conducting polymers coated on SS bipolar plates with respect to corrosion . The chemical treatment for Al plates has been studied by Lee et al. (2005). Hung and Tawfik (2005) coated with a proprietary material which shows superior performance to graphite in fuel cells. Further, they demonstrated that hydrogen consumption could be reduced by 22% by using these plates. Fe based amorphous alloys have been studied for their corrosion properties under the fuel cell environment by Jayaraj et al. (2005).

2.5.1 Reactant Flow Field

In order to supply the fuel and oxidant to the fuel cell, the bipolar plates have flow fields which are macroscopic channels running along the bipolar plates. Computer simulation studies are being carried out extensively in recent times to optimize the design of the flow field. The feed channels can vary from 'through'-channels, to dead-end channels where all the reactants are expected to be consumed within the cell, except for reformate fuel and air oxidant. Channels may have a single or parallel meander geometry, interdigitated geometry or a structure of parallel meanders with differential gas pressure. Some of the known designs of flow field plates are given in Table 4. Each of these designs have advantages and disadvantages.

In the simplest case at constant inlet flux, ideal humidification of membrane, and negligible losses on hydrogen side, the lower the stoichiometry ratio on the cathode side, the more oxygen will be consumed close to the inlet and less left to consume close to the outlet. This will result in a highly nonuniform current distribution, an aspect which is overlooked in most fuel cell literature in which average current densities are reported. If the downstream part of the channel is thus starving, then this later region of the cell will underperform or not perform at all. Similarly higher stoichiometry will lead to a more uniform reactant distribution, but this has the consequence of reducing the overall efficiency of the system. There needs to

Table 4. Some characteristics of different flow channels

Characteristics	Serpentine	Parallel	Fractal	Interdigitated
Length of the channel	More	Less	Less	Less
Pressure drop	More	Less	Less	More
Pumping power	More	Less than serpentine	Less than parallel	No data
Methanol crossover	Less than interdigitated	No data	No data	More
Reactant molecule Entry	Diffusion mechanism	Diffusion mechanism	Diffusion mechanism	Forced convection
Humidification	—	—	—	Good
Cell resistance	0.12 ohm cm^2	—	—	0.08 ohm cm^2
Contact area	Less	—	—	More
Short circuit current density	1.25 A/cm^2			1.8 A/cm^2
Methanol crossover rate @130 C, 2M	5.6 × 10^{-6} mol/min cm			11.5 × 10^{-6}
Fuel utilization	High			Less

be a balance between the overall reactant utilization, and the current distribution. Increasing the reactant utilization seems to require greater nonuniformity of current distribution, and hence greater energetic losses within the fuel cell. Minimization of energetic losses within the fuel cell requires higher stoichiometry which in turn lowers the overall fuel utilization. Understanding the laws of fuel consumption helps to rationalize the working regime of the fuel cell and provides hints towards promising cell and stack structural modifications (Kulikovsky et al. 2004).

The flow field design by CFD has been a major activity to arrive at the best possible configuration. He et al. (2000) studied the effects of various electrode and flow field design parameters on the performance of the cathode of a PEMFC. It was found that higher differential pressure between inlet and outlet channels will enhance the electrode performance. Hontanon et al. (2000) found from their 3D simulation study that fuel utilization increased when decreasing the permeability of the flow distributor. In particular, fuel consumption increased significantly when the permeability of the porous material decreased to values below that of the anode. This effect was not observed in the grooved plate, where permeability was higher than that of the anode. Even though the permeability of the grooved plate can be diminished by reducing the width of the channels, values lower than 1 mm are difficult to attain in practice. The simulation shows that porous materials are more advantageous than grooved plates in terms of reactant gas utilization Glandt et al. (2002) have carried a CFD analysis on two flow-field configurations and studied their effects on the PEM fuel cell performance. The local current density, temperature, and liquid water concentration for single pass and double pass patterns were analyzed. It is concluded that changing flow-field configuration of the PEM fuel cell can affect the current density distribution and also its performance. The double pass flow-field gives more uniform current density distribution and less condensed liquid water on the cathode than the single pass flow-field. This liquid water concentration profile is also dependent on not only inlet humidity and gas diffusion layer property but on pressure distribution created by flow-field pattern and inlet flow rate.

Kumar and Reddy (2003) studied the optimization of the channel dimensions and shape in the flow-field of bipolar/end plates. Single-path serpentine flow-field design was used for studying the effect of channel dimensions on the hydrogen consumption at the anode. They carried out simulations from 0.5 to 4 mm for different channel width, land width and channel depth and found that for high hydrogen consumptions (80%), the optimum dimension value for channel width, land width and channel depth was close to 1.5, 0.5 and 1.5 mm, respectively. Studies on the effect of channel shapes showed that triangular and hemispherical shaped cross-section resulted in increase in hydrogen consumption by around 9% at the anode.

Ganesh Mohan et al. (2004) have used CFD modeling to predict mal distribution of reactants in fuel cell stack. The parametric studies revealed that flow rates and port dimensions play a major role in distribution of the gases for a given geometry of the flow field design (Fig. 11).

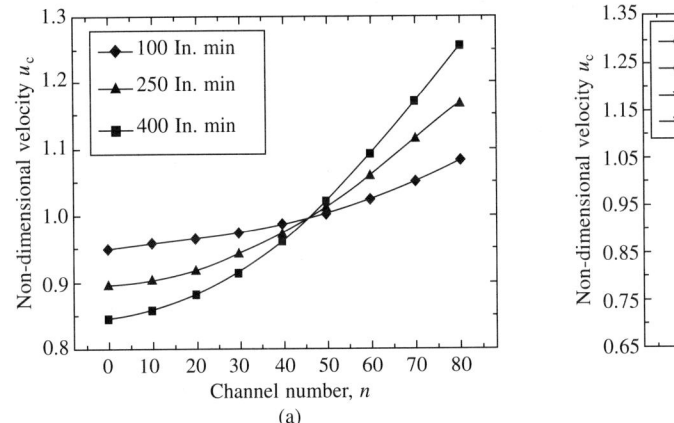
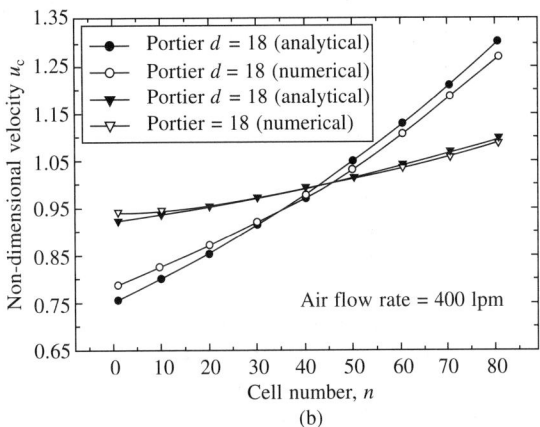

Fig. 11 (a) Channel velocity profile for the flow of air at 400 lpm in a 80 channel PEMFC of different port dimensions; (b) Channel velocity profile for the flow of air through 18/16 mm, 80 channel PEMFC of different flow rates (Ganesh Mohan et al., 2004).

The flow pattern around a 180° turn in a square section 2 mm gas channel of an optically accessible operating PEM fuel cell has been investigated using the non-invasive laser Doppler anemometry technique (Kucernak et al. 2005). The results reveal the symmetrical Ekman vortex pair expected to be present while operating at Reynolds and Dean numbers of 80 and 54, but also that additional complexity due to the combination of the operating MEA and the abrupt change in wall direction. Measured velocities normal to and near the GDL surface are found to exceed these diffusion velocities by several orders of magnitude implying that flow into the GDL is not inconsequential, contrary to the assumption made in many models.

Kim et al. (2005) have shown that the dynamic behavior of a proton exchange membrane fuel cell subjected to rapid changes in the voltage depends on the type of flow field and also on the voltage range of the voltage change. The studies have been carried out when the flow rates are constant and at levels that result in operation between fuel stoichiometries of 1.2 and 1.1. The results show that the undershoot of the final steady-state current that follows the resulting overshoot was observed with a standard triple-path serpentine flow field (SFF) but not with a single-path SFF. The results also show that the dimensionless peak current and the percentage of the overshoot current depend on the starting cell voltage and on the range of the voltage change. Evidence is presented to indicate that these peak heights are limited primarily by oxygen mass transfer, even though the PEMFC is operating at close to fuel-starved conditions.

Liu et al. (2005) found that the reactant transport and cell performance can be enhanced by the presence of the baffles in the flow channel of the bipolar plate, especially at the operating conditions of low voltage. The beneficial baffle effects become increasingly remarkable with increasing width and/or number of baffles in the tandem array.

The conventional ribbed serpentine and parallel flow distributors exhibit limited reactant/product mass transfer to and from the part of the diffusion and catalyst layers which is not covered by the flow channels (i.e., under the current collectors "shoulders"). This effect is more distinct on the cathode side due to the low diffusivity of oxygen. These deficiencies can be overcome using porous flow distributors (Senn and Poulikakos, 2004).

Kumar and Reddy (2005) also have carried out simulation studies to find the effect of gas flow-field design in the bipolar/end plates on the steady and transient state performance of the PEMFC. Simulations were performed with different flow-field designs, viz. (1) serpentine, (2) parallel, (3) multi-parallel and (4) discontinuous. The steady-state voltage at fixed current density of 5000 Am^{-2} was highest for discontinuous design. For studying the transient response, the average current density was increased suddenly from 5000 to 8000 Am^{-2}. It was observed that when the load level was increased, the voltage level suddenly dropped and then with time leveled off to a value slightly higher than the dropped value. This time for serpentine, parallel, multi-parallel and discontinuous flow-fields were 9.5, 7.5, 8.0 and 16.5 s, respectively. While it was seen that the steady-state performance of the discontinuous type of design was the maximum, its transient response was slow. On the other hand in case of parallel type of design the steady-state performance was low, but the transient response was high. The multi-parallel design offers a unique advantage of both of these properties, viz. high steady-state performance with good transient response, and is expected to perform better.

In a fuel cell the current density distribution on all scales is of high interest for the understanding and the optimization of cell structures with respect to power density. The size and ratio of the flow field ribs and the anisotropic electric conductivity of the GDL are the major issues on a sub millimeter scale. To examine and optimize these parameters, the processes occurring on this scale have to be investigated. The problem is not trivial, because the fuel cell local current generation is not constant over the MEA area covered by flow field ribs and the area covered by channels. Local current density on this scale can be modeled by accounting for the reactant flow limited by diffusion and the electrochemical reaction limited by electric resistance, as done in numerous scientific publications (Meng and Wang, 2004). These calculations are mostly based on assumptions and neglect, for example, the anisotropy of the GDL conductivity. It is therefore highly necessary to conduct experiments for gaining input information on simulations. Factors that lead to inhomogeneities in current density distribution are limitations due to electric resistance in the GDL and reactant gas diffusion. Furthermore, the presence of liquid water at the electrodes adds to the complexity.

2.6 PEMFC Stack Construction Methods

PEMFC stacks are constructed by connecting a number of cells in series or parallel depending on the voltage and current requirement with gas distributors (bipolar plates) that feed the fuel to the anode and the oxidant to the cathode. Fuel cell stack construction using bipolar plates gives very good electrical connection between one cell to the next. However, it does have the problem that there are many joints and potential problems or reactant gases and coolant leaks. The entire edge of each anode and cathode is also a potential leak. Careful manufacturing practices have to be followed. The non-repeat components in the stack assembly are: stack housing, end plates, tie rods, current collectors, insulators etc. End plates serve as structural members for the fuel cell stacks made from cast aluminum alloy, and can be subsequently either hard anodized or powder coated. Two electrical insulators, die cut from elastomer sheet are required to isolate the end plates from the electrically charged current collectors of the stack. Tie rods with associated lock washers are needed to compress the fuel cell stacks. The active area of the electrode decides the number of tie rods required for uniform compression. The repeat units are: gasket-MEA-gasket-bipolar plate (with or without cooling). The reactants to each cell can be supplied from a common gas manifold externally or internally. In micro fuel cells there are different methods employed to connect the cells in series or parallel which many times avoid using a bipolar plate. In another topology, especially useful in micro fuel cells, is using a single compartment for the air and another compartment for fuel, the compartments being made from low cost plastics. On a single sheet of polymer electrolyte, a number of electrodes matching each other on each side are placed and laminated. Edge connections using very thin platinum or gold wire is used in such cases to build the voltage. These are known as flat fuel cells. These banded structure concept is explained by Henzel et al. (1998) (Fig. 12a,b).

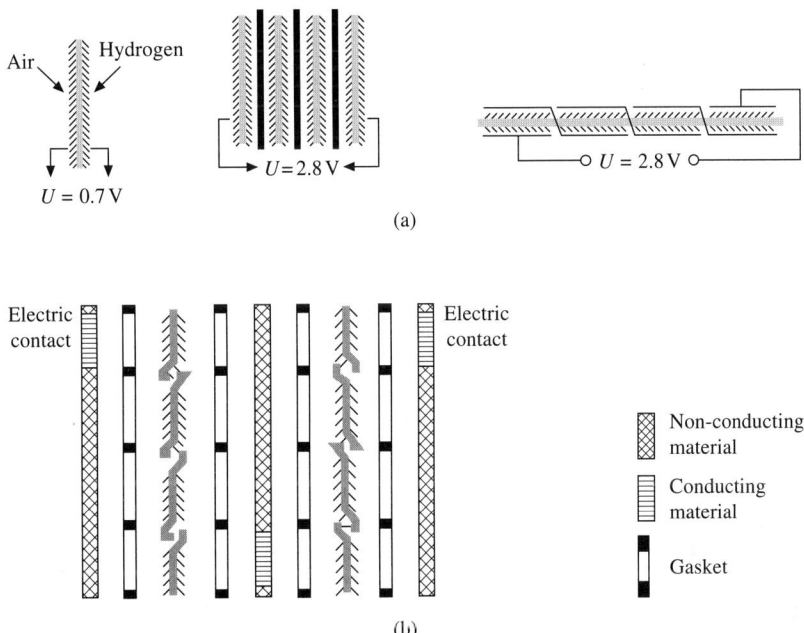

Fig. 12 (a) PEM Fuel cell concepts: single cell (left), cell stack (middle) and banded structure membrane (right) (from Heinzel et al., 1998); (b) Banded structure stack with two membranes having 4 cell units each (from Heinzel et al., 1998).

The end plates offer considerable scope for improvement. The main requirements of fuel cell end plates are:

(a) Enhance the current in the stack by reducing the contact resistance between the bipolar plates.
(b) Supply sufficient sealing forces for the different media flows. These effects may lead to increased stack pressure along the boundary of the bipolar plates, since the compaction pressure is different from sealing pressure.
(c) Stabilize the stack to resist external forces and moments in real conditions (e.g. vehicle accelerations).
(d) Stack fixation to the surrounding system.
(e) Connection of the fuel, oxidant and cooling supply lines.

The conventional end plates made of aluminum, to overcome the deformation, are made necessarily thick. Two new concepts namely (a) bomb shaped end plates and (b) 'Dbow-concept' have been demonstrated at Tribecraft AG (Evertz 2002)) which improve the specific power of the fuel cell stack (Fig. 13 (a, b, c)).

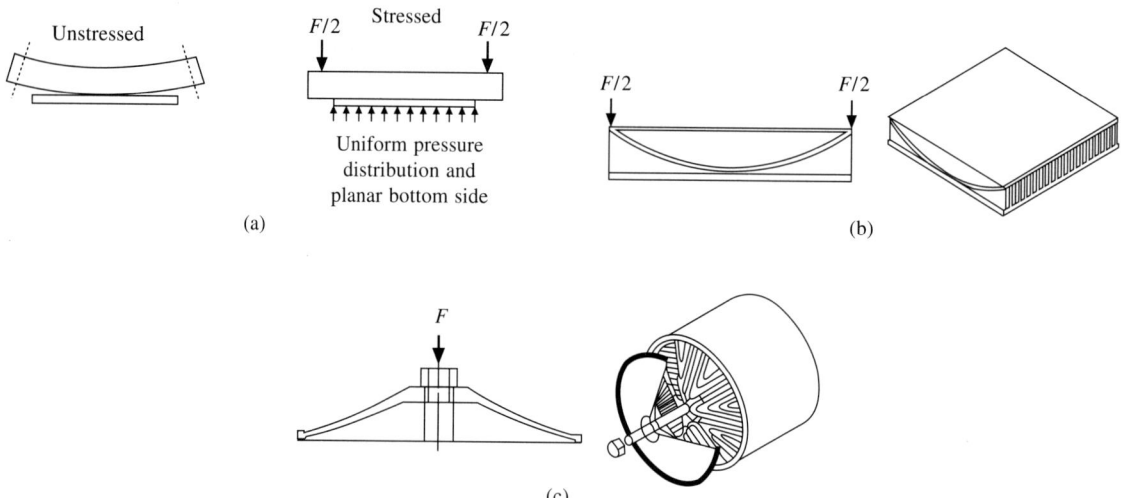

Fig. 13 (a) Bomb shaped end plate, uniform pressure distribution (from Evertz, 2002), (b) End plate based on Dbow-concept, head plate, schematic view (from Evertz, 2002), (c) End plate for axisymmetric fuel cell stack based on Dbow-concept, footplate, schematic view (from Evertz, 2002).

2.7 Water Management

Water management is a central issue in PEMFC technology. There must be sufficient water content in the electrolyte, otherwise the conductivity will decrease. However there must not be so much water that the electrodes which are bonded to the electrolyte flood, blocking the pores in the electrodes or gas diffusion layer. A balance is needed. The most important properties of membrane for ionic conduction in PEMFCs are:

1. Presence of negatively charged sites.
2. Ability to be hydrated.
3. Ability to transport protons between the charged sites with a 'Grotthus type' mechanism, in the presence of water.

The hydration of the membrane is critical to the performance of the PEMFC because the proton conductivity of the membrane is strongly correlated with the water content, defined as the number of water molecules per sulfonic acid group. The hydration originates in the humidity of the reactants and in water produced by the cathodic reaction. Indeed, some phenomena establish a water concentration gradient in the membrane (Fig. 14). They are the ion transfer phenomenon, such as:

(a) Electro osmotic water transport from the anodic side to the cathodic one (due to water "dragging" by the hydrated protons that are transferred from the anodic side). Typically between 1 and 2.5 water molecules are 'dragged' for each proton (Springer et al. 1991, Zawodzinski et al. 1993). As a result, the ion exchanged can be envisioned as a hydrated proton, $H(H_2O)_n^+$. The water drag increases at high current density, and this makes the water balance a potential concern. This means that especially at high current densities, the anode side of the electrolyte can become dried out even if the cathode side is well hydrated.

(b) Water diffusion, that moves from the cathode side, where the water is produced, to the anodic one where a small concentration occurs. Hydrogen ions are involved in a counter-diffusion mechanism (from the anode to the cathode).

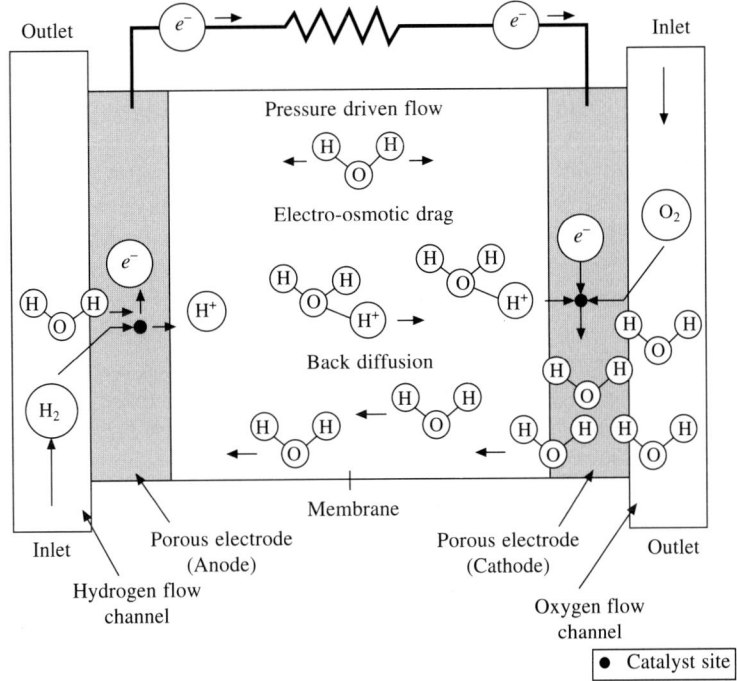

Fig. 14 Diagram showing three transport mechanisms of water vapor within the fuel cell membrane: pressure driven flow, electro-osmotic drag, and back diffusion. Transport due to pressure driven flow can occur from either electrode, electro-osmotic drag occurs solely from the anode to the cathode, and back-diffusion typically occurs from the cathode to the anode since the water concentration is usually higher at the cathode (from Evans, 2003).

(c) Pressure driving transfer, which happens when a pressure difference is created between the anodic and the cathodic regions.

To maintain a high water flux through the membrane, inlet gas conditions are often manipulated. By using in situ small angle neutron scattering microscope on an operating 4.5 cm^2 fuel cell, Mosdale et al. (1996) were able to determine the water profile within the membrane when the inlet conditions changed. When both gases were dry, the water profile appeared as a front moving from the cathode to the anode with time (Fig 15). In addition, they noted that dry hydrogen carried away three and a half times more molecules of water from the anode than the air carried away from the cathode. Hence the anode was losing water both due to evaporation into the unused hydrogen and electro-osmotic drag to the cathode side. When hydrogen was supplied in the fully saturated or "wet" state, the water profile appeared as a parabolic curve between the two electrodes. The low point of the profile was the center of the membrane. When both inlet gases were humidified, a more uniform water profile was observed between

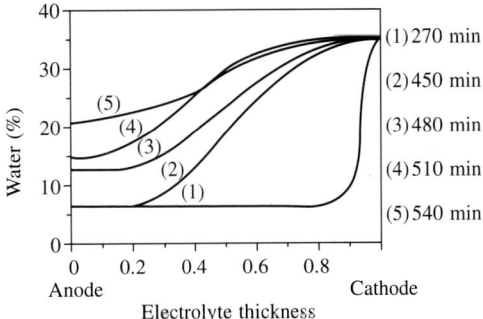

Fig. 15 Evolution of the water concentration profile during the experiment performed with a dry membrane and dry gases (from Mosdale, 1996).

the electrodes. Therefore, humidifying the inlet gases helps to balance the effects of the two dominant phenomena of water transport on membrane water content. As a result, the resistance of the membrane remains low, and higher cell voltages can be achieved.

In their two-dimensional PEMFC model, Nguyen and White (1993) stated that at high current density the transport from the anode by electro-osmotic drag exceeds transport to the anode by back diffusion from the cathode and the membrane will dry out. As the membrane becomes dehydrated, the membrane pores shrink, which further limits the back diffusion of water. For this reason, Nguyen and White (1993) concluded that water transport due to back-diffusion is not sufficient to prevent membrane dehydration. Yi and Nguyen (1998) verified these findings in their 'Along-the-Channel' fuel cell model. They found that as the hydrogen moved down the channel, the local anode moisture content decreased. This reduction in moisture content due to water transport from the anode to the cathode eventually caused the membrane to dry out. They concluded that the net water flux from the anode to the cathode was directly proportional to the current density. Dhathathreyan et al. (2001) experimentally confirmed these findings in their study of internal and external humidification methods in a fuel cell stack. At high current densities, the transport from the anode due to electro-osmotic drag is quite large, which causes the anode to dry out and the cathode to flood. It was also found that the kinetics of the reduction reaction at the cathode were adversely affected by the increase in water content due to flooding, which reduces the cell voltage. A simplified model for water management in a PEMFC operating under prescribed current has been developed by Berg et al. (2004). The sulfonic groups easily dissociate into SO_3^- and H^+ in these membranes, in the presence of water. Under this condition, the proton can be considered as a mobile charge that encounters a low resistance when moving across a potential gradient.

To achieve the humidification, many approaches have been tried for precisely humidifying inlet gas streams. Fig. 16 shows conventional fuel cell humidification methods used in research: (a) the self-humidifying fuel cell with no active external humidification of the reactant streams, (b) the setup featuring liquid water injection into an inactive portion of the fuel cell, (c) a typical dew point humidifier, (d) the evaporation setup, (e) the steam injection system and (f) the flash evaporation method (Evans 2003). Self-humidification is another concept reported by Watanabe (1996). In their 'along-the-channel' model of a PEMFC, Yi and Nguyen (1998) used direct liquid-injection within the fuel cell to humidify the gases. The water was injected into the flow channels, where it was evaporated into the gas streams. However, Yi and Nguyen concluded that external humidification at a higher temperature was preferable to direct liquid-injection, since the higher temperature allowed for more water to be absorbed by the gas stream and carried into the fuel cell.

There have also been attempts to control the water in the cell by using porous graphite plates or by external wicking connected to the membrane to either drain or supply water by capillary action. More reliable forms of water management also are being developed based on continuous flow field design and appropriate operating conditions. A temperature rise can be used between the inlet and outlet of the flow field to increase the water vapor carrying capacity of the gas streams.

The water management issue in PEMFC is also related to the air flow management (vide infra). Except for the special case of PEMFC operating on pure oxygen, it is general practice to remove the product water using air that flows through the cells. The air will also always be fed through the cell at a rate faster than that needed just to supply the necessary oxygen, which can be done by using a stoichiometry of at least 2. The air flow rate, power of fuel cell and the stoichiometry are closely related. Further complication arises as the drying effect of air is very non-linear in its relationship to temperature. It is known that at temperatures of above 60°C, the relative humidity of the exit air from the cell is below 100% necessitating using external humidification for air above this temperature. Buchi and Srinivasan (1997) have shown that even below 60°C, with counter flow of reactants, the maximum power reduces by about 40% if no external humidification

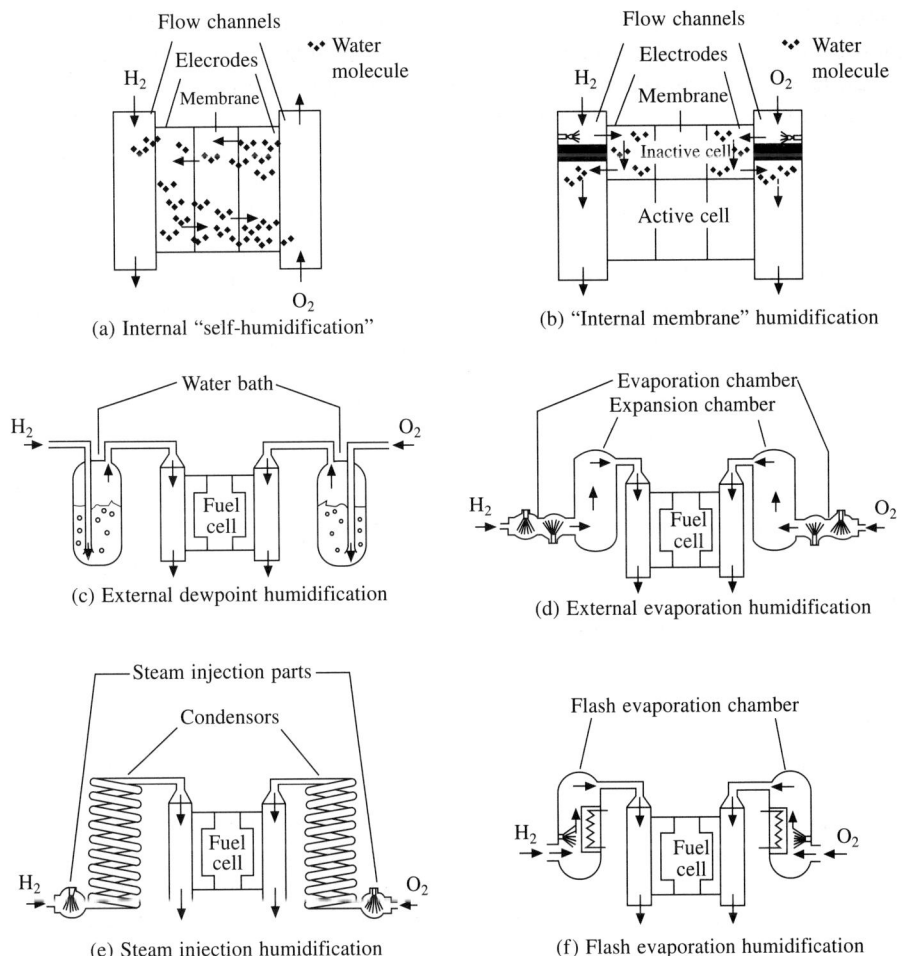

Fig. 16 Various types of humidification (six diagrams showing conventional fuel cell humidification methods used in research: (a) the self-humidifying fuel cell with no active external humidification of the reactant streams, (b) the setup featuring liquid water injection into an inactive portion of the fuel cell, (c) a typical dew point humidifier, (d) the evaporation setup, (e) the steam injection system and (f) the flash evaporation method (from Evans, 2003).

is used. In a smaller system, this kind of loss may be acceptable as the overall efficiency does not decrease too much. PEMFC can be operated without external humidification if the relative humidity of the exhaust air is about 100% which can be achieved by adjusting the air stoichiometry.

Internal humidification (actually a misnomer!) is another approach which has been successfully used. In this concept, a portion of the membrane is set aside to humidify the inlet gases and liquid water is injected directly into this inactive portion of the stack. In another method, Chow et al. (1995) developed an internal membrane humidification scheme for a PEMFC stack, where dry gas was run through a separate section of the stack to condition the gas before the electrochemically active portion of the cell. The advantage is that the gases are conditioned inside the stack, and the gas temperatures will be very close to the temperatures of the membrane itself. However, a portion of the electrochemically active section of the stack must be set aside for humidification purposes, which reduces the power density of the stack. Johnson et al. (2001) have quantified the heat flux required for internal humidification to be 1300 W from the water vapor to produce 2750 W of electric power from a 3 kW stack. In their comparison of external and internal humidification, Dhathathreyan et al. (2001) showed that self-humidification is limited by the membrane diffusion properties. At higher temperatures (90°C), the back diffusion rate of water through the membrane becomes a limiting factor. They also showed that as temperature increases, the water uptake by the reactant gases increases and the current density increases, which also increases the water transport due to electro-osmotic drag. Since the membrane limits the back diffusion rate, the electro-osmotic drag becomes the dominant mechanism and water is transported away from the anode and hence, the anode will dry out and the cathode will become flooded. Due to this phenomenon, use of external humidification at higher temperatures was recommended. This conclusion reinforces the work of Buchi and Srinivasan (1997) on internal humidification of membranes. As noted, the main problem with external humidification is that if the gas cools after the humidifier, the excess water condenses out of the gas and enters the fuel cell in droplet form, which floods the electrodes near the inlet, thereby preventing the flow of reactants.

Another widely used method of humidification involves evaporation of liquid water into the gas stream. The latent heat of evaporation required to evaporate the water into gas will cause the gas temperature to decrease. Davis (2000) documented this phenomenon in the discussion of his humidification method. He used an atomizing nozzle to inject water into the gas flow streams of hydrogen and air. In both cases, the temperature dropped because evaporation required the addition of a heater around the injection chamber. In addition, to fully evaporate the liquid water, a long length pipe was added after the humidifier. The temperature loss in this pipe was also significant for both gases, and additional heaters were added to the sections following the injection chamber. In spite of the heaters, the humidity was difficult to control and the transient response of the system was very slow. One of the recommendations that Davis made was to improve the controllability of the evaporative humidification design. In their research, Ihonen et al. (2001) point out a significant flaw in using dew point humidification in test equipment. They stated that the nominal gas temperatures (temperatures in the center of the gas stream within the pipe) were not the real dew points of the gas. Heat losses between the humidifier and the fuel cell resulted in a drop in gas temperature. To prevent this temperature loss, they recommended that the section between the humidifier and the fuel cell be heated. Unfortunately, this is not possible in most commercial test equipment due to the proximity of sensitive electronics within the test stand enclosure. Ihonen et al. concluded that a custombuilt humidification system would yield more accurate and reproducible results for experimental purposes.

Self-humidification concepts involve introducing dry gases into the anode and cathode flow channels, where they absorb water from the porous electrode. A novel self humidifying membrane has also been reported recently. While Watanabe's method (1996) involved using highly dispersed nanometer size Pt and/or metal oxides prepared by an electroless process, the new method involves multilayer composite polymer electrolytes wherein the humidifying layer has platinum catalyst particles (Liu et al., 2003). *In situ* analysis of performance degradation of PEMFC under non-saturated humidification has been reported by Yu et al. (2005). Simultaneous measurements of species and current distributions in a PEMFC under Low-Humidity Operation has been studied by Yang et al. (2005).

A novel humidification method has been reported by Santis et al., (2004). In this method each cell in the stack has a humidification section (Fig. 17) and the dry reactant air gets humidified by the exhaust air and is therefore independent of the stack size.

Understanding the importance of the humidification in PEMFC, many companies have started marketing humidification systems (the Tenney Environment Chamber which uses evaporation at ambient pressures and chillers to control the humidity, Arbin's dew point humidifier where the gas is premixed with steam and then injected into the water, Fideris humidification system which uses flash evaporation of water and the water flow rate is controlled to achieve the required dew point, Permapure humidifier uses a Nafion tubing in its humidification system).

2.8 Thermal Management

Thermal management of PEMFC is key to ensure high cell performance and efficiency. The irreversibility of electrochemical reactions and joule heating are the most important factors causing heat generation inside PEM fuel cells. The temperature distribution in the cell has a strong impact on the cell performance. It influences the water distribution by means of condensation and affects the multi-component gas diffusion transport characteristics through thermo capillary forces and thermal buoyancy. Also, the kinetics of electrochemical reactions directly depends on the temperature. Excessive local cell temperature due to insufficient or non-effective cell cooling may cause membrane dehydration, shrinking or even rupture. Hence, the thermal and water

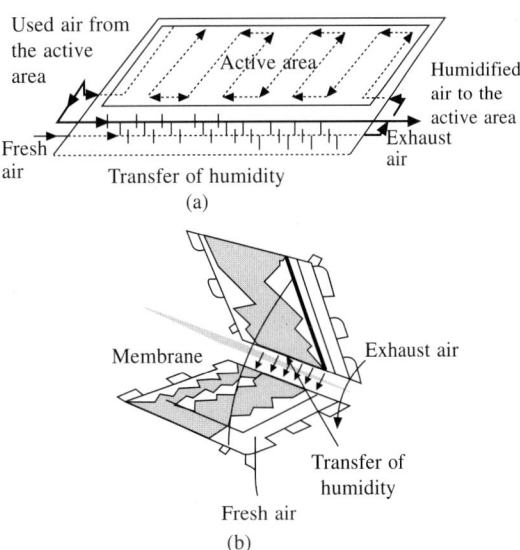

Fig. 17 Concept of internal process air humidification by transfer of humidity from the exhaust air to fresh air in an additional humidification section within each cell. (a) schematics of air flow through humidification section and active area and (b) View of opened real cell with air flow indicated (from Santis et al., 2004).

management issues are strongly coupled and they have a direct impact on cell performance.

Thermal management includes the removal of the generated heat from inside the cell to the outside. Further, a temporally and spatially uniform temperature distribution must be provided, hot spots need to be avoided. The pumping power required for the coolant circulation has to be minimized in order to ensure high overall cell efficiency. Therefore, pressure drop must be minimized while maximizing the heat transfer capability at the same time. The method employed to remove heat from the fuel cell stack depends on its size. With fuel cells of less than 100 watts, it is possible to use purely convected air to cool the cells and provide sufficient air flow to evaporate the water without using any fan (Daugherty, 1999). In the case of fuel cell stacks of higher capacity, cooling circuit need to be incorporated in the stack for thermal management. Computer simulations have been carried out to study the thermal management in a fuel cell by many groups along with the water management studies. Dumercy et al. (2005) have developed a 3D Steady State Thermal Modeling for a fuel cell stack which is helpful in defining the geometry of the fluids ducts. While a number of models assume a constant temperature of the fuel cell stack, Shan and Choe (2005) have carried out dynamic analysis especially the temperature response to the dynamic load.

2.8.1 Cooling by Air

Many systems have been reported wherein a single air blower is used to feed the reactant gas as well as supply the air to cool the stack. The thermal load can be managed simply by using fans without any water cooling system, the air-cooled PEMFC is widely used in sub kW and around 1kW systems. The performance of an air cooled system is highly dependent on ambient temperature and humidity. Air cooled systems are expensive to build as each cell has to have channels for the anode and cathode plates for the cooling air to flow. In order to reduce the cost, novel methods are being developed and one such method is reported by Ruge and Hoekel (2005) who have used a edge air cooling integrated with a fan.

2.8.2 Cooling by Liquid Coolants

When the stack capacity is more than 1.0 kW liquid cooling is preferred. In case of tropical and sub tropical countries, air cooling concept has to be thought of very seriously as the average temperature is about 35°C on an average. Water cooling becomes ideal if the hot water can be used (albeit low quality) in a small domestic combined heat and power system as has been demonstrated by Mitsubishi recently with the over all efficiency reported being a whooping 83%. In such applications, design of the cooling plate is also important. Serpentine or meander cooling patterns have been used. For overall stack functionality reasons, the number of fluid inlets and outlets must be small, making parallel cooling patterns usually inappropriate. These circumstances call for a flow geometry with minimum flow resistance between a volume subjected to two constraints: fixed total volume and fixed channel volume, has to be established. Fuel cell systems demonstrated by many leading industries use glycol/water based coolants.

Although there have been a number of studies on heat and mass transfer in the reactant gas channels, there have been very limited studies on optimizing the cooling process of a fuel cell. Musser and Wang (2000) employed a two-dimensional code to predict the temperature variation in the fuel cell. However, the two dimensional analysis could not reflect on the real cooling arrangement which includes complicated configurations such as serpentine type structures. Chen et al. (2003) have used a three dimensional CFD code to investigate the coupled cooling process involved in fluid flow and heat transfer between the solid plate and the coolant flow. They investigated six different cooling modes in their analysis and have arrived at the conclusion that serpentine type flow mode is better than the parallel type mode. Among the various serpentine modes, the mode-3 in Fig. 18 seems to offer the best solution. For practical fuel cell systems especially transportation applications, there are five different concepts for water and thermal management: an absorbent wheel, a membrane humidifier, a porous metal foam humidifier, a cathode recycle compressor, and a water injection pump. Selection of the system is done in the following descending order: size, reliability, cost, power consumption and weight. Most fuel cell vehicle designs have been based on the use of several discrete heat transfer circuits, with each system having its own set of components like plumbing, pumps, valves, etc. results in the additional weight and cost for the vehicle. The use of graphite-based heat exchangers represents a unique opportunity for using advanced materials in the design of integrated thermal management systems for fuel cell vehicles.

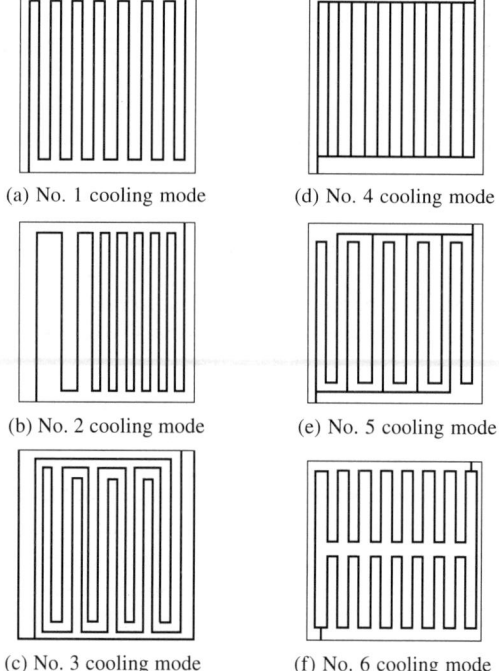

(a) No. 1 cooling mode   (d) No. 4 cooling mode

(b) No. 2 cooling mode   (e) No. 5 cooling mode

(c) No. 3 cooling mode   (f) No. 6 cooling mode

Fig. 18 Structure of the water channel of a cooling plate in a fuel cell. (a), (b) and (c) are serpentine type passages; (d), (e) and (f) are parallel type passages (from Chen et al., 2003).

2.8.3 Cooling Modules for Fuel Cell Vehicles

Cooling of the fuel cell drive is one of the key tasks for vehicle applications. The exhaust gas of the fuel cell, unlike the exhaust gas of the combustion engine, does not drag along any significant heat flow. Thus, the heat flow to be dissipated by the cooling system is approximately double as compared to that of the combustion engine (Walter et al., 2000). The other reason is the low coolant temperature in PEM fuel cells, which are always used in vehicle drives. Therefore, the difference in temperature available for heat transfer is reduced by approximately 50% compared with a combustion engine. The end result is that the cooling power required, related to the temperature difference, is four times higher for a fuel cell drive compared with the corresponding combustion engine. The cooling power installed in this case is equivalent to that of a heavy-duty truck. A solution of this kind is neither practical nor economic for serial application. Further, secondary conditions result from requirements relating to reliability, lifetime, and comfort-relevant characteristics, such as noise or A/C power, which have to meet the standard level for cooling of combustion engines. In addition, it is necessary that several electric and electronic components in the fuel cell drive be cooled down to a temperature level considerably lower than that of the fuel cell itself (Rogg et al., 2003).

2.8.4 Technological Limits of Cooling
It is crucial that a sufficient air mass flow be available to accomplish the required cooling. Besides the exchange rate of the heat exchanger (ideal exchange rate = 1) a significant criterion is the expenditure of flow rate required to achieve this exchange rate. Development of heat exchangers for engine cooling in vehicle is aimed at the most favorable design of pressure loss/heat transfer ratio possible. A mechanically assembled aluminum radiator has a lower power than brazed aluminum radiator whose performance is surpassed by brazed copper/brass radiators. It is also known that radiators with counter current is better than cross current flow mode for the air. However space limitations in a vehicle do not allow counter current flow mode.

A fuel cell driven vehicle uses more than one coolant flow to cool down: fuel cell stack coolant and a coolant for the electronic components. There are wide varieties of possible configurations. One such possibility is given in Fig. 19.

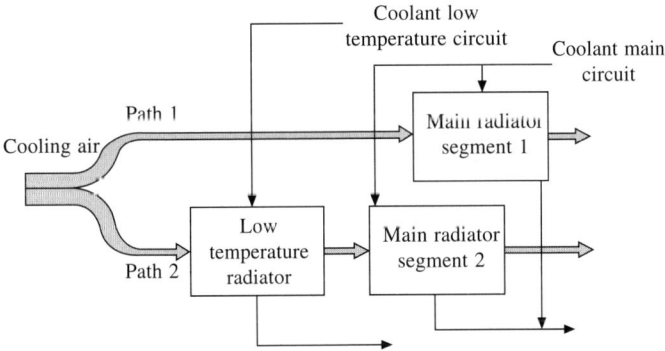

Fig. 19 Basic configuration for analysis of interaction between the main and low temperature radiator (from Rogg et al., 2003).

When selecting a fan arrangement for high-capacity cooling modules, contradictory requirements have to be linked. On the one hand, at low driving speeds, with the associated low support by dynamic pressure for air circulation, a relatively high fan speed is required to circulate a large volume of air. A typical situation is the uphill ride of a fully loaded vehicle. The operating point, and thus the airflow, results from the intersection of the characteristic of the cooling unit and that of the fan. Therefore a steep fan characteristic is desirable.

On the other hand, great dynamic pressure results from high driving speeds. This leads to a displacement of the cooling unit characteristic, which will now intersect with the fan characteristic in an area where the fan is building up low pressure only or is even overblown. In order to obtain the highest possible airflow under these conditions, the fan characteristic should be as flat as possible, or even better, no fan should be available at all.

An attempt to reconcile these contradictory requirements is normally made by combining the cooling surfaces impinged by one or more fans of steep characteristic and additional cooling surfaces impinged by dynamic pressure only. The 'fan characteristic' of the section of the cooling surface impinged by dynamic pressure is given by a static pressure increase of 0. Therefore, the aim should be to design the air circuit through the sections of the cooling surface, which are impinged by dynamic pressure only in such a way that the characteristic of the cooling unit is as flat as possible. This can be realized by directing the airflow towards the wheel housing or under body, if possible.

It is clear that a sufficient mass of cooling air is indispensable even in the case when all the heat exchangers present are operating at their theoretical maximum exchange rate. Therefore, with the existing high requirements in cooling power, it is necessary to utilize the most powerful fans and fan drives, which can be housed in the space available. For the range of possible diameters, speeds and structural depths, single-stage axial blowers are particularly suitable.

In vehicles with a fuel cell drive, high voltage supply systems are available and the disposal of electrical energy is not as limited as in vehicles with a combustion engine, where the electrical power has to be provided by a mechanically driven generator. Thus, the prerequisites for high fan drive performances are fulfilled by means of special electric motors.

An improvement in fan performance, exceeding the current level of electric fans for vehicles, can be obtained by an increase in speed, or in the number of blades, or by the utilization of fan blades with stronger deviation effect or by a combination of these. The increase of air volume by more powerful fans and fan drives is limited, however, by the fact that the required strength of the fan drive is growing by the third power of the increase in the air circulation volume.

The dissipation of nearly double the waste heat at half the temperature drop against ambient temperature still presents limits for the operation of fuel cell vehicles. Under certain conditions, limited cooling restricts driving power. A further increase of the cooling power is required. This cannot be achieved using heavy-duty cooling systems alone; at least not in serial vehicles where the available housing space as well as the shape and weight of the cooling systems is limited. The power required by radiator fans increases exponentially and will thus reduce the total efficiency of a vehicle. For this reason, an increase in the operating temperature of PEM fuel cells would be indispensable in keeping the cost of the cooling system within economically sensible limits. The cooling module in a DC Sprinter with a fuel cell engine is shown in Fig. 20. The cooling module uses the counter cross current circuit for the 35 dm^2 coolant radiator in the main circuit. For consistency the principle of an additional surface impinged by dynamic pressure only is applied here. A second coolant radiator in the main circuit creates an additional 17 dm^2 of surface. The low temperature radiator overlaps by 60% with the first main radiator and therefore is nearly arranged at its optimum. A double blower provides the air supply.

2.9 Fuel Options for PEMFC

Basically three fuel supply and storage approaches for fuel cell systems can be differentiated (Thomas et al., 2000):

1. Hydrogen storage in compressed gaseous (high

Fig. 20 Cooling module for the DC sprinter with a fuel cell drive (from Rogg et al., 2003).

pressure tanks, metal hydrides, graphite nanostructures) or liquid form (cryogenic liquid, cryo-adsorption) (potentially from renewable sources - potentially without any emissions).

2. Hydrogen storage in hydrogen rich liquid methanol (potentially from renewable sources - potentially with very low emissions).

3. Hydrogen storage in hydrogen rich hydrocarbons such as gasoline- or diesel-type fuels (from fossil sources – with emissions rather high emissions, or synthesized from renewable energy sources with still moderate emissions).

Concept 3 does not require any infrastructural changes for its supply and thus would be the ideal starter scenario. On the other hand, it prevents the introduction of renewable energy sources into the transport sector as true zero emission alternative.

Concept 2 requires only moderate infrastructural changes (exchange of some gasket and hose materials, corrosion resistive tank and tube materials, spill collection, vapor collection). This concept would also allow the introduction of renewable energy from biomass or via hydrogenation of CO_2.

Concept 1 requires major infrastructural changes but provides the possibility for a very flexible introduction of renewable energy sources into the transport sector. Since the infrastructure adaptation will require several years, concept 1 is only suitable as long term scenario at large scale. However most of the demonstration programs have followed this option.

The selection of the fuel depends on the costs associated with infrastructure and fuel costs, safety, environmental implications, ease of use, acceptance by the manufacturers, oil companies, government, and the public.

When operated with pure hydrogen gas, the PEMFC system is simpler in construction: appropriate load following controls can be placed which simply control the flow of the gas, humidification is simpler, the channels in the flow field can be designed as the volume of the gas going into the anode chamber is less than the case where reformate is supplied. Further, the electrode structure is simpler and the amount of platinum catalyst used can be reduced substantially. The disadvantage is that one needs source of high quality hydrogen gas.

Reformate gas from natural gas (methane), methanol, kerosene, ethanol is expected to be a major fuel source for PEMFCs for various applications. Steam reforming, partial oxidation (POX) and auto thermal reforming are the three major reforming technologies. Plasma Reforming (PR) is also known. However these reformation processes are required to be followed by a series of purification trains to remove mainly CO which is a poison to PEMFC in concentrations above 10 ppm. In addition the fuel cell electrodes have to be made with CO tolerant catalysts which increases the cost of the electrode and thus the stack. Further, if carbon dioxide is not completely eliminated, the reformate gas will have only 40-75% of hydrogen depending on the process of reforming used. This means that the flow field plates have to be suitably designed to accommodate higher amount of reactants. Using reformers as a source of hydrogen, load following gas flow controllers are difficult to program as the reformation step is endothermic and the response time is not fast enough for rapidly changing power requirements. While CO impedes the anodic hydrogen oxidation reaction by poisoning the Pt catalyst, dilution of the hydrogen stream by CO_2 leads to mass transport limitations in the anode catalyst layer and backing, analogous to those observed at air electrodes (Rajalakshmi et al., 2003). Zhu et al. (2005) identified the critical flow rate for the fuel gas by anode mass balance method. The PEM stack NEXA module was successfully operated with up to ca. 7% nitrogen or carbon dioxide in the absence of a palladium-based hydrogen separator at ca. 200 W power level. The system maintained a fuel efficiency of 99% at a manual purge rate of 2.22 ml/s and no auto purge. The fuel cell stack efficiency was 64% and the stack output efficiency was 75%. The overall system efficiency was 39%.

In recent times most of the fuel cell vehicles especially the buses are operated using compressed hydrogen in high pressure cylinders. This method of operation also allows the FCV to be characterized as true zero emission vehicle. However, it suffers from certain disadvantages like handling of hydrogen, safety precautions, and large volume to store enough hydrogen on board to give a vehicle an adequate driving range. One of the object of constant innovation in hydrogen supply for the fuel cell in a vehicle is to improve the amount of hydrogen available for the vehicle so that the range can be extended. One solution that has been used is to move to higher onboard gas pressures from 200 to 350 bars and most likely to 700 bars in the future. This has resulted in a series of investigations on to developing suitable storage devices which can withstand such high pressures at the same time reducing the weight of such cylinders. Novel concepts such as 'conformable hydrogen storage cylinders' (Fig. 21) have also been attempted (Aceves et al., 2004). Research into other fuels for FCVs, however continues, although most of the major car companies are investing less in this area.

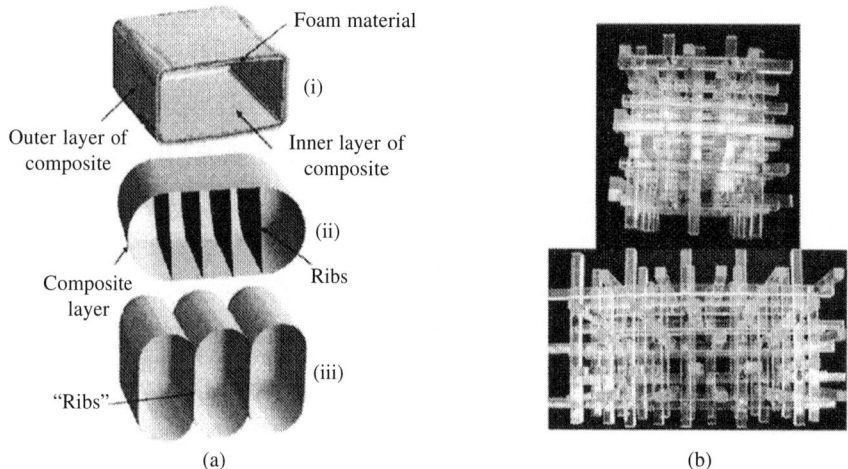

Fig. 21 (a) Three proposed designs for continuous fiber pressure vessels: (i) sandwich design, (ii) ribbed design and (iii) bucking design and (b) two views of the octahedral lattice identified as optimum for replicant conformable pressure vessels (from Aceves et al., 2004).

2.10 Oxidant Options

The oxidant can be either air or oxygen. If oxygen is the oxidant used, supply from compressed cylinder is the only option. Except in cases where fuel cell is used in closed environment or where availability of air is limited (submarine and high altitude), air is commonly used to provide the oxygen required by the fuel cells. In addition, if the stack is designed to operate with air cooling, additional air is required to be moved around the fuel cell system. Buchi et al. (2005) have shown that overall efficiency of fuel cell system would increase if oxygen is used in place of air.

The serpentine or interdigitated flow for oxygen needs to be modified to free flow of air to accommodate higher flow of air. While operation of the stack with compressed air has several advantages like high fuel cell efficiency, low volumetric flow rate, ease of humidification, less mass transport limitation, tolerance to large pressure drops, the major advantages of using blowers for air supply is the low parasitic power, lower cost and self cooling. The other advantage of the blowers being the availability of blowers with variable speed drives for flow control.

Apart from the flow management of reactants care has to be taken to avoid impurities in the reactant feeds. It has been established that impurities such as SO_2, NO_2, H_2S, and O_3 in the air stream affect the cell performance. It was estimated that the concentration of SO_2 and NO_2 in the air feed should not exceed 0.05 ppm, O_3 concentration should not be higher than 0.20 ppm, and H_2S is limited to less than 1 ppb.

The options that have been used for supplying air are: air compressor, fan and blower, membrane or diaphragm pumps. For a stationary application of fuel cells, commercially available air compressors can be easily used. In case of transport applications, the common compressors are not suitable. Extensive research is being carried out to develop air supply system where the energy requirement is less and large volume air can be supplied and compact in design. Root compressors, Lysholm or screw compressor, centrifugal or radial type compressors, axial flow compressors have all been tried each having its own set of advantages and disadvantages. For fuel cell cooling fans and blowers can be used and they are available in a huge range. The blowers have limited pressure range and the flow rate can drop to zero if the back pressure rises to even 50 Pa. (0.5 cm of water) For slightly higher pressure requirement centrifugal fans can be used. These draw air in and throw out in sideways. They are somewhat similar to a centrifugal compressor but turn at much slower speeds and have much larger blades. These compressor fan/blower can generate at best 3-10 cm of water pressure and have limited application in supplying air to PEM fuel cell stacks (small to medium sized). Another option a fuel cell engineer has to move air through a PEM fuel cell stack of medium size is the membrane/diaphragm pump which are suitable for 10-20 kPa needs (Popelis 1999).

Cunningham (2001) in his analysis of air supply options for FCV found that:

1. The peak power P_{net} can be achieved with both a blower (low pressure) and a compressor (high pressure), but the required fuel cell stack sizes are different. For the same peak P_{net} of 86 kW, 16.3% more operating PEM cells were needed in the stack for the blower application (500 vs. 430 cells with a constant active area of 490 cm^2). The blower system was able to obtain the same net power by operating just above ambient pressure at the stack and providing sufficiently higher air mass flow rates compared to that of the compressor for much of the P_{net} range.

2. The parasitic loads for the blower are significantly less than that of the compressor at the high P_{net} region. The ratio of P_{as_motor}/P_{stack} was 14.1% for the compressor vs. 3.2% for the blower at a peak P_{net} of 86 kW (though these occur at different P_{stack} values).

3. Overall, the net system efficiencies over the P_{net} range were very similar for both the blower and the compressor. However, the blower system did maintain a net efficiency 1.5-2.0 percentage points higher than the compressor system over most of the net power range.

4. High pressure application results would differ if an expander were to be included. P_{net} would be achieved at reduced P_{stack} powers and thus different air pressure and mass flow schemes. Stack size would be further reduced, potentially increasing overall power density and reducing costs. Net system efficiency may improve as well.

A critical need in fuel cell systems for vehicles is an efficient, compact, and cost effective air management system to supply air preferably at about 2.5 atm. Because no off-the-shelf compressor technologies are available to meet the stringent requirements of fuel cell air management, several compressor and blower systems are currently being developed. The efficiency, reliability and durability of compressors depend on effective lubrication or friction and wear reduction in critical components such as bearings and seals. Conventional oil or grease lubrication of compressor components is not desirable because such lubricants can contaminate and poison the fuel cell stack. Zhao et al. (2005) studied the theoretical and experimental investigations of a water injection scroll compressor (Fig. 22) for use in fuel cell systems to supply clean air. The water is used as both the lubricant and coolant in the compressor. The results show that the scroll compressor has nearly isothermal compression when injecting water in it. Increasing the compressor rotation speed increases the discharge loss and the volumetric efficiency of the scroll compressor. The difference between the calculated power and the isothermal power increases as the compressor rotation speed rises, which means the efficiency of the compressor decreases. Increasing the flow rate of water

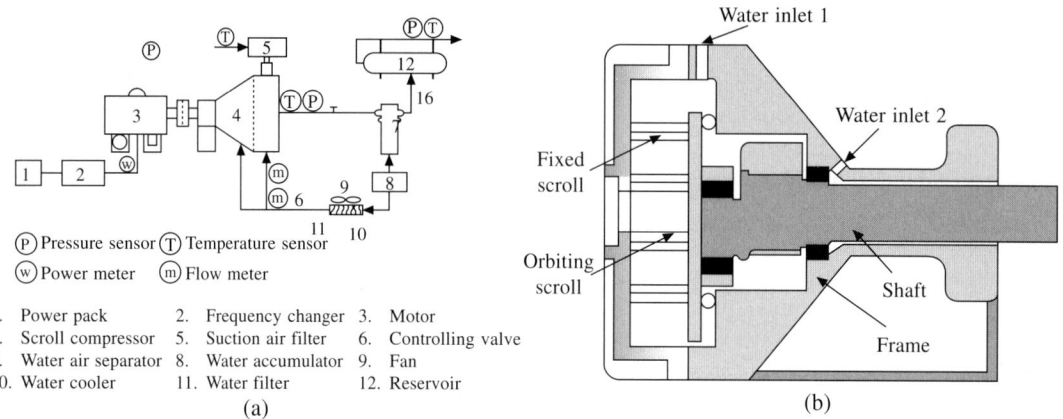

(P) Pressure sensor (T) Temperature sensor
(w) Power meter (m) Flow meter

1. Power pack   2. Frequency changer   3. Motor
4. Scroll compressor   5. Suction air filter   6. Controlling valve
7. Water air separator   8. Water accumulator   9. Fan
10. Water cooler   11. Water filter   12. Reservoir

(a)   (b)

Fig. 22 (a) Water injection scroll compressor test system, (b) Schematic of water injection scroll compressor (from Zhao et al., 2005).

injected increases the indicated isothermal efficiency and decreases the discharge temperature. Under the condition studied, the mass flow rate of water has the greatest effect on the discharge temperature.

The complexity of the air management system can be understood from the description given for the air system modules for a CITARO fuel cell bus under the CUTE program. A super charger, driven by a electric motor is used to supply pressurized air to the fuel cell stack. After leaving the stack, the pressurized air is exhausted through a turbo charger which recovers energy from the exhaust and provides a second stage of air compression. The air supply system also includes an inter-cooler to improve compression efficiency and air filter to remove contaminants. Mufflers on the air system intake and exhaust quieten the supercharger and turbo charger in order to meet the noise requirements of the over all vehicle.

2.11 PEMFC Power Plant

A complete fuel cell power plant consists of several components: a fuel supply system (gasoline or methanol reformer or hydrogen storage tank), the fuel cell stack, an air compressor/air blower to provide air to the fuel cell, cooling system (heat exchangers) to maintain the proper operating temperature, water management system to manage the humidity and the moisture in the system, a DC/DC converter to condition the output voltage of the fuel cell stack, an inverter to convert the DC variable voltage to constant AC voltage. A battery or an ultra-capacitor is generally connected across the fuel cell system to provide supplemental power and for starting the system. The fuel cell, peak power devices, electronics, and fuel supply/storage system must be designed and arranged to have a compact system without creating safety hazards. The start up battery of considerable capacity is to provide power to the fuel processor during its warm up period. For the transport applications, the battery is not intended to power the vehicle during its warm up period. An additional battery for transient power may be required during start up propulsion. The system radiator consists of three separate circuits for tail gas condenser, process water cooler and fuel cell cooler. They can be sized as separate heat exchangers. Various sensors are required in a fuel cell systems for operation requirement and safety. The sensors are mainly for temperature monitoring, oxygen sensor, sulphur sensor, CO sensor located downstream of the PROX if used, pressure sensor located upstream of the pressure regulator, ammonia sensor located downstream of the activated charcoal bed and a pH sensor for process water system. Control valves are required at various places in the fuel cell system viz., water flow (reformate, economizer, humidification, by pass anode cooler), air flow (ATR, PROX, tail gas burner, anode inlet air bleed), back pressure regulator etc., solenoid operated three way or two way diverter valves are necessary to accommodate transient and upset condition. The fuel cell system has different control systems besides a master control system. The control system architecture gets complicated when moving from pure hydrogen supply option to a reformer option.

The individual control systems are linked to:

(a) Controlling the fuel supply system with a fuel processor.
(b) DC-DC control to generate a constant DC voltage form the variable DC voltage that is commonly produced by the fuel cell stack.
(c) A DC-AC inverter control system following the load.

The role of the master control is supervisory in nature overlooking the function of the other controls and taking corrective action. Detailed software and control algorithms for these functions have been developed by various developers for the fuel cell systems and are proprietary in nature.

3. PEM Fuel Cell System Optimization and Analysis

Design and optimization of a PEMFC system is very complex because of the number of required systems, components, and functions. Many possible design options and trade-offs affect capital cost, operating cost, efficiency, parasitic power consumption, complexity, reliability, availability, fuel cell life, and operational flexibility (Lomax, 1997; James, 1999; Little, 2000; On, 2002).

Fig. 23 shows that the fuel cell itself has many trade-off options. A fundamental trade off is determining where along the current density voltage curve the cell should operate. As the operating point moves up in voltage by moving (left) to a lower current density, the system becomes more efficient but requires a greater fuel cell area to produce the same amount of power. That is, by moving up the voltage current density line, the system will experience lower operating costs at the expense of higher capital costs. Many other parameters like pressure, temperature, fuel composition and utilization, and oxidant composition and utilization can be varied simultaneously to achieve the desired operating point. The system design has a fair amount of freedom to manipulate design parameters until the best combination of variables is found, e.g., Heinzel et al. (2005) showed that the electrical efficiency of a 2.5 kW system can be improved by re-circulating the anode gas.

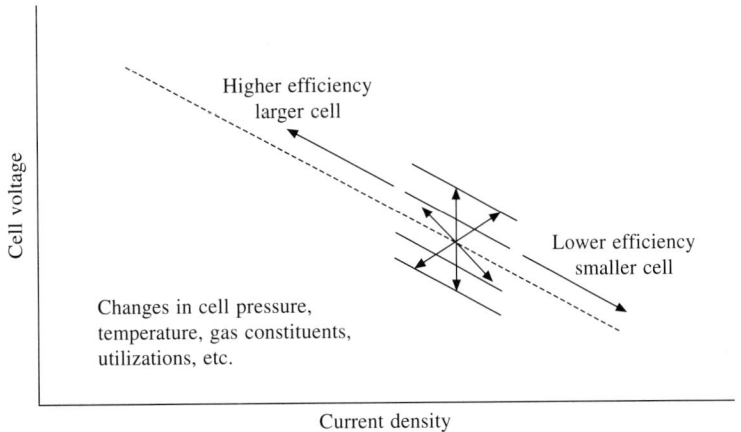

Fig. 23 Optimization trade off in a PEMFC system (from Fuel Cell Hand Book, 2000).

Ersoza et al. (2005) studied a 100 kW net electrical power PEM fuel cell system consisting of an autothermal reformer, high and low temperature shift reactors, a preferential oxidation reactor, a PEM fuel cell, a combustor and an expander. Intensive heat integration within the PEM fuel cell system was necessary to achieve acceptable net electrical efficiency levels. The fuel cell stack efficiency has been calculated as a function of the number of cells (500-1250 cells). The obtained net electrical efficiency levels are between 30% (500 cells) and 37% (1250 cells) and they are comparable with the conventional gasoline based internal combustion engine systems, in terms of the mechanical power efficiency.

In another study by Wu et al. (2005), the optimal operating conditions based on validated multi-resolution fuel cell simulation tool has been developed with four control parameters including cell temperature, cathode stoichiometry, pressure, and humidity. The study shows that different optimal solutions exist for different system assumptions, as well as different current loading levels, classified into small, medium, and large current densities. This design can be readily applied to a larger number of control parameters and further to the fuel cell design optimizations.

Yoneda and Ohno (2005) analyzed the gas reaction processes in PEMFC. In this stochastic approach they could show that the behaviours of the PEMFC gas reaction processes could be considered to be consisted of the two probability processes of the gas reactions at the reaction sites situated at both electrode/electrolyte surfaces and the random walkings of the proton ions in the electrolyte membrane. The Poisson distribution converts to Gaussian distribution of corresponding local gas reaction site on the electrode. Then the individual Gaussian distribution at each single reaction site can be added to construct one Gaussian distribution. Also, the electrolyte membrane plays role to combine both distributions at fuel electrode and air electrode.

3.1 Reactant Utilization

Utilization of the fuel and oxidant involve trade offs with respect to the optimum utilization for a given system. High utilizations are considered desirable (particularly in smaller systems) because they minimize the required fuel and oxidant flow, for a minimum fuel cost and compressor/blower load and size. However, utilizations that are pushed too high result in significant voltage drops. Low utilizations can be advantageous in large fuel cell power cycles with efficient bottoming cycles because the low utilization improves the performance of the fuel cell and makes more heat available to the bottoming cycle.

High fuel utilization is desirable in small power systems, because in such systems the fuel cell is usually the only power source. However, because the complete utilization of the fuel is not practical, except for pure H_2 fuel, and other requirements for fuel exist, the selection of utilization represents a balance between other fuel/heat requirements and the impact of utilization on overall performance. In cases where pure hydrogen is used as fuel, recirculation of the anode exhaust using ejector, pump or blower is found to improve the total system efficiency (Karnik and Sun, 2005). Similarly for oxidant utilization, a trade off between oxygen and air is involved. In addition to this, another obvious trade off occurs between cell performance and compressor or blower auxiliary power.

3.2 Cost Benefit Analysis

The key barrier to commercialization of the PEMFC is the price of the system in comparison to conventional combustion technologies. Despite the environmental advantages and the relatively high efficiency, even for a low power rated units, the capital cost must be reduced to at least the same order as reciprocating engines in order to compete in many applications and not just niche markets. PEMFC cost components are divided into fuel cell repeat and non-repeat components. Fuel cell repeat components comprise the bulk of the cost of an automotive type PEM fuel cell stack, which is generally considered to be the most costly component of a PEMFC. The repeat component is the membrane electrode assembly, in which the ion exchange membrane is very expensive and low volume product.

There is considerable scope to reduce system costs in terms of materials processing, system and component design. Some aspects of the PEMFC system are modular (stack) and costs can be expected to scale linearly with unit size, but certain components do not scale in this way, viz., pumps, heat exchangers, power conditioning equipment etc. This results in an increasing system cost/kW as the unit size decreases. In order to attain the target costs for low power rated units, it is necessary to redesign the system at the expense of efficiency and reliability.

However, the PEMFC stack has the greatest potential for cost reduction. Cell power density would have the greatest effort for the optimization. For stationary applications, lower current densities could be acceptable than used in transport applications in order to increase efficiency. A balance has to be made between system efficiency and acceptable stack current density, effectively stack cost/kW output. Another opportunity to reduce the costs lie in identifying a cheaper membrane, electrode material as well as lower platinum loadings in the electrodes.

The key area for reduction in costs of PEMFC systems by volume manufacturing involves the same stack supplier producing stacks for both stationary and automotive applications.
Power conditioning also dominates the PEMFC system cost and limit commercial opportunities. Progress in power electronics is continuous and fairly rapid, but not as dramatic as in the computer world. Gas processing also has much scope for cost reduction in terms of reducing the stages of shift reactors, which includes complicated systems with several heat exchangers.

3.3 Sensitivity Analysis
Sensitivity analysis especially analysis of the cost of manufacturing of the components and the total PEM fuel cell system is a complex subject. The analysis can be performed for each segment of the total system or for a whole lot of parameters such as cost of manufacturing of the components, and the total PEM fuel cell system, marketability of the product, availability of infrastructure for fuel, lifetime of the system, production volume and capacity, public acceptance, demand, risk factors etc. They vary dramatically depending on the applications as the operating conditions are totally different. Operating lifetimes of greater than 50,000 h (five years) are desirable for a stationary fuel cell system. Lifetimes of 5,000 h would be adequate for the life of the car. As a result, the stationary fuel cell must be more robust. In addition, the fuel cell production volume for a passenger vehicle would most likely be much larger than the production quantity for supplying distributed power to buildings. A major vehicle production line turns out 300,000 vehicles per year. A fuel cell vehicle power train would have 50 to 80 kW peak power capability. For comparison, the average electrical load buildings varies from 33 kW for hotels and stores up to 95 kW average power draw for office buildings (Little, 1995). With capacity factors near 50%, the peak power would vary between 66 kW and 200 kW, or at most three times the fuel cell peak power required for one passenger vehicle. Therefore to reach a fuel cell power production level equal to that for 300,000 vehicles per year, industry would have to install fuel cell systems of the order of 150,000 buildings per year. This comparison assumes that the stationary fuel cell has approximately the same power density as the automotive fuel cell (James et al., 1999).

Jemei et al. (2005) reported a Dynamic Recurrent Neural Network (DRNN) model of a PEMFC for a 500 W fuel cell. The proposed black box model can easily be extrapolated to more powerful fuel cell systems. For black-box models, simulation results are strongly dependant on the choice of input parameters. Thus, a sensitivity analysis is performed to assess the influences or relative importance of each input parameter on the output variable. Many different ways to perform sensitivity analysis are possible. A Multi Parameter Sensitivity Analysis (MPSA) is proposed to evaluate the relative importance of each input parameter independently on the fuel cell voltage.

An 'exergy' analysis has been carried out by Song et al. (2005) for a PEMFC system fuelled by ethanol for automotive application with different power levels, which allow an accurate allocation of the deficiencies of the subsystems of the plant and serves as a unique tool for essential technical modifications.

Cost factor basically arises from the production volume and the operating conditions in terms of its rated output. For example, in stationary power generation, it is advantageous to chose fuel cell operation at atmospheric pressure for the following reasons:

1. Stationary power generation demands 5-10 year component lifetimes.
2. Stationary systems are not weight sensitive.
3. Low pressure systems have lower parasitic loads.
4. As membrane performance improves, the cost impact between pressurized and non-pressurized operation becomes less important.

Another issue of PEM fuel cell systems is their performance degradation with time. This parameter directly affects the capital cost of the PEM stack as the membrane area and thus the MEA must be sized to supply full rated power not just at the beginning but also at the end of the rated lifetime. Thus a fuel cell stack with 10% power degradation per year and a one year design lifetime, would need to have greater membrane active area than a fuel cell stack with no performance degradation. As design lifetime increases, required membrane area can increase dramatically. Given this projected performance decrease, it is then important to analyze whether MEA replacement is a viable alternative to over sizing the fuel cell stack to ensure adequate life. Thus, when analyzing the stationary PEM fuel cell stack design, it is important to consider the impact of design features relative to the two degradation mechanisms of ion-exchange of metallic corrosion products with the polymer electrolyte and sintering of the alloy electrocatalysts.

MEA cost at low production rates is particularly difficult to estimate because of the relatively immature manufacturing techniques, the high cost and variability of the base ionomeric materials, the chemical vs. mechanical nature of the processing, and the wide production rate range considered here. However, the three cost estimation techniques are considered as most appropriate and are: (1) purchasing the components at current commercial pricing, (2) bench-top scale casting of membrane from ionomeric solutions purchased at today's commercial pricing and (3) semi-continuous automated manufacturing of the MEA in a pilot-scale plant. Techniques 1 and 3 result in the lowest finished product cost for all manufacturing rates can be considered

While the fuel cell stacks are the main functional component of a fuel cell power system, there are many other ancillary components that are needed for a complete system. Because the ancillary components tend to be 'off-the-shelf' or at least minor modifications of commercial/industrial products, a much higher level of cost estimation can be to arrive at cost estimates. The magnitude of the price discount varies depending on purchase quantity and the specific component.

James et al. (1999) have estimated the fuel cell system cost as a function of membrane area, which can be converted to the cost as a function of output power. They used the Design for Manufacture and Assembly (DFMA) technique in their analysis and have shown that the estimated cost of the fuel cell stack can be represented by the following equation:

$$C_x = M \times \left[\left(\frac{A - 105.4}{10} + \frac{17.56\,L_p C_p}{380} \right) \times \frac{P_G (1 + d)^N}{P_d} + B \right]$$

where M is a fixed cost markup (1.1 default), A the cost parameter that depends on production volume, L_p the fuel cell platinum loading for both electrodes (mg/cm^2), C_p the cost of platinum ($/troy ounce), P_G the fuel cell gross DC peak power (kW), P_d the fuel cell power density (W/cm^2), d the annual fuel cell degradation (%/year), N the planned fuel cell lifetime (years) and B a second cost parameter.

The fuel cell stack cost depends on two cost parameters, viz. *A* and *B* which in turn were developed for five different production volumes. The *A* parameter is the power-dependent term and the *B* parameter is the fixed cost for the fuel cell stack.

This analysis was done for output levels ranging from 3 to 200 kW and with the assumption that the stack operates near ambient pressure replacing an expensive compressor that provides three atmospheres of air pressure to the vehicle fuel cell cathode with a blower for the stationary system. This reduces the parasitic power required to run the fuel cell system and also reduces the ancillary costs for the system.

The cost for the ancillary components has been approximated by James et al. (1999) by a quadratic equation in fuel cell output power—the cost does not vary linearly with power. For 100 production units, the estimate ancillary cost is given by

$$C_\alpha = 3416.2 + 43.113 \times P_\alpha - 0.05_\alpha \times P_\alpha^2$$

For 10,000 production quantity, the ancillary costs are

$$C_\alpha = 3031.5 + 38.495\ 3 \times P_\alpha - 0.054 \times P_\alpha^2$$

The results of this stationary PEM fuel cell cost analysis are compared with cost projections for PEM vehicle fuel cell systems for a 50 kW stationary fuel cell system cost is projected at $490/kW for 100 units, decreasing to $310/kW for 10,000 units, while the 50 kW mobile fuel cell system cost is estimated at $36/kW in automotive production volumes.

Hailes (1999) carried out an analysis with respect to MEA manufacturing, based on the prototype manufacturing techniques and find that the manufacturing techniques adopted do not cause large environmental burden. She also found that depending on the type of material and process selected, material between 2 and 12% is lost.

A sensitivity analysis of PEMFC in transport application has been carried out after taking into consideration the leverage points, resistance points, risk analysis, value proposition, market segmentation, demand analysis and forecast (Franco, 2005).

3.4 Environmental Analysis
The low emissions in terms of both chemical and noise from PEMFC compared to conventional power generating plant would give an advantage in terms of siting flexibility. They directly influence a limited number of sales on ethical or publicity grounds and assist in creating market opportunities at present. The low emissions alone are not a selling point, but when the technology is comparable in terms of reliability and economics with conventional generating plants, the low emission characteristics may be a key factor for commercial success. Measures need to be taken in the form of non-fossil fuels obligation, whereby a specified amount of generating capacity must be secured from renewable energy technologies. In the case of PEMFC, natural gas fired plants may not be as desirable as hydrogen powered plants and renewables will be particularly attractive. In this scenario, solar derived hydrogen would be a beneficial choice of fuel, regarding integrated pollution and prevention control regulations, and in the long term, this fuel choice may be of benefit to the commercial prospects of PEMFC. From the environmental perspective, the choice would be then whether to store and transfer hydrogen or electricity produced by renewable energy technologies (Hailes, 1999). The importance of system performance was reported recently by Strachana and Farrell (2005) with respect to environmental performance in comparison with other technologies like gas turbines and micro turbines.

3.5 Thermal Efficiency and Useful Heat
The thermal efficiency of PEMFC is lower than that of combustion generator technologies. The heat to

power ratio also is not well matched to CHP applications. But they have constant thermal and electrical efficiencies and constant heat power ratio over a range of loads.

The low grade heat from PEMFC around 70°C is not enough to raise steam for steam reforming or for processing, but may find applications for specific CHP applications like brewing and in the dairy industries. The thermal efficiency is sacrificed in an attempt to keep the units compact and at low cost, which requires relatively large heat exchangers to extract useful heat.

3.6 System Dynamics (Response Time)

The dynamic response of the PEMFC has been considered both in terms of start up times from cold condition and the load following capabilities under normal operation. PEMFC has a relatively shorter start up time compared to other types of fuel cells, because of the low operating temperature. Warm up time also is shorter. However, if the fuel is natural gas or other fuels, which requires reforming, the start time is dependent on both the size of the stack as well as the time taken to heat the reformer and not the fuel cell. If the PEMFC system has hydrogen as fuel, the start up time is not affected by the unit size as it is limited only by the time taken for the gas to flow through the system.

For portable fuel cell systems a multitude of applications have been presented over the past few years. Most of these applications were developed for indoor use, and not optimized for outdoor conditions. The key problem concerning this case is the cold start ability of the polymer electrolyte membrane fuel cell which was first investigated by the automotive industry, which has the same requirements for alternative traction systems as for conventional combustion engines. The technical challenge is the fact that produced water freezes to ice after shut-down of the PEMFC and during start-up when the temperature is below 0°C was investigated by OzciPok et al. (2005) using the calculated cumulated charge transfer through the membrane which directly corresponds with the amount of produced water in the PEMFC. The charge transfer curves were mathematically fitted to obtain the parameters describing the cold start-up with the cumulated charge transfer density. The start up and failure analysis of PEMFC has been carried out by Serincan and Yèilyurt (2005) using a 2D model to study the dynamic behavior with interdigitated flow field designs. On this basis of simulations some failure modes of the PEMFC due to the auxiliary components are investigated. Isothermal operation of single fuel cells below 0°C has been investigated by potentiostatic experiments and analyzed by physical modeling by Ozcipok et al. (2005a). During potentiostatic operation a current decay is observed which is addressed to freezing of product water in the porous cathode. Thus a better understanding of freezing processes at sub-zero conditions in PEM fuel cells is gained and mechanisms proposed from earlier experimental data and statistical based interpretations are confirmed. Furthermore the transient water uptake behavior of the membrane was studied by *in situ* experiments at different temperatures.

The load following capabilities of the PEMFC, during operation are expected to be better than that of high temperature fuel cell types for stationary applications, since the electrochemical and reforming reaction kinetics are very fast under the operating conditions. They do not lead to significant temperature change which would otherwise cause thermal management problems on rapid load change.

Load following is not an issue, for grid connected operation, since fluctuations can be absorbed by the mains. In the case of stand alone applications, where rapid load change can occur. In PEMFC, it is possible to switch from zero load to higher capacity of 250 kW in less than 10 ms. There is no need to run a generator in standby mode.

However, one area of consideration in load following is the output frequency stability. Frequency stability is distinctly different between a diesel engine based system and a stand alone fuel cell. An increase in load in a diesel system causes a transient reduction in speed, and hence frequency, which is minimized by the inertia of the system and the speed of the control loop. In the case of fuel cell, the output frequency is set by the system clock and will not change on application of load. Since there is no inertia in the system, the only way to reconcile the differences between the load and the supply is by reducing the output voltage. Battery or super capacitor may be incorporated to effectively provide the inertia equivalent.

3.7 Technical Challenges

Continuous R&D efforts have addressed several technical issues with respect to better performance of PEMFC and also in reducing the cost of the components. There are still few technological challenges which are summarized as follows:

(i)   Choice of fuel (gasoline, methanol or hydrogen).

(ii)   Efficient fuel processing, with reduction of weight, volume and CO residuals.

(iii)   Finding anodic electro catalysts tolerant to CO at levels of 100 ppm (with noble metal loading lower than 0.1 mg/cm^2 or less).

(iv)   Developing a cathodic electro catalyst, to reduce the over potential encountered at open circuit and to significantly enhance the exchange current density.

(v)   Finding alternative proton conducting membranes with lower cost but same proton conductivity of the state-of-the-art perfluoro sulfonic acid membranes.

(vi)   Developing new proton conducting membranes not depending on water for high temperature operation between 333 and 473 K.

(vii)   Manufacturing low-cost bipolar plates.

(viii)   Developing an air compressor/turbine with improved performance and reduced size and cost.

(ix)   Optimizing thermal and water management.

At the system level, improvement in performance level at dynamic load conditions, performance of large area electrodes is a cause for concern. One of the reasons cited is in removal of product water which becomes more difficult in larger systems. Apart from that, high water vapor pressure in the reactant flow causes an increase in the cathode over potential. In addition, water condensation, electrode flooding and occlusion of the gas channels occur leading to operational failure. Methods like cyclic and rapid ejection of the excess water through purge valves do not satisfactorily solve this problem. Another way to overcome this difficulty to some extent is by creating a concentration gradient across the membrane, which makes the water move from the cathode to anode side and by optimizing the extent of humidification and the flow rate of fuel entering the cell. Another issue strongly related to water management is that of the thermal management in stacks. The evaluation of heat losses in PEMFC is that the power dissipated as heat is approximately equal to the electrical power supplied to the external circuit which strongly depends on the cell operating voltage.

4. Potential Applications for PEMFC

PEM fuel cells are being considered for a range of applications, from portable power generation, to distributed generation of power in the 5-100 kW range (there are couple of examples of 250 kW systems for stationary application), to use in a new generation of fuel cell powered electric vehicles. There are also other many interesting ways in which the use of fuel cells in one sector of the economy might complement in another sector. In other words the same operator can operate the fuel cells in multiple ways. There is lot of room available for improving the opportunities for fuel cell applications, by creating a central communications platform, managing an open industry database, networking associations and organizations, at different levels, within different sectors and between different countries.

4.1 Vehicular PEMFC Applications

Fuel cells are being evaluated or developed for a variety mass transit applications, including locomotives, transit buses, "people movers" and taxis. The main reason for the continued interest in fuel cell powered automobiles is the necessity to introduce Zero Emission Vehicle (ZEV)s in many western countries. In 2003, London (UK) announced imposition of a congestion tax on ICEV users, the first of such a tax in the Western world.

If a transfer from the present Internal Combustion Engine (ICE) to a future Fuel Cell (FC) transport system would be made, this will require more changes than simply 'replacing' the propulsion technology of the vehicles. A sustainable, future fuel cell transport system will require changes in user practice (different vehicle characteristics influence use of vehicle), changes in regulations (taxes, environmental standards, etc.), changes in the industrial networks (e.g. different fuel suppliers), changes in fuel infrastructure (hydrogen infrastructure), and possibly changes in symbolic meaning or culture (different experience of transportation).

For many years development work on using fuel cell as a power source in transportation was insignificant. Only a handful of vehicles were constructed until the mid-1990s. Since then prototype volumes have risen dramatically, and it is estimated that a total of 200 light duty fuel cell vehicles (FCVs) have been built and operated worldwide. This number includes: light commercial vehicles (such as vans and pick-up trucks); sedan cars; sports utility vehicles (SUVs); and smaller vehicles such as golf-buggies and go-carts. Activity in this sector has accelerated in the last few years, and the cumulative number of fuel cell vehicles (FCVs) produced reached 310 by the end of 2003 and by 2004 this number increased to 520. Since the specific energy density of PEMFC power plants (~ 1 kWh/kg) is akin to that of the present day ICEVs, comparable driving ranges may be expected. But the power density (~ 300 W/kg) of the present day PEMFCs tends to be substantially lesser than that of the ICEVs (~ 600 W/kg) (Cropper, 2004; Adamson, 2005).

4.1.1 Power and Energy Requirements of Fuel Cell-Based Automobiles

Shukla et al., (2003) have described a process to assess the energy and power requirements of a fuel cell-based automobile. They have estimated that the power plant of a modern car must be capable of delivering about 50 kW of sustained power for accessories and hill climbing, with burst-power requirement for a few tens of seconds to about 80 kW during acceleration. For a car with these performance characteristics, this sets the upper power limit required, but in common usage rarely exceeds 15 kW while cruising.

4.1.2 Fuel Cell Vehicle Systems

A complete fuel cell system for transport application consists of several components: viz., a fuel supply system: gasoline or methanol reformer or hydrogen storage tank, fuel cell stack, and an air compressor to provide pressurized air to the fuel cell, cooling system to maintain the proper operating temperature, water management system to manage the humidity and the moisture in the system, a DC/DC converter to condition the output voltage of the fuel cell stack, an inverter to convert the DC to variable voltage and variable frequency to power the propulsion motor, an AC/DC propulsion motor and transmission, a battery or an ultra capacitor is generally connected across the fuel cell system to provide supplemental power and for starting the system. While many of these component requirements are common for all fuel cell applications, the vehicular application has stringent weight, shape and volume constraints. The fuel cell, peak power devices, motor, electronics, and fuel storage system must be designed and arranged to fit into as small a space as possible, without creating safety hazards.

4.1.3 System Options

Fuel cells can be used in vehicles in the following three ways:

(a) Fuel cell along with a battery bank and DC/DC converter.
(b) Fuel cell without a battery bank but with a buck/boost controller.

(c) Fuel cell system without a battery bank and DC/DC converter.

4.1.4 Fuel Cell System with Battery and DC/DC Converter

In a fuel cell propulsion system with a battery pack and DC/ DC converter, the propulsion unit is designed at a higher voltage than the fuel cell voltage. Hence a boost type DC/DC converter is required to boost the fuel cell stack voltage to the required battery voltage and also to charge the batteries. The power required for propulsion is supplied by the batteries and the fuel cell stack. The DC/DC converter has to be sized based on the maximum power capability of the fuel cell stack. The power drawn from the fuel cell is controlled by controlling the output current of the DC/DC converter. In this type of control, since the output current of the DC/DC converter is controlled, the power can vary as a function of the battery voltage which depends on the traction power demand from the vehicle controller that is based on the acceleration and regenerative braking. This creates a wide variation in the fuel input to the stack, because the fuel cell output power is proportional to the hydrogen input. The stack controller may not be able to efficiently control the stack. The controller should have a wide bandwidth to stabilize the system under all operating conditions.

The amount of hydrogen supplied to the fuel cell stack could be better controlled, if the fuel cell stack output current is directly controlled. The reference current is demanded from the fuel cell controller and the vehicle system controller. In this control scheme, for a constant current at the DC/DC converter input, the stack voltage is also constant, and thus the power at the stack output remains constant. This control scheme avoids the wide variation in the fuel input to the stack. In addition, it enables constant current load that is ideal for fuel cell operation, and relatively constant power at the output of DC/DC converter that is optimum for hybrid vehicle operation. If the fuel cell stack system voltage is higher than the battery voltage, a buck converter needs to be used instead of a boost converter. However, designing a fuel cell stack for higher voltages is not efficient.

A range extender type fuel cell could be designed with low power to only charge the batteries. The range extender fuel cell hybrid is more like a battery operated vehicle. The battery needs to be designed to provide full power. Because of the favorable efficiency curve of the fuel cell unit in the part load range, a system with a smaller battery and a full power fuel cell stack can be more attractive.

Fuel cell system without a DC/DC converter can be used if the characteristics of the fuel cell stack and the battery are matched; it is possible to eliminate the DC/DC converter. Here the load is shared by the fuel cell and the battery. By controlling the hydrogen input to the stack, it is possible to adjust the power supplied by the battery and by the fuel cell. The fuel cell also can be operated to fully supply the load and to charge the batteries.

4.1.5 Fuel Cell System without a Battery

To reduce the cost and weight of the system, the propulsion system can be configured without using the battery also. A boost converter is required to match the low voltage output of the fuel cell stack to the high voltage requirement of the propulsion drive system. To start the fuel cell system, the necessary power can be provided from the battery of the vehicle. The voltage is boosted to the required volts using the Buck/ Boost converter. The high voltage accessories of the fuel cell stack are powered from the output of the boost converter. Depending on the system, it may take about 15 to 20 seconds to start the fuel cell unit. During this time, the power required by the fuel cell unit compressor, is about 10% of its peak power, which can be provided by the normal battery. Under normal operating conditions, the fuel cell unit charges the battery. The vehicle system without using the battery has to have direct hydrogen type fuel cell system to obtain good dynamic response as at present state of reformer technologies, they will not be able to produce the needed hydrogen fast enough to meet the sudden load changes.

4.1.6 Fuel Cell System without a Battery and DC/DC Converter

Fuel cell system can be developed to suit the required power without the batteries, if the fuel cell stack output voltage is compatible with the voltage required for the propulsion drive system and the DC/DC converter could also be eliminated. This increases the total efficiency and reduces the cost of the system. This is the ideal configuration for a fuel cell electric vehicle system. This type of fuel cells is less expensive, more efficient, and will have a better fuel economy due to the reduced weight of the total system. However, the challenges for practical implementation are: the fuel cell output is not as regular as the battery and the inverter sees a wide variation in the DC input voltage which needs to be compensated in the motor control system, otherwise leading to stability problems in the drive system. If the fuel cell system is based on direct hydrogen, it may be possible to start the system using a 12 volt battery and a boost converter. If the system has a reformer, the power available from the 12 volt battery is not adequate to start the system. In this configuration, the regenerative braking energy cannot be captured.

4.1.7 Fuel Cell Vehicle-System Integration

The development of fuel cell electric vehicles requires the on-board integration of fuel cell systems and electric energy storage devices, with an appropriate energy management system. The issues involved can be broadly classified into those related to the fuel cell system, those involving the electric drive including safety issues such as leak current, grounding, electromagnetic interference, coordination between subsystems for optimum operation, isolation of the fuel cell stack from the drive system, connecting the battery and the fuel cell stack, charging of battery from the fuel cell stack, coordination of the power delivered from the battery and from the fuel cell stack, particularly if DC-DC converter is not used, supplying the power to the accessory loads of the vehicle and to the accessory loads of the fuel cell stack, matching the fuel cell output characteristics with the characteristics of the battery and the drive system, advanced sensors for fuel flow measurement, temperature/pressure regulation.

The optimization of performance and efficiency needs an experimental analysis of the power train, which has to be effected in both stationary and transient conditions (including standard driving cycles). Corbo et al. (2005) have carried out such a study with a 2.5 kW PEMFC stack coupled to an electric propulsion chain of 3.7 kW. The control unit of the system allowed the main stack operative parameters (stoichiometric ratio, hydrogen and air pressure, temperature) to be varied and regulated in order to obtain optimized polarization and efficiency curves. Experimental runs effected on the power train during standard driving cycles have allowed the performance and efficiency of the individual components (fuel cell stack and auxiliaries, DC-DC converter, traction batteries, electric engine) to be evaluated, evidencing the role of output current and voltage of the DC-DC converter in directing the energy flows within the propulsion system. The stack characteristics are given in Fig. 24 and details of the energy losses are shown in Fig. 25, where the absorbed power of different components of the fuel cell system are reported as function of the power entering the DC-DC converter. It can be observed that the major energy consumption is due to the air compressor (about 120 Watts

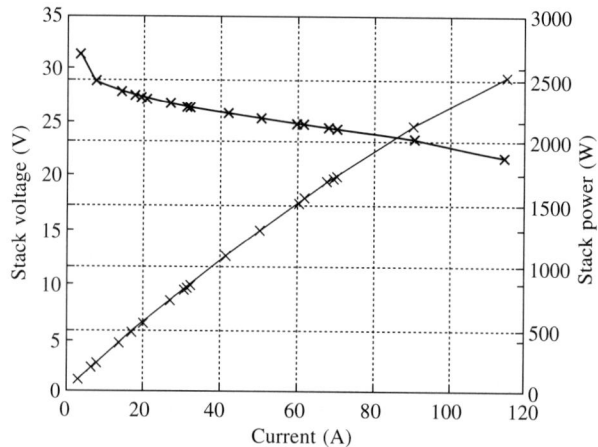

Fig. 24 Stack characteristic curves (R = 2-6, T = 333 K, P_{H_2} < 50 kPa, P_{air} < 20 kPa) (from Corba et al., 2005).

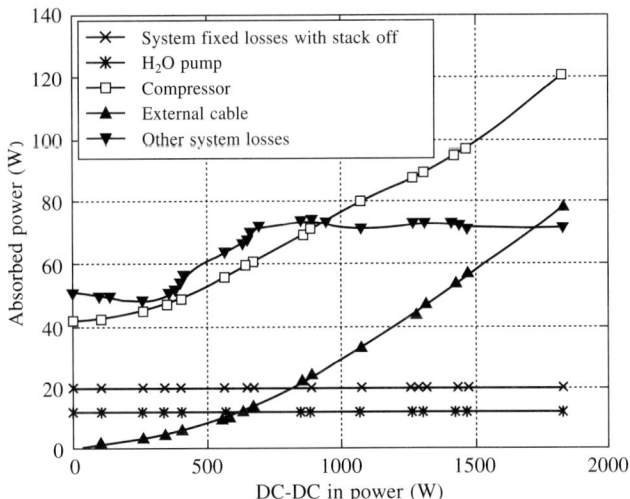

Fig. 25 Power losses associated to the main fuel cell system components vs. DC–DC converter (from Corba et al., 2005).

1.8 FCS power), while minor losses are associated to the cooling and humidification water pumps (about 10 W for each one, constant with respect to load).

In fuel cell vehicle systems, for best performance and highest efficiency, the fuel cell stack, the DC/DC converter, battery, power inverter, and the motor should be treated like one system. If the amount of hydrogen flow to the stack is higher than that required by the electrical load, then energy is wasted in the exhaust. If the fuel flow is less than that required by the electrical load, then the impedance of the stack increases thus overheating the stack, and possibly leading to cell reversal. Hence, it is necessary to match the amount of hydrogen flow to the stack to meet the desired electrical load at the output.

While the buses require a fuel cell module of 100-200 kW, small passenger vehicles are likely to need a power output of around 85 kW. To compete successfully with the internal combustion engine, stack power density will need to be over 1.6 kW/litre, while the complete system will need to provide more than 0.5 kW/litre to minimize intrusion into the passenger space. High overall thermal efficiency implies a cell voltage in excess of 0.750 V, with a decay rate of less than 2 mV/1000 h for adequate lifetimes. Performance will need to be maintained over 17,000 start-stop cycles despite adverse operating conditions, such as less than 60% relative humidity at the air inlet. In addition, stacks will need to have the ability to start up quickly from the frozen state for upto 1000 times without damage.

4.1.8 World Scenario

Almost all the major automobile manufacturers around the world are developing fuel cell powered vehicles the bulk of it being passenger cars. These development ranges from fuel cell-battery to pure fuel cell to fuel cell-super capacitor hybrid, fuel cell-ICE hybrid drive systems.

One of the largest fuel cell demonstration project is the Clean Urban Transport for Europe (CUTE) project spanning from year 2001–2006 (www. Fuel-cell-bus-club.com). CUTE is a demonstration and test of 27 fuel cell buses in public transit operation in nine participating cities. These cities are Madrid and Barcelona (Spain), Porto (Portugal), Luxembourg, London (UK), Amsterdam (the Netherlands), Hamburg and Stuttgart (Germany) and Stockholm (Sweden). Daimler Chrysler built the fuel cell city buses on the CITARO platform. The Mercedes-Benz Citaro fuel cell buses are equipped with a HY-205 P5-1 fuel cell engine developed by Ballard Power Systems, Canada. The fuel cell engine consists of a fuel cell system, based on Ballard's Mk9 generation fuel cell stacks, that supplies power to a central electric motor with a maximum power of 205 kW (Scheme 3 and Table 5). The buses use compressed hydrogen as fuel, stored in Dynetek high-pressure cylinders mounted on the roof of the vehicle. The total hydrogen storage capacity

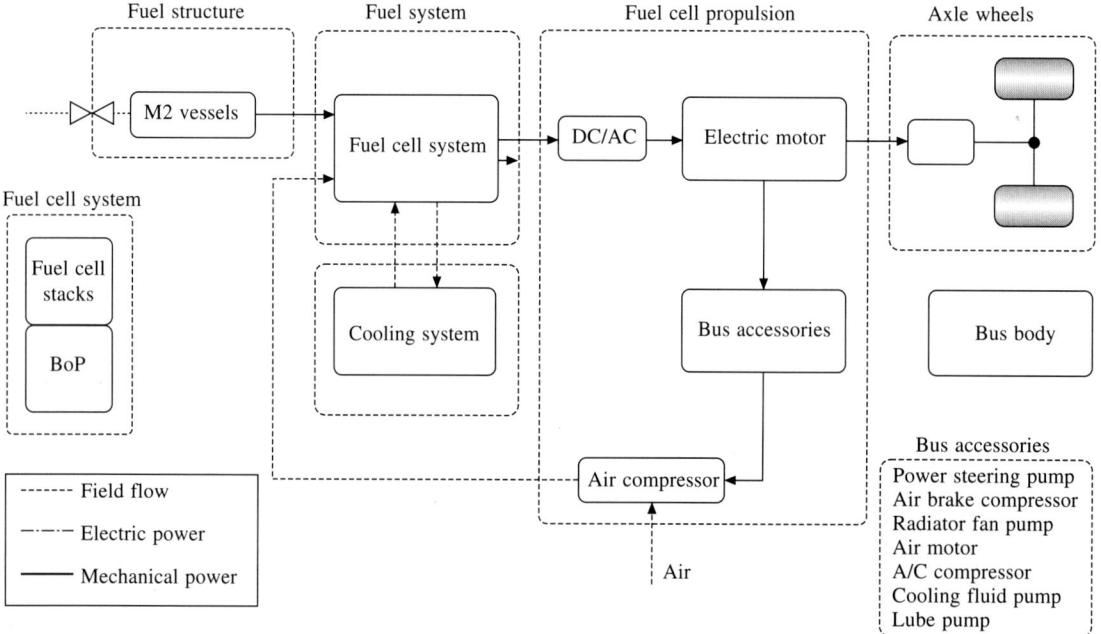

Scheme 3 Schematic overview of the electric driveline of the Mercedes-Benz Citaro fuel cell buses in the CUTE project, based on (BoP, balance of plant components) (from Haraldsson et al., 2005).

Table 5. Characteristics of the fuel cell buses in the Stockholm CUTE project according to the Swedish vehicle registration certificate

Vehicle	Description
Dimensions (L × W × H) (m)	11.95 × 2.55 × 3.69
Gross weight 9 kg)	18,000
Curb weight (kg)	13,890
Maximum frontal axle load (kg)	7245
Maximum rear axle load (kg)	11,500
No. of passengers: maximum	57
No. of passengers: seated	32
Maximum speed (km h^{-1})	80 (limited)
Driveline	
Fuel cell system (kW)	>250 (2 stacks of 150 kW gross each)
Central electric motor (kW)	205
Hydrogen storage	
Total capacity (l)	1845 (9 cylinders)
Maximum pressure per cylinder (bar)	350

is 40 kg at (15°C, 350 bar) providing a bus operation range of about 200 km (Haraldsson et al., 2005). The characteristics of a typical drive cycle are shown in Fig. 26.

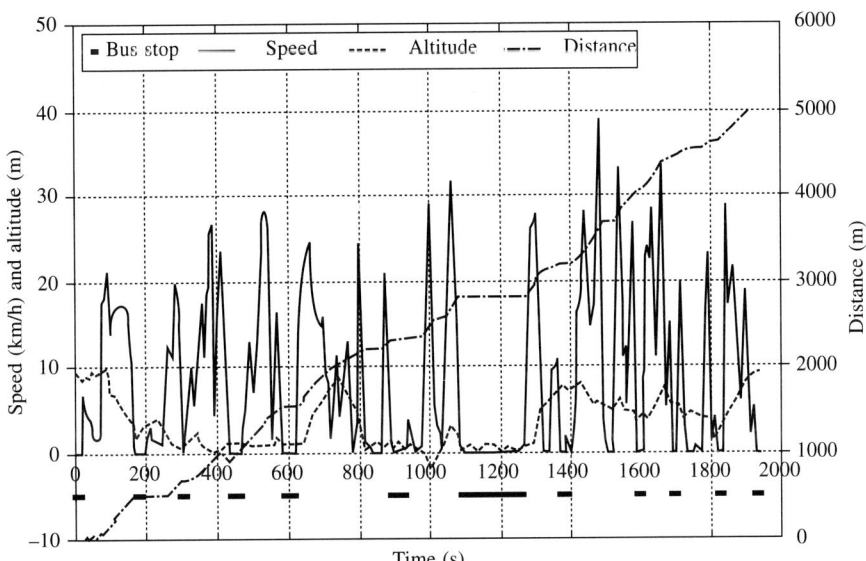

Fig. 26 The characteristics of a typical drive cycle on the demonstration route (from Haraldsson et al., 2005).

The other major bus demonstration programs are: Georgetown University (100 kW stack from XCELLSIS, 40 ft., Novabus RTS platform, Lockheed martin control system electric drive, vehicle system controller by Booz-Allan & Hamilton, Comp. hydrogen, 2000), MAN (120 kW stack from Siemens, 36 ft, Comp. hydrogen 1999), NeoPlan (80 kW DeNora stack, 100 kW flywheel, 36 ft, 2000), Toyota and Hino Motors (63 seater, low floor Comp. hydrogen).

While the number of fuel cell bus demonstration is limited, almost all the major car manufacturers across the world have demonstrated fuel cell powered cars: NECAR 1-5, F-Cell, Natrium, Jeep Commander (Daimler Chrysler), Hydro Gen 1-3, AUTOnomy using HyWire concept, Percept concept car, Opel Zaffira minivan (General Motors), P2000 Prodigy, P2000 SUV, THINK FC5 (Ford Motor Company) Bora HyMotions (Volkswagen), MOVE FCV-K-II (Daihatsu), FCX-V1, FCX-V2 and FCX-V3 (Honda), FCHV-4 and 5 (Toyota). Details are available at www.fuelcells2000.org.

There are many other aspects of recent FCV development that are worth commenting on, including a trend towards greater hybridization with batteries, which can make an FCV more responsive, and save money as a less powerful fuel cell might be required as a result. There have also been improvements in cold start-up (Honda is at the forefront here), and in the development of slim-line fuel cell systems that can fit into a car's chassis. Ju et al. (2005) have unveiled a control algorithm for active power distribution in a fuel cell/battery hybrid power source to improve system efficiency and battery life with load following capability.

4.1.9 Other Transportation Applications

Fuel cell systems for rail vehicles has been analyzed by Daimler Chrysler with the P4/P5 fuel cell system of Ballard, Canada and the selected pressurized hydrogen tank system. They are then configured into a propulsion system and laid out in different Adtranz platforms to check the technical feasibility of each application. The Adtranz product Unit for each relevant platform has selected a representative vehicle whose performance profile was then defined. A kinetic calculation model specifically developed for this study was then used to give information on the required number of FC-systems to match the desired performance as well as the attainable range and available operation time. Results of this study show that INCENTRO (LRVs), ITINO (Regional Trains) and INNOVIA (Automated Guided Transit Systems) are suitable vehicle product platforms for the application of fuel cell systems of medium performance (Niehues and Edwards, 2000).

PEM fuel cells are an attractive option for two wheeler vehicles (Laven, 1999; Lin, 1999). Extensive research is being carried out in China and Taiwan in this application and few prototypes have been demonstrated. The Vectrix electric scooter (www.zevltd.com) is a fuel cell/electric hybrid with a fuel cell (methanol powered) to continuously recharge the battery that powers the motor. This hybrid approach offers many advantages: Battery size/fuel cell combination is optimized for typical urban commuter trip profile to top-off battery during periodic stops. Battery pack life is extended by minimizing deep discharge cycles, and it is achieved as the fuel cell continuously recharges the battery pack, minimizes the size and cost of the fuel cell since peak power is delivered by the battery pack, maximizes efficiency and life of the fuel cell as it operates at a relatively constant output, recharging does not require public access to a fixed power supply, methanol based fuel eliminates the safety and fuel availability concerns of pressurized, high purity gaseous hydrogen.

The other transportation application is in ships and submarines for auxiliary power. A new submarine developed by German company Howaldts werke- Deutsche Werft AG (HDW) uses a fuel cell system (from Siemens) that allows the submarine to loiter at low speeds for extended periods of time without surfacing. During such operations, the fuel cell produces limited amounts of noise and exhaust heat, which assists in making the submarine undetectable. An autonomous underwater vehicle "DeepC®vehicle" has been powered by 2 PEMFC stacks of 1.8 kW capacity operated on neat hydrogen and oxygen. Since neat oxygen atmosphere is rather aggressive to the active components of the fuel cell, care has to be taken to find appropriate operating conditions for the system. Furthermore, since the fuel cell systems operate in a closed environment, measures have to be taken to prevent the formation of combustible gas mixtures inside the pressure hull. This has been accomplished by membrane recombination device. The fuel cell systems are placed in each compartment of the twin pressure hull of the vehicle. Each system has a nominal power of 1.8 kW. Nominal operating current will be 50 A at a stack voltage above 60 V. The systems are configured as hybrids. Additional lead acid batteries will buffer load peaks and provide surge power in case of emergency manoeuvering (Hornfield et al., 2005).

Miller et al. (2005) identified the technical challenges for incorporating fuel cells in large vehicles above 20 tonne or 1 MW of power. These challenges derive principally from the effects of large mass, density, and power on heat transfer. The two large fuel cell vehicles analyzed are: (1) a fuel cell-battery hybrid mine loader of 23 tonne and maximum power rating of 160 kW and (2) a fuel cell road-switcher locomotive with weight of 109 tonne and continuous power rating of 1.2 MW. Heat dissipation from the compact underground loader requires water cooling of the stacks, traction battery, and hydraulic systems. Operation of the system cooling fan alone requires 19 kW. The hydraulic system, which operates the bucket and steering of the articulated vehicle, requires a peak power in excess of 100 kW. The principal heat-transfer challenge of the locomotive is removal of heat from the 25-tonne hydrid bed during refueling. Allowed time for refueling with 250 kg of hydrogen is 30 minutes. Since heat removal is rate-limiting, refueling at this rate necessitates removal of heat at the rate of 1.9 MW.

4.1.10 Micro Fuel Cell Vehicles

In recent times a new category of fuel cell vehicles are being developed. This has been possible with the commercial availability of small fuel cell power modules, notably Ballard's Nexa PEMFC system. Many small companies, universities and even schools have eagerly exploited and have made working vehicles even though the power output of systems on offer is, in general, very low. Prototypes are likely to dominate this segment. The advantage of such development is that they will raise public awareness and will be a considerable asset in education, introducing fuel cells to students around the world. In Japan, even now there is a race for low powered FCVs made by schools and universities. Such vehicles can be built today at relatively low cost and without specialist technical input.

4.2 PEMFC for Stationary Power Generation

PEM fuel cells were not originally considered suitable for stationary power generation for a number of reasons most importantly for the low grade heat that is generated by PEMFC. However, in recent times PEMFC are found suitable for application in market niches. In addition, environmental concerns are motivating the development of residential and building PEMFC plants for co-production of electricity and heat. It has been evaluated (Wolk, 1999) that commercial buildings, rather than homes, are the segment of market where fuel cells have the best chance to make their entry, at an initial price higher than the target cost (e.g. US$ 800-1200/kW installed, whether at home or central station) where it is affordable. Development of such small units is in progress at a number of US companies, such as American Power Corp., Plug Power-General Electric Power Systems, Avista Laboratories and Northwest Power Systems. For example, the Plug Power-General Electric Power Systems Module is able to supply 7 kW continuously, and is able to support short-time peaks (i.e. 10 kW for 30 min and 15 kW for 0.5 s) assisted by back-up batteries, being charged during the periods when the power generated by the fuel cell exceeds the demand. This system has the size of a small refrigerator (volume around 1 m^3), and can be operated with natural gas, LPG or methanol. The package includes the fuel-processor, the fuel cell stack and the power-conditioning unit (batteries and inverter), which is the largest component. The price of the power plant is in the range US$ 7500 to 10,000 when it is first introduced into the market, with the possibility of decreasing to US$ 3500 with mass production.

In the future, the stationary applications of fuel cells will also be in decentralized power supply systems. For example, PEM cells with an output of 250 kW could be used to generate all the electrical power and part of the thermal energy required by high-rise complexes and hospitals. Ballard Power Generation Systems is very active in this field, and is developing a PEMFC stationary energy conversion plant of 250 kW power. The plant fits into a 50 m^3 volume, including fuel processor and power-conditioner units together with the fuel cell stacks. Methane, propane, hydrogen and anaerobic digester from waste water treatment facilities are being considered as the fuels. The first unit was operated in 1997; a field trial was in 1999-2000 and the first commercial unit was ready by end of 2001.

Hamada et al. (2004) reviewed the field performance of a PEMFC for residential system. The electrical efficiency and heat recovery efficiency for a rated output operation were quite high 42.5 and 49.2%, respectively and characteristics of partial load, water temperature for heat recovery, start-up time, load following and exhaust gas were clarified and it was proved that sufficient performance can be obtained even under continuous operation. An integrated system framework for fuel cell-based distributed energy applications including a physical energy system application, a virtual simulation model, a distributed coordination and control, a human system interface and a database were developed by Wu et al. (2005). The integrated system framework provides a means to optimize system design, evaluate its performance and balance supplies and demands in a hydrogen assisted renewable energy application. Through integration with an available renewable energy profile database, the developed system efficiently assists in selecting, integrating and evaluating different system configurations and various operational scenarios at the application site. The simulation results provide a solid basis for the next phase of demonstration projects.

In January 2004, Ned Stack Fuel Cell Technology and Akzo Nobel have started the PEM Power Plant Project, with the final prospect to convert industrial hydrogen into electric power on a scale of 50 MWe. The fuel cell heating appliance for the (use) application in detached and semidetached houses has been developed by Gummett et al. (2005) for the European Fuel Cell GmbH (efc) in Hamburg. After the systematic development and laboratory tests, a field test with 15 plants is planned to be carried out during 2005.

4.2.1 Standby/UPS Application

PEMFC systems appear as strong contenders to replace batteries in the emergency back up or uninterrupted power systems (UPS) market, especially in grid connected applications where good quality, reliable power supply is required and where interruptions could last several hours. One such market is that of telecommunications and in particular providing emergency back up for mobile phone repeater stations that are connected to the electricity grid, but are in remote locations where interruptions are common. According to an IFC report, such applications could include fiber optic repeating stations, multiplexing stations, cellular towers, internet backbone computing facilities. The power requirements for such telecommunication applications could range from 1 to 10 kW but can rise to 50 or even 100 kW for multi-purpose sites with many suppliers. Such telecommunication systems require an autonomy of 1-2 h according to Teledyne or 24 h, according to IFC. Conventional UPS, on the other hand, typically have an autonomy of 12 min at full load or 30 min at half load, while for longer interruptions an uninterruptible battery system is installed, consisting of a genset feeding the batteries. UTC Fuel Cells has recently employed their proprietary PEMFC technology in a prototype 5-kW back-up power unit, which provides a seamless power transition without the use of batteries (Perry and Kotso, 2004).

A power system suitable for remote application based on PEM was investigated by Argumosa et al., (2005) with concern about climate change and energy security which is required for telecom centers (GSM, radio, fibre optic equipment) in places without the possibility of grid connection due to environmental or cost factors. In the frame of a European project called FIRST (Fuel Cell Innovative Remote System for Telecom) an international consortium is developing power system based on fuel cells for remote telecom equipment. The consortium is formed by the following partners: INTA (Spain, ascoordinator), AIR LIQUIDE (France), FRAUNHOFER ISE (Germany), CIEMAT (Spain), ICP-CSIC (Spain), NUVERA FCE (Italy), WÜRTH (Germany), INABENSA (Spain), CHLORIDE SPAIN (Spain) and ISOFOTON (Spain). The objective of the project was to develop and evaluate two prototypes: the first one has a fuel cell as an auxiliary power unit and the hydrogen should be provided to the installation periodically; the second prototype includes an electrolyzer and a hydride based hydrogen storage system, being an autonomous system in which the hydrogen is "manufactured" in summer and "used" in winter in the fuel cell. Britz et al. (2005) also studied Fuel cells for telecommunication with a power output of 2 kW of 48 volts with P21 Premion fuel cell systems.

4.3 Portable Applications

Today's portable electronic devices perform an ever increasing number of complex tasks. The increased functionality of these devices is a major challenge to manufacturers who must supply them with batteries capable of meeting their power demands. The inherently higher energy density of small fuel cells, in comparison to batteries, would lead to longer operational times and serve the power demands of next generation portable electronics. As a result, many device manufacturers are seriously examining the technology potential of fuel cells. Casio, HP, NEC and Motorola all have ambitious fuel cell programs and are developing miniature fuel cell solutions to power portable electronics called Micro fuel cells. There are quite a lot of operational/functional differences with a battery pack versus micro fuel cells in terms of potential performance, safety and regulatory implications, and markings/instructions for end products. Each of the companies have their own strategies. NEC has demonstrated an experimental direct methanol micro fuel cell for portables that uses nano carbon material for its electrodes. NEC expects that the device, which measures 40 mm × 50 mm and 5 mm thick, will be more powerful by an order of magnitude than lithium ion batteries, currently the best available power sources for portable devices such as laptop computers. The energy capacity of current Li-ion batteries is around 130 Wh/kg and it is estimated about 1,300 Wh/kg or more for a well-designed portable direct methanol fuel cell based on carbon nano tube technology.

While consumer applications such as cell phones, notebook computers, camcorders, and cordless tools will be the obvious mass market application for portable fuel cells at a power level upto 100 W, there should be an earlier market for PEM fuel cells in a niche <1 kW market. The type of device that fits this class includes weather stations, signal units, APUs, gas sensors, and security cameras. Nevertheless, portable fuel cells are in their infancy. Investment costs to generate 1 kW are still very high, around $10,000 to $100,000 per produced kilowatt. However, portable fuel cells should become cost competitive (~$5,000 per kW) with lithium-ion batteries, commonly used in laptop computers, within the next five years. Voller energy group (Turpin, 2005) has recently developed a 100 W portable power generator called Voller Energy VE100 and developed a software to standardize the components to avoid poor security of supply chain. The use of a software based processor means that the VE100 can be configured to accommodate different build standards and can be remotely reconfigured to accommodate field maintenance. This is claimed to be the first product of this type to get CE certification. To validate the certification, the unit had to undergo rigorous testing in controlled conditions while working at full power using the internal hydrogen supply. This included EN55014 category 4 tests for radio frequency immunity, and compliance testing to EN55022 for radio frequency emissions.

4.4 Renewable Energy PEM Applications

A renewable energy application of PEMFC includes a source of energy such as solar or wind, hydrogen generator (such as a water PEM based water electrolyser), a hydrogen storage device coupled with PEM fuel cell to produce the electricity on demand. Early in 2000, PEM technology was selected to provide night-time power for the solar-powered Helios, a long-duration aircraft. The goal was to make the unpiloted aircraft fly continuously for upto six months. Photovoltaic panels during the day ran electric motors and electrolyzed water. At night, the fuel cell ran the motors by converting the hydrogen and oxygen back into water. The Schatz solar/PEM fuel cell project has been aerating the aquarium at Humboldt State University since 1994. It uses energy from the sun to generate and produce hydrogen that will be used in a PEM fuel cell when the sun is not available. The result is that the fish have enjoyed solar and fuel cell powered air bubbles twenty-four hours a day for over 7 years. Regenerative fuel cells are covered elsewhere in the book.

4.5 Other Applications

There are many interesting and novel applications of fuel cell technology that are being explored. For example, the impending electrification of a fraction of the light-duty vehicle fleet has led some analysts to suggest that the electricity stored in batteries or generated with fuel cells could be used to provide peak power, or even base-load power, to the electrical grid (Kempton and Letendre, 1997; Kissock, 1998; Lovins and Williams, 1999). The current motor vehicle fleet in the U.S. (about 146 million vehicles) has a total shaft power generating capacity of about 14 TW, which if connected to generators could produce about 12 TW of electric power (Kempton and Letendre, 1997). This is approximately sixteen times the present stationary electric generating capacity in the U.S.! Motor vehicles are driven on average about one hour a day, so the generating capacity of the vehicle fleet is idle approximately 95% of the time.

In California, a fleet of 100,000 fuel cell EVs connected in this vehicle-to-grid (or "V2G") fashion could produce about 3.8 GW of power for the grid, assuming 40 kW net fuel cell output power per vehicle and 95% vehicle availability. Even if the vehicles were only available as generating capacity 50% of the time, about 2 GW of generating capacity could be provided by each 100,000 vehicles. If, for example, 50% of today's vehicle population in San Diego alone were fuel cell vehicles by 2025, each capable of supplying 50 kW of power to the grid 50% of the time, the total generating capacity of these vehicles would be over 35% of the present level of installed generating capacity in the entire state of California. A key feature of this concept is that the automotive and light-truck fuel cell capacity will automatically be distributed to the locations where people are working by virtue of the vehicles also being used for transportation. This literally places the power by the people, paving the way for efficiency through increased distributed power management.

Use of fuel cell EVs in this way could help to reduce the need to add generating capacity to meet the expected 1.8% per year growth in electricity demand (and 1.7% per year growth in peak power demand) for California over the next decade (Goeke et al., 1998), and the forecast electricity supply deficit of 2.8 GW in 2003 and 6.7 GW in 2007 (CEC 1997).

Power capacity for current electric utilities has been purchased at a cost of about $1,000/kW, with costs of about $300/kW for modern natural gas systems (Kempton and Letendre, 1997). Current PEM fuel cell system costs are of the order of $10,000 per kW for automotive systems and at least $2,500 per kW for small stationary systems (compared with costs of about $4,000 per kW for 200-kW stationary phosphoric acid fuel cell systems), but costs for vehicle applications are projected to rapidly decline to much lower levels as production expands—perhaps even as low as $50/kW in mass production (Lomax et al., 1997).

Residential and commercial fuel cell systems could also be used in the same manner to provide distributed generating capacity, thereby delaying or eliminating the need to build new centralized electricity generating plants (Kammen, 2002).

5. Conclusions

The prevailing architecture of the PEM fuel cell is the stack with controls and balance of plant, but it has to be assembled with various peripherals required for specific applications, requiring professional (and potentially expensive) installation.

PEM fuel cell technology advances (primarily in the areas of materials and controls) in the coming years will be substantial, reflecting system wide innovation and optimization. The stack will be designed to optimize the use of hydrogen or a hydrogen-rich fuel without the need for an additional fuel processor. PEM fuel cells will primarily use hydrogen or hydrogen-rich commodity fuels. By implication, the hydrogen infrastructure will have to be in place to support PEM fuel cell commercialization.

The governments has to play a decisive role in the development and commercialization of PEM fuel cells by entrepreneurs, venture capitalists, and corporations.

PEM fuel cells will be used primarily in isolated locales unconnected, underserved, or overpriced by the utility electric grid or in especially niche applications. They will be seen as electric generators in competition with internal combustion engines or advanced batteries.

Distributed generation will grow as a cooperative and combined effort of customers and electric utilities. In this context, PEM fuel cells will increase in popularity as energy efficient and environment friendly power generators for both backup and peak-saving applications. They will compete well with advanced internal combustion engines and renewable energy forms. The prevailing customers for PEM fuel cells will be residential and light commercial. Residential customers might include a broad range of private residences, including individual homes, neighborhoods or subdivisions, apartment etc. By implication the prevailing PEM fuel cell unit successful in the marketplace will be of relatively small sizes, less than 50 kW capacity. The other customers could be light industrial with a 50-250 kW capacity demand.

The key elements (Millet and Mahadevan, 2005) of success in PEM fuel cell commercialization are:

1. The PEM fuel cell architecture must be fully integrated.
2. PEM fuel cells must be able to run continually or continuously as generators of premium power for base, backup, and peak-shaving loads in cooperation with the electric grid.
3. Technical advances must be substantial and achieve system-wide innovation and optimization.
4. Customer benefits and expectations must be consistently met.
5. System cost to come down initially to US$ 1500 for stationary application with a life time of operation of 30000-40000 hrs. In the case of transportation application, the price has to come down dramatically to US$ 100-200 range.
6. Hydrogen or hydrogen-rich fuels must be widely available and priced like commodities. This may require a hydrogen infrastructure, although other possibilities are emerging.
7. Codes and standards must be updated continually and they must be comprehensive and aligned well with changing PEM fuel cell technologies and products.
8. Government support must be substantial, including being a lead buyer and customer of PEM fuel cells.
9. Distributed generation must be a combined and cooperative endeavor of both electricity consumers and electric utilities.
10. PEM fuel cell operating costs must be comparable to present power generation systems.
11. To achieve the high range of sales, the PEM fuel cell must be attractive to residential and light commercial customers. This suggests a unit size of less than 50 kW capacity.

References

Aceves S M, Perfect S and Weisberg A (2004), Optimum utilization of available space in a vehicle through conformable hydrogen vessels, DOE Hydrogen Program-Progress report.

Adamson K A (2005), Fuel Cell Today Market Survey: Light Duty Vehicles. Fuel Cell Today, pp. 1-19 (March).

Adjemian K T, Lee S J, Srinivasan S, Benziger J, Bocarsly A B (2002) Silicon oxide Nafion composite membranes for proton exchange membrane fuel cell operation at 80-140°C. *J Electrochem Soc.*, **149**, pp. A256-A261.

Ahn S Y and Le Y C (2005) Effect of the ionomers in the electrode on the performance of PEMFC under non-humidifying conditions, *Electrochim Acta*, **50**, pp. 669 672.

Allcock H R, Hofmann M A, Ambler C M (2002) Phenylphosphonic acid functionalized polywaryloxyphosphazene as proton-conducting membranes for direct methanol fuel cells. *J Membr Sci.*, **201**, pp. 47-54.

Antolini E (2004) Recent developments in polymer electrolyte fuel cell electrodes, *J. Appl. Electrochemistry*, **54**, pp. 563-576.

Antonucci P L, Arico A S, Creti P, Ramunni E, Antonucci V (1999) Investigation of a direct methanol fuel cell based on a composite Nafion®-Silica electrolyte for high temperature operation, *Solid State Ionics*, **125**, pp. 431-437.

Argumosa M P and Pendones R B (2005), A Power System for Remote Applications Based on PEM Fuel Cell, 3rd European PEFC Forum, Lucerne, Poster 405.

Asensio J A, Borros S, Gomez Romero (2002) Proton conducting polymers based on benzimidazole and sulfonated benzimidazoles. *J PolymSci. Part A: PolymChem.*, **40**, pp. 3703-10.

Asensio J A , Borros S and Go´mez-Romeroa Pet al. (2004) Polymer Electrolyte Fuel Cells Based on Phosphoric Acid-Impregnated Poly-2,5-benzimidazole Membranes, *J. Electrochem. Soc.*, **151**, pp. A304-A310

Bae J M, Honma I, Murata M, Yamamoto T, Rikukawa M, Ogata N (2002) Properties of selected sulfonated polymers as proton-conducting electrolytes for polymer electrolyte fuel cells, *Solid State Ionics*, **147**, pp. 189-94.

Bahar B, Hobson A R and Kolde J A (1997), Integral composite membrane, US Patent No. 5,599,614.

Baturina O A and Wnek G E (2005) Characterization of Proton Exchange Membrane Fuel cells with Catalyst Layers obtained by Electro spraying, *Electrochemical and Solid-State Letters*, **8**, pp. A267-A269.

Bender G, Zawodzinski T A and Saab A P (2005) Fabrication of high precision PEFC membrane electrode assemblies, *J. Power Sources*, **124**, pp. 114-117.

Benítez R, Soler J and Daza L (2005) Novel method for preparation of PEMFC electrodes by the electro-spray technique, *J. Power Sources* (In Press).

Bernardi D M and Verbrugge M W (1992) "A Mathematical Model of the Solid-Polymer-Electrolyte Fuel Cell", *J Electrochem Soc.*, **139**, pp. 2477-2490.

Berg P, Promislow K, St. Pierre J, Stumper J and Wetton B (2004), Water Management in PEM Fuel Cells, *J Electrochem Soc.*, **151**, pp. A341-A353.

Besse S, Capron P and Diat O (2002) Sulfonated polyimides for fuel cell electrode membrane assemblies (EMA). *J New Mater Electrochem System*, **5**, pp. 109-112.

Bevers B, Wagner N and Bradke M (1998) Innovative production procedure for low cost PEFC electrodes and electrode/membrane structures, *Int. J. Hydrogen Energy*, **23**, pp. 57-63.

Bockris J O'M and Appleby A J (1986) Assessment of research needs for advanced fuel cells, in: S. Penner (Ed.), Energy **11** p. 110 (Chapter 3).

Bonnet B, Jones D J and Roziere J (2000) Hybrid organic-inorganic membranes for a medium temperature fuel cell. *J New Mater Electrochem. System*, **3**, pp. 87-92.

Bouwman P J, Dmowski W, Stanley J, Cotton G B and Swider-Lyonsa K E (2004) Platinum-Iron Phosphate Electrocatalysts for Oxygen Reduction in PEMFCs, *J. Electrochem. Soc.*, **151**, pp. A1989-A1998.

Bozkurt A and Meyer W H (2001) Proton conducting blends of poly(4-vinylimidazole) with phosphoric acid. *Solid State Ionics*, **138**, pp. 259–265.

Bozkurt A and Meyer W H (2001) Proton-conducting poly(vinylpyrrolidon)-phosphoric acid blends, *J Polym Sci. Polym Phys.*, **39**, pp. 87–94.

Britz P and Esser P (2005) Uninterutible Fuel Cell Power for Telecommunication Transmitters, Fuel Cells for a Sustainable World, Lucerne, Session A12.

Brosha E, Choi Jong-Ho, Davey J, Garzon F, Hamon C, Piela B, Ramsey J, Uribe F and Zelenay P (2005) Non-Precious Metal Catalysts, Hydrogen, Fuel Cells & Infrastructure Technologies Program, 2005 Annual Review, Washington, DC, May 23-27, 2005.

Büchi F and S Srinivasan (1997) "Operating Proton Exchange Membrane Fuel Cells Without External Humidification of the Reactant Gases", *J. Electrochem. Soc.*, **144**, pp. 2767-2772.

Büchi F N, Reum M, Freunberger S A and Delfino A (2005) On the Efficiency of Automotive H_2/O_2 PE Fuel Cell Systems, 3rd European PEFC Forum, Lucerne File No. B091.

Carter R, Wycisk R, Yoo H, Pintauro P N (2002) Blended polyphosphazeney polyacrylonitrile membranes for direct methanol fuel cells. *Electrochem Solid State Letters*, **5**, pp. A195–A197.

Cavaliere S, Raynal F, Etcheberry A, Herlem M and Perezb H (2004) Direct Electrocatalytic Activity of Capped Platinum Nanoparticles Toward Oxygen Reduction in Electrochemical and Solid-State, *Letters*, **7** pp. A358-A360.

Cavalca C A, Arps J H and Murthy M (2001) Fuel cell membrane electrode assemblies with improved power outputs and poison resistance, US Pat. No. 6,300,000.

Cha S Y and Lee W M (1999) Performance of proton exchange membrane fuel cell electrodes prepared by direct decomposition of ultra thin platinum on the membrane surface, *J. Electrochem. Soc.* **146**, pp. 4055-4060.

Chen F C, Gao Z, Loutfy R O and Hecht M (2003) Analysis of optimal heat transfer in a PEM fuel cell cooling plate, *FUEL CELLS*, **3**, pp. 181-188.

Chen S and Kucernak A (2004) Electrocatalysis under conditions of High Mass Transport: Investigation of Hydrogen Oxidation on Single Submicron Pt Particles Supported on Carbon, *J. Phys. Chem. B*, **108**, pp. 13984-13994.

Cheng X, Yi B, Han M, Zhang J, Qiao Y and Yu J (1999) Investigation of platinum utilization and morphology in catalyst layer of polymer electrolyte fuel cells, *J. Power Sources*, **79**, pp. 75-81.

Cheng X, Chen L, Peng C, Chen Z, Zhang Y and Fanc Q (2004) Catalyst Microstructure Examination of PEMFC Membrane Electrode Assemblies vs. Time, *J. Electrochem. Soc.*, **151**, pp. A48-A52.

Cho E A, Jeon U-S, Hong S-A, Oh I-H and Kang S-G (2005) Performance of a 1 kW-class PEMFC stack using TiN-coated 316 stainless steel bipolar plates, *J. Power Sources*, **142**, pp. 177-183.

Choi P, Jalani N H and Datta R (2005) Thermodynamics and Proton Transport, Nafion, III. Proton Transport in Nafion/ Sulfated ZrO_2 Nanocomposite Membranes, *J. Electrochem. Soc.*, **152**, pp. A1548-A1554.

Chow C Y and Wogniczka B M (1995) Electrochemical fuel cell stack with humidification section located upstream from the electrochemically active section, U.S. Patent No. 5,382,478.

Chun Y G, Kim C S, Peck D H and Shin D R (1998) Screen printing of electrodes, *J. Power Sources*, **71**, pp. 174-180.

Cleghorn S, Kolde J and Liu W (2003) Catalyst coated composite membranes Volume 3, Part 3, pp. 566–575. Handbook of Fuel Cells—Fundamentals, Technology and Applications, edited by Wolf Vielstich, Arnold Lamm, Hubert A. Gasteiger, John Wiley & Sons Ltd, Chichester, 2003.

Cooper J S (2004), Design analysis of PEMFC bipolar plates considering stack manufacturing and environment impact, *J. Power Sources*, **129**, pp. 152-169.

Corbo P, Corcione F E , Migliardini F and Veneri O (2005) Experimental study of a fuel cell power train for road transport application, *J. Power Sources*, **145**, pp. 610-619.

Corti C W, Holliday R J and Thompson D T (2005) Commercial aspects of gold catalysis, *Applied Catalysis A: General*, **291**, pp. 253-261.

Costamagna P and Srinivasan S (2001) Quantum jumps in the PEMFC science and technology from the 1960s to the year 2000 Part I. Fundamental scientific aspects, *J. Power Sources*, **102**, pp. 242-252.

Costamagna P, Yang C, Bocarsly A B, Srinivasan S (2002) Nafion 115 Zirconium phosphate composite membranes for operation of PEMFCs above 100 C. *Electrochim Acta*, **47**, pp. 1023-1033.

Cropper D (2004) Fuel Cell Today Market Survey: Light Duty Vehicles, Fuel Cell Today, pp. 1-13 (April).

Cunningham J M (2001) Air System Management for Fuel Cell Vehicle Applications, M S Thesis, University of California, Davis, USA.

Curtin D E, Lousenberg R D, Henry T J, Tangeman P C and Tisack M E (2004) Advanced materials for improved PEMFC performance and life, *J. Power Sources*, **131**, pp. 41-48.

Daugherty M, Haberman D, Stetson N, Ibrahim S, Lokken D, Dunn D, Cherniak M, Salter C (1999) Modular PEM fuel cell for outdoor applications, Proceedings of the European Fuel cell Forum Portable Fuel Cells Conference, Lucerne, pp. 205-213.

Davis M W, Development and Evaluation of a Test Apparatus for Fuel Cells (2000) Electronic Thesis, Mechanical Engineering Department, Virginia Tech.

Davies D P and Adcock P L (2002) Light weight high power density fuel cell stack, DTI/Pub URN 02/643.

De Castro E S (2005) Integrated Manufacturing for Advanced MEAs DOE Hydrogen and FC Program Review-2005, DE-FC04-02AL67606.

Debe M K (2005) Advanced MEA's for Enhanced Operating Conditions, Amenable to High Volume Manufacture, DOE Hydrogen and FC Program Review-2005, DE-FC04-02AL67606.

Dhar H P (1994) Near ambient unhumidified solid polymer fuel cell US Patent No. 5,318,863.

Dhar H P, Lee J H, Lewinski K A (1996) Proceedings of Fuel Cell Seminar, November 1996, Orlando, FL, USA, p. 583.

Dhathathreyan K S, Ramya K and Vishnupriya B (2001) A Blend Membrane, Indian Patent Application No. 303/MAS/2001.

Dhathathreyan K S, Sridhar P, Perumal R, Rajalakshmi N and Raja M (2001) Humidification Studies on Polymer Electrolyte Membrane Fuel Cell, *J. Power Sources*, **101**, pp. 72-78.

Ding J, Chuy C and Holdcroft S (2002) Solid polymer electrolytes based on ionic graft polymers: Effect of graft chain length on nano-structured, ionic networks, *Adv Funct. Mater.*, **12**. pp. 389-394.

Divisek J, Oeijen H-F, Peinecke V, Schmidt V M and Stimming U (1998) Components for PEM fuel cell systems using hydrogen and CO containing fuels, *Electrochim. Acta*, **43**, pp. 3811-3815.

Dong Z (2001) PEM Fuel Cell Stack Development and System Optimization, Ph. D Thesis, University of Victoria.

Doyle M, Choi S K, Proulx G (2000) High temperature proton conducting membrane based on perfluorinated ionomer membrane – Ionic fluid composites, *J. Electrochem. Soc.* **147** pp. 34-37.

Dumercy L, Glises R and Kauffmann J M (2005) 3D Steady state thermal modeling of a three cells PEMFC stack, 3rd European PEFC Forum, Lucerne, File No. P208.

Eikerling M, Kornyshev A A, Kuznetsov A M, Ulstrup J and Walbran S (2001) Mechanisms of proton conductance in polymer electrolyte membranes, *J Phys Chem B*, **105**, pp. 3646-62.

Eikerling M, Kornyshev A A (2001a) Proton transfer in a single pore of a polymer electrolyte membrane. *J Electroanal Chem.*, **502**, pp. 1-14.

Eikerling M, Paddison S J, Zawodzinski T A (2002) Molecular orbital calculations of proton dissociation and hydration of various acidic moieties for fuel cell polymers, *J New Mater Electrochem Sys.*, **5**, pp. 15-23.

Eisman G A, in: J W, Van Zee, R E White K, Kinoshita H S, Burney (Eds.) (1986) Proceedings of the Symposium on Diaphragms, Separators and Ion-Exchange Membranes, PV 86-13, *Electrochem Soc. Proc. Ser.*, Pennington, NJ, p. 156.

Ersoza A, Olguna Hand Ozdogan S (2005) Simulation study of a proton exchange membrane (PEM) fuel cell system with autothermal reforming, Energy, pp. 1-11.

Evans J P (2003) Experimental Evaluation of the Effect of Inlet Gas Humidification on Fuel Cell Performance, Electronic Thesis (MS), Mechanical Engineering Department, Virginia Tech.

Evertz J and Guenthart M (2003) Structural Concepts for light weights and cost effective end plates for fuel cell stacks, European Fuel cell forum, Lucerne, pp. 1-8.

Faubert G, Cote R, Doodle J P, Lefevre M and Bertrand P (1999) Oxygen reduction catalysts for polymer electrolyte fuel cells from the pyrolysis of FeII acetate adsorbed on 3,4,9,10 –perylene tetracarboxylic dianhydride, *Electrochim. Acta*, **44**, pp. 2589-2593.

Fedkiw P and Her W (1989) An impregnation – reduction method to prepare electrodes on Nafion SPE, *J. Electrochem. Soc.*, **136**, pp. 899-900.

Ferna´ndez J L and Bard A J (2003) Scanning Electrochemical Microscopy. 47. Imaging Electrocatalytic Activity for Oxygen Reduction in an Acidic Medium by the Tip Generation-Substrate Collection Mode, *Anal. Chem.*, **75**, pp. 2967-2974.

Fernández R, Ferreira-Aparicio P and Daza L (2005) PEMFC electrode preparation: Influence of the solvent composition and evaporation rate on the catalytic layer microstructure, *J. Power Sources* (In Press).

Ferna´ndez J L, Walsh D A and Bard A J (2005a) Thermodynamic Guidelines for the Design of Bimetallic Catalysts for Oxygen Electroreduction and Rapid Screening by Scanning electrochemical Microscopy. M-Co (M: Pd, Ag, Au), *J Amer. Chem. Soc.*, **127**, p. 9.

Foster S, Mitchell P and Mortimer R (1994) Proceedings of the Fuel Cell—Program and Abstracts on the Development of a Novel Electrode Fabrication Technique for Use in Solid Polymer Fuel Cells, pp. 442-443.

Franco A V "Fuel cells automotive industry" (2005) Procyan Report.

Frey T H and Linardi M (2005) Effects of membrane electrode assembly preparation on the polymer electrolyte membrane fuel cell performance, *Electrochimica Acta*, **50**, pp. 99-105.

Fuel Cells (2000) Breakthrough Technologies Institute, Status of Fuel Cell Technology for Distributed and Portable Power Generation, www.fuelcells.org.

Fuel Cell Hand Book (2000) (Fifth ed.) EG&G Parsons Inc., Science applications International Corporation under Contract No. DE-AM26-99FT40575 Oct.

Ganesh Mohan, Prabhakara Rao B, Das S K, Pandiyan S, Rajalakshmi N and Dhathathreyan K S (2004) Analysis of Flow Maldistribution of Fuel and Oxidant in a PEMFC, Journal of Energy Resources Technology, Transactions of ASME, **126**, pp. 262-270.

Gasteiger H A, Panels J E and Yan S G (2004) Dependence of PEM fuel cell performance on catalyst loading, *J. Power Sources*, **127**, pp. 162-171.

Gasteiger H A, Kocha S S, Sompalli B and Wagner F T (2005) Activity benchmarks and requirements for Pt, Pt-alloy, and non-Pt oxygen reduction catalysts for PEMFCs, *Applied Catalysis B: Environmental*, **56**, pp. 9-35.

Genies C, Mercier R, Sillion B, Cornet N, Gebel G and Pineri M (2001) Soluble sulfonated naphthalenic polyimides as materials for proton exchange membranes, *Polymer*, **42**, pp. 359-73.

Genies C, Mercier R, Sillion B (2001a) Stability study of sulfonated phthalic and naphthalenic polyimide structures, aqueous medium, *Polymer*, **42**, pp. 5097–5105.

Genova-Dimitrova P, Baradie B, Foscallo D, Poinsignon C and Sanchez J Y (2001) Ionomeric membranes for proton exchange membrane fuel cell (PEMFC): sulfonated polysulfone associated with phosphatoantimonic acid, *J Membr Sci.*, **185**, pp. 59-71.

Glandt J, Shimpalee S, Lee w -K and van Zee J W (2002) Modeling the effect of flow field design on PEM fuel cell performance, 2002 Spring National Meeting, New Orleans, LA.

Glises R, Hissel D, Harel F and P´era M C (2005) New design of a PEM fuel cell air automatic climate control unit, *J. Power Sources* (In Press).

Goeke K et al. (1998) Baseline Energy Outlook, P300-98-012, California Energy Commission, Sacramento, August.

González-Huerta R G, Chávez-Carvayar J A and González-Huerta O S (2005) Electro catalysis of oxygen reduction on carbon supported Ru-based catalysts in a polymer electrolyte fuel cell, *J. Power Sources* (In Press).

Gottesfeld S and Wilson M (1992) High performance catalysed membranes of ultra low Pt loadings for polymer electrolyte fuel cells, *J. Electrochem. Soc.*, **139**, pp. L28–30.

Gottesfeld S and Wilson M (1992a) *J. Appl. Electrochem*, **22**, pp. 1-7.

Gottesfeld S (1993) "Polymer Electrolyte Fuel Cells: Potential Transportation and Stationary Applications", No. 10, An EPRI/GRI Fuel Cell Workshop on Technology Research and Development, Stonehart Associates, Madison, Connecticut.

Gottesfeld S and Zawodzinski T A (1997) in: R.C. Alkire, H. Gerischer, D.M. Kolb, C.W. Tobias (Eds.), Advances in Electrochemical Science and Engineering., Vol. 5, Wiley–VCH, Weinheim, Germany, p. 195.

Gottesfeld S, Zawodzinski T A (1998) PEFC Chapter, Advances in Electrochemical Science and Engineering, Volume 5, edited by R. Alkire, H. Gerischer, D. Kolb, C. Tobias, pp. 197-301.

Grune H (1992) Fuel Cell Seminar Program and Abstracts, November 29 - December 2, 1992, Tucson, Arizona, p. 161.

Grubb W T (1957) Proceedings of the 11th Annual Battery Research and Development Conference, PSC Publications Committee, Red Bank, NJ, p. 5.

Grubb W T (1959) Fuel Cell U.S. Patent No. 2,913,511.

Gummert and Winkelmann T (2005) Fuel Cell Co-generator for Residential Applications, Fuel Cells for a Sustainable World, Lucerne, File No. A114.

Guo X, Fang J, Watari T, Tanaka K, Kita H, Okamoto K (2002) Novel sulfonated polyimides as polyelectrolytes for fuel cell application. 2. Synthesis and proton conductivity of polyimides from 9,9-bis (4-aminophenyl)fluorene-2,7-disulfonic acid. *Macromolecules*, **35**, pp. 6707-6713.

Guo Q, Sethuraman V A and White R E (2004) Parameter Estimates for a PEMFC Cathode, *J. Electrochem. Soc.*, **151**, pp. A983-A993.

Guzlow E, Schulze M , Wagner N, Kaz T, Reissner R, Steinhilber G and Schneider A (2000) Dry layer preparation and characterization of polymer electrolyte fuel cell components, *J. Power Sources*, **86**, pp. 352-362.

Guzlow E and Kaz T (2002) New results of PEFC electrodes produced by the DLR dry preparation technique, *J. Power Sources*, **106**, pp. 122-125.

Hagiwara R, Nohira T, Matsumoto K and Tamba Y (2005) A Fluorohydrogenate Ionic Liquid Fuel Cell Operating Without Humidification, *Electrochemical and Solid-State Letters*, **8**, pp. A231-A233.

Hailes R (1999) Fuel cells for transportation, An inventory analysis of environmental interventions associated with a prototype stack component manufacturing route, Thesis, Imperial College, London, UK.

Hamada Y, Nakamura M, Kubota H, Ochifuji K, Murase M and Goto R (2004) Field performance of a polymer electrolyte fuel cell for a residential energy system, Renewable and Sustainable Energy Reviews, **9**, pp. 345-362.

Haraldsson K, Folkesson A and Alvfors P (2005) Fuel cell buses in the Stockholm CUTE project—First experiences from a climate perspective, *J. Power Sources*, **145**, pp. 620-631.

He W, Yi J S and Nhuyen T V (2000) Two-phase flow model of the cathode of PEM fuel cells using interdigitated flow fields, *AIChE J.*, **46**, pp. 2053 2064.

Henzel A, Nolte R, Ledjeff-Hey K and Zedde M (1998) Membrane fuel cells—concepts and design, *Electrochim. Acta,* **43**, pp. 3817-3820.

Heinzel A, Roes J and Brandt H (2005) Increasing the electric efficiency of a fuel cell system by re-circulating the anodic off gas, *J. Power Sources*, **145**, pp. 312-318.

Hofmann M A, Ambler C M, Maher A E (2002) Synthesis of polyphosphazenes with sulfonimide side groups. *Macromolecules*, **35**, pp. 6490-6493.

Hontanon E, Escudero M J, Bautista C, Garcıa-Ybarra P L and Daza L (2000) Optimisation of flow-field in polymer electrolyte membrane fuel cells using computational fluid dynamics techniques, *J. Power Sources*, **86**, pp. 363-368.

Hickner M, Kim Y S, Wang F, Zawodzinski T A and McGrath J E (2001) Proton exchange membrane nanocomposites for fuel cells., *Intern. SAMPE Tech. Conf.*, **33**, pp. 1519-1532.

Hogarth M and Glipa (2001) High temperature membranes for solid polymer fuel cells, ETSU F/02/00189/REP, DTI/Pub URN 01/893.

Holze R, Vogel I and Vielstich W (1986), New oxygen cathodes for fuel cells with organic fuels, *J. Electroanal. Chem.* *J. Electroanal. Chem*, **210**, pp. 277-286.

Honma I, Nomura S and Nakajima H (2001) Proton conducting organic-inorganic nanocomposites for polymer electrolyte membrane. *J. Membr Sci.*, **185**, pp. 83-94.

Hornfield W and Joerissen L (2005) PEM-Fuel Cell Power in the Autonomous Underwater Vehicle DeepC®, Fuel Cells for a Sustainable World, Lucerne, File No. A124.

Hung Y and Tawfik H (2005) Testing and Evaluation of aluminum coated bipolar plates of PEM fuel cells operating at 70°C, Proceedings of FUEL CELL 2005 Third International Conference on Fuel Cell Science, Engineering and Technology, May 23-25, 2005, Ypsilanti, Michigan-paper 74018.

Ihonen J, Jaouen F, Lindbergh F and Sundhom F (2001) A Novel Polymer Electrolyte Fuel Cell For Laboratory Investigations and In-Situ Contact Resistance Measurements, *Electrochim Acta*, 46, pp. 2899-2911.

International Fuel Cells (1991) Investigation of Design and Manufacturing Methods for Low-Cost Fabrication of High Efficiency, High Power Density PEM Fuel Cell Power Plant, Final Report FCR-11320A, June 10.

Iojoiu C, Chabert F, Mar'echal M, Kissi N E, Guindet J and Sanchez J Y (2005) From polymer chemistry to membrane elaboration—A global approach of fuel cell polymeric electrolytes, *J. Power Sources* (In press).

Ishihara A, Lee K, Doi S, Mitsushima S, Kamiya N, Hara M, Domen K, Fukuda K, and O'Hayre R, Lee S J, Cha S W and Prinz F B (2002) A sharp peak in the performance of sputtered platinum fuel cells at ultra-low platinum loading, *J. Power Sources*, **109**, pp. 483-493.

Ismagilov Z R, Kerzhentsev M A, Shikina N V, Lisitsyn A S, Okhlopkova L B, Barnakov Ch. N, Sakashita M, Iijima T and Tadokoro K (2005) Development of active catalysts for low Pt loading cathodes of PEMFC by surface tailoring of nanocarbon materials, Catalysis Today, 102-103, pp. 58-66.

James B D, Lomax Jr. F D and (Sandy) Thomas C E (1999) Manufacturing Cost of Stationary Polymer Electrolyte Membrane (PEM) Fuel Cell Systems, DTI.

Jemeï S, Hissel D, Péra M C and Kauffmann J M (2005) Multi-Parameter Sensitivity Analysis of a Proton Exchange Membrane Fuel Cell Model, 3rd European PEFC Forum, Lucerne, File No. P205.

Jannasch P, Recent developments in high-temperature proton conducting polymer electrolyte membranes (2003) Current Opinion, Colloid and Interface Science, **8**, pp. 96-102.

Jayaraj J, Kim Y C, Kim K B, Seok H K and Fleury E (2005) Corrosion studies on Fe-based amorphous alloys in simulated PEM fuel cell environment, *Science and Technology of Advanced Materials*, **6**, pp. 282-289.

Jayaraman S and Hillier A C (2001) Construction and Reactivity Mapping of a Platinum catalyst Gradient Using the Scanning Electrochemical Microscope, Langmuir, **17**, pp. 7857-7864.

Johnson R, Morgan C, Witmer D and Johnson T (2001) Performance of a Proton Exchange Membrane Fuel Cell Stack, *Int. J. Hydrogen Energy*, **26**, pp. 879-887.

Jonissen L, Gogel V, Kerres J, Garche J (2002) New membranes for direct methanol fuel cells. *J Power Sources*, **105**, pp. 267-73.

Joseph S, McClure J C, Chianelli R, Pich P and Sebastian P J (2005) Conducting polymer-coated stainless steel bipolar plates for proton exchange membrane fuel cells (PEMFC), *Int. J. Hydrogen Energy*, **30**, pp. 1339-44.

Ju R A, An J H, Lee J K and Lee S H (2005) DSP-Based Actively Controlled Fuel Cell/Battery Hybrid DMFC System, 3rd European PEFC Forum, Poster 417.

Kammen D M, Lipman T E and Edwards J (2002) Economic and Environmental Analysis of PEM Fuel Cell System Performance Using the Clean Energy Technologies Economic and Emissions Model (CETEEM) A publication from Renewable and Appropriate Energy Laboratory (RAEL).

Karnik A Y and Sun J (2005) Modeling and control of an ejector based anode recirculation system for fuel cells, Proceedings of FUELCELL, Third International Conference on Fuel Cell Science, Engineering and Technology May 23-25, 2005, Ypsilanti, Michigan, Paper - 74102.

Kato H (2000) Ion Exchange and electrode assembly for an electrochemical cell US Patent No. 6,054,230.

Kawahara M, Rikukawa M, Sanui K and Ogata N (2000) Synthesis and proton conductivity of sulfopropylated poly(benzimidazole) film, *Solid State Ionics*, 136-137 pp. 1193-1196.

Kempton W and Letendre S E (1997) Electric Vehicles as a New Power Source for Electric Utilities, *Transportation Research - D*, **2**(3), pp. 157-175.

Kerres J, Ullrich A, Meier F and Häring T (1999) Synthesis and characterization of novel acid-base polymer blends for application in membrane fuel cells, *Solid State Ionics*, **125**, pp. 243-249.

Kerres J A (2001) Development of ionomer membranes for fuel cells. *J. Membr Sci* , **185**, pp. 3-27.

Kiefer J, Brack H-P, Huslage J, Buchi F N, Tsakada A, Geiger F, Scherer G G (1999) Radiation grafting: A versatile membrane preparation tool for fuel cell applications. Proceedings of the European Fuel Cell Forum Portable fuel cells conference, *Lucerne*, pp. 227-235.

Kim J D and Honma I (2005) Anhydrous solid state proton conductor based on benzimidazole/monododecyl phosphate molecular hybrids, *Solid State Ionics*, **176**, pp. 979-984.

Kim S, Shimpalee S and van Zee J W (2005) Effect of Flow Field Design and Voltage Change Range on the Dynamic Behavior of PEMFCs, *J. Electrochem. Soc.*, **152**, pp. A1265-A1271.

Kissock J K (1998) "Combined Heat and Power for Buildings Using Fuel-Cell Cars", Proceedings of the ASME International Solar Energy Conference, Albuquerque, NM, June 13-18, pp. 121-132.

Kohler J, Starz K -A, Wittphal S and Diehl M (2002) Process for Producing a Membrane Electrode Assembly for Fuel Cells, US Patent 2002/0064593 A1.

Kreuer K D, Fuchs A, Ise M, Spaeth M and Maier J (1998) Imidazole and pyrazole-based proton conducting polymers and liquids, *Electrochim Acta*, **43**, pp. 1281-1288.

Kreuer K D (2001) On the development of proton conducting membranes for hydrogen and methanol fuel cells. *J Membr Sci.*, **185**, pp. 29-39.

Krumpelt M, Kumar R, Miller R and Christianson C (1992) Fuel Cell Seminar Program and Abstracts, November 29 - December 2, Tucson, Arizona, p. 35.

Krumpelt M and Myles K M (1993) An EPRI/GRI Fuel Cell Workshop on Technology Research and Development, April 13-14, Stonehart Associates, Madison, Connecticut.

Kucernak A, Ladewig B, Blewitt R and Shrimpton J (2005) Laser Doppler Anemometry Study of Reactant Flow in Fuel Cell Channels, 3rd European PEFC Forum, Lucerne, Poster 21.

Kumar A and Reddy R G (2003) Effect of channel dimensions and shape in the flow-field distributor on the performance of polymer electrolyte membrane fuel cells, *J. Power Sources*, **113**, pp. 11-18.

Kumar A and Reddy R G (2005) Effect of gas flow-field design in the bipolar/end plates on the steady and transient state performance of polymer electrolyte membrane fuel cells, *Journal of Power Sources* (In Press).

Kulikovsky A A, Kucernakb A and Kornyshevb A A (2004) Feeding PEM fuel cells in Electrochim Acta, **50**, pp. 1323-1333.

Lakeman J B and Scott K (2003) Abstract, Conference on High Energy Density Electrochemical Power Sources, Nice, France, September 17-20.

Laven A (1999) Development of a Prototype Fuel Cell Powered Motor Scooter, Masters Thesis, University of Nevada.

Lebedeva N P and Janssen G J M (2005) On the preparation and stability of bimetallic PtMo/C anodes for proton-exchange membrane fuel cells, *Electrochim Acta*, June (In Press).

Lee S-J, Huang C -H, Chen Y -P, Chen Y -M (2005) Chemical Treatment Method for the Aluminum Bipolar Plates of PEM Fuel Cells, *J Fuel Cell Sci & Tech.*, **2**, pp. 208-212.

Lee S-J, Huang C-H, Lai J-J and Chen Y –P (2004) Corrosion-resistant component for PEM fuel cells, *Journal of Power Sources*, **131**, pp. 162-168.

Lee et al. (2005a) The Development of a Heterogeneous Composite Bipolar Plate of a Proton Exchange Membrane Fuel Cell, Ming-San Lee, Long-Jeng Chen, Zheng-Ru He, Shih-Hong Yang, *J Fuel Cell Sci & Tech.*, **2**, pp. 14-19.

Li T, Wlaschin A, Balbuena P B (2000) Theoretical studies of proton transfer in water and model polymer electrolyte systems, *Ind. Eng Chem Res.*, **40**, pp. 4789-800.

Li Q, He R, Jensen J O and Bjerrum (2003), Approaches and Recent Development of polymer electrolyte membrane for Fuel Cells operating above 100°C, *Chem. Mater.*, **15**, pp. 4896-4915.

Li M, Luo S, Zeng C, Shen J, Lin H and Cao C (2004) Corrosion behavior of TiN coated type 316 stainless steel in simulated PEMFC environments, *Corrosion Science*, **46**, pp. 1369-1380.

Lin B (1999) Conceptual design and modeling of a fuel cell scooter for urban Asia, Masters Thesis. Princeton PU/CEES Report 320.

Lindermeir A, Rosenthal G, Kunz U and Hoffman U (2004) Improvement of MEAs for direct methanol fuel cells by tuned layer preparation and coating technology, *FUEL CELLS*, **4**, pp. 78-85.

Little A D (1995) Fuel Cells for Building Cogeneration Applications – Cost/Performance Requirements and markets, prepared for the Office of Building Technologies, U.S. Department of Energy, January.

Little A D (2000) Cost Analysis of Fuel Cell System for Transportation, Baseline System Cost Estimate, Ref 49739, Department of Energy, March 2000.

Litster S and McLean G (2004) PEM fuel cell electrodes, *J. Power Sources*, **130**, pp. 61-76.

Liu B and Bard A J (2002) Scanning Electrochemical Microscopy: Study of the Kinetics of Oxygen Reduction on Platinum with Potential Programming of the Tip, *J. Phys. Chem. B*, **106**, pp. 12801-12806.

Liu F, Yi b, Xing D, Yu J, Hou Z, Fu Y (2003) Development of novel self-humidifying composite membranes for fuel cells, *J. Power Sources*, **124**, pp. 81-89.

Liu H –C, Yan W -M, Soong C –Y and Chen F (2005) Effects of baffle-blocked flow channel on reactant transport and cell performance of a proton exchange membrane fuel cell, *J. Power Sources*, **142**, pp. 125-133.

Iliev I, Kaisheva A and Gamburzev S (1991) Proceedings of the Intersociety Energy Conversion Engineering Conference, Boston, MA, USA, 3 p. 469.

Lomax F D, James B D, Baum G N and Thomas C E (1998) Detailed Manufacturing Cost Estimates for Polymer Electrolyte Membrane (PEM) Fuel Cells for Light Duty Vehicles, Directed Technologies, Inc., Arlington, August.

Lovins A B and Williams B D (1999) A Strategy for the Hydrogen Transition. 10th Annual U.S. Hydrogen Meeting, Vienna VA, National Hydrogen Association.

Lundblad A, Kiros Y, Onsten A, Jaouen F and Lindbergh G (2005) Porphyrin-Based Catalysts for PEFC Cathodes, 3rd European PEFC Forum, Lucerne July, 2005, File No. B044.

Ma Y -L, Wainright J S, Litt M H and Savinell R F (2004) Conductivity of PBI Membranes for High-Temperature Polymer Electrolyte Fuel Cells, *J. Electrochem. Soc.*, **151**, pp. A8-A16.

Mathias M, Roth J, Flemimg J and Lehnert W (2003) Diffusion media materials and characterization, Chapter 46, Handbook of Fuel Cells – Fundamentals, Technology and Applications, edited by Wolf Vielstich, Hubert A. Gasteiger, Arnold Lamm. Volume 3: Fuel Cell Technology and Applications, John Wiley & Sons, Ltd.

Matsubayashi T, Hamada A, Taniguchi S, Miyake Y and Saito T (1994) Proceedings of the Fuel Cell–Program and Abstracts on the Development of the High Performance Electrode For PEFC, pp. 581-584.

Matsushita Electric Ind. Co., (2002) Electrode catalyst for fuel cells, Japanese Patent Application JP 305001.

Matsuoka K, Iriyama Y, Abe T and Ogumi Z (2004) 2004 ECS Joint International Meeting, Tapa, October 3–8, (Abstract 1518).

Maye M M, Luo J, Han L, Kariuki N L and Zhong C –J (2003) *Gold Bull.*, **36**, pp. 75-78.

Mehta V and Cooper J S (2003), Review and analysis of PEM fuel cell design and manufacturing, *J. Power Sources*, **128**, pp. 32-53.

Meng H and Wang C Y (2004) Electron transport in PEFC's, *J. Electrochem. Soc.*, **151**, pp. A358-A367.

Middelman E (2002) Improved PEM fuel cell electrodes by controlled self-assembly, Fuel Cells Bulletin, November 2002, pp. 9-12 Miyake N, Wainright J S, Savinell R F (2001) Evaluation of a sol-gel derived Nafion silica hybrid membrane for proton electrolyte membrane fuel cell applications-I. Proton conductivity and water content. *J Electrochem. Soc.*, **148**, pp. A898–A904.

Middelman E, Pek J and Verhage A (2005) The PEM Power Plant Project, 3rd European PEFC Forum, Lucerne, File No. B102.

Millet S and Mahadevan K (2005) Commercialization scenarios of polymer electrolyte membrane fuel cell applications for stationary power generation in the United States by the year 2015, *J. Power Sources* (In press).

Miller A R and Barnes D L (2005) Technical Challenges of Large Fuel Cell Vehicles, Fuel Cells for a Sustainable World, Lucerne, File No. A122.

Misubhshi (2004) Mitsubhishi electric achieves domestic top level energy efficiency of 83% with PEFC cogeneration system using a Lossnay humidifier, www.global.mitsubishielectric.com, Release No. 2322.

Mizuhata H, Nakao S and Yamaguchi T (2004) Morphological control of PEMFC electrode by graft polymerization of polymer electrolyte onto platinum-supported carbon black, *J. Power Sources*, **138**, pp. 25-30.

Morikawa H, Tsuihiji N, Mitsui T and Kanamura K (2004) Preparation of Membrane Electrode Assembly for Fuel Cell by Using Electrophoretic Deposition Process, *J. Electrochem. Soc.*, **151**, pp. A1733-A1737.

Mosdale R, Gebel G and Pineri M "Water Profile Determination in a Running Proton Exchange Membrane Fuel Cell Using Small-Angle Neutron Scattering" (1996) *J. Membrane Science*, **118**, pp. 269-277.

Mukerjee S and Srinivasan Soriaga M P (1995) Role of structural & electronic properties of Pt and Pt alloys on electrolcatalysis in oxygen reduction, *J. Electrochem. Soc.*, **142**, pp. 1409-1422.

Mukerjee S, Urian R C, Lee S J, Ticianelli E A and McBreen J (2004) Electrocatalysis of CO Tolerance by Carbon-Supported PtMo Electro catalysts in PEMFCs, *J. Electrochem. Soc.*, **151**, pp. A1094-A1103.

Munch W, Kreuer K D, Silvestri W, Maier J, Seifert G (2001) The diffusion mechanism of an excess proton in imidazole molecule chains: First results of an *ab initio* molecular dynamics study. *Solid State Ionics*, **145**, pp. 437-443.

Murphy O J, Hitchens G D and Manko D J (1994) High power density proton-exchange membrane fuel cells, *J. Power Sources*, **47**, pp. 353-368.

Musser J and Wang C Y (2000) Heat transfer in a fuel cell engine proceedings of NHTC'00, 34th National Heat Transfer Conference, Pittsburgh, Aug. **20**, 2000.

Neyerlin K C, Gasteiger H A, Mittelsteadt C K, Jorne J and Gua W (2005) Effect of Relative Humidity on Oxygen Reduction Kinetics in a PEMFC, *J. Electrochem. Soc.*, **152**, pp. A1073-A1080.

Nakajima H, Nomura S, Sugimoto T, Nishikawa S, Honma I (2002) High temperature proton conductive organic-inorganic nanohybrids for polymer electrolyte membrane. Part II. *J Electrochem. Soc.*, **149**, pp. A953-A959.

Nguyen T V and White R E "A Water and Heat Management Model for Proton-Exchange-Membrane Fuel Cells" (1993) *J. Electrochem. Soc.*, **140**, pp. 2178-2186.

Niehues M and Edwards T H (2000) Fuel Cells for Rail Vehicles. Proceedings of the UIC Energy Efficiency Conference, Paris, 2000.

Niyogi S, Kumar R and Myers D (2005) High-Temperature Polymer Electrolyte Membranes, DOE Hydrogen program review, May 2005.

O'Hayre R, Lee S J, Cha S W and Prinz F B (2002) A sharp peak in the performance of sputtered platinum fuel cells at ultra-low platinum loading, *J. Power Sources*, **109**, pp. 483-493.

On Isa Bar, Kirchain R and Roth R (2002) Technical cost analysis of PEM Fuel cells, *J Power sources*, **109**, pp. 71-75.

Oosthuizen P H, Sun L and McAuley (2005) The effect of channel-to-channel gas crossover on the pressure and temperature distribution in PEM fuel cell flow plates, *Applied Thermal Engineering*, 25, pp. 1083-1096.

Oszcipok M, Riemann D, Kronenwett U, Kreideweis M and Zedda M (2005) Statistic analysis of operational influences on the cold start behaviour of PEM fuel cells, *J. Power Sources*, **145**, pp. 407-415.

Oszcipok M, Hakenjos A, Riemann D and Hebling C (2005a) Freezing Processes in PEM Fuel Cells, 3rd European PEFC Forum, Lucerne, File No. P114.

Otaa K (2005) Tantalum Oxynitride for a Novel Cathode of PEFC, Electrochemical and Solid-State Letters, **8**, A201-A203.

Paddison S J, Paul R and Kreuer K D (2002) Theoretical computed proton diffusion coefficients in hydrated PEEKK membranes. *Phys Chem. Chem Phys.*, **4**, pp. 1151-1157.

Pan Mu, Tang H, Jiang and Liu Z (2005) Fabrication and Performance of Polymer Electrolyte Fuel Cells by Self-Assembly of Pt Nanoparticles, *J. Electrochem. Soc.*, **152**, pp. A1081-A1088.

Park J -H, Kim J -H, Lee H -K, Lee T -H and Joe Y (2004) A novel direct deposition of Pt catalysts on Nafion impregnated with polypyrrole or PEMFC, *Electrochim Acta*, **50**, pp. 765-771.

Pehnt M, Fuel cells for Distributed Power: Benefits, Barriers and Perspectives, www.panda.org/EPO.

Perry M L and Kotso S (2004) A Back-up Power Solution with No Batteries, INTELEC 2004 Proceedings, pp. 210-217.

Pharkya P, Alfantazi A and Farhat Z (2005) Fabrication Using High-Energy Ball-Milling Technique and Characterization of Pt-Co Electrocatalysts for Oxygen Reduction in Polymer Electrolyte Fuel Cells, *J Fuel Cell Science and Technology*, **2**, pp. 171-177.

Popov et al., (2005) Novel Non-Precious Metals for PEMFC: Catalyst Selection through Molecular Modeling and Durability Studies, Hydrogen, Fuel Cells and Infrastructure Technologies Program, 2005 Annual Review, Washington, DC, May 23-27.

Popelis I, Tsukada A and Scherer G (1999) "12 Volt 300 Watt PEFC power pack", Proceedings of the European Fuel Cll Forum Portable Fuel Cells Conference, Lucerne, pp. 147-155.

Poppe D, Frey H, Kreuer K D, Heinzel A, Mulhaupt R (2002) Carboxylated and sulfonated poly(arylene-co-arylene sulfone)s: thermostable polyelectrolytes for fuel cell applications. *Macromolecules*, **35**, pp. 7936-41.

Prasanna M, Ha H Y, Cho E A, Hong S -A and Oh I (2004) Investigation of oxygen gain in polymer electrolyte membrane fuel cells, *J. Power Sources*, **137**, pp. 1-8.

Prater K (1990) "The renaissance of the solid polymer fuel cell", *J. Power Sources*, **29**, pp. 239-250.

Preischel C, Hedrick P and Hahn A (2001) Continuous Method for Manufacturing a Laminated Electrolyte and Electrode Assembly, US Patent 6,291,091 B1.

Rajalakshmi N, Jayanth T T and Dhathathreyan K S (2003) Effect of carbon dioxide and ammonia on polymer electrolyte membrane fuel cell performance, *FUEL CELLS*, **3**, pp. 177-180.

Rajalakshmi N, Ryu H and Dhathathreyan K S (2004) Platinum catalysed membranes for proton exchange membrane fuel cells – higher performance, *Chemical Engineering Journal*, **102**, pp. 241-247.

Rajalakshmi N, Ryu H, Shaijumon M M and Ramaprabhu S (2005) Performance of polymer electrolyte membrane fuel cells with carbon nanotubes as oxygen reduction catalyst support material, *J. Power Sources*, **140**, pp. 250-257.

Rajalakshmi N, Velayutham G, Ramya K, Subramanyam C K and Dhathathreyan K S (2005) Characterisation and Optimization of low cost activated carbon fabric as a substrate layer for PEMFC electrodes, Proceedings of Fuel Cell 2005, Third International Conference on Fuel Cell Science, Engineering and Technology May 23-25, Ypsilanti, Michigan, USA, Paper-74182.

Ramani V, Kunz H R and Fenton J M (2004) Investigation of Nafion®/HPA composite membranes for high temperature/ low relative humidity PEMFC operation, *J. Membrane Science*, **232**, pp. 31-44.

Rao C R K and Trivedi D C (2005) Chemical and electrochemical depositions of platinum group metals and their applications, Coordination Chemistry Reviews, **249**, pp. 613-631.

Reichman S, Duvdevani T, Aharon A, Philosoph M, Golodnitsky D and Peled E (2005) A novel PTFE-based proton-conductive membrane, *J. Power Sources* (In Press).

Reiner A, Hajbolouri F, Döbeli M, Wokaun A and Scherer G G (2005) Co-Sputtering: A Novel Platinum-Carbon Catalyst Preparation Method, 3rd European PEFC Forum, Lucerne, Poster 109.

Ren X, Wilson M S, Gottesfeld S (1996) High performance direct methanol polymer electrolyte fuel cells, *J. Electrochem. Soc.*, **143**, pp. L12-L15.

Reeve R W, Eweka I E and Mepsted G O (2003) Eighth Grove Fuel Cell Symposium, London, September 24-26, 04B.6.

Rogg S, Höglinger M, Zwittig E, Pfender C, Kaiser W and Heckenberger T (2003) Cooling Modules for Vehicles with a Fuel Cell Drive, *FUEL CELLS*, **3**, pp. 153-158.

Roser J, Dyck A, Gogel V, Bauer B, Holdik H, Dohle H, Müller M, Felber S and Wilde P (2005) Flexible Graphite Foil Solutions for Low-Cost PEM Fuel Cell Systems, 3rd European PEFC Forum, Lucerne, Poster 123.

Ross P N (2005) New Electrocatalysts For Fuel Cells, DOE hydrogen program review 2005.

Rosso I, Galletti C, Saracco G, Garrone E and Specchia V (2004) Development of A zeolites-supported noble-metal catalysts for CO preferential oxidation: H_2 gas purification for fuel cell, *Applied Catalysis B: Environmental*, **48**, pp. 195-203.

Ruge M and Hoekel M (2005), Air-Cooled Fuel Cell Stack Made of Foil Materials, 3rd European PEFC Forum, Lucerne File No. B096.

Santiago E I, Batista M S, Assaf E M and Ticianelli E A (2004) Mechanism of CO Tolerance on Molybdenum-Based Electrocatalysts for PEMFC, *J. Electrochem. Soc.*, **151**, pp. A944-A949.

Santis M, Schmid D, Ruge M, Freunberger S and Buechi F N (2004), Modular Stack—Internal Air Humidification Concept—Verification in a 1 kW stack, *FUEL CELLS*, **4**, pp. 214-218.

Savadogo O, Xing B (2000) Hydrogen oxygen polymer electrolyte membrane fuel cell (PEMFC) based on acid-doped polybenzimidazole (PBI). *J New Mater Electrochem. Sys.*, **3**, pp. 345–349.

Savinell R, Yeager E, Tryk D, Landau U, Wainright J, Weng D, Lux K, Litt M, Rogers C (1994) A polymer electrolyte for operation at temperatures upto 200 C, *J. Electrochem. Soc.*, **141**, pp. L46-L48.

Sawai K and Suzuki N (2004) Highly Active Nonplatinum Catalyst for Air Cathodes, *J. Electrochem. Soc.*, **151**, pp. A2132-A2137.

Shukla A K et al., (2003) The promise of fuel-cell based automobiles, *Bull Mat Sci.*, **26**, pp. 207-214.

Schmidt T J, Gasteiger H J and Behm R J (1999) Rotating disc electrode measurements on CO tolerance of high surface area Pt/Vulcan, *J. Electrochem. Soc.*, **146**, pp. 1296-1304.

Schuster M, Meyer W H, Wegner G (2001) Proton mobility in oligomer-bound proton solvents: imidazole immobilization via flexible spacers. *Solid State Ionics*, **145**, pp. 85-92.

Senn S M and Poulikakos D, Polymer Electrolyte fuel cells with porous materials as fluid distributors and comparisons with traditional channeled systems, Transactions of the ASME, **126**, pp. 410-418.

Serincan M F and Yeilyurt S (2005) An Analysis of a Proton Electrolyte Membrane Fuel Cell (PEMFC) at Start-ups and Failures, 3rd European PEFC Forum, Lucerne, File No. P203.

Shan Y and Choe S-Y (2005) A high dynamic PEM fuel cell model with temperature effects, *J. Power Sources*, **145**, pp. 30-39.

Shin S-J, Lee J-K , Ha H-Y, Hong S-A, Chun H-S and Oh I-H (2002) Effect of the catalyst ink preparation method on the performance of polymer electrolyte membrane fuel cells, *J. Power Sources*, **106**, pp. 146-152.

Shin J H, Passerini S, Shin J H and Passerini S (2004) PEO-LiN.SO$_2$CF$_2$(CF$_3$)$_2$ Polymer Electrolytes V. Effect of Fillers on Ionic Transport Properties, *J. Electrochem. Soc.*, **151**, pp. A238-A245.

Siroma Z, Sasakura T, Yasuda K, Azuma M and Miyazaki Y (2003) Effects of ionomer content on mass transport in gas diffusion electrodes for proton exchange membrane fuel cells, *Journal of Electroanal. Chem*, **546**, pp. 73-78.

Smith R A (1984) Coextruded Multilayer Cation Exchange Membranes, US Patent 4,437,952.

Smitha B, Sridhar S and Khan A A (2005) Solid polymer electrolyte membranes for fuel cell applications—A review, *J. Membrane Sci.*, **259**, pp. 10-26.

Song S, Douvartzides S and Tsiakarasm P (2005) Exergy analysis of an ethanol fuelled proton exchange membrane (PEM) fuel cell system for automobile applications, *J. Power Sources* (In Press).

Souzy R and Ameduri B (2005) Functional fluoropolymers for fuel cell membranes, *Prog. Polymer Sci.*, **30**, pp. 644-687.

Springer T E, Zawodzinski T A and Gottesfeld S (1991) Polymer Electrolyte Fuel Cell Model, *J. Electrochem. Soc.*, **138**, pp. 2334-2341.

Srinivasan S, Ferreira A, Mosdale R, Mukerjee S, Kim J, Hirano S, Lee S, Buchi F and Appleby A (1994) Proceedings of the Fuel Cell—Program and Abstracts on the Proton Exchange Membrane Fuel Cells for Space and Electric Vehicle Application, pp. 424-427.

Staiti P, Arico A S, Baglio V, Lufrano F, Passalacqua E, Antonucci V (2001) Hybrid Nafion-silica membranes doped with heteropolyacids for application in direct methanol fuel cells. *Solid State Ionics*, **145**, pp. 101–107.

Staiti P (2001a) Proton conductive membranes constituted of silicotungstic acid anchored to silica-polybenzimidazole matrices, *New Matter Electrochem. Syst.*, **4**, pp. 181-186.

Strachana N and Farrell A (2005) Emissions from distributed vs. centralized generation: The importance of system, performance, *J. Energy Policy* (In Press).

Stangar U L, Groselj N, Orel B, Schmitz A and Colomban P (2001) Proton-conducting sol–gel hybrids containing heteropoly acids. *Solid State Ionics*, **145**, pp. 109-118.

Staudt R (2005) Development of Polybenzimidazole-based High Temperature Membrane and Electrode Assemblies for Stationary and Automotive Applications, 2005 DOE Hydrogen, Fuel Cells & Infrastructure Technologies.

Steele B C H and Heinzel A (2001) Materials for fuel-cell technologies. *Nature*, **414**, pp. 345-52.

Subramanian C K, Rajalakshmi N, Ramya K and Dhathathreyan K S (2000) *Bull. Electrochem.*, **16**, pp. 350-354.

Swider-Lyons K, Bouwman P, Urgate N and Dmowski W (2003) Low Platinum hydrous metal oxides for PEMFC cathodes, DoE Review, Berkeley C A, May 2003.

Tang H and Pintauro P N (2001) Polyphosphazene membranes IV. Polymer morphology and proton conductivity in sulfonated poly bis(3-methylphenoxy) phosphazene films. *J Appl. Polym. Sci.*, **79**, pp. 49-59.

Taylor E J, Anderson E B and Vilambi N (1992) Preparation of high platinum utilization gas diffusion electrodes for proton exchange membrane fuel cells, *J. Electrochem. Soc.*, **139**, pp. L45–L46.

Teranishi K, Tsushima S and Hirai S (2005) Study of the Effect of Membrane Thickness on the Performance of Polymer Electrolyte Fuel Cells by Water Distribution in a Membrane, Electrochemical and Solid-State Letters, **8**, A281-A284.

Thangamuthu R and Lin C W (2005) Membrane electrode assemblies based on sol–gel hybrid membranes—A preliminary investigation on fabrication aspects, *J. Power Sources* (In Press).

Thomas C E, James B D, Lomax Jr F D and Kuhn Jr I F (2000) Fuel options for the fuel cell vehicle: hydrogen, methanol or gasoline?, *Int. J. Hydrogen Energy*, **25**, pp. 551-557.

Travitsky N, Ripenbein T, Golodnitsky D, Livshits V, Rosenberg Y, Lereah Y, Burstein L and Peled E (2005) Nanometric Platinum and Platinum-Alloy-Supported Catalysts for Oxygen Reduction in PEM Fuel Cells, 3rd European PEFC Forum, Lucerne, Poster 105.

Turpin M (2005) Power from Portable PEFC Generators, Fuel Cells for a Sustainable World, Lucerne, File No. A052.

Uda T and Haile S M (2005) Thin-Membrane Solid-Acid Fuel Cell, Electrochemical and Solid-State Letters, **8**, A245-A246.

Venkataraman R, Kunz H R and Fenton J M (2004) CO-Tolerant, Sulfided Platinum Catalysts for PEMFCs, *J. Electrochem. Soc.*, **151**, pp. A710-A715.

Vickers C (1996) Proceedings of Fuel Cell Seminar, Orlando, FL, USA, 1996.

Villers D, Jacques-Bédard X and Dodelet J (2004) Fe-Based Catalysts for Oxygen Reduction in PEM Fuel Cells, Pretreatment of the Carbon Support, *J. Electrochem. Soc.*, **151**, pp. A1507-A1515.

Walker M, Baumgärtner K M, Kaiser M, Kerres J, Ullrich A and Räuchle E (1999) Proton-conducting polymers with reduced methanol permeation, *J. Appl. Polym. Sci.*, **74**, pp. 67.

Walter J (2000) Modellierung eines Brennstoffzellen Antribesstrangs VDI Beriche Nr. 1565: p. 355.

Wang F, Hickner M, KimY S, Zawodzinski T A and McGrath J E (2002) Direct polymerization of sulfonated poly(arylene ether sulfone) random (statistical) copolymers: candidates for new proton exchange membranes. *J Membrane Sci.*, **197**, pp. 231-42.

Wang H, Brady M P, Teeter G and Turner J A (2004) Thermally nitrided stainless steels for polymer electrolyte membrane fuel cell bipolar plates: Part 1: Model Ni–50Cr and austenitic 349™ alloys, *J. Power Sources*, **138**, pp. 86-93.

Wang H, Brady M P, More K L, Meyer III H M and Turner J A (2004a) Thermally nitrided stainless steels for polymer electrolyte membrane fuel cell bipolar plates: Part 2: Beneficial modification of passive layer on AISI446, *J. Power Sources*, **138**, pp. 79-85.

Wang H and Turner J A (2004b), Ferritic stainless steels as bipolar plate material for polymer electrolyte membrane fuel cells, *J. Power Sources*, **128**, pp. 193-200.

Warshay M and Prokopius P R (1990) "The fuel cell in space yesterday, Today and Tomorrow", *J. Power Sources*, **29**, pp. 193-200.

Watanabe M, Uchida H, Seki Y, Emori M and Stonehart P (1996) Self humidifying polymer electrolyte membrane for fuel cells, *J. Electrochem. Soc.*, **143**, pp. 3847-3852.

Wei Z D, Chan S H, Li L L, Cai H F, Xia Z T and Sun C X (2005) Electrodepositing Pt on a Nafion-bonded carbon electrode as a catalyzed electrode for oxygen reduction reaction, *Electrochim Acta*, **50**, pp. 2279-2287.

Wells P, Wiltshire R, King C, Thompsett D, Crabb E M and Russell A E (2005) Preparation of Cr/Pt/C Catalysts by the Controlled Surface Modification of Pt/C Using an Organometallic Precursor, 3rd European PEFC Forum, Lucerne, Poster 119.

Williams M V, Leonard E B, Bonville H, Kunz R and Fentona J M (2004) Characterization of Gas Diffusion Layers for PEMFC, *J. Electrochem. Soc.*, **151**, pp. A1173-A1180.

Wilson M S, Springer T E, Zawodzinski T A and Gottesfeld S (1993) Proceedings 28th Intersociety Energy Conversion Engineering Conference, 1, 1993, Atlanta, GA, p. 1203.

Wolk R H (1999) Fuel cells for Homes and Hospitals, IEEE Spectrum, **36**, pp. 45-52.

Wood D L, Yi Y S, Nguyen T V (1998) Effect of direct liquid water injection and interdigitated flow filed on the performance of proton exchange membrane fuel cells, *Electrochim. Acta*, **43**, pp. 3795-3809.

Wu S H, Kotak D B and Fleetwood M S (2005) An integrated system framework for fuel cell-based distributed green energy applications, Renewable Energy, **30**, pp. 1525-1540.

Wu J, Liu Q and Fang H (2005) Toward the optimization of operating conditions for hydrogen polymer electrolyte fuel cells, *J. Power Sources* (In Press).

Xie Z, Navessin T, Shi K, Chow R, Wang Q, Song D, Andreaus B, Eikerling M, Liu Z and Holdcroft S (2005) Functionally Graded Cathode Catalyst Layers for Polymer Electrolyte Fuel Cells II. Experimental Study of the Effect of Nafion Distribution, *J. Electrochem. Soc.*, **152**, pp. A1171-A1179.

Xie J, Wood D L, Wayne D M, Zawodzinski T A, Atanassov P and Borup R L (2005) Durability of PEFC's at high humidity conditions, *J. Electrochemical Soc.*, **152**, pp. A104-A113.

Yamada Y, Ueda A, Shioyama H and Kobayashi T (2004) High-throughput screening of PEMFC anode catalysts by IR thermography, *Applied Surface Science*, **223**, pp. 220-223.

Yang C, Costamagna P, Srinivasan S, Benziger J, Bocarsly A B (2001) Approaches and technical challenges to high temperature operation of proton exchange membrane fuel cells. *J. Power Sources*, **103**, pp. 1-9.

Yang T -H, Yoon Y -G, Park G -K, Lee W -Y and Kim C S (2004) Fabrication of a thin catalyst layer using organic solvents, *J. Power Sources*, **127**, pp. 230-233.

Yang X-G, Burke N, Wang C -Y, Tajiri K and Shinoharab K (2005) Simultaneous measurements of species and current distributions in a PEFC under low-humidity operation, *J. Electrochem. Soc.*, **152**, pp. A759-A766.

Yi J S and Nguyen T V (1998) "An Along-the-Channel Model for Proton Exchange Membrane Fuel Cells", *J. Electrochem. Soc.*, **145**, pp. 1149-1159.

Yoneda S and Ohno Y (2005) Stochastic Approach for the Analysis of PEFC Behavior, 3rd European PEFC Forum, Lucerne, File No. B075.

Yoon C B, Meyer H W and Wegner G (2001) New functionalized polyurethane with proton conductivity. *Synthetic Met.*, **119**, pp. 465-466.

Yoon Y-G, Park G-G, Yang T-H, Han Y-N, Lee W-Y and Kim C S (2003) Effect of pore structure of catalyst layer in a PEMFC on its performance, *Int. Journal of Hydrogen Energy*, **28**, pp. 657-662.

Yu T L, Lin Hsiu-Li , Shen Kun-Sheng, Huang Li-Ning, Chang Yu-Chen, Jung Guo-Bin and Huang J C ((2004) Nafion/PTFE Composite Membranes for Fuel Cell Applications, *Journal of Polymer Research*, **11**, pp. 217-224.

Yu J, Yoshikawa Y, Matsuura T, Islam M D and Hori M (2005) Preparing Gas-Diffusion Layers of PEMFCs with a Dry Deposition Technique, Electrochemical and Solid-State Letters, **8**, pp. A152-A155.

Yu J, Matsuura T, Yoshikawa Y, Islam M D and Hori M (2005) In Situ Analysis of Performance Degradation of a PEMFC under Nonsaturated Humidification, Electrochemical and Solid-State Letters, **8**, pp. A156-A158.

Yu P, Pemberton M and Plasse P (2005) PtCo/C cathode catalyst for improved durability in PEMFCs, *J. Power Sources*, **144**, pp. 11-20.

Zawodzinski T A, DeRouin C, Radzinski S, Herman R J, Smith V T, Springer T E and Gottesfeld S (1993) Water uptake by and transport through Nafion 117 membrane, *J. Electrochem. Soc.*, **140**, pp. 1041-1047.

Zawodzinski T A, Development of New Polymer Electrolytes for Operation at High Temperature and Low Relative Humidity, DOE Hydrogen program review, May 2005.

Zen J -M and Wang C W (1994) Oxygen reduction on ruthenium oxide pyrochlore produced in a proton exchange membrane, *J. Electrochem. Soc.*, **141**, pp. L51-L52.

Zhao et al. (2005) Theoretical and experimental studies of water injection scroll compressor in automotive fuel cell systems, Yuanyang Zhao, Liansheng Li, Huagen Wu, Pengcheng Shu, Energy Conversion and Management, **46**, pp. 1379-1392.

Zhong C-J and Maye M M (2001) Core Shell assembled nanoparticles as catalysts, *Adv. Mater.*, **13**, pp. 1507-1511.

Zhong C-J, Luo J, Maye M M, Han L and Kariuki N N (2003) Nanostructured Gold and Alloy Electrocatalysts, in: B Zhou, S Hermans, G A Somorjai (Eds.), Nanotechnology in Catalysis, *Kluwer Academic/Plenum Publishers*, (Chapter 11).

Zhong C-J, Luo J, Maye M M , Han L and Kariuki N (2003) Proceeding of the GOLD 2003, Vancouver, Canada, September-October.

Zhu W H, Payne R U and Tatarchuk B J (2005) Critical flow rate of anode fuel exhaust in a PEM fuel cell system, *J. Power Sources* (In Press).

Recent Trends in Fuel Cell Science and Technology
Edited by S. Basu
Anamaya Publishers, New Delhi, India

4. Fundamentals of Gas Diffusion Layers in PEM Fuel Cells

Virendra K. Mathur and Jim Crawford*
Department of Chemical Engineering, University of New Hampshire,
Durham, NH 03824, USA

*Crawford Associates, Rye, NH 03870, USA

1. Introduction

The main components of a proton exchange membrane (PEM) fuel cell are the gas diffusion layer, membrane and catalyst. Gas diffusion layers (GDLs) are commercially available in various forms such as carbon paper or woven carbon fabrics. These are placed on either side of the membrane in a fuel cell. A GDL should allow the flow of reactant gases H_2, air/oxygen and product gases to pass through it. The water formed in a cell should not choke the pores of this paper or fabric and so they are pre-coated with polytetrafluoroethylene (PTFE), which changes the GDL material from hydrophilic to hydrophobic. PTFE is also commonly called Teflon. The platinum catalysts for both the anode and cathode side can be coated on the surface of a GDL. The hydrogen or oxygen reacts in three phases: interface-gas phase (hydrogen or oxygen), liquid phase (water) and solid phase (catalyst). The carbon paper or fabric serves as a structural support for the electrocatalyst layer as well as current collector. Some of the GDLs available on the market are not coated with PTFE and are thinner in size. In such cases, the user has to coat the GDL with PTFE.

Carbon papers and carbon fabrics have both been successfully used as GDLs. Most of the carbon papers that are used in making demonstration fuel cells are too rigid or too fragile to be wound on typical rolls, making them less suited for mass production. The carbon fabrics, on the other hand, are inherently more flexible and can withstand higher compression loads. Continuous, roll to roll processing is key to accomplishing the ultimate goal of fuel cell commercialization. There are several brand name GDLs available in the market. The Toray paper is the most extensively used, primarily because of its low cost. These GDLs can be used with PEM fuel cells operating on either hydrogen or methanol as fuels. Similar porous carbon paper GDLs are also used in the phosphoric acid fuel cells (PAFC) requiring a different choice of carbon and heat treatment. A GDL can be evaluated by using it in a fuel cell system and studying its performance.

2. Gas Diffusion Layer

Gas diffusion layer is a slightly misleading name because it plays more than one role. It is an electrical conductor between the carbon supported catalyst and the current collector plates. Thin gas diffusion layers with less resistance to electricity are desirable. They allow the easy flow of the reactant gases H_2, and air/oxygen and the product gases through. To improve the mass transport they need to be made porous at some loss of electrical conductivity (Larminie, 2003). Often GDLs are used as the base substrates for the deposition of the catalysts. GDLs are made of porous conductive materials such as a carbon cloth or carbon paper with a thickness in the range of 100-400 μm. Generally, GDLs are wet proofed with a hydrophobic material like PTFE or other special material to keep them from getting wet. Wet proofing rejects excess liquid water so that the catalyst layer will not be flooded by water thus improving the catalyst performance. It also helps maintain the water balance in the membrane by allowing the appropriate amount of water to reach the membrane and draining the excess water.

In the dry state, the gas diffusion layers are characterized by the thickness, gas permeability and electrical conductivity. Thinner GDLs, providing less resistance to mass flow (i.e. high gas permeability) and high electrical conductivity, are generally better (Fuel Cell Handbook, 1994; Larminie, 2003). According to researchers, too thin GDLs cannot provide good electrical contact between the current collecting plates and the catalyst layer while too thick ones exhibit high electrical resistance. The thickness needs to be optimized and it differs for different GDL materials. The base for GDL is made of carbon fibers either in the form of non-woven paper or woven fabric. Several authors have reported on the use of fabric GDLs such as ELAT® (Buchi and Srinivasan, 1997; Gamburzev and Appleby, 2002; Qi and Kaufman, 2003; Wilson et al., 1995) and Centre for Electrochemical and Energy Research (CEER) (Kumar et al., 1995). Paper GDLs such as Toray (Gasteiger et al., 2004; Giorgi et al., 1998), Kureha Chemical Co. (Passalacqua et al., 1998), and SGL Carbon, Japan (Sasikumar et al., 2004) were also used in several experiments. Toray (THGP-90 and THGP-120) is the most extensively used GDL paper because of its low price. Moreira et al. (2002) investigated both carbon fabric and carbon paper from Electrochemical Inc. and concluded that membrane electrode assemblies (MEAs) with carbon paper gave slightly better results than with fabric at low current densities but MEAs with fabric GDLs gave better results at high current densities and withstood flooding.

Wetproofing the GDL with a hydrophobic substance, usually PTFE, is essential for proper water management. However, PTFE is not an electrical conductor and hence should be applied in careful measure. The common PTFE content cited in the literature is about 33% by weight. But it differs with GDL materials. Qi and Kaufman (2002) explained that carbon paper, which is the most commonly used GDL, may differ from one manufacturer to another and may possess different characteristics. Experiments need to be performed to find the optimal PTFE amounts for each type. Toray carbon papers (THGP-90 and THGP-120) with different thicknesses have different wet proofing requirements. In a recent study by Park et al. (2004), the effect of PTFE content in the two Toray GDLs on single cell performance was investigated under various operating conditions. Similarly, different fabric GDLs have different wetproofing requirements.

The GDLs used initially were single layered with a wetproofed carbon base. Later developments included the introduction of a separate hydrophobic carbon layer to both sides or to one side only. Paganin et al. (1996) fabricated a typical gas diffusion layer by applying a layer of PTFE and carbon powder suspension to both sides of the base carbon cloth and then drying and sintering it. Glora et al. (2001) used carbon aerogels to form a porous substrate. Micro thin structures were formed by using this substance. These thin layers on both sides of the gas diffusion layer decreased the contact resistance between the electrode and the membrane, as well as bipolar plates. But this GDL when tested produced a maximum power of only about 1/6th of a typical fuel cell. The bad performance was, however, attributed to the poor catalyst layer preparation.

A further development in GDL manufacture was the introduction of a micro porous hydrophobic layer applied to one side of the carbon backing which faced the catalyst layer (Lister and McLean, 2004; Song et al., 2001). This layer typically consisted of PTFE and carbon black. Carbon black imparts good electrical properties and improves the gas transport to the catalyst layer. The type of carbon powder used also impacts the cell performance. Antolini et al. (2002) investigated two different carbon powders (oil furnace carbon black and acetylene black) and reported that acetylene black gave better cell performance than oil furnace carbon black. Carbon is a critical material in fuel cells since it has the necessary properties of electrical conductivity, corrosion resistance, surface properties and low cost factors which could make an inexpensive fuel cell a reality. Carbon is a nonmetallic substance with a wide range of crystalline and amorphous structures and exhibits large variation in chemical stability (Mathur et al., 2005). The role of PTFE in the hydrophobic layer is to enhance the water management property of the GDL and hence it has a very important role in cell performance. PTFE content of about 30–35% in this layer was reported to be optimum by some authors (Qi, Kaufman, 2002; Song et al., 2001) while some did not find significant change in performance with the PTFE variation (Chu et al., 2003). A study of the PTFE content in the two-layered GDLs was presented by Giorgi et al. (1998). PTFE content in the hydrophobic diffusion layer was varied from 10 to 40% and applied to 35% wet proofed Toray TGPH-90 paper. At high current densities, the lowest PTFE content gave the best results. Prasanna et al. (2004) reported an optimum PTFE content of about 20%. The difference in the optimum PTFE levels in the microporous layer, as reported by the above mentioned authors, could be because of the different base carbon cloth and paper used. Hence, each material can have its own optimum level of PTFE requirements and should be experimentally determined for various materials.

Several authors attempted to find the correlation between thickness, porosity and gas permeability of GDLs and their performance. Prasanna et al. (2004) used scanning electron microscopy to study the morphology and Micrometrics Auto Pore IV to obtain the porosimetry data of two carbon paper based GDLs. It was concluded that gas permeability and pore size of the GDL played a crucial role in cell performance compared to all other physical properties. In a recent publication, theoretical mathematical models were presented by Chu et al. (2003) and several conclusions on the influence of porosity on GDL performance were discussed. It was also stated that optimization of GDL depends not only on the porosity morphology but also on the thickness and chemical content of the GDL. Kong et al. (2002) reported the influence of pore size distribution on mass transport characteristics. It was also reported that pore size distribution is a more critical parameter than total porosity and that mass transport limitations can be reduced by enlarging the macro pore volume in the GDL. A correlation between the GDL properties and GDL performance could not be established. A clear guideline for choosing or manufacturing the GDL has not been presented by any author. Another difficulty in the GDL study is that most of the companies have proprietary techniques with patents for the GDL manufacture and characterization (Allen et al., 1981; Kato, 2000).

2.1 GDL Preparation

As mentioned earlier, a GDL consists of a carbon substrate that can either be woven or non-woven, on to which is coated a carbon catalyst layer, usually consisting of a mixture of carbon black and PTFE-controlling hydrophobicity while ensuring conductivity. The various substrates used to produce GDLs result in very different physical characteristics and therefore in the performance of the fuel cell. Probably the most well-known GDL substrate used during the development of the PEM fuel cell was that produced by Toray of Japan. This substrate is produced by pyrolysing a non-woven carbon-fiber sheet which gives a sheet very good conductive properties and low elongation but limited resistance to brittle fracture. In general, the carbon/PTFE layer is applied by the end-user with the amount depending on the hydrophobicity required.

Another approach to GDL manufacture is to use a woven carbon cloth as the substrate, the best known example being Elat® from Etek, a US company which is part of the Italian de Nora group. The cloth is produced using highly conductive fibers, which are then "loaded" with carbon black. The carbon/PTFE coating can be applied either by the substrate manufacturer or the end-user. Cloth based GDLs have the advantage of being very flexible and durable but their tendency to stretch can make automated assembly of a stack rather difficult. Woven substrates also tend to be more expensive than similar non-woven sheets.

A rather different approach is being taken by Lydall with its LyFlex® GDL where the substrate is again a non-woven carbon sheet but does not need to be pyrolysed and remains very flexible. In addition, as there are no high-temperature post-processes, the carbon/PTFE can be integrated directly during the substrate manufacturing process – eliminating a costly production step. The advantage of such an approach is not only the low cost of the GDL itself but also its ability to be supplied in roll-form for use in industrial stack-assembly processes. This type of material is, however, inherently more compressible than a pyrolysed sheet and will experience structural failure if too high a pressure is applied (Xie et al., 2004). Much know-how is proprietary to the GDL developers.

3. GDL and PEM Fuel Cell

A fuel cell experimental set-up is needed to evaluate GDL performance. Design features of a PEMFC are relatively simple. A schematic of a single cell is presented in Figure 1 showing the locations of the two GDLs. Hydrogen gas from a storage tank enters machined or molded channels in a bipolar or field plate where it is distributed over the surface area of the cell. Field plates are typically machined from bulk graphite although lower cost materials such as moldable or injectable compounds are also being developed. Important features for this component include high thermal conductivity, in-plane and through the thickness, and it must be relatively impervious to gas and water. On the opposite side of the cell is another similar field plate through which oxygen/air is passed. These plates are separated by a membrane (electrolyte), which allows hydrogen ions (H^+) to flow through, and in-turn keeps oxygen from penetrating into the anode (hydrogen side). The most commonly used membrane is Nafion (made by DuPont), a solid electrolyte perflourosulfonic acid ionic exchange polymer. Using a solid electrolyte eliminates the need for electrolyte management because it will not leak. However, the electrolyte must be kept wet by pre-wetting and feeding humidified gases. On either side of the membrane, gas diffusion layers (GDLs) are used to further control the flow of gas and support a carbon-based coating. The electrochemical reactions occur on highly dispersed electrocatalyst particles supported on the carbon coating. Platinum (Pt) deposited on carbon is generally used as the catalyst. This combination forms anode and cathode electrodes. There are two options for introducing the platinum catalyst: one is to coat both the surfaces of the membrane or second adding platinum catalyst to the carbon black particles coated on the GDLs facing the two sides of the membrane. In Fig. 1, the catalyst is shown to be coated on the membrane.

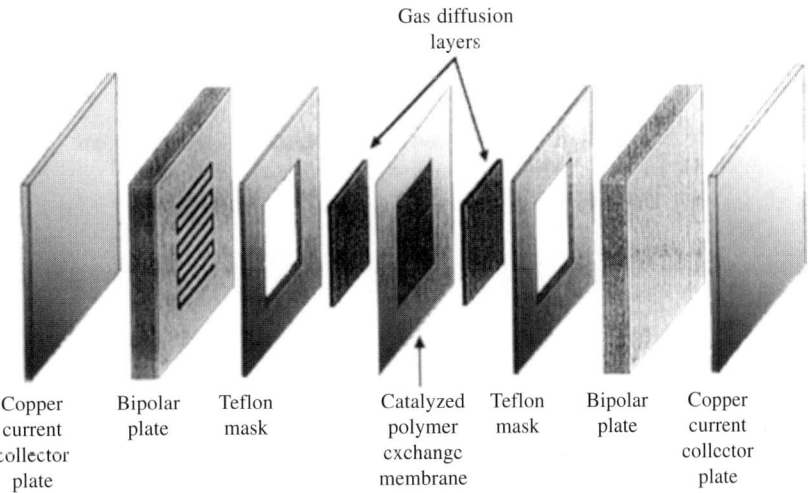

Fig. 1 Conventional design for single cell hardware.

Electrochemical reactions

Anode reaction (negative electrode): $H_2 \rightarrow 2H^+ + 2e^-$

Cathode reaction (positive electrode): $^1/_2 O_2 + 2H^+ + 2e^- \rightarrow H_2O$

A labor intensive step in the process of making a PEMFC is fabricating the membrane-electrode assembly which consists of a membrane sandwiched between the two GDLs. Demonstration MEAs are typically made in small hand sheet quantities by highly trained scientists in a laboratory. There is no standard procedure for an MEA preparation. To make MEAs that allow reproducible processing at lower manufacturing costs, a large scale continuous process will be required. Where platinum is the single most expensive material in a PEM fuel cell, the assembly of the MEA is the most expensive labor contributor.

4. GDL and Cell Performance

There are several experimental parameters for which a GDL should be evaluated to study its performance characteristics. Some of the important experiments conducted are discussed subsequently for the optimization of a fabric gas diffusion layer called GDL-UNH (Koppula, 2004; Koppula et al., 2004) developed at the University of New Hampshire and a local company. Hydrogen was used as the fuel. The data obtained are compared with other commercially available GDLs termed A, B and C. The identity of manufacturing companies is protected to avoid the possibility of promoting one product over the other.

The GDL-UNH is woven out of carbon fibers and wetproofed with a proprietary technique by the local company. Experiments were conducted to study the effect of weaving patterns on the cell performance and the best pattern was adopted in making the GDL used in this study. Efforts were made to optimize the amounts of various materials used in the GDL and electrode preparation starting from the base GDL. The PTFE amount used in the microporous carbon layer, the types of coating used, and the Nafion and catalyst amounts used in the catalyst layer have been studied with respect to cell performance. Some of the results from these experiments are discussed as below. Details are given elsewhere (Koppula, 2004; Koppula et al., 2004).

5. Effect of PTFE Content in the Hydrophobic Carbon Layer of the GDL on Cell Performance

PTFE is used as a binder to apply the carbon layer to the GDL. The function of PTFE is to provide bonding to the carbon black and also to impart a hydrophobic nature to the GDL. The amount of PTFE needed on the GDL-UNH for maximum cell performance was studied. Three samples of GDLs with PTFE contents of 7 mg/cm^2, 12 mg/cm^2 and 26 mg/cm^2 were evaluated. The carbon content was kept constant.

Useful amounts of work (electrical energy) are obtained from a fuel cell only when a reasonable amount of current is drawn, but the actual cell potential is less than equilibrium potential because of losses. The performance of a cell can be characterized by measuring cell voltage and current density (A/cm^2) for a particular set of conditions under which a cell is operated. The voltage-current (V-I) and power density-current (P-I) data showing the performance curves are presented in Fig. 2. The MEA made with the GDL containing the lowest amount of PTFE of 7 mg/cm^2 was found to give better cell performance than the one with higher PTFE content of 12 mg/cm^2. At about 0.9 A/cm^2 current density, the former gave a 7% greater power output. The GDL with 26 mg/cm^2 showed a sudden drop in voltage even at a low current density of 0.4 A/cm^2.

6. Effect of Coating Types on Cell Performance

In the preparation of GDL electrodes and MEAs, there are several steps involved such as coating, drying, sintering and hotpressing that had significant impact on the overall cell performance. The catalyst can be applied on the GDL as a thick single layer or can be coated as thin multi-layers resulting in different

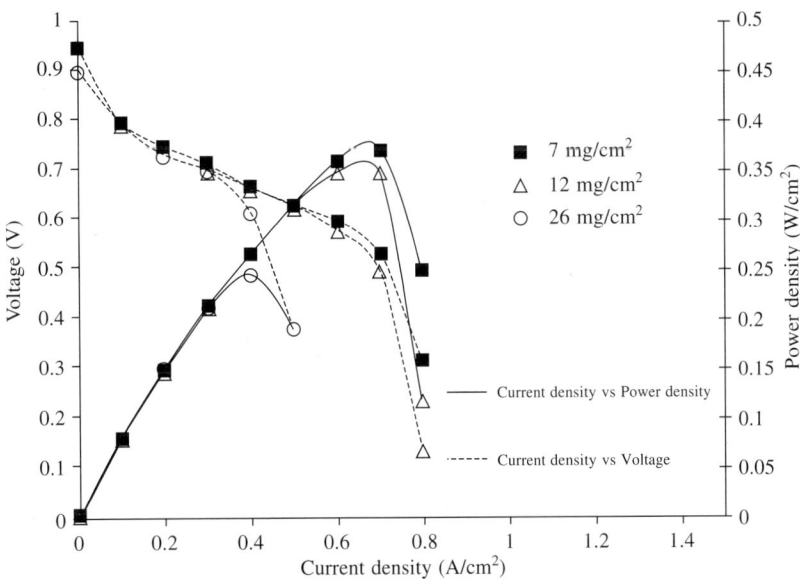

Fig. 2   Effect of PTFE content in carbon layer.

material distribution and surface characteristics. Quick drying and sintering at high temperatures may result in cracks all over the surface while slow drying for a longer period of time may limit the formation of cracks. Pressure and temperature of hot pressing affects the bonding of the electrodes with the membrane during MEA preparation. A slight variation in any of the above steps can change the cell performance even with all the other parameters remaining constant.

Two types of coating techniques were used to coat catalyst on the GDL-UNH samples to prepare electrodes and study cell performance. Two MEAs were prepared with these samples with the same catalyst loadings. The cell was operated under the same conditions of temperature (60°C), pressure (1 atm) and gaseous humidity (H_2 sat @ 75°C and dry air). The performance curves are shown in Fig. 3. A maximum power output of about 0.37 W/cm^2 at a current density of 0.7 A/cm^2 was obtained with Type 2 multi-layered coating, while Type 1 coating provided a maximum power of only about 0.28 W/cm^2 at a current density of 0.6 A/cm^2. Type 2 coating showed an increase in the power output by about 30% at a current density of 0.6 A/cm^2. The SEM scans showed that in Type 2, the surface exhibited unidirectional hairline cracks limited to mainly the top layer. Type 1 coating showed deep cracks separating the surface into blocks.

7. Catalyst Effect on Cell Performance

The three-catalyst loadings of 1, 2 and 4 mg/cm^2 (20% Pt on carbon) were coated onto the GDL surfaces facing the membrane and their performance was compared. The performance curves for various catalyst amounts are presented in Fig. 4. The optimum catalyst loading for maximum cell performance was found to be 2 mg/cm^2 (0.4 mg/cm^2 Pt).

8. Comparative Studies of GDL-UNH with Commercial GDLs

Gas diffusion layer is of vital importance to the performance and durability of a PEM fuel cell. For portable systems like cell phones and laptops, flexible and more durable fabric GDLs are preferred over the paper GDLs. GDL-UNH is a fabric GDL that is robust, flexible and durable. Several commercial GDLs are also

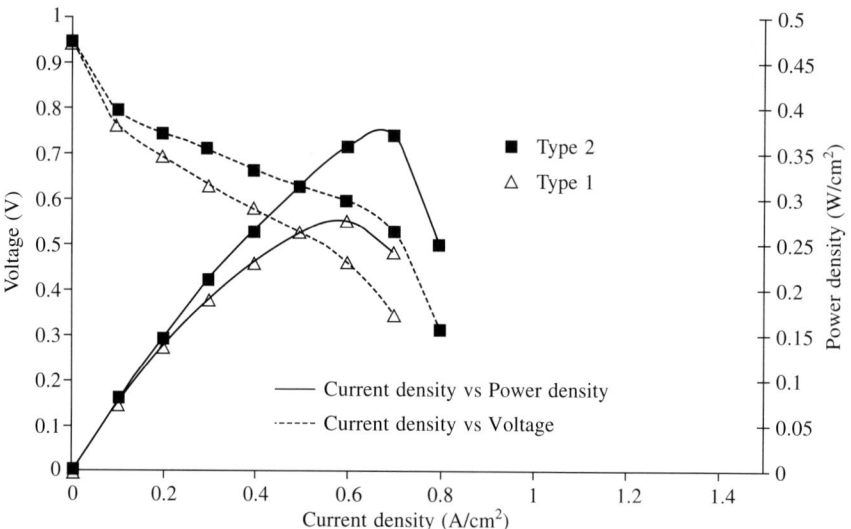

Fig. 3 Effect of coating types.

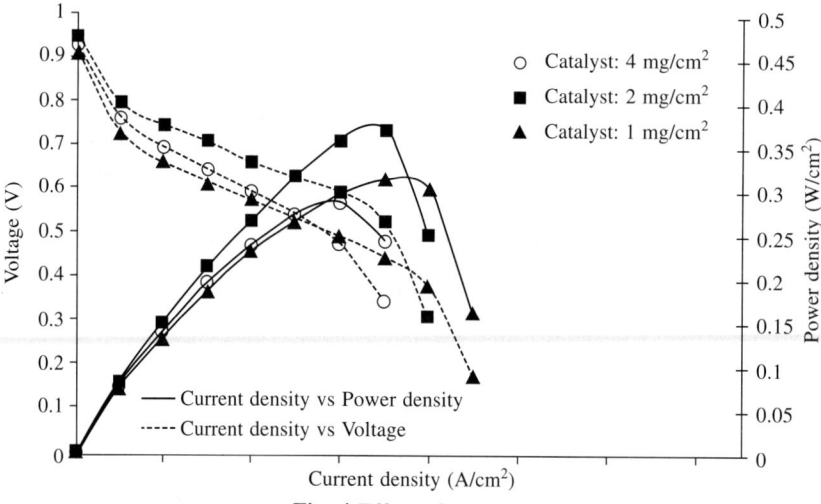

Fig. 4 Effect of catalyst.

available, catering to the specific needs of the users. Efforts were made to compare and characterize the various GDLs and the results are discussed as follows.

8.1 *V-I* and *P-I* Data

The performance of MEAs prepared with GDL-UNH has been compared with some of the commercially available GDLs: GDL-A (fabric), GDL-B (paper) and GDL-C (paper). The catalyst loadings were kept constant at 2 mg/cm^2 for all the MEAs prepared with the above GDLs. Hydrogen saturated at 75°C was used as anode gas and dry air was used as cathode gas. The cell temperature was maintained at 60°C for all the experiments. Fig. 5 shows the cell performance curves for MEAs made from four GDLs and evaluated in our laboratory.

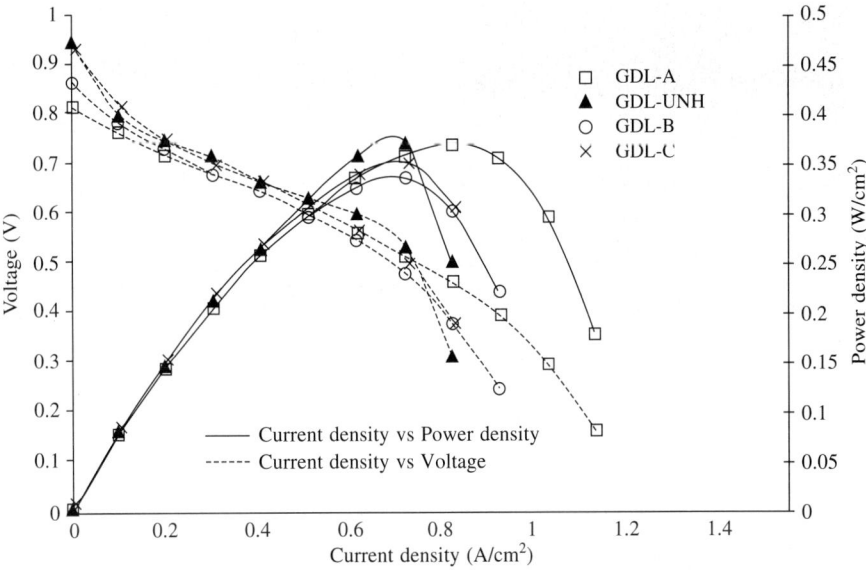

Fig. 5 Performance of MEAs with GDL-UNH and commercial GDLs.

8.2 SEM Scans of GDL Surfaces

A scanning electron microscope (Amray 3300 FE) was used to study the surface characteristics of the various GDLs. Surface scans of both the sides, the carbon coated side that faced the membrane and the uncoated side that faced the gas flow channels, for all the GDLs were taken. The SEM scans of the surfaces of GDLs of the fabric and paper GDLs are presented in Figs. 6 and 7, respectively. GDL-A (fabric), GDL-B (paper) and GDL-C (paper) are commercial products available in the market. The GDL-UNH is under development.

GDL-A and GDL-UNH were made of carbon fabrics with a carbon layer on the side facing the membrane. The scans showed the weaving patterns of the fabric and the characteristics of the coated surface. GDL-A surface showed no cracks. The scan of the uncoated surface facing the reactant gases, showed that even the small openings between the weaving patterns were blocked by the carbon layer coating. GDL-UNH showed cracks on the carbon coated surface indicating the presence of large crevices in the layer. Both the GDLs gave about the same power output. However, the cell performance with GDL-A was distributed over a wide range of current densities. The performance curve with GDL-UNH showed diffusion overpotential at high current density with a sudden drop in voltage.

GDL-B and GDL-C were made of carbon paper. GDL-B had a carbon layer on one side whereas GDL-C did not have any. The cell performance of both GDLs is almost the same as shown in Fig. 5.

These SEM scans can also provide a better understanding of the performance data (*V-I* and *P-I*) for every GDL. The details of such an analysis can be found elsewhere (Koppula, 2004; Koppula et al., 2004).

9. Smaller Fuel Cells, a Shorter Wait

We will have to wait 20 years, if not longer, for cars powered by fuel cells to become a familiar sight. But much smaller fuel cells may well power electronic devices such as laptop computers, video cameras and cell phones by the end of this decade. Prototypes of long-lasting fuel cells that can replace batteries are being tested in laboratories in the United States and overseas. Small cells have several economic advantages

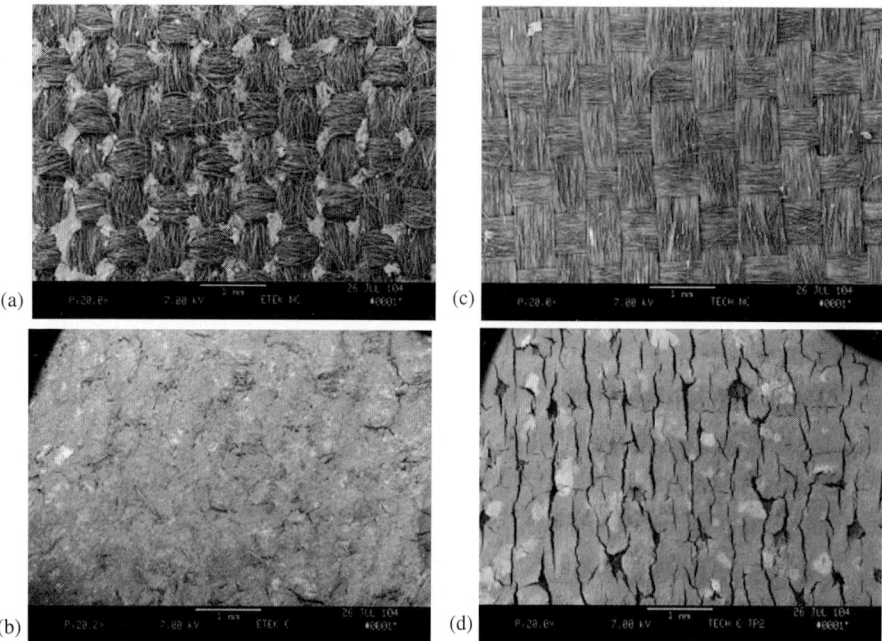

Fig. 6 Surface SEM scans of GDL-A and GDL-UNH: (a) Uncoated surface view of GDL-A, (b) Carbon coated surface view of GDL-A, (c) Uncoated surface view of GDL-UNH and (d) Carbon coated surface view of GDL-UNH.

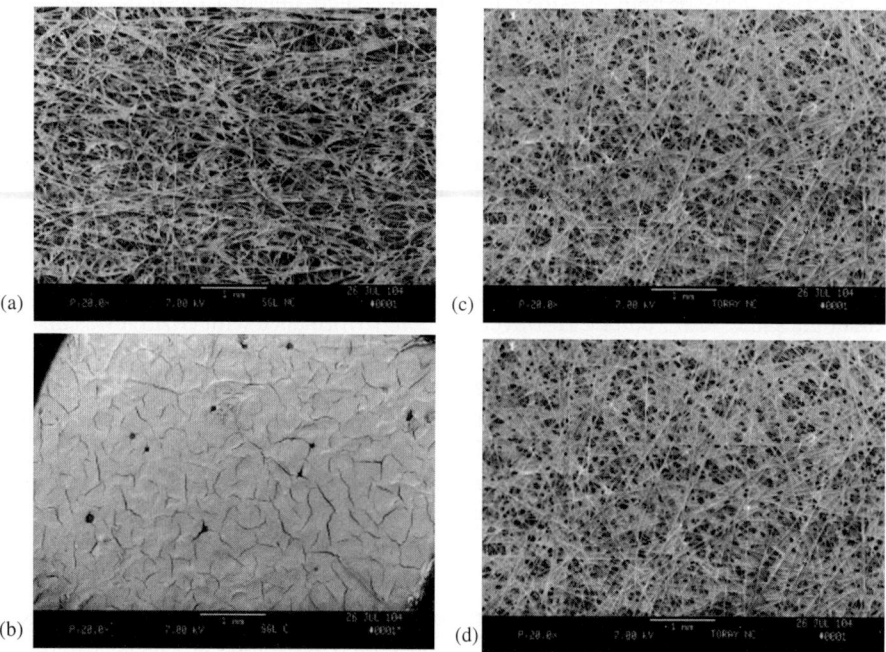

Fig. 7 Surface SEM scans of GDL-B and GDL-C: (a) Uncoated surface view of GDL-B, (b) Carbon coated surface view of GDL-B, (c) Uncoated surface view of GDL-C and (d) GDL-C does not have carbon layer.

over the higher power cells in the race to commercialization. It is expected that small cell production costs can be reduced to be competitive with those of batteries long before larger cells can be manufactured at anything close to the cost of internal combustion engines. But the biggest reason for the smaller cells to become popular sooner is their appeal as a convenience—something that consumers have shown a willingness to pay for—and not as an answer to energy and environmental problems. Half of the interest in fuel cells is out of frustration with batteries.

10. Commercialization

The cost of MEAs for applications up to 1 kW is commercially acceptable to the fuel cell industry today. The focus of MEAs for power applications greater then 1 kW, approaching 10-50 kW for stationary power (a typical space fuel cell is 15-30 kW and 75-100 kW for automotive fuel cells), is to drive the cost of the MEA lower by a factor of 10. This will be accomplished by developing the technology for manufacturing MEAs in a low cost, continuous lamination process and by finding alternate catalyst materials to platinum.

It is believed that the raw material cost for membrane and GDL manufacture will follow a typical commercial cost reduction curve as the technology is commercialized. Therefore significant advances needed to meet this cost target will be achieved by developing manufacturing processes to automate MEA production.

The first commercial application for high volume MEA production will be for small portable electronics, which is driven by ever increasing power requirements and for longer duration energy sources. The US Fuel Cell Council, which has over 115 corporate members, has forecasted that the fuel cell market for small portable electronics will be $ 2 billion by 2011 with the potential of a 70% market penetration into the lithium ion battery market by 2007. To get a better understanding of this market size it is also forecasted that there will be 1.6 billion cell phone subscribers in 2007 worldwide, 76 million laptop shipments in 2007 and 37 million shipments of PDAs (personal digital assistants) in 2007.

The commercial market for larger fuel cells, greater then 1 kW, will evolve as the portable market and manufacturing technologies are established. Presently there are over 500 stationary fuel cell systems operating in North America and this number will grow as lower cost MEAs are realized and the need for power located outside of an electrical grid system grows. The military will also benefit from fuel cells as they retool the combat foot soldier. The US Army has a goal of deploying a soldier into the field for 2 weeks with an unlimited source of power for his combat electronics without resupply. This can only be accomplished with the use of fuel cells.

Presently the United States consumes 26% of the world's oil production. Automotive fuel cells that are commercially accepted will be the ultimate use for fuel cells as the hydrogen economy and infrastructure is developed.

11. Continuous MEA Fabrication

One of the most significant steps to commercialization of fuel cells will be the development of a continuous process to manufacture membrane electrode assemblies. The process envisioned will include coating, sintering and laminating steps and will have a continuous roll of MEA as its output. Different fuel cell applications being developed require MEAs with different characteristics. For example, different carbon/PTFE ratios and different catalyst concentrations are tailored to the application. Therefore, the ideal MEA manufacturing process must be able to adjust every step of the process to be useful to the broadest range of applications. The process must also be compatible with the various sizes of fuel cells. MEAs currently in use range from 5 sq. cm to 1 sq. m. It is unlikely that one MEA production line would be able to accommodate this entire range. Hence, there are several challenges that must be overcome. The most important of these is the quality and consistency of the resulting MEA.

At one end of the process line, a roll of carbon fabric or paper (GDL) would be mounted. The carbon material will have been heat-treated to the appropriate temperature to enhance the thermal properties and also to remove any contaminants and/or convert the same to carbon, such as polymeric binders. The next step in the process is the plannerization of the carbon substrate. This entails the application of one or more layers of an emulsion consisting of PTFE, carbon particles, and water to both sides of the carbon substrate. The coating(s) must be dried and sintered slowly to minimize, or ideally avoid, the formation of cracks in the coating due to shrinkage. These are often referred to as "mud cracks". Multiple thin coatings are often preferred to one thick coating for at least two reasons. First, thin coatings are less prone to shrinkage cracks and second, they are less stiff and therefore more robust in later handling operations. The sintering temperature must be selected to adhere the carbon particles to the carbon substrate.

The final step in the process is the lamination of two GDL layers of plannerized carbon to one layer of membrane film. One GDL layer of carbon is laminated to each side of the membrane. The membrane is very fragile and must not be compromised during the lamination process. Heat and pressure must be applied in a manner that will not induce curvature into the lamination. However, the final material should be able to be taken-up on a roll. The stiffness of the assembly will dictate the minimum roll diameter that can be realized. A continuous roll can then be fed into a die-cutting process to remove precise MEA components to be used in the fuel cell stack. Fig. 8 illustrates one approach for a continuous MEA manufacturing plan. This approach was presented by Lydall Filtration/Separation Group, Rochester, NH at a recent AIChE Meeting (Xie et al., 2004).

Lydall, a manufacturer of roll-good carbon paper type GDLs, claims that this procedure will be highly cost effective both from the logistics and process standpoints. The GDL is basically micro porous and

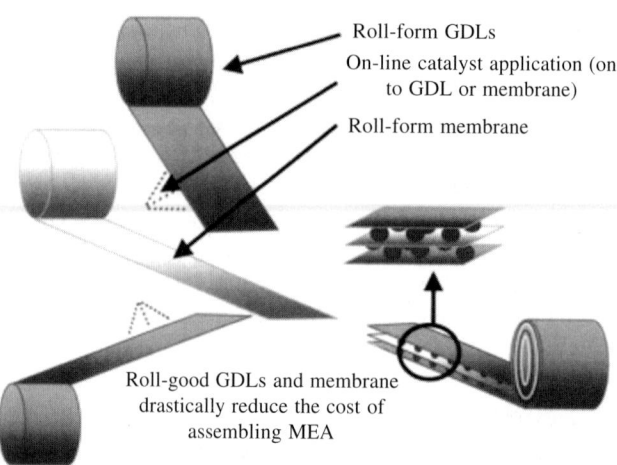

Roll-form GDLs
On-line catalyst application (on to GDL or membrane)
Roll-form membrane

Roll-good GDLs and membrane drastically reduce the cost of assembling MEA

Fig. 8 Lydall's concept of continuous MEA manufacture.

hydrophobic. This allows the diffusion of gases uniformly to the surface of the catalyst. This layer also helps in proper water management in the membrane and catalyst layers by rejecting the excess water formed at the cathode or by retaining water for proper membrane hydration.

Acknowledgement

We are thankful to our students Karuna K. Koppula, Michael C. Johnston and Zongyuan Chen for their contributions towards the experimental work.

References

Allen, R.J., Lindstrom, R. and Juda, W. (1981) Thin Carbon-Cloth-Based Electrocatalytic Gas Diffusion Electrodes, and Electrochemical Cells Comprising the Same, U.S. Patent **4**, 293 396, Oct. 6.

Antolini, E., Passos, R.R. and Ticianelli, E.A. (2002) Effect of Carbon Powder Characteristics in the Cathode Gas Diffusion Layer on the Performance of Polymer Electrolyte Fuel Cells, *J. Power Sources*, **109**, 477-482.

Buchi, F.N. and Srinivasan, S. (1997) Operating Proton Exchange Membrane Fuel cells without External Humidification of the Reactant Gases, *J. Electrochemical Society*, **144**, 2767-2772.

Chu, H., Yeh, C. and Chen, F. (2003) Effects of Porosity Change of Gas Diffuser Performance of Proton Exchange Membrane Fuel Cell, *J. Power Sources*, **123**, 1-9.

Fuel Cell Handbook, 1994, fourth edition, B/T Books, CA, USA.

Gamburzev, S. and Appleby A.J. (2002) Recent Progress in Performance Improvement of Proton Exchange Membrane Fuel Cell (PEMFC), *J. Power Sources*, **107**, 5-12.

Gasteiger, H.A., Panels, J.E. and Yan, S.G. (2004) Dependence of PEM Fuel Cell Performance on Catalyst Loading, *J. Power Sources*, **127**, 162-171.

Giorgi, L., Antolini, E., Pozio, A. and Passalacqua, E. (1998) Influence of PTFE content in the Diffusion Layer of Low-Pt Loading Electrodes for Polymer Electrolyte Fuel Cells, *Electrochimica Acta*, **43**, 3675-3680.

Glora, M., Wiener, M., Petricevic, R., Probstle, H. and Fricke, J. (2001) Integration of Carbon Aerogels in the PEM Fuel Cells, *J. Non-Crystalline Solids*, **285**, 283-287.

Kato, H. (2000) Gas Diffusion Layer for Solid Electrolyte Fuel Cell, U.S. Patent, **6**, 127 059, Oct. 3.

Kong, C.S., Kim, D., Lee, H., Shul, Y. and Lee, T. (2002) Influence of Pore Size Distribution of Gas Diffusion Layer on Mass Transport Problems of Proton Exchange Membrane Fuel Cells, *J. Power Sources*, **108**, 185-191.

Koppula, K.S. (2004) "Study of Gas Diffusion Layers in PEM Fuel Cells", M.S. Thesis, University of New Hampshire, Durham, NH, Aug.

Koppula, K.S., Johnston, M.C. and Mathur, V.K. (2004) Study of Gas Diffusion Layers in PEM Fuel Cells, AIChE Annual Meeting, Austin, TX, Nov. 7-12.

Kumar, G.S., Raja, M. and Parthasarathy, S. (1995) High Performance Electrodes with very Low Platinum Loading for Polymer Electrolyte Fuel Cells, *Electrochimica Acta*, **40**, 285 290.

Larminic, J., Dicks A. (2003) Fuel Cell Systems Explained, John Wiley & Sons, NY, USA.

Lister, S. and McLean, G. (2004) PEM Fuel Cell Electrodes, *J. Power Sources*, **130**, 61-76.

Mathur, V.K., Xie, X. and Crawford, J. (2005) Role of Carbon in a PEM Fuel Cell System, AIChE Spring National Meeting, Atlanta, GA, April 10-14.

Moreira, J., Sebastian, P.J., Ocampo, A.L., Castellanos, R.H., Cano, U. and Salazar, M.D. (2002) Dependence of PEM Fuel Cell Performance on the Configuration of the Gas Diffusion Electrodes, *J. New Materials for Electrochemical Systems*, **5**, 173-175.

Paganin, V.A., Ticianelli, E.A. and Gonzalez, E.R. (1996) Development and Electrochemical Studies of Gas Diffusion Electrodes for Polymer Electrolyte Fuel Cells, *J. Applied Electrochemistry*, **26**, 297-304.

Park, G., Sohn, Y., Yang, T., Yoon, Y., Lee, W. and Kim, C. (2004) Effect of PTFE Contents in the Gas Diffusion Media on the Performance of PEMFC, *J. Power Sources*, **131**, 182-187.

Passalacqua, E., Lufrano, F., Squadrito, G., Patti, A. and Giorgi, L. (1998) Influence of the Structure in Low-Pt Loading Electrodes for Polymer Electrolyte Fuel Cells, *Electrochimica Acta*, **43**, 3665-3673.

Prasanna, M., Ha, H.Y., Cho, E.A., Hong, S. and Oh, I. (2004) Influence of Cathode Gas Diffusion Media on the Performance of the PEMFCs, *J. Power Sources*, **131**, 147-154.

Qi, Z. and Kaufman, A. (2003) Low Pt Loading High Performance Cathodes for PEM Fuel Cells, *J. Power Sources*, **113**, 37-43.

Qi, Z. and Kaufman, A. (2002) Improvement of Water Management by a Microporous Sublayer for PEM Fuel Cells, *J. Power Sources*, **109**, 38-46.

Sasikumar, G., Ihm, J.W. and Ryu, H. (2004) Dependence of Optimum Nafion Content in Catalyst Layer on Platinum Loading, *J. Power Sources*, **132**, 11-17.

Song, J.M., Cha, S.Y. and Lee, W.M. (2001) Optimal Composition of Polymer Electrolyte Fuel Cell Electrodes Determined by the AC impedance Method, *J. Power Sources*, **94**, 78-84.

Wilson, M.S., Valerio, J.A. and Gottesfeld, S. (1995) Low Platinum Loading Electrodes for Polymer Electrolyte Fuel Cells Fabricated using Thermoplastic Ionomers, *Electrochimica Acta*, **40**, 355-363.

Xie, X., Koppula, K.S., Hamblin, T. and Mathur, V.K. (2004) Materials—Key to Hydrogen Economy, AIChE Spring Meeting, New Orleans, LA, April 24-29.

Recent Trends in Fuel Cell Science and Technology
Edited by S. Basu
Anamaya Publishers, New Delhi, India

5. Water Problem in PEMFC

Kohei Ito

Department of Mechanical Engineering Science, Graduate School of Engineering, Kyushu University, Japan

1. Introduction

A new generator performing highly-efficient with low (or no) pollutant emission is necessary to resolve the problem of exhaustion of fissile fuel and spreading environmental pollution. Fuel cell has been expected as one candidate of the generator. Among several types of fuel cells polymer electrolyte membrane fuel cell (PEMFC), can operate at low temperature (below 100°C) and generate specific power and a power density higher than any other fuel cell with relatively simple constitution. These benefits make us enthusiastic to apply PEMFC to the power source of transportation, portable and resident (Larminie et al., 2003).

In spite of this large expectation of PEMFC, its commercialization has not been spread out. One reason is that the cost and the durability of PEMFC is worse than that of conventional engines competing with PEMFC. Another reason is that water problem occurs during the practical operation of PEMFC, and this problem makes its performance worse.

The water problem is an intrinsic problem of PEMFC caused by its low-temperature operation and by the characteristics of polymer electrolyte membrane used. This problem can be divided into two parts as shown in Fig. 1: (1) flooding occurred in gas diffusion layer (GDL) and flow channel, (2) drying polymer

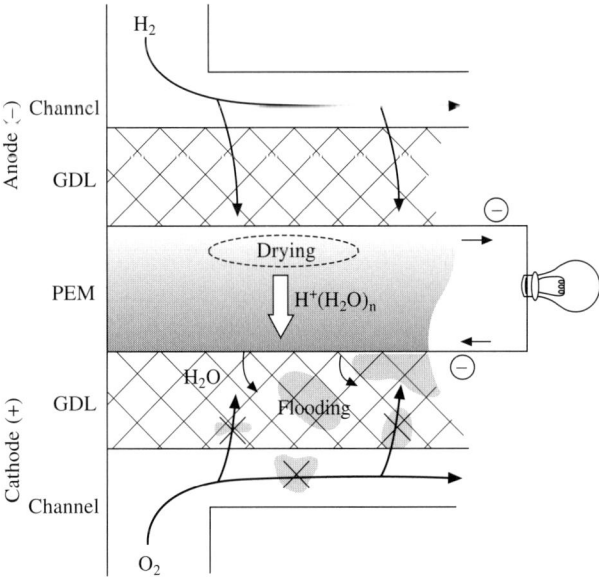

Fig. 1 Schema of flooding and drying in PEMFC.

electrolyte membrane (PEM). The flooding is caused by the following mechanics: water or vapor generates at catalyst layer in proportion to load current, and a portion of vapor condenses in GDL and flow channel when vapor pressure is larger than the saturated vapor pressure. If the condensation progress, flooding occurs, and then the flooding blocks the supplied gases resulting in the decrease of cell voltage and stopping electricity generation at worse. The drying is caused by the so-called electro-osmotic flow, which is the water molecules immigration in PEM with the proton conduction from anode to cathode resulting in decreasing water content especially near anode. Once PEM dries up, the ionic resistance increases and the cell voltage decreases.

The water problem, which is thus represented by the flooding and drying, becomes more important issue related to the recent trend of high current-density operation that is aiming for higher performance. For spreading the practical use of PEMFC, we must solve the water problem, and must construct the optimization of design and operation so that the flooding and the drying is difficult to occur. For solving water problem we must develop two tools: (a) *in situ* measurement or observation of the flooding and the drying, (b) reliable numerical simulation that has good agreement with the measurement result. Then we must optimize the design and operation based on these two tools. Following shows the recent reports of measurement of the flooding and drying in PEMFC, and the reports of the numerical simulation considering that a two-phase flow occurred.

2. Measurement of Flooding and Drying in Cell

For the sake of understanding the water problem in PEMFC, it is necessary to measure the flooding distribution in flow channel and gas diffusion layer (GDL), and to measure the water content in PEM. There are a lot of experimental reports to observe the flooding in flow channel. The majority of them are transparent cell experiments. In Tüber et al. (2003) experiment, a digital camera recorded the flooding distribution in the channel through transparent cell cover. The camera is placed outside the cell. Due to an optimized adjustment of digital camera, the PEMFC is positioned 70° to the horizontal. In addition to this flooding observation, Hakenjos et al. (2004) measured simultaneously the temperature distribution by IR camera through zinc selenide covered cell and the current distribution with segmented electrode. Then they discussed the relation between flooding, temperature, and current distribution. Ogawa et al. (2004) compared the cell voltage fluctuation to the observed flooding change. Hottinen et al. (2004) observed the flooding on the cathode GDL surface of the planar free-breathing cell, which is recently developing. Note that Mench et al. (2003) measured not the flooding but the water vapor distribution in the flow channel using a gas chromatograph.

It is challenging to measure the flooding appeared in GDL. A recent paper reports that Satija et al. (2004) achieved success to measure the flooding in GDL by using a novel and interesting method. The method is the *in situ* neutron imaging technique that utilized the high sensitivity of neutron beam to water. This idea led to the success of measuring the flooding in GDL though the measurement system is relatively large.

The drying distribution, that is, the water content distribution in PEM was measured by use of magnetic resonance imaging (MRI). Ito et al. (2000) measured the water content distribution in PEM by using heavy water and chemical shift imaging technique. They measured the water content distribution and its time evolution, and estimated the water diffusion coefficient in PEM, though they used the simple cell where only PEM is supported between two separators made of acrylic resin without load current. Tushima et al. (2004) also measured the water content in PEM. They obtained the water content distribution estimated by the conversion of NMR intensity distribution. It is worth mentioning that they acquired the *in situ* water content distribution in PEM while generating electricity with unit cell placed in a cylinder-shaped RF coil. Ouriadov et al. (2004) also measured the water content distribution in PEM by MRI with single point imaging technique employing surface-shaped RF coil. All measurement methods explained here are summarized in Table 1.

Table 1. Method for *in situ* measurement flooding and drying in PEMFC

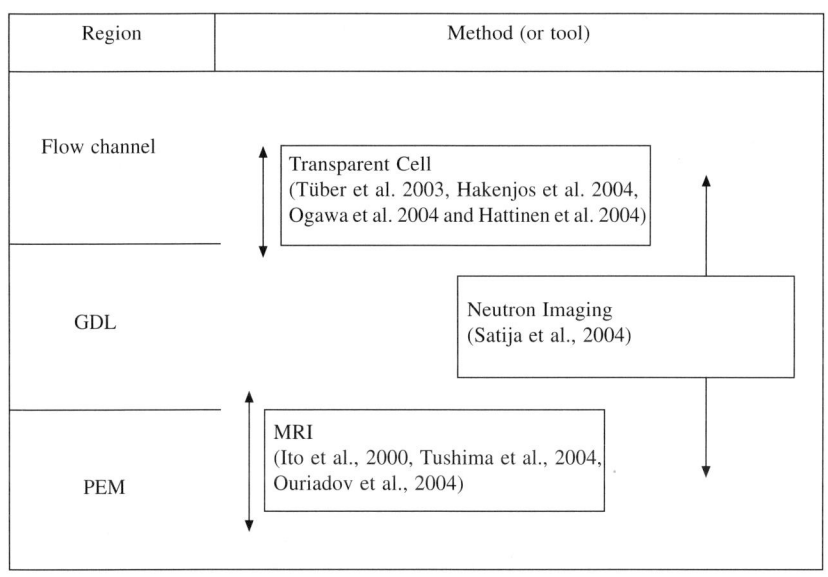

Region	Method (or tool)
Flow channel	Transparent Cell (Tüber et al. 2003, Hakenjos et al. 2004, Ogawa et al. 2004 and Hattinen et al. 2004)
GDL	Neutron Imaging (Satija et al., 2004)
PEM	MRI (Ito et al., 2000, Tushima et al., 2004, Ouriadov et al., 2004)

Though there are a lot of experiments to measure the flooding and drying, it is plausible to mention that quantitative understanding of the water problem has not been obtained yet. As for the measurement of the flooding in channel, we have not shown the quantitative correspondence of the flooding distribution to that of temperature and current: how much does this size water droplet located here cause cell-voltage to decrease? In addition we have not shown the quantitative correspondence of the flooding change to the cell-voltage change: does this frequency of flooding change correspond to that of the cell-voltage to change? As for the measurement of the water contents distribution in PEM by using MRI, more efforts seem to be necessary to obtain more reliable data aiming the higher resolution of time and space, and overcoming many difficulties in the piping and wiring in MRI equipment. Moreover, it is important to develop a new *in situ* measurement tool for PEMFC, and to compare quantitatively between the measurement and simulation, explained as follows.

3. Numerical Simulation

The fundamental equation of PEMFC simulation consists of mass, momentum, energy conservation, and equivalent electric circuit that is composed of the elements: open circuit voltage, resistance overvoltage, and charge transfer overvoltage. For understanding the flooding and drying that occurred in the cell, we must consider the model of gas/liquid two-phase flow in the fundamental equations.

The mathematical model of two-phase flow in PEMFC is often based on the multiphase mixture model (MMM) developed by Wang et al. (1996), Chang et al. (1996) and Wang et al. (1997). The key idea of this model is to focus not on the level of separate phases, but on the level of a multiphase mixture such as mass-averaged mixture velocity. Hence the model need not to track phase interface separating one from two-phase region. The developed formulation based on the MMM is as follows (Wang et al., 2001 and You et al., 2002). In GDL, continuity equation is

$$\frac{\partial}{\partial t}(\varepsilon\rho) + \nabla \cdot (\varepsilon\rho\mathbf{u}) = 0 \qquad (1)$$

and momentum conservation equation is,

$$\frac{\partial}{\partial t}(\varepsilon\rho\mathbf{u}) + \nabla \cdot (\varepsilon\rho\mathbf{uu}) = -\varepsilon\nabla P + \nabla \cdot (\varepsilon\mu\nabla\mathbf{u}) + \varepsilon\rho\mathbf{g} - \varepsilon^2 \frac{\mu}{K}\mathbf{u} \tag{2}$$

and species conservation equation is

$$\frac{\partial}{\partial t}(\varepsilon\rho C^\alpha) + \nabla \cdot (\varepsilon\gamma_\alpha\rho\mathbf{u}C^\alpha) = \nabla \cdot (\varepsilon\rho D^\alpha\nabla C^\alpha) + \nabla \cdot \{\varepsilon[\rho_l s D_l^\alpha\nabla C_l^\alpha + \rho_g(1-s)D_g^\alpha\nabla C_g^\alpha - \rho D^\alpha\nabla C^\alpha]\}$$

$$- \nabla \cdot [(C_l^\alpha - C_g^\alpha)\mathbf{j}_l] \tag{3}$$

The coordinate system and calculation mesh for these fundamental equations is shown in Fig. 2. The x-axis and y-axis are taken as being parallel and horizontal to the flow channel direction. The two-phase mixture quantities in the above formulation are defined as

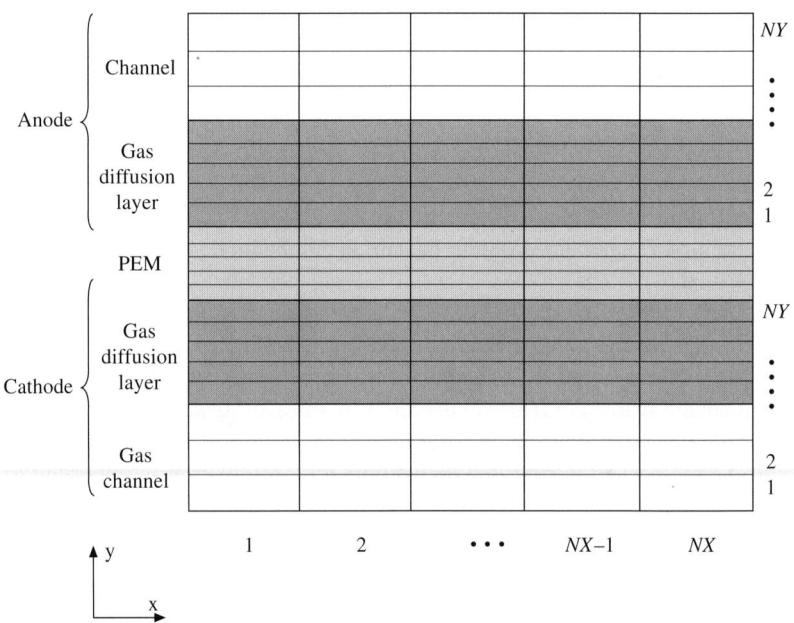

Fig. 2 Coordinate system for the mathematical model of two-phase flow in PEMFC.

Density: $\qquad\qquad\qquad\qquad \rho = (1-s)\rho_g + s\rho_1 \tag{4}$

Velocity: $\qquad\qquad\qquad\qquad \rho\mathbf{u} = \rho_g\mathbf{u}_g + \rho_l\mathbf{u}_l \tag{5}$

Concentration: $\qquad\qquad\qquad \rho C = \rho_g C_g(1-s) + \rho_l C_i s \tag{6}$

In addition to these equations, further equations, such as constitutive equations, are imposed in the MMM formulation. Some of them are:

Individual velocity of liquid phase: $\quad \varepsilon\rho_l\mathbf{u}_l = \mathbf{j}_l + \lambda_l\varepsilon\rho\mathbf{u} \tag{7}$

Individual velocity of gas phase: $\qquad \varepsilon \rho_g \mathbf{u}_g = \mathbf{j}_g + \lambda_g \varepsilon \rho \mathbf{u}$ $\qquad\qquad$ (8)

The other constitutive equations and the nomenclature in the equations denoted are summarized in Table 2. For more details, see Wang et al., 2001 and You et al., 2002. In simulation, all the equations mentioned are simultaneously solved, and important physical quantities, such as water saturation and gas-phase velocity, can be obtained as shown in Fig. 3 (Masuda et al., 2004). It is worth to mention that, in the formulation based on the MMM, the governing equations in flow channel are same as those in GDL with the porosity being unity and with the permeability being infinity.

Table 2. Nomenclature used to describe the two-phase flow in PEMFC

Symbols

C_g^α	Mass fraction for species α in k-phase	$C_k^\alpha = \dfrac{\rho_k^\alpha}{\rho_k}$
C^α	Mass fraction for species α	$\rho C^\alpha = \rho_l s C_l^\alpha + \rho_g (1-s) C_g^\alpha$
D^α	Diffusion coefficient for species α	$\rho D^\alpha = \rho_l s D_l^\alpha + \rho_g (1-s) D_g^\alpha$
$\mathbf{j}_l$	Diffusive mass flux of liquid phase	$\mathbf{j}_l = \dfrac{K \lambda_g \lambda_l}{\nu} [\nabla P_C + (\rho_l - \rho_g)\mathbf{g}]$
$\mathbf{j}_l$	Diffusive mass flux of gas phase	$\mathbf{j}_l + \mathbf{j}_g = 0$
k_{rl}	Relative permeability of liquid	$k_{rl} = s^3$
k_{rg}	Relative permeability of gas	$k_{rg} = (1-s)^3$
K	Permeability of porous GDL	
P	Pressure	
P_C	Capillary pressure between gas and liquid given by a empirical model	$P_C = \sigma \cos \theta_C \left(\dfrac{\varepsilon}{K}\right)^{\frac{1}{2}} J(s)$
		$- \sigma \cos \theta_C \left(\dfrac{\varepsilon}{K}\right)^{\frac{1}{2}} [1.417(1-s) - 2.120$
		$(1-s)^2 + 1.2263 (1-s)^3]$
$\mathbf{u}$	Liquid-gas mixture velocity	
ε	Porosity	
λ_l	Relative mobility of liquid phase	$\lambda_l = \dfrac{k_{rl}/v_1}{k_{rl}/v_1 + k_{rg}/v_g}$
μ	Liquid-gas mixture viscosity	$\mu = \dfrac{\rho}{k_{rl}/v_1 + k_{rg}/v_g}$
v_k	Kinetic viscosity of k-phase	
ρ	Liquid-gas mixture density	
θ_C	Contact angle	
σ	Surface tension	
Subscripts		
g	Gas phase	
l	Liquid phase	

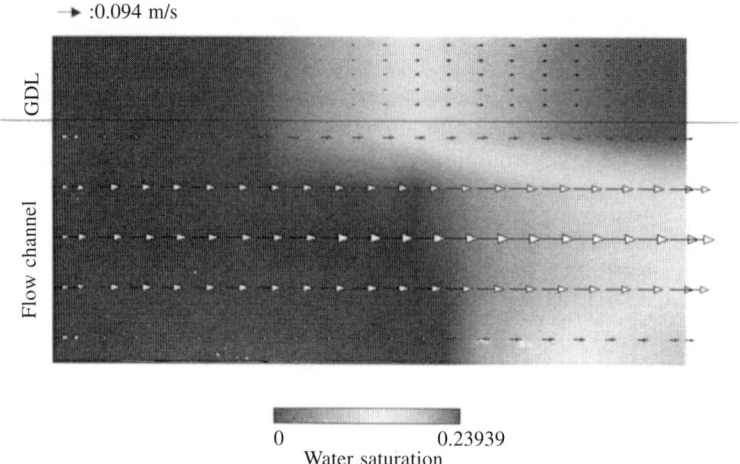

→ :0.094 m/s

GDL

Flow channel

0   0.23939
Water saturation

Fig. 3 A two-phase flow simulation in PEMFC based on MMM. Water saturation distribution and gas phase velocity are expressed by color scale and vector, respectively. The typical water saturation distribution, where a large water saturation region appeared in the GDL near channel outlet, is caused by the water production at cathode catalytic layer, and by its accumulation along flow channel. The gas phase velocity in the flow channel is larger than that in the GDL, and is almost constant in the stream-wise. The calculation condition used are: cell temperature is 80°C, average current density is 1.0 A/cm^2, hydrogen and oxygen utilization ratio is 0.5, humidifier temperature for anode and cathode is 60 and 25°C, respectively.

Wang et al. (2001) numerically analyzed it, and showed the liquid water saturation distribution in the flow channel and GDL at cathode, and showed the critical current density in the case where the two-phase region appeared. Explicitly adding the catalytic layer region to the Wang et al. (2001) work, You et al. (2002) executed the two-phase simulation, and showed the good agreement of current-voltage characteristic between simulation and measurement. Hu et al. (2004 a, b) expanded the Wang et al. (2001) work to three-dimensional simulation, and showed the detail distribution of the velocity and the species-concentration especially under separator rib. They also estimated the performance of the interdigitated flow channel PEMFC. In the interdigitated cell, the reactant gases are forced to flow into the electrode by making the inlet and outlet gas channel dead-ended, and have the characteristics of three dimensional flow. Their three dimensional simulation is available to capture the three dimensional phenomena appeared in such an interdigitated cell. Masuda et al. (2004), Ferng et al. (2003) and Yuan et al. (2004) also reported the two-phase flow simulation in PEMFC.

In addition to these simulation developed by the academic researchers mentioned above, software houses also developed the two-phase flow PEMFC simulator based on the MMM, and commercialized it. Mizuho Information and Research Institute (http://www.mizuho-ir.co.jp) and CFD Research Corporation (http://www.cfdrc.com) supply the two-phase-flow PEMFC module. On their websites, we can see some examples of the numerical results such as the distribution of water saturation. CD-adapco Japan Co. Ltd (http://www.cdaj.co.jp) and Fluent Inc. (http://www.fluent.com) also supply the two-phase PEMFC simulator. The soft houses' simulator developed with their proven simulation technology, seems to have merits like helpful graphic user interface and high execution-speed especially for three-dimensional system that has large calculation domain.

Thus there are a lot of two-phase flow simulators developed. However the evaluation of the simulators is insufficient, because the comparison between practical cell operation and its simulation has been focused on the level of only current-voltage characteristics. To get the reliability on the simulation, a comparison at the level of space distribution and time evolution between simulation and measurement is necessary. Reflecting this detail comparison, we should evaluate the mathematical model, and refine it if necessary.

4. Concluding Remarks

Solving the water problem occurred in PEMFC operation is a key issue to spread and commercialize it. In the frame of this context, recent reports on the measurement of flooding and drying in cell and on two-phase flow simulation for unit cell were shown. Then the remarkable progress of the measurement and the simulation were also shown. On the other hand, we understood that the several subjects to them still remained. As for the measurement, quantitative comparison remains: (a) the change of cell potential vs. flooding and drying-up; (b) the distribution of current and temperature distribution vs. flooding and drying up; (c) the numerical result vs. the measurement result focused on the distribution level. In addition, developing new measurement method, which has higher resolution of time and space, is expected. As for the two-phase flow simulation, it is necessary to obtain the higher reliability in comparison with the measurement in detail. Once we have solved the water problem based on the measurement and the simulation, that is, once we have optimized the cell design and operation that is free from the flooding and the drying, the practical use of PEMFC spreads with good acceleration.

References

Cheng, P. and Wang, C.Y., A multiphase mixture model for multiphase, multicomponent transport in capillary porous media–2. Numerical simulation of the transport of organic compounds in the subsurface, *Int. J. Heat and Mass Transfer*, **39**, pp. 3619-3632, 1996.

Fering, Y.M., Sun, C.C. and Su, A., Numerical simulation of thermal-hydraulic characteristics in a proton exchange membrane fuel cell, *Int. J. Energy Res.*, **27**, pp. 495-511, 2003.

Hakenjos, A., Muenter, H., Wittstadt, U. and Hebling, C., A PEM fuel cell for combined measurement of current and temperature distribution, and flow field flooding, *J. Power Sources*, **131**, pp. 213-216, 2004.

Hottinen, T., Himanen, O. and Lund, P., Effect of cathode structure on planar free-breathing PEMFC, *J. Power Sources*, **138**, pp. 205-210, 2004.

Hu, M., Gu, A., Wang, M., Zhu, X. and Yu, L., Three dimensional two phase flow mathematical model for PEM fuel cell: Part 1. Model development, energy conversion and management, **45**, pp. 1861-1882, 2004a.

Hu, M., Zhu, X., Wang, X., Gu, A. and Yu, L., Three dimensional two phase flow mathematical model for PEM fuel Cell: Part 2. Analysis and discussion of the internal transport mechanism, Energy Conversion and Management, **45**, pp. 1883-1916, 2004b.

Ito, K. and Ogawa, K., Investigation of Water Molecule Distribution and Transport Mechanism in Polymer Electrolyte Membrane by Magnetic Resonance Imaging, Proceeding of the 4th JSME-KSME Thermal Engineering Conference, Kobe, Japan, **3**, pp. 355-360, 2000.

Larminie, J. and Dicks, A., Fuel Cell Systems Explained, John Wiley & Sons Ltd., 2003.

Masuda, H., Ito, K., Masuoka, T. and Kakimoto, Y., Investigation of Water Blocking Phenomena in PEMFC by Numerical Analysis, *Proceedings of Thermal Eng. Conf.* 2004, Japan, pp. 31-32, 2004.

Mench, M.M., Dong, Q.L. and Wang, C.Y., *In situ* water distribution measurement in a polymer electrolyte fuel cell, *J. Power Source*, **124**, pp. 90-98, 2003.

Ogawa, T., Nohara, N., Kikuta, K., Chikahisa, T. and Hishinuma, Y., Observation of Water Production and Temperature Distribution in PEM Fuel Cell, 41st National Heat Transfer Symposium of Japan, Toyama, pp. 235-236, (2004).

Ouriadov, A.V., MacGregor, R.P. and Balcom, B.J., Thin film MRI-high resolution depth imaging with a local surface coil and spin echo SPI, *J. Magnetic Resonance*, **169**, pp. 174-186, 2004.

Satija, R., Jacobson, D.L., Arif, M. and Werner, S.A., *In situ* neutron imaging technique for evaluation of water management system in operating PEM fuel cell, *J. Power Sources*, **129**, pp. 238-245, 2004.

Tüber, K., Pccza, D. and Hebling, C., Visualization of water buildup in the cathode of a transparent PEM fuel cell, *J. Power Sources*, **124**, pp. 403-414, 2003.

Tushima, S., Teranishi, K. and Hirai, S., Magnetic Resonance Imaging of the Water Distribution within a Polymer Electrolyte Membrane in Fuel Cell, *Electrochemical and Solid-State Lett.*, **7**, No. 9, A269-A272, 2004.

Wang, C.Y. and Cheng, P., A multiphase mixture model for multiphase, multicomponent transport in capillary porous media–1. Model development, *Int. J. Heat and Mass Transfer*, **39**, pp. 3607-3618, 1996.

Wang, C.Y. and Cheng, P., Multiphase Flow and Heat Transfer in Porous Media, *Adv. Heat Transfer*, **30**, pp. 93-196, 1997.

Wang, Z.H., Wang, C.Y. and Chen, K.S., Two-phase flow and transport in the air cathode of proton exchange membrane fuel cell, *J. Power Sources*, **94**, pp. 40-50, 2001.

You, L. and Liu, H., A two-phase flow and transport model for the cathode of PEM fuel cell, *Int. J. Heat Mass Transfer*, **45**, pp. 2277-2287, 2002.

Yuan, J., Sunden, B., Hou, M. and Zhang, H., Three-dimensional analysis of two-phase flow and its effect on the cell performance, *Numerical Heat Trans*. Part A, **46**, pp. 669-694, 2004.

Recent Trends in Fuel Cell Science and Technology
Edited by S. Basu
Anamaya Publishers, New Delhi, India

6. Micro Fuel Cells

S. Venugopalan

Battery Division, Power Systems Group, ISRO Satellite Centre, Bangalore, India

1. Introduction

The advent of microelectromechanical systems (MEMS), micromachines, microsystems, integrated passive components and low power electronics for various types of functionalities has resulted in a surge of research in the development of power sources of matching configuration to meet the device requirements (Chan et al., 2005; Yen et al., 2003). The increasing demand for a wide variety of commercial, portable, electronic appliances such as cellular phones, laptop computers, personal organizers, digital cameras, portable radios, notebook computers, personal digital assistants, embedded monitors, military devices and specialised devices such as autonomous sensors, clinical and diagnostic test devices, micro-analytical systems, global positioning systems etc., has spurred the development of micro power sources (Lu et al., 2004; Choban et al., 2004). Since energy is stored in a fuel cell as a reservoir of fuel rather than as an integral part of the power source, fuel cells are expected to provide higher total energy for a given size or weight than batteries, calculated as Watt-hours per liter (Wh/l) or Watt-hours per kilogram (Wh/kg) due to the high amount of energy that can be stored in fuels such as hydrogen, sodium borohydride, methanol, ethanol, hydrocarbon. A second advantage of fuel cells is that once the fuel is consumed, it can be replenished instantly and the system continues to provide power. In fact, a fuel cell will continue to supply power indefinitely as long as fuel and air (O_2) are supplied. As a result along with the evolution of various electronic devices, the miniature fuel cells have emerged as a possible power source through the endless pursuit of ever-higher levels of performance, density and functionality.

The fuel cell industry is experiencing a very dynamic development especially in the area of micro fuel cells. In parallel with the drive for improved fuel cells for electric vehicle of macroscopic dimensions, the high speed integrated circuits, require incorporation of fuel cells of microscopic dimensions at nodal points within the circuit in order to minimise the power losses associated with power distribution along interconnects of ever decreasing dimensions. Such fuel cells would be based on micro electronic (thin film solid state) technology. However, to date, microscale systems research has focused mostly on miniaturization of functional components (Choban et al., 2004). Polymer electrolyte membrane fuel cell (PEMFC), direct methanol fuel cell (DMFC) and solid oxide fuel cell (SOFC) are all witnessing a rapid design evolution towards miniaturization to diminish size, weight and complexity while improving performance and reducing cost (Sarkar, 2005; Blakely, 2005; Goto, 2005; O'Hayre et al., 2003; Shimshon, 2005; Barton et al., 2005).

2. Direct Methanol Fuel Cell

The fuel cells are classified based on the fuel, operating temperature, electrolyte type, physical state of fuel cell components and the fabrication technology. The polymer electrolyte membrane fuel cell (PEMFC) operating on a methanol-water mixture as a fuel is called a Direct Methanol Fuel Cell (DMFC). DMFC uses sulphonated fluoropolymer, such as NAFION 117 as the electrolyte membrane. Nafion membranes require high levels of humidification and can operate comfortably only within a narrow temperature range of 25°C to 80°C. The leakage of methanol from the anode through the membrane to the cathode is called methanol crossover. However, recently developed hydrocarbon based DMFC membranes have substantially improved performance with respect to operating temperature range, humidity level requirement and methanol cross over rate when compared to the fluorocarbon-based Nafion membrane. Due to the low operating temperature, it is necessary to include noble metals such as platinum (Pt) and ruthenium (Ru) as electrocatalysts in the electrodes. Methanol is fed either as a liquid or a gas, depending on operating conditions and application. One drawback of the DMFC is the more complicated and much slower anodic oxidation of methanol compared to hydrogen. Ruthenium is added to the anode as a binary catalyst to prevent carbon monoxide poisoning during the reaction. The electrochemical reduction reaction of oxygen and the oxidation of methanol are:

Anode oxidation of methanol: $\qquad CH_3OH + H_2O \rightarrow CO_2 + 6H^+ + 6e^-$

Cathode reduction of oxygen: $\qquad 1\frac{1}{2}O_2 + 6H^+ + 6e^- \rightarrow 3H_2O$

Total DMFC reaction: $\qquad CH_3OH + 1\frac{1}{2}O_2 \rightarrow CO_2 + 2H_2O$

DMFC is similar to the PEMFC in that the electrolyte is a polymer and the charge carrier is the hydrogen ion (proton). However, the liquid methanol (CH_3OH) is oxidized in the presence of water at the anode generating CO_2, hydrogen ions and the electrons that travel through the external circuit as the electric output of the fuel cell. The hydrogen ions travel through the electrolyte and react with oxygen from the air and the electrons from the external circuit to form water at the anode completing the circuit.

Though the power density of the DMFC is lower than the PEMFC, easy storage of high energy density fuel and simple system design makes it very attractive as a substitute for rechargeable batteries in portable applications.

3. Micro Fuel Cell Fundamentals

Micro fuel cell (MFC) refers to small fuel cells that are designed to power portable electronics equipment, and in many cases can be integrated into the electronic device itself. A micro fuel cell is a portable source of energy that converts chemical energy of a fuel into useable electrical energy under ambient conditions. In general it generates power through the electrochemical oxidation of a fuel on a catalytic surface. The oxidant is generally air or oxygen. The fuel in a microfuel cell generally refers to methanol, a form of alcohol, a safe, non-combustible, inexpensive, renewable, plant based energy source. Micro fuel cells offer cheap and easy means to power electronic devices by simple refuelling, enabling continuous use when travelling with no access to the power grid.

Conventional fuel cell stack mainly comprises of: (a) membrane electrode assemblies (MEAs) for achieving the electrochemical energy conversion process, (b) bipolar plates for the supply of reactant (fuel and oxidant) gases to MEAs in addition to providing cell to cell electronic conduction path and removal of heat and (c) auxiliary components for the reactant supply and product removal. Table 1 provides some of the essential differences between DMFC and micro fuel cell.

In fact a fuel cell is a mini chemical plant with a number of pumps, valves and recirculation loops. Obviously it is hard to miniaturize it to a pocket size system. Miniaturizing fuel cells is not a simple matter of reducing physical dimensions as macro-sized (conventional) fuel cell components are limited by characteristic fabrication constraints (Cha et al., 2004, Lee et al., 2002). The materials and the manufacturing processes used in the conventional fuel cells have restrictions on dimensions and scalability. The machining of flow structures, for example, is constrained by the brittleness of graphite, and molding is limited in deep narrow

Table 1. Micro fuel cell vs. DMFC

Parameter	Conventional DMFC	Micro fuel cell
1. Power level Power density	Few watts to few kW 100 mW.cm^{-2} to 500 mW.cm^{2}	Few mW to few watts 10 mW.cm^{-2} to 50 mW.cm^{-2}
2. Components of the cell	The individual components of the stack — electrolyte membranes, gas diffusion layers, bipolar plates, gaskets end plates etc., are separately manufactured and assembled into the stack.	The components of the cell such as current collectors, electrolyte, gas diffusion layers, catalyst layers etc., are deposited using thick film technology
3. Manufacturing technology	Cutting, molding, fastening, etc.	Microelectronic manufacturing processes such as photolithography, focused ion beam (FIB) etching/deposition, deep reactive ion etching, physical vapor deposition, chemical vapor deposition, electron beam evaporation, spin coating, sputtering and electroplating etc.
4. Fuel: Methanol concentration	Approximately 4% with Nafion membrane electrolyte	Cells are being designed to work with 100% methanol
5. Electrolyte Membrane	Nafion 117	Membraneless cells utilizing characteristics of fluid flow at microscale are under development (microfluidics)
6. Peripherals/auxiliaries	Generally requires water management system, forced air system, fuel delivery/regulating system, thermal management system etc.	Devices without peripheral units such as pumps, valves etc., that are self activated by electrochemical reactions are being designed

channels. Rather, new designs, materials and manufacturing approaches must be employed. Micromachining processes has opened an avenue for developing fuel cells in small dimensions and for observing scaling effects.

Furthermore, as the raw size of an engineering product becomes smaller, it becomes more essential to manufacture units in parallel or continuous processes considering economical aspects. Therefore, micro electronic manufacturing processes such as photolithography, focused ion beam (FIB) etching/deposition, deep reactive ion etching, physical vapor deposition, chemical vapor deposition, electron beam evaporation, spin coating, sputtering and electroplating may be more appropriate than cutting, molding, fastening, etc. The production of integrated-circuit chips is a particularly strong example that takes advantage of parallel fabrication (Lee et al., 2000)

3.1 Microfabrication Technology

Microfabrication is a series of thin film material application (deposition) and removal (etch) steps, that are used to create mechanical or electronic structures in and/or on a substrate. Photolithography is the method used to transfer the two dimensional pattern from the master layout, which defines the required surface pattern of material. Photoresist, a photosensitive polymer, is used as an etch or deposition mask. First a thin film material is deposited uniformly over the entire substrate. Then, photoresist is applied by spin coating. The cured photoresist is then exposed to UV light through a photomask. After the latent image is developed, the polymer is thermally cured to harden. Next, the photoresist masked aluminum film is exposed to the etchant. After etching of the unprotected aluminum, the transfer of the pattern from the photomask (the master) to the thin film has been accomplished, and the photoresist is removed. This process is repeated for each layer of material that comprises the structure. Therefore, when finished, each individual structure on the substrate is made from the same set of thin film materials, deposited in the same order, but not all structures need have all layers and each layer of each structure can have a different shape.

Physical Vapor Deposition (PVD) Coatings: PVD coatings involve atom-by-atom, molecule-by-molecule, or ion deposition of various materials on solid substrates in vacuum systems.

Chemical Vapor Deposition (CVD): It is a chemical reaction which transforms gaseous molecules, called precursor, into a solid material, in the form of thin film or powder, on the surface of a substrate. Conventional CVD coating process requires a metal compound that will volatilize at a fairly low temperature and decompose to a metal when it contacts a substrate at higher temperature. Contrasting to the PVD coating in the "line of sight", the CVD can coat all surfaces of the substrate.

Thermal evaporation: It uses the atomic cloud formed by the evaporation of the coating metal in a vacuum environment to coat all the surfaces in the line of sight between the substrate and the target (source).

Sputtering: This is a PVD method involving the removal of material from a solid cathode. This is accomplished by bombarding the cathode with positive ions emitted from a rare gas discharge. When ions with high kinetic energy are incident on the cathode, the subsequent collisions sputter atoms from the material. The process of transferring momentum from impacting ions to surface atoms forms the basis of sputter coating. The film deposited by this method is uniform in thickness and capable of covering areas usually shadowed by other deposition methods.

The sputter deposition system consists (in a general case) of a vacuum chamber, sputter sources, a substrate holder and a pumping system. The control panel selects the target source and substrate position and controls the sputtering power. The vacuum system contains a mechanical pump, a turbo pump, throttle valve, and chamber. The separate RF power supply is remote. It supplies the RF power needed to generate plasma in the deposition chamber.

Sputtering applies high-technology coatings such as ceramics, metal alloys, organic and inorganic compounds by connecting the work piece and the substance to a high-voltage DC power supply in an argon vacuum system (10^{-2}-10^{-3} mmHg). The plasma is established between the substrate (work piece) and the target (donor) and transposes the sputtered off target atoms to the surface of the substrate. When the substrate is non-conductive, e.g., polymer, a radio-frequency (RF) sputtering is used instead.

Plasma Enhanced Chemical Vapor Deposition (PECVD): When energy is applied to a solid, it becomes a liquid. Apply more energy to a liquid and it becomes a gas. If further energy is applied to a gas, it then becomes a plasma. When used for surface treatment and critical cleaning applications, ions and electrons in the plasma react with the surface of materials placed within the plasma chamber. The result is a complete removal of organic contamination. On polymers, a permanent chemical modification of the surface reactive chemical functionalities may be imparted to the surface resulting in a dramatic increase in bond strength and other properties, without affecting the bulk properties of the material.

Microfluidics: It is the science of designing, manufacturing, and formulating devices and processes that deal with volumes of fluid of the order of nanoliters. Microfluidics hardware requires construction and design that differs from macroscale hardware. It is not generally possible to scale conventional devices down and then expect them to work in microfluidics applications. When the dimensions of a device or system reach a certain size as the scale becomes smaller, the particles of fluid, or particles suspended in the fluid, become comparable in size with the apparatus itself. This dramatically alters system behaviour. Capillary action changes the way in which fluids pass through microscale-diameter tubes, as compared with macroscale channels. In addition, there are unknown factors involved, especially concerning microscale heat transfer and mass transfer, the nature of which only further research can reveal.

4. Recent Developments in Micro Fuel Cell

For the realisation of micro fuel cell, a combination of one or more of the following solutions have been tried by commercial industries and academic institutions involved in the active research and development activity.

1. Substrate material compatible with micro electronic fabrication: Silicon wafer, Glass, ceramic, glass epoxy laminates (PCB), organic polymers as substrate materials instead of traditional high density non-porous graphite material (Chan et al., 2005; Yen et al., 2003; Lu et al., 2004; Apanel et al., 2004; Wozniak et al., 2004; Choban et al., 2005; Chachuat et al., 2004; Meyers et al., 2002; Kelley et al., 2002; Min et al., 2002; Savinell et al., 2000).
2. New electrolyte formulation: Non-fluorinated ionomer membranes such as phosphazene, arylene, poly(benzimidazole) or PBI, hydrocarbon polymers, nanoporous silicon membranes to serve as proton conductors (Apanel et al., 2004; Chu et al., 2005).
3. Microfluidic fuel cell: liquid-liquid interface-elimination of polymer electrolyte membrane (Choban et al., 2004; Chu et al., 2005; Cohen et al., 2005).
4. Micro-scale fabrication technologies (Chan et al., 2005; Yen et al., 2003; Kelley et al., 2000; Press Release, 2004; Wainright et al., 2003).
5. Elimination of plumbing (Press Release, 2004, Wainright et al., 2003).
6. Enhanced catalytic activity for the desired electrode reactions—use of carbon nanohorn (Yoshimi, 2004).
7. Ceramic based fuel reformer, micro reactors (Masaharu, 2004; Pattekar et al., 2004).
8. Selection of fuel (Apanel et al., 2004).

The manufacturing process/technology for the microfuel cells is being evolved. Laboratory demonstration models and prototypes have been built by several industries. However, the technology is not optimised for simultaneous achievement of energy density, power density and durability required for a commercial product. Microfluidics is the science of designing, manufacturing, and formulating devices and processes that deal with volumes of fluid of the order of nanoliters. Microfluidics hardware requires construction and design that differs from macroscale hardware. It is not generally possible to scale conventional devices down and then expect them to work in microfluidics applications. When the dimensions of a device or system reach a certain size as the scale becomes smaller, the particles of fluid, or particles suspended in the fluid, become comparable in size with the apparatus itself. This dramatically alters system behavior. Capillary action changes the way in which fluids pass through microscale-diameter tubes, as compared with macroscale channels. In addition, there are unknown factors involved, especially concerning microscale heat transfer and mass transfer, the nature of which only further research can reveal. Typical *V/I* characteristics of a micro fuel cell is shown in Fig. 1.

A representative process for the substrate and electrode preparation is given in Fig. 2. The exploded view of the micro fuel cell assembly developed by Manhattan Scientific Inc., is shown in Fig. 3.

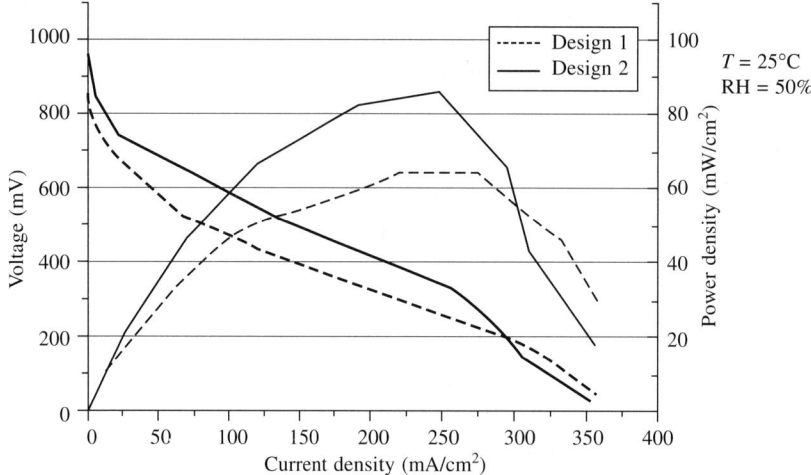

Fig. 1 *V/I* characteristics of a typical micro fuel cell.

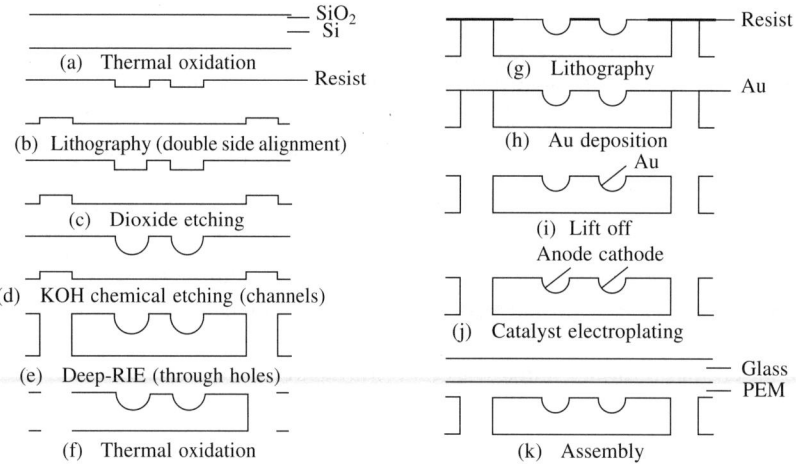

Fig. 2 Micro fabrication process (Osaka et al., 2005).

The schematic cross sectional view of the direct methanol micro fuel cell developed by Manhattan Scientific Inc., is shown in Fig. 4.

The exploded view of the cylindrical micro fuel cell is given in Fig. 5.

An exploded view of the micro fuel cell manufactured by Mechanical Technology Incorporation, using micro electronic fabrication technology, under the trade name of Mobion is shown in Fig. 6.

5. Status of Development

Fuel cells have been used in portable applications since the 1960s (Cropper, 2002). However, progress of development work was insignificant until 1995. Now increasing number of organizations are turning their attention to the development of portable micro fuel cells. Most portable fuel cell systems built to date have used proton exchange membrane fuel cell (PEMFC) technology. The ease with which methanol can be handled, stored and distributed, makes DMFC attractive for consumer electronic applications.

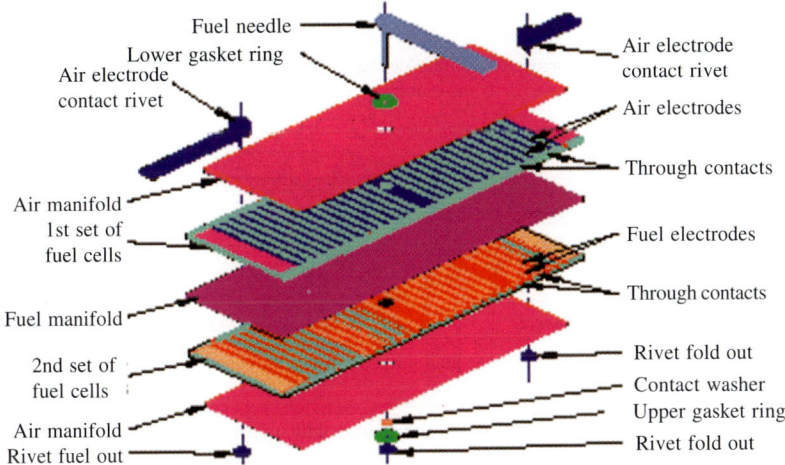

Fig. 3 Exploded view of the micro fuel cell assembly.

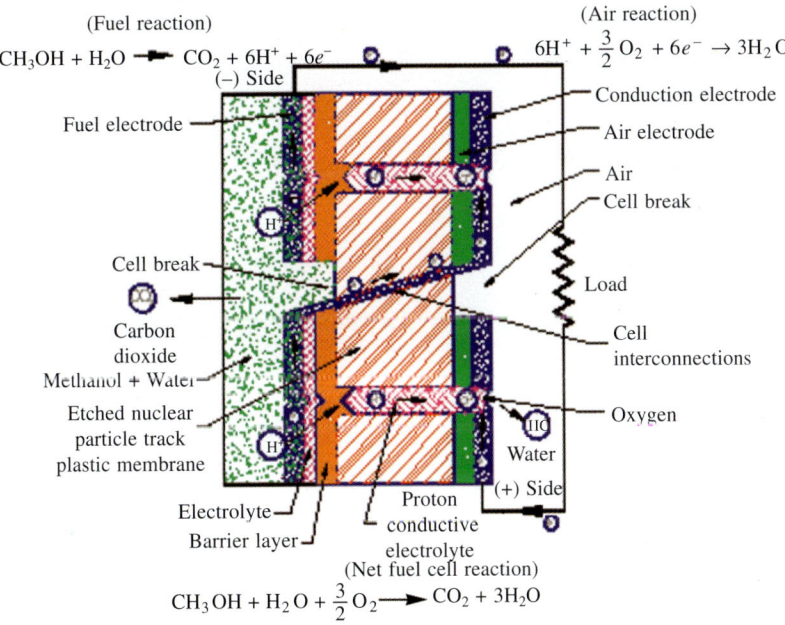

Fig. 4 Schematic of cross sectional view of the DMFC micro fuel™ cell developed by Manhattan Scientific Inc.

The development of portable fuel cells had been underpinned by US space and military programs. Though good number of companies have brought out their prototype and demonstration model, commercial production for consumer electronic application is way off due to high cost and bulkiness of the state of the art product. Details of microfuel cells developed by some of the organizations are described as follows.

Manhattan Scientific (MSI) is commercialising the micro fuel cell developed by Bob Hockaday. Here disposable ampoules are used to supply either methanol or H_2 gas generated by sodium borohydride/water

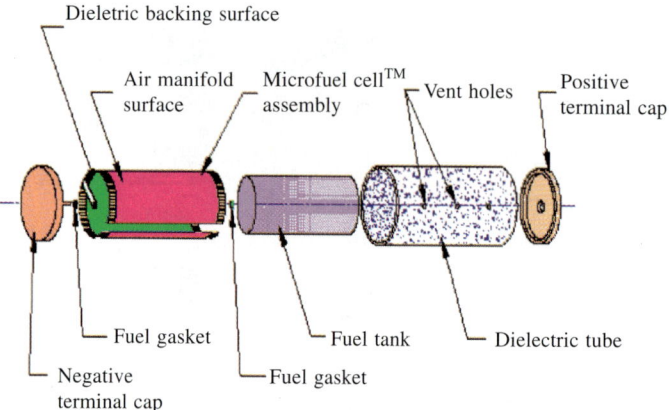

Fig. 5 Exploded view of the cylindrical micro fuel cell™.

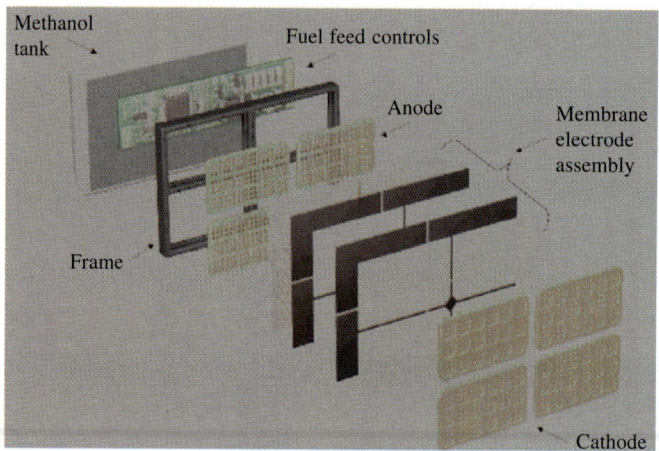

Fig. 6 Exploded view of MTI's Mobion fuel cell.

reaction. MSI's micro fuel cell array unit consisting of 10 cells in 120 mm × 46 mm × 0.5 mm package generates approximately 1 W of electric power. The cells are formed by paper thin plastic sheets that can be mass produced by roll to roll technology.

The fuel cell developed by Medis Technologies Ltd., has a very simple design and architecture, consisting of an anode, a cathode, a chamber for the liquid electrolyte and a fuel chamber and does not require reformer, heat control system, water management system, forced air system, fuel regulating and delivery system etc. Non-noble metals are used as catalysts for the reduction of oxygen at the cathode. A non-flammable and non-toxic alkaline solution of borohydride combined with a mixture of alcohols is used as the liquid fuel. Fig. 7 shows the medis fuel cell power pack that measures 80 × 55 × 30 mm, weighing 200 g when filled with fuel, designed to operate and charge advanced portable devices such as cell phones digital cameras, PDAs, MP3 players, hand-held video games etc.

A simplified, effectively packed device integrated 30 Wh DMFC power source has been developed at MTI Microfuel Cells under the trade-named Mobion that allows direct feed by 100% methanol fuel without any need of water collection and for recirculation. The Mobion™ technology architecture uses a proprietary

Fig. 7 Medis fuel cell pack powering a cell phone and digital camera.

approach to manage water that is produced at the cathode to flow to the anode, internal to the fuel cell without the need for a complex "micro-plumbing." Mobions technology allows ease of manufacture, provides more room for the fuel, permits use of concentrated fuel, has compact size, gives excellent opportunity for the adoption of DMFC technology by hand-held electronics. Fig. 8 shows the RFID tag reader manufactured by Intermec Inc., integrated with Mobion micro fuel cell. Fig. 9 shows MTI Micro's Mobion™ technology prototype powering a PDA.

Fig. 8 RFID tag reader manufactured by Intermec Inc.

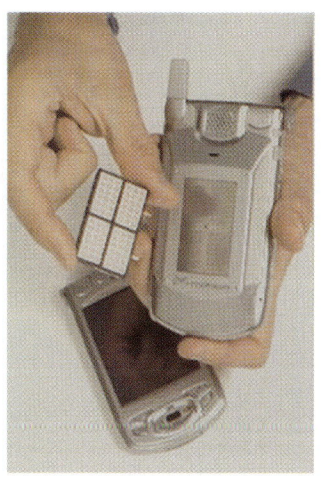

Fig. 9 MTI Micro's Mobion™ technology prototype powering a PDA.

Fujitsu has developed a MEA based on an aromatic hydrocarbon solid electrolyte material, coated with a high density of highly active platinum-based nano-particle catalyst, having less than one-tenth of the methanol crossover rate encountered with typical fluorinated polymers. Fig. 10 shows Fujitsu's 15 W micro fuel cell based on a new hydrocarbon solid electrolyte material that enables use of 30% methanol powering a note book PC. The basic specification of a 3.78 W prototype micro fuel cell is given in Table 2.

Casio (Japan) has developed a prototype methanol reformate PEM fuel cell to power notebook computers and PDAs. It could power a Casio notebook computer (the Cassiopeia FIVA) for more than twenty hours, four times longer than a lithium ion battery. Casio's fuel cell uses a proprietary methanol reformer, which

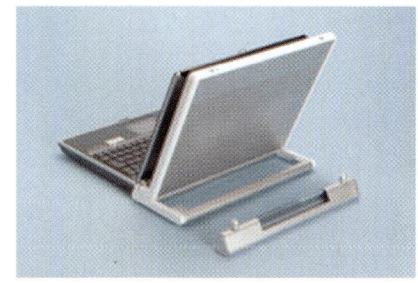

Fig. 10 Fujitsu's 15 W micro fuel cell powering a notebook PC.

Table 2. Main specifications of prototype micro fuel cell

Height × width × thickness	152 × 57 × 16 mm
Size (volume)	180 cc
Weight	190 grams
Fuel	Methanol (18 cc, 30% concentration)
Number of Recharges	1 per methanol cartridge
Design	Desktop holder
Output	5.4 V 700 mA (same as other FOMA rechargers)

is a highly efficient microreactor of the size of a postage stamp. The reformer is made by depositing a catalyst (specially developed in conjunction with Kagakuin University) on an etched thin silicon wafer, using Casio's own semiconductor micro fabrication techniques. Figs 11, 12 and 13 show laptop with micro fuel cell and reformer, details of the fuel cell and reformer, respectively.

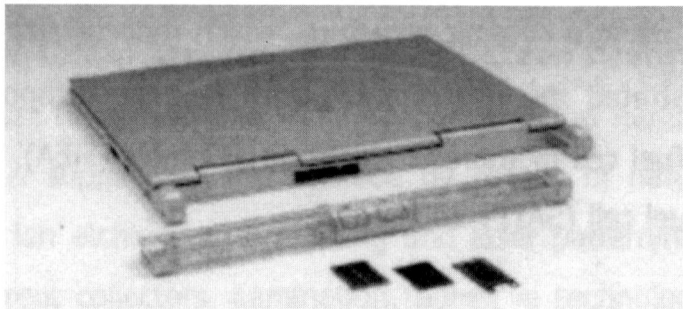

Fig. 11 Casio's micro fuel cell with reformer.

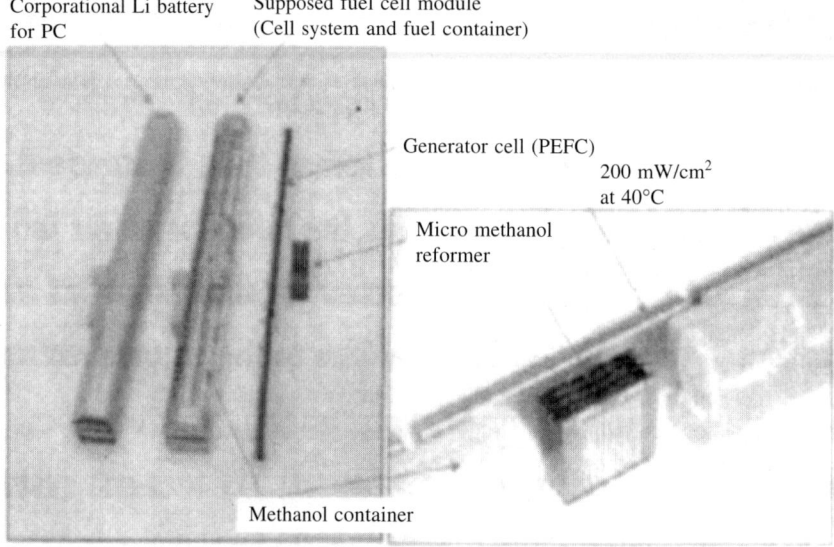

Fig. 12 Details of Casio's micro fuel cell.

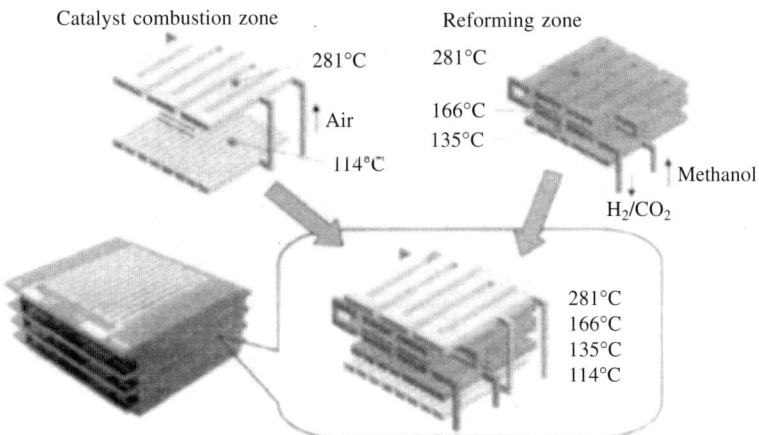

Catalyst combustion zone

Reforming zone

Fig. 13 Reformer for fuel cell.

Polyfuel (USA), a spin-off of SRI International, is developing direct methanol fuel cells to replace Lithium ion batteries in wireless, handheld and portable devices, based on patented, proprietary technology. PolyFuel's membranes are based on hydrocarbon polymers, rather than perfluorinated and are considered to be best-in-class for portable direct methanol fuel cells (DMFC) designed for portable electronic devices such as laptops, PDAs or cell phones.

Motorola Labs (USA) is developing fuel cells to power a range of portable electronic devices, including cell phones, two way radios, PDAs, and notebook computers. It has developed unique direct methanol fuel cell technology using a ceramic structure, which reduces size and cost. Motorola has now successfully demonstrated the use of multi-layer ceramic technology for processing and delivering fuel and air to the fuel cell membrane electrode assembly (MEA). This fuel delivery system can be built into a miniature fuel cell (SAITT 2004).

The prototype, combines fuel mixing and microchannels for delivery, substrate for MEA mounting, and electrical contact in just two ceramic pieces. The lower ceramic piece handles the liquid fuel processing while the upper piece provides for passive air delivery (air-breathing). The MEA is sandwiched between the two ceramic layers, making for simple assembly.

This ceramic technology also simplifies the interconnection of multiple fuel cells. In this implementation, they are arranged in a planar layout rather than a standard vertical stack. This simplifies the design of the fuel cell system and eliminates the need for an air fan or pump since all of the fuel cells are exposed to air. Several cells are connected together in series electrically to increase the output voltage of the system. This simplifies the interface to the actual electronic system.

The system, highlights Motorola's expertise in miniaturization, microfluidics, energy conversion systems, electronic packaging and interface. Making a practical 1 W fuel cell system (Fig. 14) requires not only a small-sized fuel cell stack, but also miniature, low-power peripheral components.

Energy Visions (Canada) is developing direct methanol fuel cell technology in conjunction with Alberta Research Council Inc. The production of prototype systems (including 20 W units for the US and Canadian military) has been put on hold while its basic fuel cell has been reconfigured to incorporate a new electrode design which offers greater power and efficiency at a lower cost. Next step involves the creation of a hybrid portable power system using DMFCs in conjunction with nickel zinc batteries.

The Fraunhofer Initiative (Germany) was created in 2000 by five institutes to develop innovative energy systems based on portable, miniature fuel cells. Partners include the Fraunhofer Institute, where several prototypes have been developed and demonstrated, including fuel cell powered notebook computers at the

Fig. 14 The 1 W DMFC system inside modified Motorola IMPRES charge.

Hannover Fair in April 2000. It is developed in conjunction with LG Caltex Oil (Korea) and the Korean-American joint venture Clean Energy Technologies Inc. (CETI). The system consists of just three foils and is independent of the number of individual planar cells connected in series. Air is supplied to the cathode by means of natural convection. Wafer level technologies like reactive ion etching, electroplating and laser patterning has been used to develop micro flow fields and current collectors. Lamination, adhesive technologies and screen printing are used for assembly and interconnection of the cells. A demonstrator of a planar PEM fuel cell, as shown in Fig. 15 with an active area of $0.54\ cm^2$ was developed, which consisted of three serial interconnected single cells. Serial interconnection of individual cells is attained with gang bonding technologies. Fine distribution of gases is achieved by a micro patterned flow field which avoids extra gas diffusion layers. Stable operation at a power density of $100\ mW/cm^2$ has been proven.

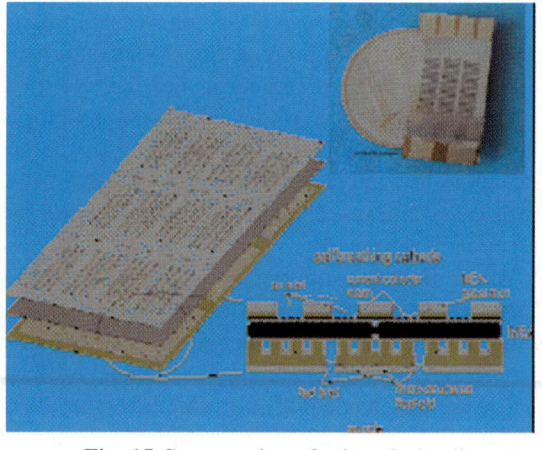

Fig. 15 Cross section of micro fuel cell.

The cross-section of the micro fuel cell is shown in Fig. 15. The performance characteristics of cells is shown in Fig. 16. A dispensing technique or screen printing is used for integrated planar sealing. The system is mechanically flexible for easy integration into the surface of electronic devices and low volume consumption. Fig. 17 provides the fuel cell modules produced by roll to roll process.

Wafer processing reduces costs and guarantees a high quality standard. The demonstrator cell consists of three individual cells which are serial interconnected. The anodic micro flow field has been patterned with dimensions down to 10 μm. It is powered by a hydrogen fueled PEMFC. The Fraunhofer Institute is also working on fuel cell powered camcorders, a 50 W external power unit, and a very flat fuel cell in the sub-watt range. In addition, it is working on technology to mass produce small fuel cells. It exhibited a prototype of an automated assembly line at the 2002 Hannover Fair, the first in the world designed to assemble miniature fuel cells.

H-tec (Germany), a leading supplier of small PEM fuel cells for use in education and training, is diversifying into the development and production of fuel cells for industrial applications in the range 0.5 to 100 W. These could be used in a range of applications, including powering computers, cell phones and power tools.

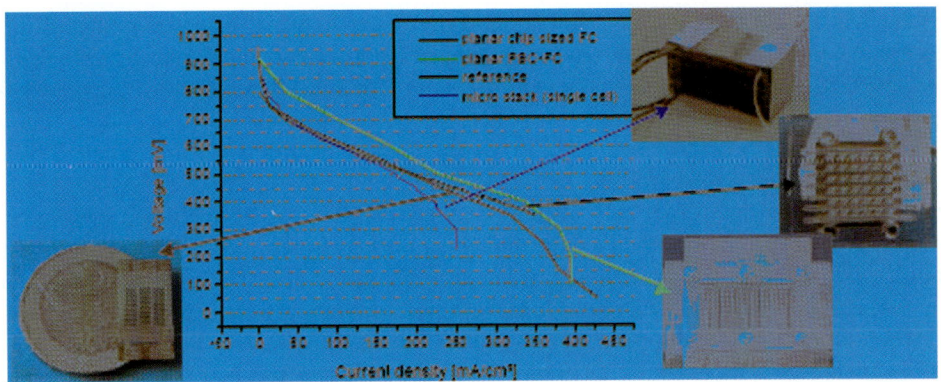

Fig. 16 Performance characteristics of cells.

Patterning of micro flow field

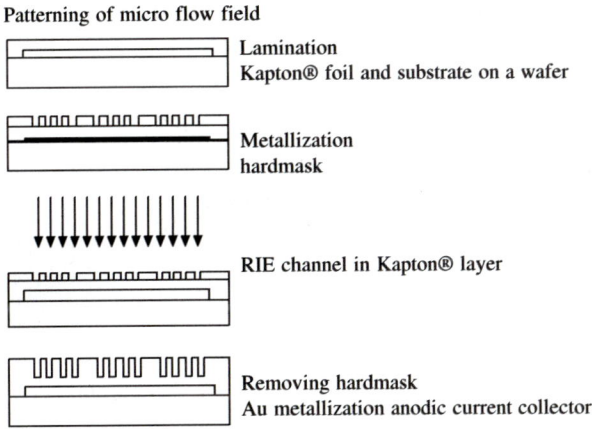

Lamination
Kapton® foil and substrate on a wafer

Metallization
hardmask

RIE channel in Kapton® layer

Removing hardmask
Au metallization anodic current collector

Fig. 17 Fuel cell modules produced by roll-to-roll process.

NEC (Japan) has developed miniature direct methanol fuel cells using electrodes made from a type of carbon nanotube called a nanohorn. Several prototypes have been built to power mobile devices. The miniature energy source has an energy capacity ten times that of current lithium ion batteries, and it generates 20% more current than conventional fuel cells based on activated carbon. By using carbon nanohorns (CNHs), a type of carbon nanotube (CNT), having a complex surface morphology to support fine platinum catalyst dispersion for fuel cells, NEC has succeeded in the development of high-performance fuel cells with the world's highest power density of 70 mW/cm². In addition, NEC was the world's first to demonstrate laptop PC operations using micro fuel cells as the main power supply (Apanel et al., 2004). The laptop PC powered by NEC micro fuel cell is given in Fig. 18.

Samsung Advanced Institute of Technology (SAIT)

Fig. 18 Laptop PC powered by micro fuel.

(Korea) is the R & D centre for the Samsung group. It has been active in the development of fuel cells to power electronic devices for some time, having developed a notebook computer powered by a 40 W PEM fuel cell in 1999. It is active in the development of membranes and MEAs for direct methanol fuel cells, as well as complete systems. In May 2002 it revealed that it had developed a mobile phone with a built in direct methanol fuel cell which is roughly the size of a credit card (6 × 8 × 1 cm). Fuelled by methanol stored in a container like an ink cartridge, the unit has an output of 2 W (SAITT, 2004).

For the realization of micro fuel cell, Toshiba has developed a system including new materials that allows a higher concentration of methanol to be diluted by the water produced as a by-product of the power generation process. This technology allows methanol to be stored at a much higher concentration, and achieves a fuel tank less than 1/10 the size of that required for storing the same volume of methanol in a 3 to 6% concentration. The current prototype can operate for approximately five hours on 50 cc of high concentration methanol. The essential technologies incorporated in the fuel cell include interface and electric circuits to assure efficient control of power supply, sensors to monitor methanol concentration and liquid level, and a remaining quantity sensor to tell users when they need to change the methanol fuel cartridge. All these components, and low power liquid and air transmission pumps, are controlled by a super small DC-DC converter (SFC, 2004). Fig. 19 gives the micro fuel fuel cell developed by Toshiba powering a laptop. The specification of Toshiba's micro fuel cell is given in Table 3.

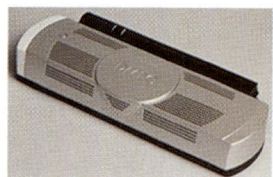  

Fig. 19 Toshiba's micro fuel cell powering a laptop.

Table 3. Specifications of Toshiba micro fuel cell

Product	Methanol fuel cell directly connected to the PC
Output	Average 12 W
	Maximum 20 W
Voltage	11 V
Size	275 × 75 × 40 mm (825 cc)
Weight	900 g
Operating hours	Approximately five hours with 50 cc, and 10 h with 100 cc, of high concentration methanol fuel
Cartridge weight	120 g (100 cc), 72 g (50 cc) (Approximate)
Cartridge size	100 cc: 50 × 65 × 35 mm 50cc: 33 × 65 × 35 mm
Fuel	Methanol

Hitachi and Tokai develop direct methanol fuel cell (DMFC) prototype for use in handheld electronic devices. The DMFC has a methanol concentration of approximately 20 percent, which is expected to be increased by 10 percent once the units are commercially available (www. hitachi.com).

Neah Power Systems has replaced the typical thin polymer membranes with stacked, porous silicon layers. This could improve performance and make the fuel cells easier to manufacture via microelectronic fabrication techniques. In a fuel cell, the number of electrons produced is directly related to the membrane's surface area. Therefore, increasing power requires expanding the size of the polymer membrane and thus the fuel cell itself, which causes problem to systems designed for small devices. However, by stacking porous silicon layers, depending on the number of layers the surface area can be increased to the desired level keeping the overall size of the fuel cell same. Atmospheric air contains only 20% O_2. This can be enhanced to 90% by using 30 percent hydrogen peroxide solution, rather air from the atmosphere (Apanel et al., 2004).

Companies like Poly Fuel, Neah Power Systems, Medis Technologies, MTI MicroFuel Cells, Alberta Research Council, Toshiba, Hitachi, NEC, CASIO, AIST and Honda Technical Research Laboratory are exploring innovative ways for reduction in component count, elimination of auxiliaries such as pumps and valves, alternatives to membranes and packaging technologies to realize an elegant, simpler micro fuel cell product. Hitachi, Toshiba, NEC and NTT DoCoMo have all announced plans to sell methanol-powered devices (Bostaph et al., 2002, Dillon et al., 2004).

Researchers at the University of Illinois, Urbana-Champaign, (Mitrovski et al., 2004; Paul Kenis et al., 2005; Choban et al., 2002; Choban et al., 2003; Jayashree, 2004); Lawrence Livermore National Laboratory (LNNL); Morse et al., 2002; Upadhye, 2003; Brown University, Palmore et al., 2002, Cornell Nanoscale Facility, Cornell University, Cohen et al., 2005; Rapid Prototyping Laboratory, Stanford University, Lee et al., 2000, 2002, 2003, Centre for Electrochemical Sciences, Case Western Reserve University, Wainright et al., 2003) have developed and demonstrated a laboratory prototype miniature thin-film fuel cell power source, which provides portable electrical power for a range of consumer electronics. The miniature fuel cell technology incorporates a thin film fuel cell and microfluidic fuel processing components integrated into a common package.

LLNL's miniature fuel cell product incorporates integrated microfluidic fuel distribution architecture within a miniature fuel cell package. This feature enables the fuel cell to operate from highly concentrated methanol supplied from a replaceable fuel cartridge.

The heart of this miniature power source utilizes a thin layer of electrolyte material sandwiched between electrode materials containing appropriately proportioned catalyst materials. Microfluidic control elements distribute methanol through a silicon chip over one electrode surface while air is simultaneously distributed over the other electrode. Catalytically generated protons from the fuel supply cross the electrolyte to the air breathing electrode and combine with oxygen to generate electrical current. Optimization of current output through control of the catalyst and electrode surface area, and microfluidic fuel distribution, offer a miniature energy source providing continuous power for greater than three times longer than existing rechargeable batteries.

In PEM fuel cells, the polymer electrolyte membrane (PEM) serves as a barrier to prevent the fuel and oxidizer from mixing without generating electricity. Ideally, electrons and protons from a fuel are liberated at a catalyst-coated anode and travel via separate routes to a cathode where they react with an oxidizer. Protons migrate through an electrolyte medium—the membrane—while electrons travel through an external circuit to provide electrical power.

Much of the research on PEM cells focuses on improving the membrane to improve overall cell performance. For example, scientists are trying to find robust substitutes for the commonly used membranes that can tolerate operating temperatures higher than about 80°C. Raising the temperature would speed up reaction kinetics but may dehydrate the membrane, which would reduce proton conduction and cell performance. Fuel permeating through the membrane is another problem.

Kenis et al. (2004), are circumventing the problems with PEM, by devising a fuel cell that omits the membrane. By pumping aqueous fuel and oxidizer solutions through a micrometer-sized Y-shaped channel, the group induces laminar flow in which the reagents proceed along the length of the channel in parallel streams with no mixing except for a small amount of diffusion that is controlled by the channel dimensions and flow rates. The side view of the membraneless microfuel cell is given in Fig. 20.

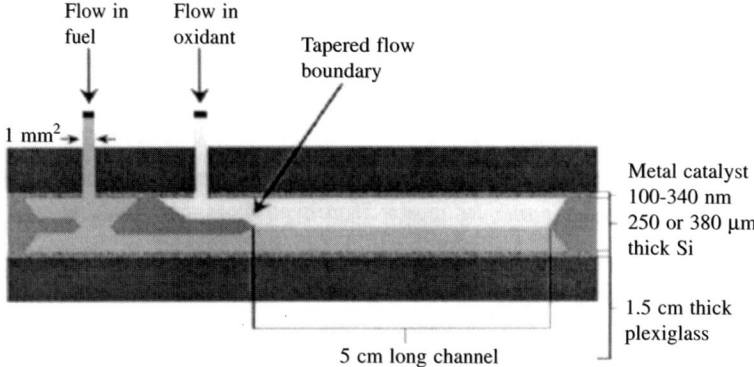

Fig. 20 Side view of the planar microfluidic membraneless micro fuel cell.

Chan et al. (2005), have realised micro fuel cells through an approach that combines thin film materials with MEMS (micro-electro-mechanical system) technology. The membrane electrode assembly was embedded in a polymeric substrate (PMMA) which was micromachined through laser ablation to form gas flow channels. The micro gas channels were sputtered with gold to serve as current collectors. This cell utilized the water generated by the reaction for the humidification of dry reactants $(H_2$ and $O_2)$. The peak power density achieved was 315 $mW \cdot cm^{-2}$ (901 $mA \cdot cm^{-2}$ at 0.35 V) for the H_2-O_2 system with 20 ml min^{-1} O_2 supply and H_2 at 10 psi in dead ended mode of operation. A Y shaped microfluidic channel is depicted in Fig. 21.

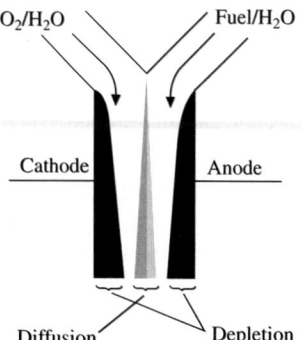

Fig. 21 The fuel cell system uses a Y-shaped microfluidic channel where two liquid streams containing fuel and oxidant merge and flow between catalyst-covered electrodes without mixing.

6. Potential Operational and Performance Issues of Micro Fuel Cells

Traditionally, fuel cells have been too big and expensive for smaller devices. Micro fuel cell faces lot of challenges, which researchers are trying to overcome. The performance of micro fuel cells should approach or exceed that of its macro counter part for it to be attractive. Micro fuel cell technologies, should preferably be self sustaining in its operation and shall be self regulating without the need for complex plumbing for fluid circulation for thermal and water management for it to be compact and autonomous.

One of the critical component of the fuel cell is proton conducting electrolyte membrane/polymer electrolyte membrane (PEM). The PEM is a complex polymer membrane, often intimately associated with the anode and cathode catalysts, creating a complex system that must be decreased in size without diminishing its effectiveness. With its inherent problem of dryout, tear, deterioration, fuel cross over, the thickness has to be reduced without compromising its performance. Fuel cells run more efficiently at higher temperatures due to faster kinetics, however, this results in dry out of membrane reducing the effectiveness of proton conduction due to associated water management issues. Fuel crossover through the membrane results in a mixed potential at the cathode and thereby lowers the cell performance (Choban et al., 2004).

The relatively sluggish kinetics for the anodic oxidation of methanol, the resistive drop across the membrane, formation of CO_2 bubble in the anode structure, flooding of cathode structure with water, dryout of the membrane etc., all result in poor high pulse power capability of the micro fuel cell (DMFC) for start-up of today's high-tech electronic devices, such as laptop computers and PDAs.

Since the design and manufacturing process parameters influence performance, a clear understanding of the design variables such as electrolyte thickness, electrode and flow channel geometry, catalyst particle size, shape and microstructure, which, in turn, control the length of the tri-phase boundaries are important for the electrical behavior and the efficiency of fuel cells. One of the challenges is the development of highly conductive, tough, dimensionally stable, thin electrolyte membrane which can be operated under fairly dry condition and yet provide long term stability under high operating temperatures, pressures and current density with high resistance to fuel cross over. Despite extensive efforts, these problems still remain major issues preventing PEM-type fuel cells from being applied in wide-scale portable applications (Apanel et al., 2004).

Some of the design issues involve the potential heat and noise/vibration generated by the fuel cell system. Since all fuel cell systems are less than 100% efficient in converting chemical (fuel) energy into electricity, the fuel cell will generate excess heat that must be removed from the system. When the fuel cell is operating at its maximum power density the efficiency falls to less than 50%. The heat output is expected to be in the range of 80% to 100% of its rated power and has to be removed efficiently to overcome its negative impact on system performance (Hahn et al., 2003). This can be partially accomplished by the oxidant/air supply and exhaust, but additional heat sinks/spreaders and cooling may be needed. Many of the fuel cell systems being considered are active systems that use pumps/fans to move fuel and air into and out of the fuel cell stack. These will generate some level of noise and possible vibration in the fuel cell system. A part of the generated power is consumed by the auxiliary system. System efficiency can be improved dramatically by the elimination of auxiliary system that needs power for its operation.

Since the electrochemical reaction, heat transfer etc. are surface phenomena significant improvements in power density and efficiency is expected to be achieved due to orders of magnitude increase in the surface to volume ratio in micro fuel cells. Correspondingly, however, new innovations are required to offset miniaturization penalties such as high pressure drops in small channels and high electrical resistance through thin-film current collectors/conductors.

One of the fundamental requirement of the fuel cell is uninterrupted power support and hence the micro fuel has to be designed preferably with a fuel reservoir that can be filled periodically or may be provided with a fuel cartridge that can be replaced periodically. A user friendly micro fuel cell system should have a fuel gauge that provides an estimate of the remaining capacity and alert the user regarding replacement of fuel cartridge. To be compliant with environmental pollution control act (EPA) the exhaust of the micro fuel cell system should not contain any appreciable levels of unconsumed fuel or other potentially flammable or toxic materials.

To be robust or rugged, the system design should take into account possible forced discharge/short circuit of the fuel cell power source. Though fuel cells will not be able to source very high peak currents similar to batteries, adequate precautions must be taken to limit the current drawn from the fuel cell through inclusion of short circuit protection. The system design shall include logic controls to sense an excessive load or short circuit condition that stops the flow of fuel and/or oxidant to effectively shutdown the fuel cell. Note also that since an air supply is generally required for operating and cooling the fuel cell, air inlet/outlets and fans/air pumps must not be blocked during operation of the fuel cell. This could cause a shutdown of the fuel cell due to loss of oxidant and/or overheating.

7. Conclusions and Recommendations

The multifunctional hand held devices with enhanced features demand more and more power for longer periods of operation between recharges. The striking feature of microfuel cell system, with its capability for independent sizing of power and energy capacity, is their inherent potential of higher energy densities compared to the best of the secondary battery systems available today. To fulfil the present and possible future requirements the microfuel cell system should be designed to have a simple architecture and should operate reliably at room temperature. There should be minimum amount of peripheral elements. The main challenge for the successful realisation of the microfuel cell is elimination of active components such as fans, pumps, compressors, humidifiers, coolers, heaters etc. The fabrication process should be compatible with other microelectronic, microelectromechanical or microfluidic devices.

Micro fuel cell system faces numerous challenges for its widespread application in mobile devices. The electrical interfaces and the size of the fuel cell system should be maintained same as that of batteries. Fuel cells lack size-related standards similar to the one available for batteries. There is no standard fuel cell product that can be used as a power pack for all brands of the same electronic device. Currently micro fuel cells are quite expensive due to the use of noble metal catalyst and tiny pumps, fans, valves etc. for the reactant supply, recirculation, thermal, water management and product handling. Air traffic regulation could prohibit use and transportation devices fitted with micro fuel cells due to the flammable nature of most of the fuels used in these devices. The peak power rating of many of the micro fuel cells are inferior to that of batteries resulting in their inability to meet some of the functional requirement of microelectronic devices.

No smart solution is available to store the gaseous hydrogen used in miniature fuel cell (PEMFC) and hence DMFC is receiving enormous interest due to system simplicity. However, specialised air-breathing DMFC components have to be developed. New materials have to be developed in addition to optimisation of structure and operating conditions to take care of performance decay modes. New membrane/electrode assemblies appropriate for the microscale to be developed exploiting the enhanced heat and mass transfer on the microscale for improved performance, and developing microfluidic components for micro fuel cells.

The micro fuel cell technology is progressing at such an astonishing rate that its commercial breakthrough is only a matter of time. It is anticipated that miniature fuel cells will be power source of choice for a wide range of consumer/portable electronic produces in the near future.

References

Apanel, G. and Johnson, E. (2004) Fuel Cells Bulletin, November 2004.

Barton, P.I., Mitsos, A. and Chachuat, B. (2005) Optimal start-up of micro power generation processes, Presented at the ESCAPE 15, Barcelona, Spain, 29 May-1 June 2005.

Blakely, K.A. (2005) The Revolution 50™: The First Commercial Portable Solid Oxide Fuel Cell, Presented at the 7th Annual SMALL FUEL CELLS (sm) 2005—Small Fuel Cells for Portable Applications, Washington, DC., April 27-29, 2005.

Bostaph, J., Xie, C.G., Pavio, J., Fisher, A.M., Mylan, B. and Hallmark, J. (2002) Design of a 1-W Direct Methanol Fuel Cell System, Proceedings of the 40th Power Sources Conference, June 10-13, 2002, 211-214.

Cha, S.W., Lee, S.J., Park, Y.I. and Prinz, F.B. (2003) Investigation of Transport Phenomena in Micro Flow Channels for Miniature Fuel Cells, Proceedings of 1st International Conference on Fuel Cell Science Engineering and Technology, American Society of Mechanical Engineers, Rochester, NY, April 2003.

Cha, S.W., O'Hayre, R. and Prinz, F.B. (2004) The influence of size scale on the performance of fuel cells, *Solid State Ionics*, **175**, 789-795.

Chachuat, B., Mitsos, A. and Barton, P.I. (2004) Optimal operation and design of micro power generation processes, Presented at the AICHE Annual Meeting, Austin, Texas, USA, November 2004.

Chan, S.H., Nguyen, T., Xia, Z. and Wu, Z. (2005) Development of a polymeric micro fuel cell containing laser-micromachined flow channels, *J. Micromech. Microeng.*, **15**, 231-236.

Choban, E.R., Markoski, L.J., Stoltzfus, J., Moore, J.S. and Kenis, P.A.J. (2002) Power Sources Proceedings, 40, 317-320.

Choban, E.R., Waszczuk, P., Markoski, L.J., Wieckowski, A. and Kenis, P.A.J. (2003) ASME Fuel Cell Science, Engineering and Technology Proceedings, 2003, 261-265.

Choban, E.R., Markoski, L.J., Wieckowski, A. and Kenis, P.A.J. (2004) Microfluidic fuel cell based on laminar flow, *J. Power Sources*, **128 (1)**, 54–60.

Choban, E.R., Royer, M., Waszczuk, P., Markoski, L.J., Wieckowski, A. and Kenis, P.A.J. (2005) Extended Abstracts, 207th meeting of the Electro Chemical Society, Quebec, Canada, May 15-20, 2005.

Chu, K., Gold, S., Ravi Subramanian, Shannon, M.A. and Masel, R.I. (2005) Extended Abstracts, 207th meeting of the Electro Chemical Society, Quebec, Canada, May 15-20, 2005.

Cohen, J.L., Volpe, D.J., Westly, D.A., Pechenick, A. and Abruna, H.D. (2005) Dual Electrolyte H_2/O_2 Planar Membraneless Microchannel Fuel Cell System with Open Circuit Potentials in Excess of 1.4 V, *J. Langmuir*, **21(8)**, 3544-3550.

Cohen, J.L., Westly, D.A., Pechenick, A. and Abruna, H.D. (2005) Fabrication and preliminary testing of a planar membraneless microchannel fuel cell, *J. Power Sources*, **139**, 96-105.

Cropper, M. (2002) Fuel Cell Market Survey: Portable Applications, Fuel Cell Today, 18 September 2002.

Dillon R., Srinivasan S., Aricò S.A. and Antonucci, V. (2004) International activities in DMFC R&D: Status of technologies and potential applications, *J. of Power Sources*, **127**, 112-126.

Goto, Y. (2005) Toshiba DMFC for Portable Applications, Presented at the 7th Annual SMALL FUEL CELLS (sm) 2005—Small Fuel Cells for Portable Applications, Washington, DC., April 27-29, 2005.

Hahn, R., Krumm, M. and Reichl, H. (2003) Thermal Management of Portable Micro Fuel Cell Stacks in 19th IEEE SEMI-THERM symposium, San Jose, CA, USA, March 11-13, 2003.

Jayashree, R.S., Yeom, J., Mozsgai, G.Z., Choban, E.R., Spendelow, J., Kenis, P.A.J. and Shannon, M.A. (2004) Palladium-Nanoparticles on Platinum-Black Catalysts Integrated into a Microfabricated Si-Based Micro-Fuel Cell, Solid-State Sensors and Actuators Workshop, Hilton Head Island, SC, June 2004, 266-269.

Kelley, S.C., Deluga, G.A. and Smyrl, W.H. (2002) Miniature fuel cells fabricated on silicon substrates. *American Institute of Chemical Engineers Journal*, **48(5)**, 1071-1082.

Kelley, S.C., Deluga, G.A. and Smyrl, W.H. (2000) A Miniature Methanol/Air Polymer Electrolyte Fuel Cell, Electrochem. *Solid-State Letters*, 3(9), 407-409.

Lee, S.J., Cha, S.W., Liu, Y.C., O'Hayre, R. and Prinz, F.B. (2000) High Power-Density Polymer-Electrolyte Fuel Cells by Microfabrication, Micro Power Sources, K. Zaghib and S. Surampudi (eds.), Proceedings Volume 2000-3, The Electrochemical Society Proceeding Series, Pennington, NJ, 2000.

Lee, S.J., Chang-Chien, A., Cha, S.W., O'Hayre, R., Park, Y.I., Saito, Y. and Prinz, F.B. (2002) Design and fabrication of a micro fuel cell array with "flip-flop" interconnection, *J. Power Sources*, **112(2)**, 410-418.

Lee, S.J., Cha, S.W., Liu, Y.C., O'Hayre, R. and Prinz, F.B. (2000) Micro-Power Sources, H.Z. Massoud, I. Baumvol, M. Hirose, E.H. Poindexter, Editors, PV 2000-3, The Electrochemical Society Proceeding Series, Pennington, NJ (2000).

Lee, S.J., Cha, S.W., O'Hayre, R., Chang-Chien, A. and Prinz, F.B. (2000) Miniature Fuel Cells with Non-Planar Interface by Microfabrication, Power Sources for the New Millenium, M. Jain, M.A. Ryan, S. Surampudi, R.A. Marsh, and G. Nagarajan (eds.), Proceedings Volume 2000-22, The Electrochemical Society Proceeding Series, Pennington, NJ, 2000.

Lu, G.Q., Wang, C.Y., Yen, T.J. and Zhang, X. (2004) Development and characterization of a silicon-based micro direct methanol fuel cell, *Electrochimica Acta*, **49**, 821–828.

Masaharu, S. (2004) ATIP SCOOP, Technology: Micro Fuel Cell, Energy, Power, 27 September 2004.

Meyers, J. and Maynard, H. (2002) Design considerations for miniaturized PEM fuel cells, *J. Power Sources*, **109** (1), 76-88.

Min, K.B., Tanaka, S. and Esashi, M. (2002) Mems-based polymer electrolyte fuel cell, *Electrochemistry*, **70** (12), 924.

Mitrovski, S.M., Elliott, L.C.C. and Nuzzo, R.G. (2004) Microfluidic Devices for Energy Conversion: Planar Integration and Performance of a Passive, Fully Immersed H_2-O_2 Fuel Cell, *Langmuir*, **20**, 6974-6976.

Morse, J.D. (2002) MEMS-Based Proton Exchange Membrane Fuel Cells, presented at the 2002 SMALL FUEL CELLS (sm) - 4th Annual International Conference for Portable Power Applications, Washington, DC April 21, 2002.

O'Hayre, R., Braithwaite, D., Herman, W., Lee, S.J., Fabina, T., Cha, S.W., Saito, Y. and Prinz, F.B. (2003) Development of portable fuel cell arrays with printed-circuit technology, *J. Power Sources*, **124**, (2) 459-472.

Osaka, T., Motokawa, T., Mohamedi, M., Momma, T. and Shoji, S. (2005) Extended Abstracts, 207th meeting of the Electro Chemical Society, Quebec, Canada, May 15-20, 2005.

Palmore, G.T.R. (2002) Fabrication of Microbiofuel Cells Using Soft Lithography, presented at the SMALL FUEL CELLS (sm)—4th Annual International Conference for Portable Power Applications, Washington, DC April 21, 2002.

Pattekar, A. and Kothare, M. (2004) A microreactor for hydrogen production in micro fuel cell applications, *J. Microelectromechanical Systems*, **13**, 7-18.

Press release, www.mtimicrofuelcells.com, Albany, N.Y. U.S.A, December 14, 2004.

SAITT (2004) Samsung presents new 10 hour fuel cell notebook, press release fuel cell today, April 2004.

Sarkar, P. (2005) Micro Solid Oxide Fuel Cell for Portable Applications, Presented at the 7th Annual SMALL FUEL CELLS (sm) 2005—Small Fuel Cells for Portable Applications, Washington, DC., April 27-29, 2005.

Savinell, R., Wainright, J., Dudik, L., Yee, K., Chen, L., Liu, C.C, Zhang, Y. and Litt, M. (2000) Microfabricated fuel cells for portable power, Micro-Power Sources, The Electrochemcial Society Proceeding Series, 2000.

SFC, Toshiba show DMFC prototypes at CeBIT, press release Fuel Cells Bulletin, May, 2004.

Shimshon, G. (2005) DMFC Portable Power Products at MTI MicroFuel Cells: Present and Future, Presented at the 7th Annual SMALL FUEL CELLS (sm) 2005 - Small Fuel Cells for Portable Applications, Washington, DC., April 27-29, 2005.

Upadhye, R. (2003) Power Plant on a Chip Moves Closer to Reality, Science & Technology Review, November 2003.

Wainright, J.S., Savinell, R.F., Liu C.C. and Litt, M. (2003) Microfabricated fuel cells, *Electrochimica Acta*, **48**, 2869-2877.

Wozniak, K., Johansson, D., Bring, M., Sanz-Velasco, A. and Enoksson, P. (2004) Micro direct methanol fuel cell demonstrator, *J. Micromech. Microeng.* **14**, No. 9 (September 2004), S59-S63.

www.hitachi.com, Press release, Hitachi fuel cell PDA, Tokio (Japan), April 2004.

Yen, T.J., Fang, N., Zhang, X., Lu, G.Q. and Wang, C.Y. (2003) A micro methanol fuel cell operating at near room temperature, *Applied Physics Letters*, **83**, 4056-4058.

Yeom, J., Mozsgai, G.Z., Flachsbart, B.R., Choban, E.R., Asthana, A., Shannon, M.A. and Kenis, P.A.J. (2005) Microfabrication and characterization of a silicon-based millimetre scale PEM fuel cell operating with hydrogen, methanol or formic acid, *Sensors and Actuators B: Chemical*, **107(2)**, 882-891.

Yoshimi, K., (2004) ATIP SCOOP, Technology: Micro Fuel Cell, Energy, Power, 13, December 2004.

Yu, J.R., Cheng, P. and Ma. Z.Q. (2003) Fabrication of a miniature twin-fuel-cell on silicon wafer, *Electrochemica Acta*, **48** (11), 1537-1541.

Recent Trends in Fuel Cell Science and Technology
Edited by S. Basu
Anamaya Publishers, New Delhi, India

7. Direct Alcohol and Borohydride Alkaline Fuel Cells

Anil Verma[†] and Suddhasatwa Basu

Department of Chemical Engineering, Indian Institute of Technology, Delhi, New Delhi-110 016, India
(email:sbasu@chemical.iitd.ac.in)

[†]Presently, Department of Chemical Engineering, Indian Institute of Technology, Guwahati, Assam-781039, India

1. Introduction

The alkaline fuel cell (AFC) was the first fuel cell technology used in many practical services like Apollo space mission and running an automobile (Kordesch et al. 1999, Carrette et al. 2001). It is a low temperature fuel cell technology, which uses hydrogen as a fuel, oxygen or air as oxidant and alkaline solution as electrolyte. The development of AFC technology has reached its peak in the beginning of 1980s (Schulze et al. 2004) but its further development is stopped due to many technical, commercial and safety issues. The research and development on fuel cell is gaining momentum as a new way power generation technology, which is environmental friendly. The main focus of technology development of fuel cells is to increase the power output per unit area of electrode at low cost. The interest in AFC increased again due to more favorable oxygen reduction (Burchardt et al. 2002, Yu et al. 2004) and fuel oxidation reactions (Wang et al. 2003, Rahim et al. 2004) in alkaline condition. Apart from these, the cost, simplicity, efficiency and the possibility to use the non-noble metal catalyst (Schulze et al. 2004, Cifrain et al. 2003a) compared to other low temperature fuel cell technology have given an impetus to the AFC research. A detailed comparison of AFC with polymer electrolyte membrane fuel cell (PEMFC) is given by McLean et al. (2002).

Alkaline fuel cells (Fig. 1) generally operate on hydrogen and oxygen gases and at a temperature of 150-200°C. A solution of potassium hydroxide in water caged in a matrix is used as the electrolyte. During cell operation, the hydroxyl ions (OH⁻) work as charge carrier in electrolyte and they migrate from the cathode to the anode. At the anode (eq. (1)), hydrogen gas reacts with the OH⁻ to produce water and release electrons. Electrons generated at the anode, conduct through the load connected to the external circuit and migrate to the cathode. These electrons react with oxygen and water at cathode to produce hydroxyl ions (eq. (2)) that diffuse into the electrolyte for further reaction. The reactions in the AFC are (McLean 2002):

$$\text{Anode: } 2H_2 + 4OH^- \rightarrow 4H_2O + 4e^- \quad (1)$$

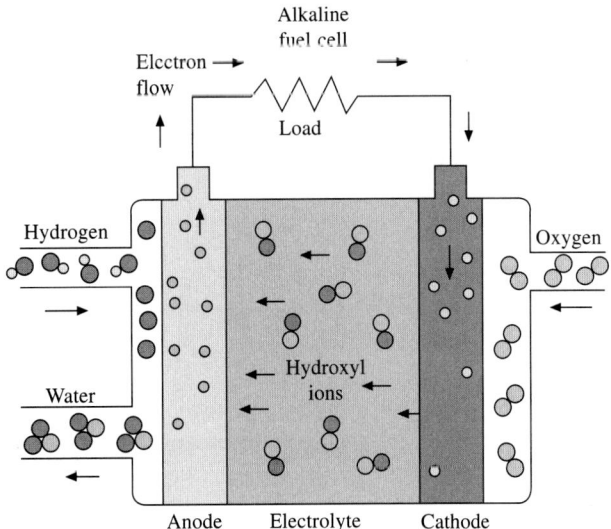

Fig. 1 Schematic diagram of an alkaline fuel cell.

$$\text{Cathode: } O_2 + 2H_2O + 4e^- \rightarrow 4OH^- \tag{2}$$

$$\text{Overall: } 2H_2 + O_2 \rightarrow 2H_2O \tag{3}$$

Though, the hydrogen as a fuel has many advantages but many obstacles have given thrust to look at direct use of hydrogen-rich liquid and solid fuels in fuel cells. Some of the obstacles associated with the use of hydrogen as fuel are the generation of hydrogen gas in huge amount cost effectively vis-à-vis gasoline cost, on board storage of low density gas, safety issues and low power output per unit weight of the fuel cell and fuel processor (Kordesch et al. 1999). The investigators are working on direct feeding of hydrogen rich liquid fuels e.g., alcohols, ethers, and hydrogen rich gaseous compounds or solid compounds in solution, e.g. N_2H_4, NH_3, and $NaBH_4$ into the alkaline fuel cells (Lee et al. 2002, Verma et al. 2005a, b, c). They are appropriately named direct alcohol or borohydride alkaline fuel cell. These fuel cells produce low power density and operated at a lower temperature (20-60°C) and thus thought to be suitable for use in powering portable electronic equipments.

Among the fuels, methanol is an attractive liquid fuel because it is relatively cheap, readily available, easily stored and handled, and soluble in aqueous electrolytes. Energy density of methanol is about 6 KWh/Kg (5 KWh/l). Ideally, the electrochemical oxidation of methanol produces 6 electrons per molecule of methanol. Fig. 2 (a) shows photograph of a flash light equipped with methanol/air fuel cell and the performance curve is shown in Fig. 2 (b) (Koscher et al. 2003). However, methanol has some disadvantages, for example, it is toxic, has a low boiling point (65°C) and it is not a primary fuel. Generally, methanol is synthesized from natural gas via its incomplete combustion, producing synthesis gas (CO + H_2). Synthesis gas is catalytically converted to methanol. Therefore other alcohols, particularly those derived from biomass resources are considered, as they are renewable in nature. In this category, ethanol is one of the potential fuels and it can be easily produced in great quantity by fermentation of biomass. The energy density of ethanol is about 7.44 KWh/kg (5.9 KWh/l). Ideally, the electrochemical oxidation of ethanol can produce 12 electrons per molecule of ethanol. It is non-toxic and its boiling point is relatively high compared to methanol. Ethanol and the higher alcohols have the drawback that their C-C bond rupture for 100% electro-oxidation to carbon dioxide is difficult in the presence of electro-catalyst in the temperature range of 25-90°C (Lamy et al. 2001).

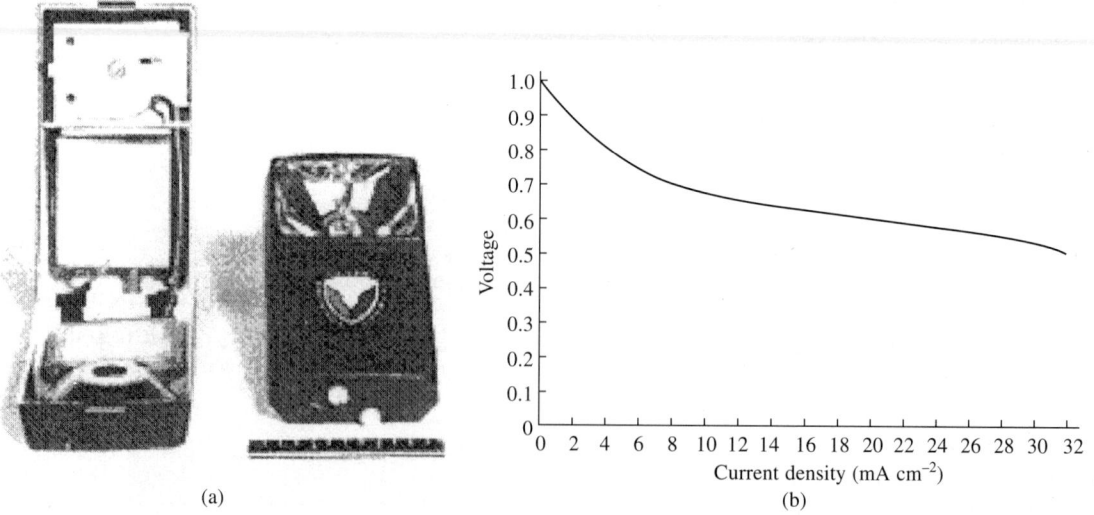

(a)   (b)

Fig. 2 (a) Working methanol/air fuel cell equipped with flashlight and (b) voltage-current characteristics of the methanol/air fuel cell.

The existence of the reactive intermediate and the inactive intermediates (poisoning species) generated during electro-oxidation of alcohols (e.g. methanol, ethanol, *n*-propanol and *n*-butanol) suggests a dual-path mechanism (eqs. (4) and (5)) in an alkaline solution. This is summarized as (Tripković et al. 2001):

$$\text{Alcohol} \rightarrow \text{Reaction intermediate} \rightarrow \text{Acid (in anionic form)} \tag{4}$$
$$\downarrow$$
$$\text{Poisoning species (CO)} \rightarrow CO_2 \tag{5}$$

The path shown by eq. (4) is the main reaction path and assumes the formation of reactive intermediates, weakly bonded to the surface, in contrast to the poisoning intermediates, which are strongly bonded (Morallón et al. 1995). Tripković et al. (2001) proposed the following general mechanism for the alcohol electro-oxidation (eqs. (6)-(9)) on noble metal catalyst in alkaline medium:

$$R - COH + OH^- \leftrightarrow R - CO_{ad} + H_2O + e^- \tag{6}$$

$$OH^- \leftrightarrow OH_{ad} + e^- \tag{7}$$

$$R - CO_{ad} + OH_{ad} \rightarrow R - COOH \tag{8}$$

$$R - COOH + OH_{ad} \rightarrow R - COO^- + H_2O \tag{9}$$

The adsorption of alcohol on catalyst site initiates dehydrogenation reaction where alcohol reacts with hydroxyl ion to produce water molecule and generates electrons. The adsorbed OH_{ad} reacts with adsorbed alcohol molecule to produce acid. The acid resides as anion in the electrolyte. Recently, investigators have (Gupta et al. 2004, Torresi et al. 2003) reported the cleavage of C—C bond of ethanol by electrochemical oxidation in the presence of new catalyst. Gupta et al. (2004) reported the cleavage of C—C bond in ethanol electro-oxidation by CuNi/PtRu. They have observed the presence of slight amount of acetaldehyde and evolution of CO_2 in alkaline condition at 20–60°C. Torresi et al. (2003) reported the presence of soluble oxidation products of ethanol when polycrystalline gold was used as working electrode after long hours of electrolysis (over 60 h) at different applied potentials in acid and alkaline solution. The ethanol concentration diminished with time and only acetate is detected as reaction product. The comparison of Pt/Ru and Pt catalyst on oxidation products of methanol, ethanol and formic acid are summarized in Table 1 (Torresi et al. 2003).

Table 1. Comparison of the Pt/Ru and Pt catalyst on the oxidation products of methanol, ethanol and formic acid (Torresi et al. 2003)

Product formation	Methanol	Ethanol	Formic acid
CO_2 formation	Pt/Ru is catalytically more active	Effects uncertain, but small if present	Pt/Ru is catalytically more active
Aldehyde formation	Pt/Ru is catalytically more active	Pt/Ru is catalytically more active	Not available
Selectivity towards CO_2 formation	Pt/Ru is more selective	No differences in selectivity found	Not available

Among solid fuels, sodium borohydride in alkaline fuel cell is gaining attention from the scientists developing fuel cell technology. The reasons are that the sodium borohydride ($NaBH_4$) contains 10.6-wt % hydrogen and 7 KWh/g of energy density, much more than most of the hydrogen storage alloys. The solution of $NaBH_4$ is noninflammable and quite stable at high pH values. The electro-oxidation reaction products of $NaBH_4$ are environmentally safe and can be recycled. Although hydrogen as a fuel can be easily obtained from the hydrolysis reaction of borohydrides, the direct anodic oxidation of borohydride provides more negative potential than that of hydrogen gas. There are two possible routes for oxidation of $NaBH_4$. In the first route $NaBH_4$ is directly oxidized (Li et al. 2003a, b and Amendola et al. 1999)

$$NaBH_4 + 8OH^- \rightarrow NaBO_2 + 6H_2O + 8e^- \tag{10}$$

In the second route, hydrogen is liberated at high temperature or pH less than 7 and then the hydrogen reacts with OH^- to produce electrons

$$NaBH_4 + 2H_2O \rightarrow NaBO_2 + 4H_2 \tag{11}$$

$$4H_2 + 8OH^- \rightarrow 8H_2O + 8e^- \tag{12}$$

Because of hydrolysis reaction, 7 electrons are utilized per molecule of borohydride electro-oxidation instead of theoretically 8 electrons (Amendola et al. 1999).

This chapter attempts to provide a critical review of the work carried out on alkaline fuel cell, which directly uses hydrogen rich liquid fuel and oxygen or air as an oxidant. The subjects covered are electrode materials, electrolyte, half-cell analysis and single cell performance in alkaline medium. Koscher et al. (2003) brought out elaborate review work on direct methanol alkaline fuel cell. Earlier Parsons et al. (1988) reviewed literature on anode electrode where, the oxidation of small organic molecules in acid as well as in alkaline conditions was considered. A review work on electro-oxidation of boron compounds was done by Morris et al. (1985). However, in this chapter use of three specific fuels, e.g., methanol, ethanol and sodium borohydride in alkaline fuel cell is discussed.

2. Cell Components

2.1 Electrode Material
The cell components are discussed in detail as follows.

2.1.1 Anode Catalyst
The anode electrode-catalyst is one of the important components of the alkaline fuel cell as it helps in the electro-oxidation of fuel. It is desirable that the anode electrode-catalyst provides faster reaction kinetics and 100% oxidation of fuels to CO_2 and H_2O. The most widely used catalyst, without doubt, is platinum. Platinum seems to be the best choice for acidic solutions, but other metallic alloy with platinum or other metals can match its performance in alkaline medium because of the favorable fuel oxidation in alkaline medium. Different anode materials based on Pt (Prabhuram et al. 1998, Morallón et al. 1995, Tripkivić et al. 1996), Pt-Ru (Wang et al. 2003, Manoharan et al. 2001), Co-W alloys (Shobba et al. 2002), sintered Ag/PdO (Koscher et al. 2003), spent carbon electrodes impregnated with Fe^+, Fe^{++} or Ag^+ (Verma 2000), nickel impregnated silicate-1 (Khalil et al. 2005) and nickel dimethylglyoxime complex (Golikand et al. 2005) are some of the catalysts studied for the electro-oxidation of methanol in alkaline medium.

The electro-oxidation of ethanol is studied recently by few investigators in alkaline medium. Tripković et al. (2001) used Pt (111) for the electro-oxidation of ethanol, methanol, propanol and butanol in alkaline solution. Gupta et al. (2004) used CuNiPt and CuNiPtRu alloys for the electro-oxidation of ethanol in alkaline medium. The cell performances using these anodes are mediocre.

Various noble and non-noble metal catalysts have been used for the study of electro-oxidation of sodium borohydride like Zr-Ni alloy (Li et al. 2003a and 2003b), Ag or Ag/Pt (Amendola et al. 1999), Ni (Liu et al. 2003), $ZrCr_{0.8}Ni_{1.2}$ alloy (Lee et al. 2002), LaCeNdPrNiAlMnCo, ZrTiVMnCrCoNi alloys (Chaudhury et al. 2005) and Pt (Verma et al. 2005b).

2.1.2 Cathode Catalyst
It is well known that the reaction kinetics of oxygen reduction is superior in an alkaline medium compared

to that in acidic medium. The reduction of oxygen in alkaline medium generally proceeds by either of two pathways (Ortiz et al. 2003, Yang et al. 2003, Verma et al. 2005a). They are described below as direct oxygen reduction to OH^- ions, i.e., two electron pathway:

$$O_2 + 2H_2O + 4e^- \rightarrow 4OH^- \qquad (13)$$

or, an oxygen reduction to HO_2^- ions, i.e., two electron pathway:

$$O_2 + H_2O + 2e^- \rightarrow HO_2^- + OH^- \qquad (14)$$

with subsequent reduction of peroxide ion to OH^- ions, i.e., two electron pathway:

$$HO_2^- + H_2O + 2e^- \rightarrow 3OH^- \qquad (15)$$

Equations (14) and (15) collectively produce 2 + 2 electron mechanism (Ortiz et al. 2003). In the case of direct alcohol and sodium borohydride alkaline fuel cell, the fuel and electrolyte mixture comes in contact with the cathode as well as with the anode. A non-noble electro-catalyst should be used at cathode such that no fuel oxidation takes place at cathode. There are non-noble metal catalysts, where oxygen reduction is active and unwanted fuel oxidation is prevented. Thus, the advantage of alkaline fuel cell is the possibility of use of non-noble metals in place of platinum as electro-catalysts at cathode. The electro-catalysts alternative to Pt were studied as active catalysts for oxgyen reduction reaction in alkaline medium like iron tetramethoxyphenyl porphyrin chloride (Gojković et al. 1999), Ni/Co, Ni/Co spinel (Heller-Ling et al. 1997, Ponce et al. 2001 and Rashkova et al. 2002), Ag (Wagner et al. 2003 and Demarconnay et al. 2005), and manganese dioxide based catalysts (Mao et al. 2002, Yang et al. 2003, Klápště et al. 2002, Bezdička et al. 1999 and Verma et al. 2005a).

2.2 Electrode Fabrication Method and Analyses

The fabrication method of electrode for alkaline fuel cell has significant impact on the performance of the overall cell. Literature on fabrication of electrode comprising catalyst for alkaline fuel cell is very scanty. Generally, the anodes and cathodes are manufactured by a wet fabrication followed by sintering or by a method of dry fabrication through rolling and pressing of components into the electrode structure. In all cases the resulting electrode consists of a catalyzed layer on top of a gas diffusion layer and the other side is bonded to a perforated supporting material which is usually metallic. The metallic plate works as current collector and generally it should have excellent electron conducting properties to avoid losses. The best results appear to be achieved when the electrode structure is build up from layers.

Most of the literature describes about the catalyst properties in terms of active area per unit area of electrode surface, structural view using SEM or TEM photomicrograph, porosity and lattice structure. Normally, electrodes are analysed through half-cell analysis. The half cell is constructed with reference electrode, working electrode and counter electrode. The working electrode is electrochemically analyzed to predict the formation of intermediate products, poisoning species and reaction mechanism with the help of cyclic voltammetry, impedance spectroscopy, FTIR and liquid chromatography. Tables 2 to 4 show the electrode-catalyst fabrication methods by various investigators and specific remarks on their work for methanol, ethanol oxidation at anode and oxygen reduction at cathode in alkaline condition.

2.3 Electrolyte

The alkaline electrolytes have several distinct advantages over acid electrolytes. The most notable advantage is that the oxidation of fuel in alkaline solution by the non-noble metal catalyst and it is as active as noble metal catalyst. The other major advantage of alkaline solution is the absence of electrode poisoning from the reaction intermediates. Vielstich (Koscher et al. 2003) had chosen KOH as an electrolyte over NaOH

Table 2. Electrode-catalyst fabrication for methanol electro-oxidation at anode in alkaline condition and techniques used for the analyses of the system

Subject of investigation/Major technique	System information		Catalyst and electrode preparation technique	Remarks	References
	Catalyst	Electrolyte			
Methanol oxidation/cyclic voltammetry	Porous unsupported Pd, Pt and Pt/Ru	KOH solution	Catalysts are prepared by aqueous phase reduction method. Electrodes prepared from catalyst powder by compaction method	Possible to prevent the formation of the poisoning species with unsupported porous structure methanol anodes with a suitable combination of electrolyte/methanol mixture	Manoharan et al. (2001)
Methanol oxidation/cyclic voltammetry	Porous unsupported Pt	KOH solution	Catalysts are prepared by aqueous phase reduction method. Electrodes prepared from catalyst powder by compaction method	Highest methanol activity is found at 6 M CH_3OH/6 M KOH. Proper ratio of fuel to electrolyte is found to prevent formation of poisoning species	Prabhuram et al. (1998)
Methanol oxidation/cyclic voltammetry In-situ FTIR spectroscopy	Pt (100), Pt (110), Pt (111)	Na_2CO_3 and NaOH solutions	Pt was treated electrochemically to get the desired surface structure	FTIR result shows that CO is weakly adsorbed on Pt (111) than Pt (110) and Pt (100) which is in well agreement with cyclic voltammetry result	Morallón et al. (1995)
Reaction kinetics and mechanism of methanol oxidation	Pt (100)	$NaHCO_3$, Na_2CO_3, and NaOH solution	Electrode purchased from Metal Crystal and Oxides Ltd., Cambridge, UK	Oxidation proceeds with some poisoning species formation. Main reaction product is formate and CO. Potential range for OH_{ad} is $0.4\,V < E < 0.7\,V$	Tripković et al. (1996)
Reaction kinetics and mechanism of methanol oxidation	Pt (110)	$NaHCO_3$, Na_2CO_3, and NaOH solutions	Pt was treated electrochemically to get the desired surface structure	Oxidation proceeds with some poisoning species formation. Main reaction product is formate and CO. Potential range for OH_{ad} is $0.4\,V < E < 1.0\,V$	Tripković et al. (1998a)
Reaction kinetics and mechanism of methanol oxidation	Pt (111)	$NaHCO_3$, Na_2CO_3, and NaOH solutions	Electrode purchased from Metal Crystal and Oxides Ltd., Cambridge, UK	Oxidation proceeds with some poisoning species formation. Main reaction product is formate and CO. Potential range for OH_{ad} is $0.6\,V < E < 0.85\,V$. Less poisoning than Pt (100) and Pt (110)	Tripković et al. (1998b)

Study/Method	Catalyst	Electrolyte	Preparation	Findings	Reference
Preparation and characterization of catalyst for methanol oxidation/cyclic voltammetry, XRD	Co-W alloys	KOH solution	Alloys were electroplated potentiometrically on pretreated copper foil	Good corrosion resistance and partially amorphous heat treatment of alloy increases the catalytic activity and decreases the polarization	Strobba et al. (2002)
Influence of oxygen containing species on methanol, ethanol, n-propanol and n-butanol oxidation reaction/cyclic voltammetry quasi-steady-state analysis	Pt (111), Pt (755) and Pt (332)	NaOH solution	Pt was treated electrochemically to get the desired surface structure	General mechanism for methanol, ethanol, n-propanol and n-butanol oxidation is suggested. PtO species inhibits the oxidation reaction	Tripković et al. (2001)
Methanol oxidation/linear voltammetry, *in-situ* FTIR spectroscopy	Pt/Ru/C	$KHCO_3$ and K_2CO_3 solution	Co-deposition of fine oxides of Pt and Ru followed by reduction with hydrogen bubbling	CO_2 peak of *in-situ* FTIR spectra showed that the full oxidation of methanol is possible. Carbonate and bicarbonate have better kinetics than H_2SO_4	Wang et al. (2003)
Catalyst synthesis for methanol oxidation/cyclic voltammetry	Nickel electrode modified by nickel dimethylglyoxime complex	NaOH solution	Dimethylglyoxime ligand deposition on the mechanically pretreated nickel disc	Heat treatment of catalyst increases the catalyst activity. Methanol tolerant catalyst	Golikand et al. (2005)
Methanol electro-oxidation/cyclic voltammogram	Ni modified graphite	KOH solution	Nickel electrodeposited on pure crystalline graphite disc	Ni modified graphite is found to be a good catalyst (current density over 150 mA cm^{-2}) for the oxidation of methanol	Rahim et al. (2004)
Methanol electro-oxidation/cyclic voltammogram	Pt-black and Pt/Ru	KOH solution	Electrodes are prepared by wet fabrication method	Oxide layer formation is confirmed. Reaction mechanism is proposed.	Verma et al. (2005e)

Table 3. Electrode-catalyst fabrication for ethanol oxidation at anode in alkaline condition and techniques used for analyses of the system

Subject of investigation/ Major technique	System information		Catalyst and electrode preparation technique	Remarks	References
	Catalyst	Electrolyte			
Ethanol electro-oxidation/cyclic voltammetry, electrochemical impedance spectroscopy	Pt and Pt/Ru	NaOH solution	Electrodeposition of noble metal on CuNi alloys	Higher electrocatalytic activity is found for ethanol oxidation for the catalyst layer prepared from PTFE suspension of noble metal salts rather without PTFE suspension. The charge transfer resistance is greatly reduced in the Pt/Ru-modified CuNi electrodes	Gupta et al. (2004)
Preparation of catalyst and study of reaction of ethanol electro-oxidation/XPS, XRD, polarization curve	RuNi/C	KOH solution	Catalyst deposited on carbon support by the dissolved metallic salts in aqueous-alcohol mixture	Reaction order of 0.5 with respect to ethanol is found to be 0.5. 15% RuNi catalyst containing CO_2 Ni amount of 20–40 at% provides a higher ethanol electro-oxidation. CO_2 and acetic acid are found to be the reaction product	Tarasevich et al. (2005)
Ethanol electro-oxidation/cyclic voltammetry	Pt-black, Pt/Ru	KOH solution	Electrodes prepared by wet fabrication method	Oxide layer formation is confirmed. Pt/Ru is more active than Pt-black. C–C bond cleavage is not observed. Reaction mechanism is predicted	Verma et al. (2005e)

Table 4. Electrode-catalyst fabrication for oxygen reduction at cathode in alkaline condition and techniques used for analyses of the system

Subject of investigation/Major technique	System information		Cathode catalyst/electrode preparation technique	Remarks	References
	Catalyst	Electrolyte			
Oxygen electro-reduction and evolution reaction/cyclic voltammetry, rotating ring disc electrode (RRDE)	$Ni_xAl_{1-x}Mn_2O_4$	KOH solution	Two preparation methods are used: coprecipitation of metal hydroxides and sol-gel route using metal propionates were used. A Teflon-bound electrode technique was chosen with graphite addition	The substitution of Al, by Ni increases the catalytic activity. Maximum activity is experienced by $NiMn_2O_4$	Ponce et al. (2001)
Oxygen electroreduction/EDAX, XRD	$Ni_xCo_{3-x}O_4$ spinel	KOH solution	Electro-deposition of oxide films by spraying on Ni foil	The ratio of oxygen molecules reduced via the direct $4e^-$ pathway with respect to those reduced via indirect $2 + 2$ electron pathway depends on x. The ratio is maximum for $x = 0$ and $0.6 < x < 1.2$	Heller-Ling et al. (1997)
Oxygen electroreduction/XRD, cyclic voltammetry, polarization curve	Nonporous amorphous manganese oxide	KOH solution	Catalyst prepared by adsorption on high surface area carbon black. Electrode was prepared by catalyst ink coating	Current density of 100 mA cm^{-2} is obtained with a low-catalyst loading of 0.85 mg cm^{-2}. Amorphous morphology is advantageous for the design of effective gas diffusion layer	Gojković et al. (1999)
Synthesis and investigation of catalyst for oxygen electro-reduction/cyclic voltammogram, XRD	MnO_x/C	KOH solution	Impregnation technique for catalyst preparation. PTFE-bonded electrode	Ni doped catalyst show the best cycling stability in PTFE bonded electrode than Bi, Pb or Ti	Bezdička et al. (1999)
Lifetime test for oxygen electroreduction electrode/XPS, electrochemical impedance spectroscopy)	Ag	KOH solution	Catalyst preparation not given. Porous PTFE bonded gas diffusion layer is prepared by cold rolling method	12-15% performance loss of cell voltage in 5000 h of operation time due to decrease of the surface roughness of electrode and PTFE degradation effect.	Wagner et al. (2003)

(Contd.)

| Subject of investigation/ Major technique | System information | | Cathode catalyst/electrode preparation technique | Remarks | References |
	Catalyst	Electrolyte			
Electrochemical characterization of catalytic activities for oxygen electroreduction/cyclic voltammetry	Mn_2O_3, Mn_3O_4, Mn_5O_8 and $MnOOH$	KOH solution	Chemical and thermal oxidations are used to obtain different catalysts. Catalyst paste application on Au to obtain electrode	2 + 2 electron pathway $MnOOH$ shows the highest catalytic activity among other species examined.	Mao et al. (2002)
Oxygen electroreduction/ XRD, cyclic voltammetry	Nonporous amorphous manganese oxide	KOH solution	Catalyst synthesized by aqueous redox sol-gel route. Electrode prepared by painting catalyzed ink on carbon paper.	Amorphous MnO_2 provided much more distortion hence higher active sites than crystalline MnO_2 and higher corrosion resistance is found	Yang et al. (2003)
Oxygen evolution and electroreduction reaction/XPS, polarization curve	Mixed Co and Ni oxides	KOH solution	Vacuum thermal co-deposition method. Electrode prepared by deposition of metal oxide on gas diffusion layer	High catalytic activity for oxygen evolution and reduction with very small catalyst loading (0.07 mg cm^{-2})	Rashkova et al. (2002)
Electrochemical reduction of oxygen/RRDE	Ag/C and Pt/C	NaOH solution	Catalyst synthesis by colloidal precursor method. Electrode prepared by painting catalyzed ink on glossy carbon substrate	The mechanism of oxygen electroreduction on Ag is similar to Pt catalyst	Demarconnay et al. (2005)
Electrochemical reduction of oxygen/ cyclic voltammetry	MnO_2 on carbon paper	KOH solution	Electrode prepared by wet fabrication method	Four-electron pathway was confirmed	Verma et al. (2005a)

because of its lower overpotential, particularly at cathode side. The drawback of alkaline electrolyte is the progressive "carbonation" of the solution due to carbon dioxide produced by the reaction product of organic fuel oxidation as well as CO_2 from air. The probable electrochemical oxidation reaction of methanol in alkaline medium is given by (Koscher et al. 2003)

$$CH_3OH + 8OH^- \rightarrow CO_3^{2-} + 6H_2O + 6e^- \tag{16}$$

This reaction has the effect of reducing the number of hydroxyl ions available for reaction at the electrodes. Further, it reduces the ionic conductivity of the electrolyte solution. In a very concentrated electrolyte solution, it may also have the effect of blocking the pores of the gas diffusion layer (GDL) by the precipitation of K_2CO_3 salt. Gülzow et al. (2004) reported that, although carbon dioxide poisoning decreases the alkaline fuel cell performance, it does not cause any degradation of the electrodes. Even after thousands of hours of operation in a carbon dioxide rich atmosphere, no additional electrode degradation, like deposition of K_2CO_3, was observed. Saleh et al. (1994) showed that the concentrations of up to 1% carbon dioxide in the oxidant stream of Ag/PTFE electrodes did not affect the cell performance over a period of 200 hours. Therefore, the most probable reason for the decrease in the cell performance is the change in electrolyte composition. The conversion of the electrolyte, from KOH to K_2CO_3, by the absorption of carbon dioxide slows down the rate of oxidation of fuel at the anode (Cifrain et al. 2003b, Gülzow et al. 2004). In addition, decreased electrolyte conductivity also increases the ohmic polarization leading to lower cell efficiency.

Although the harmful effects of carbon dioxide poisoning can be partly reduced by circulating the electrolyte as discussed by Cifrain et al. (2003b), a permanent solution must be sought for possible commercialization of alkaline fuel cell.

The efforts are being made to rectify the problem of carbon dioxide poisoning in alkaline fuel cells. Molecular sieves and polymeric membranes are being used for CO_2 separation from air (Appleby et al. 1989). At present, these membranes require large area to bring down the CO_2 level from 300 ppm to 10 ppm at low gas velocity. Another alternative proposed is CO_2 management, which involves the synergistic possibility of using liquid hydrogen to condense the carbon dioxide out of the air. Ahuja et al. (1996, 1998) discussed this at length and developed a model of the heat exchanger required for this purpose. This solution presents low parasitic energy consumption, but the condensation and re-vaporization of water and CO_2 makes the structure complicated. McLean et al. (2002) mentioned that the "Removal of the 0.03% carbon dioxide from the air can be accomplished by chemical absorption in a tower filled with "soda lime" One kilogram of soda lime has the ability to clean 1000 m^3 of air from 0.03% to 0.001% CO_2. The only method commercially employed to alleviate the oxidant side carbon dioxide poisoning is CO_2 scrubbing using soda lime. Technically, soda-lime method works, but it requires periodic maintenance resulting high operational costs over the lifetime of the fuel cells.

The oxidation of alcohol, like methanol ($CH_3OH + 6OH^- \rightarrow CO_2 + 5H_2O + 6e^-$), produces carbon dioxide at the anode and, hence, contributes to the carbon dioxide poisoning in the fuel cells. Morallón et al. (1995) studied methanol electro-oxidation on platinum in Na_2CO_3 and NaOH medium and found no major differences in the deactivation of the electrode. The deactivation was due to the structural modification of the electrode surface and/or due to poisoning of the electrode surface by the adsorbed species. Tripković et al. (1996) studied the oxidation of methanol in different concentration of bicarbonate, carbonate and sodium hydroxide solution. The basic voltammograms were found similar except somewhat difference in position and peak current. The removal of carbonate from cathode by means of electrochemical method is reviewed by McLean et al. (2002). In the electrochemical method a large current is drawn from the fuel cell, which reduces the OH^- concentration at anode. The carbonate ions migrate from cathode to anode due to anion concentration gradient producing carbonic acid at the anode. With the increase in current density, the carbonic acid is electrolyzed and CO_2 escapes from the solution. Cifrain et al. (2003b), and Verma et al.

(2005b) suggested that by circulating electrolyte one not only removes the heat of reaction from the system, but it helps to remove the carbonate externally from the solution. Prabhuram et al. (1998) showed that the right combination of KOH and methanol suppresses the formation of oxide layer, intermediate organic species and/or poisoning species on the electrode surface thus improving the methanol oxidation reaction performance. They have shown that the equi-molar mixtures of methanol and KOH give the maximum current densities. Yu et al. (2004) used anion exchange membrane in direct methanol alkaline fuel cell. Wang et al. (2003) also used anion exchange membrane for the feasibility analysis of direct methanol fuel cell in the presence of carbonate and bicarbonate solution.

The direct sodium borohydride alkaline fuel cell uses electrolyte not only as an ionic conductive medium but also to stabilize the $NaBH_4$ to prevent hydrolysis reaction (eq. 11). NaOH (Liu et al. 2003, Verma et al. 2005c) or KOH (Lee et al. 2002) solution is used as electrolyte. Anion exchange membrane as a solid electrolyte is also used with NaOH solution mixed with $NaBH_4$ to prevent hydrolysis. The anion exchange membrane (AEM) must have resistance to alkali solution, mechanically stable and behave as electrical insulator. An ideal AEM must allow OH^- permeation from the cathode to the anode compartment through AEM but not permeation of BH_4^-. The crossover of BH_4^- results in oxidation at cathode thus hampers the performance significantly. Li et al. (2003a) studied the BH_4^- crossover through different Nafion membranes. They have found that Nafion 112 (51 μm) has higher borohydride crossover than Nafion 117 (178 μm). The use of thicker membrane is an effective way to reduce the borohydride crossover at the expense of ionic conductivity.

Choudhury et al. (2005) and Li et al. (2003a, b) used doped Na^+ Nafion membrane (Fig. 3) and explained the electron transfer mechanism. Li et al. (2003a, b) found that the Na^+ in Na^+ doped Nafion membrane worked as the charge carrier in the electrolyte rather than the OH^-. To verify the electron transfer mechanism in the Na^+ form of Nafion

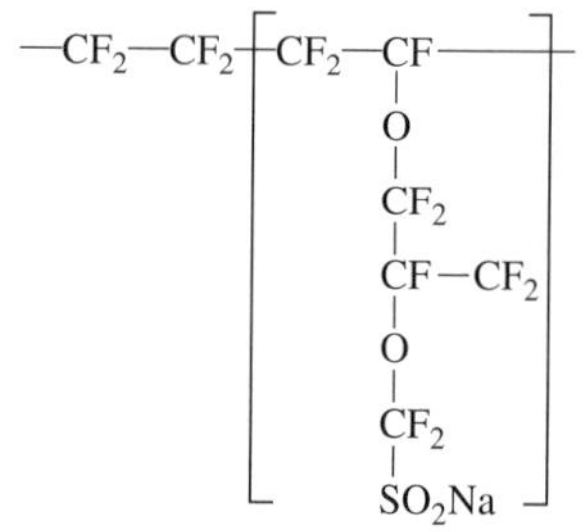

Fig. 3 Na^+ form of Nafion membrane.

membrane, Li et al. (2003a) analyzed the Na^+ distribution by the electron probe micro analyzer as well as pH measurement before and after the cell operation. Prior to the cell operation the pH was 7 and no Na^+ was found at cathode. After the cell operation, they have found the presence of Na^+ in the cathode and pH value was 14. They have concluded that Na^+ migrated from the anode compartment to the cathode compartment through the Na^+ form of Nafion membrane. The pH increases because of OH^- production at cathode during cell operation.

3. Half-cell Analysis

3.1 Anode
The comparison of various techniques used for the analyses of methanol and ethanol electro-oxidation at anode and the results are given in Tables 2 to 3 for methanol and ethanol, respectively. The most commonly used technique in electrochemical studies of fuel cell reactions has been cyclic voltammetry. The cyclic voltammetry is used to study the redox behavior of electrodes in fuel-electrolyte solutions (Bard et al. 2001). The cyclic voltammogram helps to identify the reaction intermediates, poisoning species, reaction mechanism, suitable combination of electrode material and electrolyte/fuel mixtures, such that the formation of poisoning species is prevented. Prabhuram et al. (1998) investigated methanol oxidation on unsupported platinum electrodes in alkaline condition (Table 2). The cyclic voltammograms (CVs) were recorded in

different electrolyte concentration of NaOH in the presence of different concentration of methanol. They have assumed the following mechanism for the methanol oxidation reaction (MOR) in alkaline solutions as

$$Pt + OH^- \leftrightarrow Pt\text{-}(OH)_{ad} + e^- \tag{17}$$

$$2Pt + CH_3OH \leftrightarrow Pt\text{-}H + Pt\text{-}(CH_3O)_{ad} \tag{18}$$

$$Pt\text{-}(CH_3O)_{ad} + Pt\text{-}(OH)_{ad} \leftrightarrow Pt_2\text{-}(CH_2O)_{ad} + H_2O \tag{19}$$

$$Pt_2\text{-}(CH_2O)_{ad} + Pt\text{-}(OH)_{ad} \leftrightarrow Pt_3\text{-}(CHO)_{ad} + H_2O \tag{20}$$

$$Pt_3\text{-}(CHO)_{ad} + Pt\text{-}(OH)_{ad} \leftrightarrow Pt_3\text{-}(CO)_{ad} + 2Pt \tag{21}$$

$$Pt_2\text{-}(CO)_{ad} + Pt\text{-}(OH)_{ad} \rightarrow Pt\text{-}(COOH)_{ad} + 2Pt \tag{22}$$

$$Pt\text{-}(COOH)_{ad} + Pt\text{-}(OH)_{ad} \leftrightarrow 2Pt + CO_2 \uparrow + H_2O \tag{23}$$

The cyclic voltammogram recorded for an unsupported Pt electrode in 6 M KOH/1 M CH$_3$OH is shown in Fig. 4. During the forward scan, an anodic peak, O^f, occurs at -0.6 V and the maximum peak current (i_p) is observed at $E_p = -0.04$ V. The current of this peak then declines exponentially until $E = 0.2$ V where the less active PtO monolayer film inhibits the methanol oxidation reaction. The reactions (17) to (23) appear to occur in the potential region of the peak O^f. The reactions (21) to (23) involving the release of CO$_2$ gas seem to occur in the potential range between -0.2 and 0.2 V, as indicated by the presence of a potential hump. Any residual weakly bonded CHO species do not become oxidized at higher potentials, hence another peak corresponding to the oxidation of organic species is not seen in the oxygen evolution reaction (OER) potential region ($E > 0.6$ V) shown in Fig. 4(b). A shoulder is noted in the potential range of -0.6 to -0.18 V on the left side of the O^f peak in Fig. 4(a).

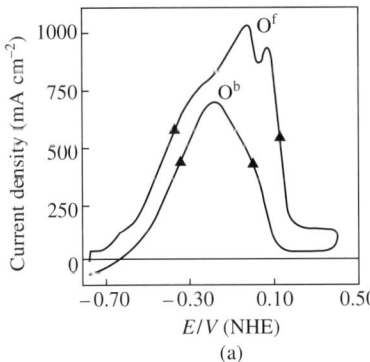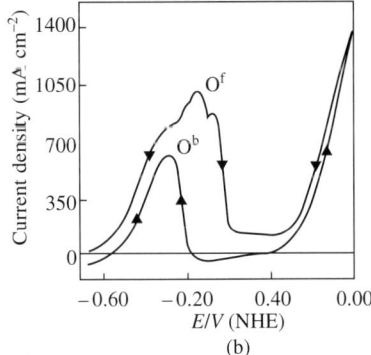

Fig. 4 CVs for the MOR on unsupported platinum electrodes in 6 M KOH/1 M CH$_3$OH solution in the potential range: (a) -0.775 to 0.4 V; (b) -0.775 to 1.0 V. Temperature: 25°C, Scan rate: 25 mV s^{-1}.

The analysis of the variation of the maximum peak current of this shoulder indicates that the associated process is diffusional controlled. It is speculated that the appearance of the shoulder could be due to oxidation of methanol arising out of the condition when OH$^-$ ions are available in excess and chemisorbed organic species availability is insufficient at the electrode-catalyst surface in the given electrolyte. During the backward sweep, an oxidation peak, O^b, is present at 0.13 V, where the PtO film is reduced. In this region the E_p of the O^b peak coincide with the potential region of the shoulder of the O^f peak. In this potential region, the electrode can acquire a large quantity of active oxygen atoms from 6 M KOH solution and use them to oxidize the residual weakly bonded CHO species that remain at the end of the forward sweep and as well as to oxidize the freshly chemisorbed methanol molecules.

When the electrolyte consists of equimolar concentration of alkali and methanol, dissociatively chemisorbed methanol and OH_{ads} species appear to cover the adjacent platinum sites simultaneously. Almost all the sites are covered in high equimolar mixtures, viz., 6 M KOH/ 6 M CH_3OH and 11 M KOH/11 M CH_3OH, and hence complete oxidation of chemisorbed organic species takes place without permitting any CHO species to remain on the electrode surface. The O^f and O^b peaks (Fig. 5) are not formed in the CVs recorded in these solutions, and only a featureless current-potential response (no peak) is observed. The high current density of 2.53 A cm^{-2} (Fig. 5a) is obtained for the 6 M KOH/6 M CH_3OH mixture as compared to 1.97 A cm^{-2} (Fig. 5b) for 11M KOH/11M CH_3OH mixture at $E=0.4$ V for a scan rate of 25 mV s^{-1}. This decrease in current density for 11M KOH/11M CH_3OH mixture may be due to the lower ionic conductivity of the mixture. The O^b peaks are not observed because CHO species does not remain on the electrode surface at the end of the forward sweep. CVs of the lower equimolar concentration mixture (1 M KOH/1 M CH_3OH) show O^f and O^b peaks (Fig. 6a) at 0.28 and 0.16 V, respectively. Although this mixed solution is an equimolar concentrated solution, the CHO species does not completely oxidize at the end of the forward

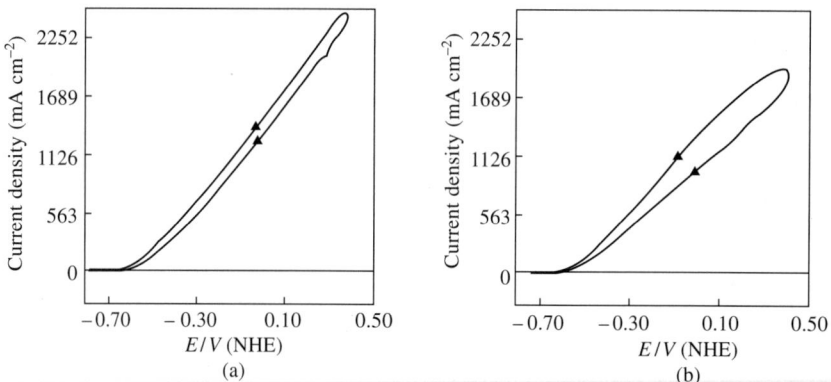

Fig. 5 CVs for the MOR on unsupported platinum electrode in: (a) 6 M KOH/6 M CH_3OH solution; (b) 11 M KOM/11 M CH_3OH solution. Temperature: 25°C, Scan rate: 25 mVs^{-1}.

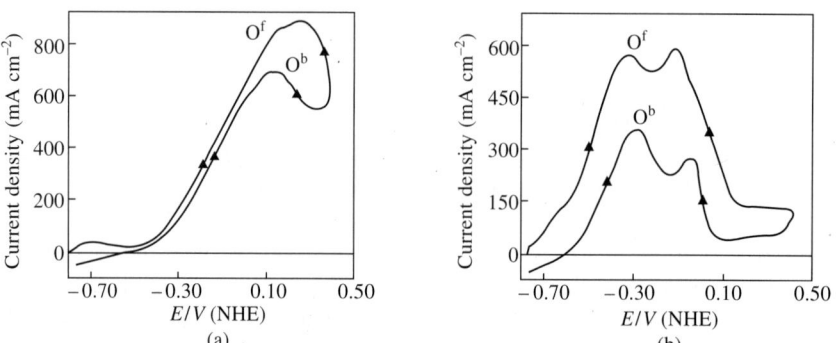

Fig. 6 CVs for the MOR on unsupported platinum electrode in: (a) 1 M KOH/1 M CH_3OH solution; (b) 11 M KOH/1 M CH_3OH solution. Temperature 25°C, Scan rate: 25 mVs^{-1}.

sweep, possibly because 1 M KOH does not provide sufficient active oxygen atoms for oxidizing the carbonaceous species. Nevertheless, only a small amount of the carbonaceous species may exist. Therefore, only a small O^b peak is observed during the backward sweep. The peaks are also observed in the CV (Fig. 6b) of 11 M KOH/6 M CH_3OH and 11 M KOH/1 M CH_3OH mixtures. This may be due to the fact that the excess coverage of OH_{ad} species on the electrode surface oxidizes the chemisorbed organic species quickly and block the chemisorption of further methanol molecules at lower potentials. As the excess OH^- ions are not utilized for oxidizing the organic species, these ions could assist the rapid formation of the PtO layer in the higher potential range. As a result, a decrease in the rate of the MOR has been found.

Manoharan et al. (2001) studied cyclic voltammograms for the methanol oxidation reaction on several smooth and carbon unsupported porous Pd, Pt and Pt-Ru alloys electrodes in different concentration of methanol and KOH mixture and assumed the reaction mechanism be given by eq. (17) to eq. (23). The CV recorded for the porous unsupported Pt electrode in 6 M KOH solutions that contain the varying concentrations of methanol (1, 3, 6, 9 and 11 M CH_3OH) are shown in Fig. 7. They have found that the complete electro-oxidation of methanol could not be carried up to the concentration of 3 M as less active PtO layer was formed. On increasing the concentration of methanol to 6 M, the complete oxidation took place without permitting any poisoning species (CHO_{ad}) to remain on the electrode surface.

In the case of porous unsupported Pt-Ru alloy electrodes, both Pt and Ru atoms supply the active oxygen atoms and the ratio of supply being governed by the alloys' compositions. Pt/Ru alloys with 9 : 1, 7 : 3 and 5 : 5 composition were used in the study. The anodic peaks appear in the CVs with 6 M KOH/1 M CH_3OH and 3 M CH_3OH mixtures. The featureless polarization curves are obtained for 6 M methanol for all the Pt-Ru compositions. CV recorded

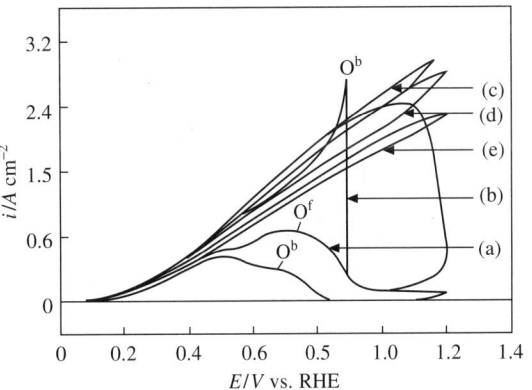

Fig. 7 CVs for the MOR reaction on porous unsupported Pt electrodes in 6 M KOH solutions that contain varying concentrations of CH_3OH: (a) 1 M; (b) 3 M; (c) 6 M; (d) 9 M; (e) 11 M. Scan rate: 25 mVs^{-1}.

for the smooth Pt electrodes showed the appearance of the forward and backward peaks and higher MOR polarizations in all KOH/methanol mixtures. It was inferred that the oxidation of the intermediate organic species (CHO_{ad}) did not take place completely on the smooth Pt electrodes. The population of the active sites present on the smooth electrodes is lower compared to those in the porous electrodes. Hence, inadequate amounts of the active oxygen atoms might have been extracted from the electrolyte on the smooth Pt electrodes to completely oxidize the CHO_{ads} species in the lower potential region and these organic species remain as the poisoning species on the surface of the electrodes. Similarly, the CVs recorded for the smooth Pd electrodes, shows that the intermediate bridge bonded CO species is not completely oxidized in the KOH/methanol mixtures of all concentrations and they remain as poisoning species on the surface of the electrodes.

To understand the various phenomena take place during electro-oxidation of methanol and ethanol on Pt-black surface, Verma et al. (2005e) studied half-cell using cyclic voltammetry. Fig. 8 shows the cyclic voltammograms with and without the fuel (methanol and ethanol) in 1 M KOH solution at the scan rate of 50 mV s^{-1}. In the forward scan, ethanol electro-oxidation shows a prominent peak at 0.03 V as compared to electro-oxidation of methanol, which shows a broad plateau from -0.4 V to 0.6 V. It implies that Pt-black electrode catalyst is more active in the case of ethanol than methanol for electro-oxidation. During reverse scan, the oxidation peaks were found for methanol (0.17 V) and ethanol (0.18 V). These peaks correspond to the electro-oxidation of unreacted methanol and ethanol left in the vicinity of the electrode during forward scan. The single oxidation peak for ethanol confirms that C—C bond could not be broken. It is further confirmed through product analyses of the methanol and the ethanol electro-oxidation by high performance liquid chromatography (HPLC) and gas chromatography (GC). In the case of methanol, formic acid was detected while for ethanol, acetic acid and acetaldehyde were detected using HPLC and GC analyses.

As stated earlier, the most serious problem for alcohol oxidation and CO_2 contaminated oxygen/ air reduction in alkaline medium is the carbonate ion formation (eq. (16)). The problem may be overcome by using circulating electrolyte

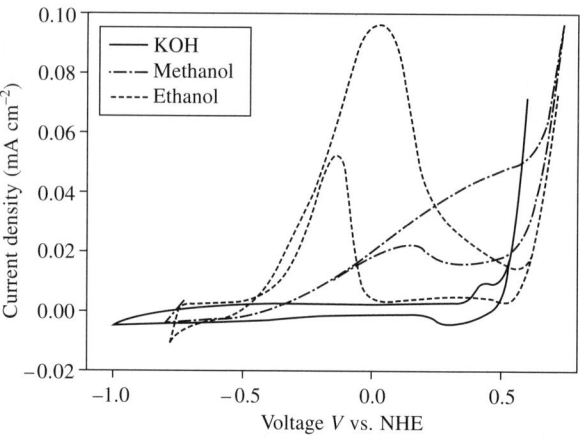

Fig. 8 CVs for electro-oxidation of methanol (1 M) plus KOH (1 M), ethanol (1 M) plus KOH (1 M) and 1 M KOH and 1 M KOH (no fuel) electrolyte solution on Pt electrode, Temperature: 25°C, Scan rate: 50 mV s^{-1}.

(Kordesch et al. 2004, Gouérec et al. 2004 and Verma et al. 2005b). In search of a suitable electrolyte Morallón et al. (1995), Tripković et al. (1996) and Wang et al. (2003) studied the methanol electro-oxidation on Pt in carbonate, bicarbonate and NaOH solutions to compare the results. They have found that the methanol oxidation on Pt (111) catalyst is lower in bicarbonate and carbonate solution than in NaOH solution. On the other hand, Morallón et al. (1995) reported the methanol electro-oxidation on Pt (110) and Pt (100) and they found the higher current density for carbonate and bicarbonate electrolytes than that for NaOH but the deactivation of the catalyst was very high (30-50%). Tripković et al. (1996) carried out cyclic voltammetry of methanol oxidation on a Pt (111) surface in different concentration of NaOH solutions. The CVs of Pt (111) electrode in three different concentrations of NaOH solution (without methanol) were showing two separate, well-defined zones. The first zone in the potential range of 0.06-0.4 V was associated with the hydrogen adsorption-desorption process. The origin of the sharp reversible peaks, which appear in the second zone at potential range of 0.6-0.85 V, indicate the formation of OH_{ad} species. The OH_{ad} species are formed at 0.55 V. Hydrogen adsorption-desorption process, appearing in the first zone, remains unchanged with changing concentration of NaOH solution. On the contrary, the second zone process undergoes changes caused by change in NaOH concentration. The CVs given in Fig. 9 demonstrate the oxidation behavior of 0.3 M CH_3OH on the Pt (111) electrode in 0.1 M NaOH. Here, methanol adsorbed in the hydrogen adsorption-desorption region, blocking sites that might be available for hydrogen adsorption (Fig. 9(a)). The reaction commences immediately after the hydrogen desorption and it proceeds further with small rates to the upper potential limit. The current decreases as shown by curves 1 and 2 of Fig. 9 (a), which corresponds to first and second sweep, respectively. The increase of surface coverage by adsorbed species in the hydrogen region may indicate surface poisoning. The current-time transient curve in Fig. 9 (b) supports the view that "poisoning species" are produced during methanol oxidation at $E = 0.55$ V. The amount of "poisoning species" formed depends on time. The maximum surface coverage by the poisoning species was reached after holding the electrode for 25 min in Methanol/NaOH solution. The CV in Fig. 9(c) correspond to the oxidation of irreversibly adsorbed species formed in the oxidation of methanol at $E = 0.55$ V. The influence of OH_{ad} and poisoning species on methanol oxidation is also studied.

The CVs of methanol oxidation in three different concentrations (0.01, 0.1, 1 M) of NaOH solutions

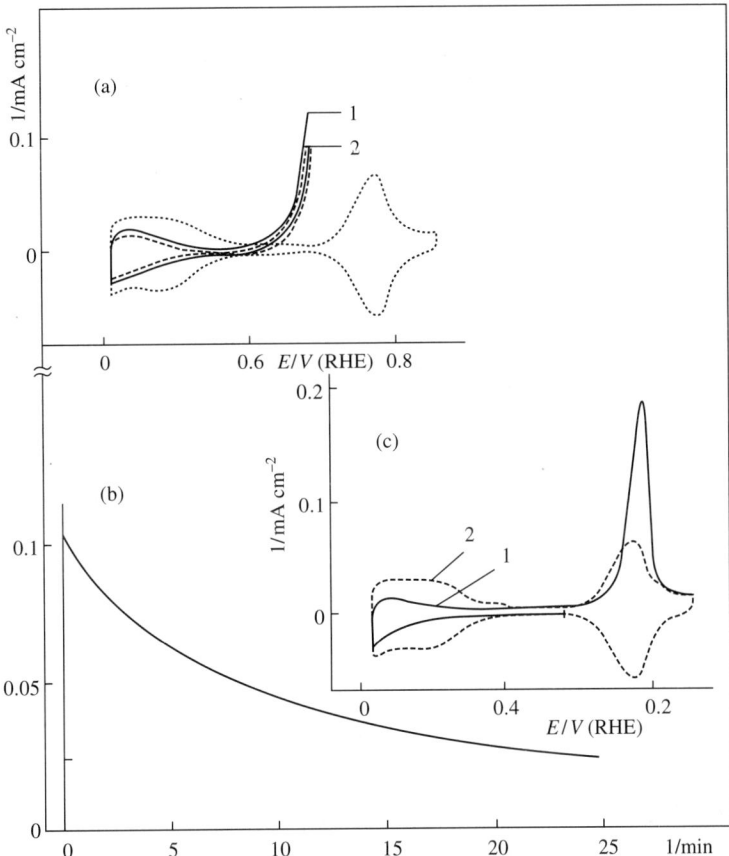

Fig. 9 (a) Cyclic voltammograms (first and second scan) for the oxidation of 0.3 M CH_3OH on the Pt (111) electrode in 0.1 M NaOH up to $E = 0.55$ V and the corresponding basic curve (dotted line). (b) Current-time transient of the methanol oxidation from the solution at $E = 0.55$ V. (c) Cyclic voltammograms for the oxidation of the irreversible adsorbed species formed in the oxidation of methanol. Temperature: 22°C, Scan rate: 50 mV s^{-1}.

were recorded and compared. Initial potential, peak potential and the potential where reaction finishes are at the same potential but the current density of the Pt (111) surface was different depending on the OH$_{ad}$ coverage. The coverage by the OH$_{ad}$ species is probably decreased by the presence of methanol in the solution, due to the competitive adsorption of OH$^-$ ions and methanol molecules. The poisoning species is formed during methanol oxidation because of decreased concentration of OH$_{ad}$. Thus, the current density should be decreased and featureless curve should be obtained. However, Morallon et al. (1995) found decrease in current density but featureless curve is not obtained. The results of Morallon et al. (1995) are contradictory in nature. Shobba et al. (2002) developed Co-W alloys by electroplating technique for the electro-oxidation of methanol in alkaline as well as acidic medium. The alloys were characterized in order to determine their prospects as effective anode materials for methanol fuel cells in both acidic and alkaline mediums. They have showed that the heat treatment of prepared catalyst increases the open circuit voltage by 5.5% (i.e. from 0.87 to 0.918 V) for 3 M CH_3OH/3 M KOH.

The catalytic activity of the nickel electrodes (Ni/C) for methanol oxidation in alkaline condition was studied by Rahim et al. (2004). The activity of the Ni/C varied with the amount of electrodeposited Ni.

They have recorded the cyclic voltammograms in 1.0 M KOH and 0.5 M methanol. The results suggested that methanol is oxidized on the nickel electrode through the reaction with NiO(OH) to form Ni(OH)$_2$ as follows:

$$Ni(OH)_2 + OH^- \leftrightarrow NiO(OH) + H_2O + e^- \tag{24}$$

$$NiO(OH) + \text{Organic compound} \xrightarrow{\text{slow}} Ni(OH)_2 + \text{Products} \tag{25}$$

They have reported that the accumulation of Ni(OH)$_2$ species evolves more stable species. The increasing amount of NiO(OH) seems to have inhibiting effects causing inactivity of the electrode. The electrode can be regenerated by periodic reactivation in the hydrogen evolution potential region. Golikand et al. (2005) modified the nickel electrode surface by nickel dimethyglyoxime complex for the electro-oxidation of methanol in alkaline medium to cope up with the inhibiting effects that causes inactivity of the nickel electrode. Khalil et al. (2005) prepared nickel impregnated zeolite-modified electrodes by mixing Ni-silicate-1 with carbon black in different ratios. The performance of Ni-zeolite electrodes is compared with the dispersed Ni and Pt on graphite. Ni-zeolite electrodes were found to be superior electrode-catalysts for methanol oxidation compared to the other electrodes in terms of current density obtained. Ni-zeolite electrodes have less blockage of active site than Pt-electrode due to repeated utilization. The comparative study of methanol electro-oxidation on different catalyst surface in alkaline solution is given in Table 2.

Gupta et al. (2004) studied the cyclic voltammograms (Fig. 10) of ethanol electro-oxidation behavior on CuNi, CuNi/Pt and CuNi/PtRu alloys electro-catalysts in 0.5 M NaOH solution. Fig. 10 (a) shows a steady rise of the anodic peak current for the CuNi/Pt electro-catalyst. The peak current increases substantially from 1st to the 50th scan. Fig. 10 (b) shows the increase in reaction kinetics for ethanol electro-oxidation when Ru is added in the alloy. They have detected the presence of acetaldehyde and CO$_2$ (as carbonate) with CuNi/PtRu electro-catalyst. Authors found carbonate ions because of the cleavage of C—C bond of ethanol molecule. The temperature of ethanol electro-oxidation was not mentioned although the experimental work was done at room temperature. Tripković et al. (2001) studied the electro-oxidation of methanol, ethanol, n-propanol and n-butanol (C$_1$—C$_4$ alcohol) in alkaline solution at the Pt (111) and vicinal stepped planes Pt (755) and Pt (332). The nature of the oxygen-containing species as well as their role in the alcohol oxidation is proposed. A dual path reaction mechanism as shown by eqs. (4) and (5) is proposed based on the assumptions that RCO$_{ad}$ is a reactive intermediate of the main reaction path, while CO$_2$ is a product of the poisoning species oxidation in a parallel reaction path.

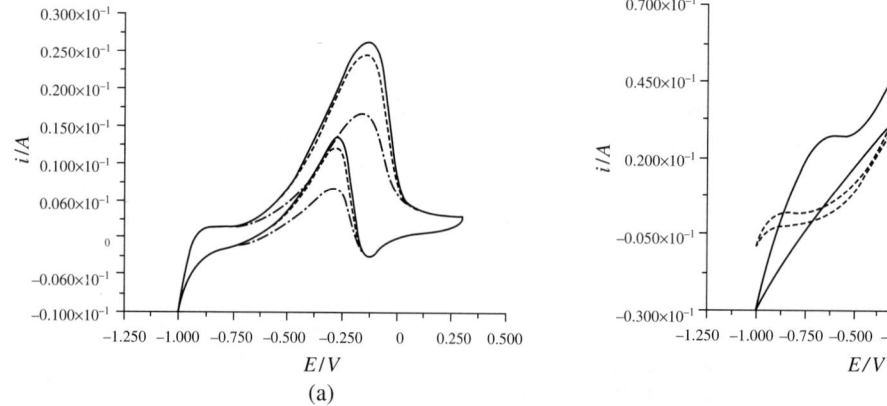

Fig. 10 CVs for 0.5 M NaoH/1.0 M ethanol: (a) CuNi/Pt, (—) 1st scan, (···) 25th scan, (——) 50th scan; (b) (···) CuNi/Pt (PTFE) and (——) CuNi/PtRu (PTFE). Temperature: 25°C, Scan rate: 50 mV s^{-1}.

Two kind of oxygen-containing species, i.e., PtOH and PtO were detected during the surface oxidation on Pt (111), Pt (755) and Pt (332) surfaces. Their formation are shown as follows:

$$Pt + OH^- \rightarrow PtOH + e^-$$ (26)

$$PtOH \rightarrow PtO + H^+ + e^-$$ (27)

They have found ethanol is the most active alcohol among C_1–C_4 alcohol on the Pt (111) surface. The reversibly adsorbed OH species were the active intermediates in alcohol oxidation but the irreversible adsorbed OH species are inactive, strongly bound intermediates, and act as a poison in the alcohol oxidation. Table 4 shows the comparative study of ethanol electro-oxidation in alkaline condition. It should be noted that Tripković et al. (2001) did not report the rupture of C—C bond on Pt catalyst whereas Gupta et al. (2004) reported the breaking of C—C bond on CuNi/PtRu catalyst.

Liu et al. (2003) studied the oxidation of BH_4^- on nickel electrocatalyst and reported that the reaction proceeds by the four-electron rather than eight-electron pathways. Normally, Na^+ or K^+ cations in the solution do not influence the four-electron reaction pathways. The four-electron reaction pathways and hydrolysis reaction leads to a decrease in efficiency of oxidation. A suitable catalyst should be identified such that the eight electron mechanism of electro-oxidation of $NaBH_4$ is followed. Otherwise, hydrogen gas generated from four-electron mechanism of electro-oxidation should be utilized to maintain higher efficiency.

3.2 Cathode

Gojković et al. (1999) proposed that active material for oxygen reduction reaction (eq. (13)) might be obtained by heat treatment of different Fe or Co and Ni precursors on high surface area carbon. They have presented a comprehensive electrochemical study of oxygen reduction on iron (III) tetramethoxyphenyl porphyrin chloride (FeTMPP-Cl) adsorbed on the black pearls (BP) carbon and heat treated up to 1000°C. The reaction kinetics in both acid and alkaline solutions using heat-treated catalyst is reported. The cyclic voltammograms (Fig. 11) were recorded with oxygen-saturated solutions of 0.1M H_2SO_4 or 0.1M NaOH in 1M methanol to study the effect of different anions and methanol on oxygen reduction. No trace of

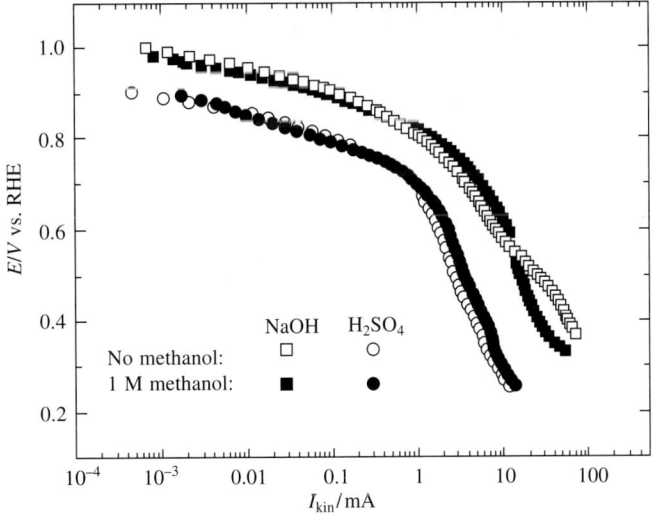

Fig. 11 Tafel plots for oxygen reduction on FeTMPP-Cl/BP heat treated at 800°C in 0.1 M NaOH or 0.1 M H_2SO_4 solution with or without methanol. Temperature 25°C, Scan rate: 10 mV s^{-1}, Rotational speed: 1000 rpm.

methanol oxidation current was observed. The reaction rate of oxygen reduction was not influenced by the presence of methanol either in acidic or alkaline solution, showing that the electrocatalyst is methanol tolerant. Further, the oxygen reduction reaction is not influenced by the presence of sulphate, perchlorate, or phosphate anions. The number of electrons liberated per molecule of oxygen molecule was estimated to be between 3.41 and 4, depending on the electrode potential. Verma et al. (2005a) studied the influence of methanol, ethanol and sodium borohydride in the reduction process of oxygen on manganese dioxide electrocatalyst in alkaline medium. The oxygen consumption at cathode in the alkaline electrolyte fuel cell was measured by gas displacement method. The rate of oxygen consumption is directly related to the number of electrons consumed per molecule of oxygen in the alkaline fuel cell (Verma et al. 2005a). It was found that 4.26 to 4.54 electrons were used in the oxygen electro-reduction reaction irrespective of fuel used. It is in well agreement with the theoretical 4 electrons requirement for the electro-reduction of oxygen. Demarconnay et al. (2005) estimated about four electrons evolved per oxygen molecule reduced on Ag catalyst. The nickel or cobalt based electrocatalysts showed poor activity of oxygen reduction reaction in alkaline medium. The oxygen reduction on nickel or cobalt material occurs via two-electron mechanism producing hydrogen peroxide as main product. On the other hand, spinel of cobalt and nickel oxide $Ni_xCo_{3-x}O_4$ $(0<x<1)$ showed a better activity towards oxygen reduction reaction in alkaline medium but via the 4-electron pathway (Heller-Ling et al. 1997).

Ponce et al. (2001) studied the oxygen reduction reaction and the oxygen evolution reaction on mixed oxides $Ni_xAl_{1-x}MnO_2$ $(0 \leq x \leq 1)$. The substitution of Al by Ni increases the catalytic activity, the maximum being exhibited by $NiMn_2O_4$. Nickel hydrous oxide and mixed Co and Ni oxides as electrocatalysts were found active towards oxygen reduction and evolution reaction (Rashkova et al. 2002).

Manganese oxides have been extensively studied and the activity of these catalysts depends on the types of oxides e.g., Mn_2O_3, Mn_3O_4, Mn_5O_8 and $MnOOH$. Normally, manganese dioxide consists of manganese in lower valence states such as Mn (II, III) together with the dominant Mn (IV) state. Mao et al. (2002), represented it as MnO_x and they studied many forms of manganese dioxide and found γ-$MnOOH$ compared well with the other varieties, e.g. MnO_2, α-Mn_2O_3, Mn_3O_4, Mn_5O_8. The cyclic voltammetric investigations of different forms of MnO_x in the presence of oxygen and argon saturated 0.1 M KOH medium showed two reduction current peaks. According to them the two reduction peaks correspond to 2 + 2 electron mechanism. Yang et al. (2003) synthesized nonporous amorphous manganese dioxide by means of aqueous redox sol-gel route. The catalyst was analyzed using oxygen or nitrogen saturated KOH solution through cyclic voltammetry. Yang et al. (2003) pointed out that only one reduction peak is obtained and the second peak could not be obtained because of the range of potential chosen. The study also investigated the suitability of the proposed manganese dioxide as an electrocatalyst to reduce oxygen in 1 M KOH. A current density of more than 100 mA/cm^2 was achieved by the half-cell at 0.45 volt using MnO_2 catalyst with a loading of 0.85 mg/cm^2. Verma et al. (2005a) studied the oxygen reduction reaction at the MnO_2 cathode in alkaline medium using cyclic voltammogram. The experiments were performed in either nitrogen or air-saturated alkaline electrolyte over the range of −0.6 to 0.2 V at a scan rate of 5 V s^{-1}. The cyclic voltammograms for the MnO_2 cathode in nitrogen or air-saturated electrolyte are shown in Fig. 12. The single reduction potential peak supports the four-electron path for the oxygen reduction reaction. The reduction peak clearly indicates that the MnO_2 is electro-active. The higher peak current is observed at −0.43 V for the air saturated than that for nitrogen-saturated alkaline electrolyte. The presence of air increases the cathodic current and expands the voltammogram. Demarconnay et al. (2005) prepared Ag/C catalyst using a colloidal route. An amount of 20 wt% Ag on carbon was found to be the best loading in terms of current density and mass activity. According to them the mechanism of oxygen electro-reduction on Ag catalyst is similar to that on Pt. Wagner et al. (2003) prepared cathode using silver catalyst on a high surface area porous electrode bonded with PTFE by a cold rolling process. They found a decrease in electrochemical performance

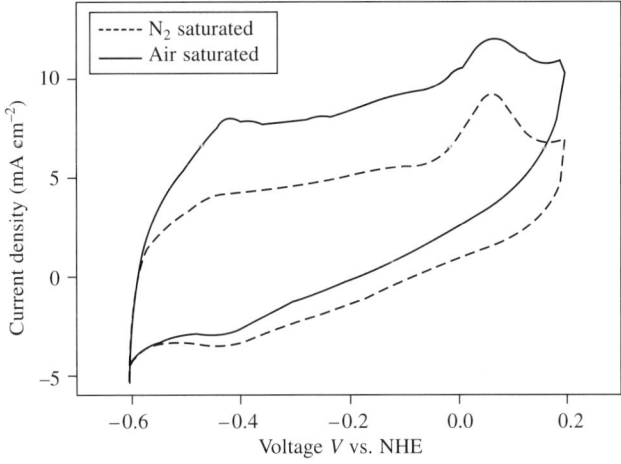

Fig. 12 CVs for MnO_2 cathode (3 mg cm^{-2}) in 3 M KOH saturated with air or nitrogen gas. Temperature: 25°C, Scan rate: 5 mV s^{-1}.

during oxygen reduction at 70°C in 30 wt% KOH. The linear decrease of the electrochemical performance has been found with a gradient of approximately 20 µV h^{-1} in an operation period of 5000 h. The total voltage loss was approximately 100 mV, which is 12-15% of the cell voltage in 5000 hours operation time. This gives a fair stability to mobile application but not for stationary application. The two reasons found out as a cause of decreased performance were the decrease of surface roughness and chemical deposition of PTFE during electrochemical operation. Choudhury et al. (2005) reported the use of H_2O_2 in an alkaline borohydride fuel cell to cope up with the problem of CO_2 contaminated air. The cell performance was highly dependent on H_2O_2 concentration and the optimum concentration of H_2O_2 was 4.45 M.

4. Cell Performance

Many investigators have started working on oxidation of methanol, ethanol or sodium borohydride and reduction of oxygen in alkaline medium in a half-cell as discussed in previous sections. But, surprisingly very few investigators are working on the single cell or stack of direct methanol or ethanol alkaline fuel cell. Table 5 shows the salient features of the different studies by several investigators on direct alcohol and sodium borohydride alkaline fuel cell. Vielstich (in Koscher et al. 2003) fabricated laboratory cell with platinum as anode electro-catalyst (2-5 mg cm^{-2}) and active carbon (250 mg cm^{-2}) as cathode electro-catalyst. The open circuit voltage of about 0.9 V was achieved when a mixture of 10 M KOH and 4.5 M methanol was fed to the cell. The terminal voltage dropped to 0.6-0.75 V at a current of 0.5 A (electrode area is not mentioned by the author). A five-fold higher current density was obtained when Pt/Ru catalyst was used instead of platinum catalyst.

In 1970, J. E. Wynn (in Koscher et al. 2003) described a fuel cell that contains three electrodes, an anode, a cathode and a grid electrode (Fig. 13). The anode consisted of an unsintered Ag/PdO catalyst in a carbon/ polytetrafluoroethylene (PTFE) bonded structure. The cathode consisted of a platinized carbon/PTFE bonded structure (15% Pt by wt.). A mixture of 9 M KOH and 6 M methanol was used as the anolyte and 9 M KOH solution was used as the catholyte. The anolyte was separated from the catholyte by an anion exchange membrane. The third electrode, a platinum catalyzed nickel grid, was placed in-between the membrane and the cathode. The grid electrode was electrically connected to the cathode through a small resistance. The three-electrode cell assembly was used to reduce the unwanted methanol oxidation at cathode. Any methanol that

Table 5. Different electro-catalyst used in direct methanol, ethanol or sodium borohydride alkaline fuel cell

Fuel/oxidant	System information			Operating temperature (°C)	Current density (mA cm^{-2})	OCV (V)	References
	Anode	Cathode	Electrolyte				
Methanol/not specified	Fe(III)-treated graphite	Ag(I)-treated graphite	6 M KOH	25	18	0.85	Verma (2000)
Methanol/air	Pt/C	Pt/C	Anion exchange membrane	60	69.3	0.75	Yu et al. (2004)
Methanol/air	Pt/C	MnO$_2$	3 M KOH	25	28.5	1.1	Verma et al. (2005a)
Ethanol/air	Pt/C	MnO$_2$	3 M KOH	25	34	1.08	Verma et al. (2005a)
Sodium borohydride/ air	Au/Pt	Not specified	Anion conducting membrane	25	1.1	152	Amendola et al. (1999)
Sodium borohydride/ oxygen	Zr-Ni alloy	Pt-black	Na$^+$ form of Nafion-117 membrane	60	1.2	300	Li et al. (2003)
Sodium borohydride/ air	Pt/C	MnO$_2$	3 M KOH	25	0.98	39	Verma et al. (2005a)
Sodium borohydride/ hydrogen peroxide	LaCeNdPrNi AlMnCo and ZrTiV MnCrC oNi alloys	Pt/C	Pretreated Nafion-117	70	1.25	500	Choudhury et al. (2003)

diffuses through the anion exchange membrane was electrochemically oxidized at the grid. The cells with grid electrode showed a better performance in the long hours of tests than cells without grids. It is because the grid reduces the oxidation of methanol on platinized carbon cathode thus increasing the cell performance and life span of cathode. Verma (2000) studied metal-ion impregnated graphite electrode for methanol-air fuel cell in alkaline medium. The spent graphite electrodes were used with and without impregnation by Fe^{3+} to prepare three kinds of anodes. The cathode was prepared by Ag$^+$ impregnated on graphite electrode. The Al^{3+} impregnated graphite electrodes performed better than Fe^{3+} impregnated electrode in methanol-air alkaline fuel cell. The maximum current density of 54 and 18 mA cm^{-2} were obtained for Al^{3+} and Fe^{3+} treated graphite anode, respectively.

Verma et al. (2005c) investigated the performance of direct methanol, ethanol and sodium borohydride

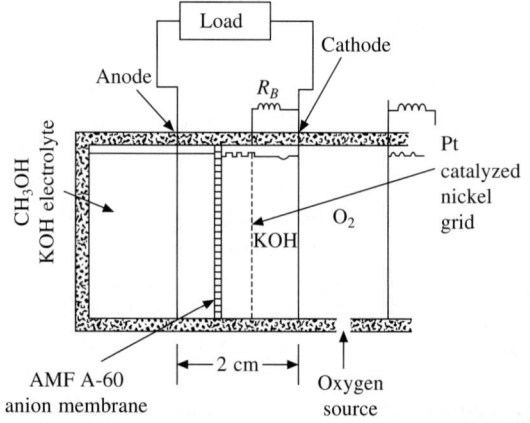

Fig. 13 Schematic diagram of methanol/air alkaline fuel cell.

alkaline fuel cell. Pt and Pt/Ru electro-catalysts were used at anode and manganese dioxide electro-catalyst was used at cathode. Nickel mesh was used as a current collector. The electrodes were prepared by wet fabrication method. Fig. 14 (a) shows the schematic diagram of direct alcohol or sodium borohydride alkaline fuel cell. The prepared electrodes were used in the alkaline fuel cell to investigate the effect of various parameters like electrolyte concentration, fuel concentration, temperature variation, catalyst loading and the use of different catalyst at anode and cathode. The photograph of the experimental set up and fuel cell is shown in Fig. 14 (b). The current density-cell voltage characteristics in Figs. 15 to 17 show the effect of different KOH concentrations for methanol, ethanol and sodium borohydride. For a given load the cell voltage increases with the increase in KOH concentration and then it decreases with further increase in KOH concentration. The reason may be the relative decrease in concentration of fuel in the presence of KOH solution, which hinders methanol to come in contact with anode electrode. Although the increase in KOH concentration has minimum effect on activation polarization, the concentration polarization increases with the increase in KOH concentration because of less availability of methanol at the anode.

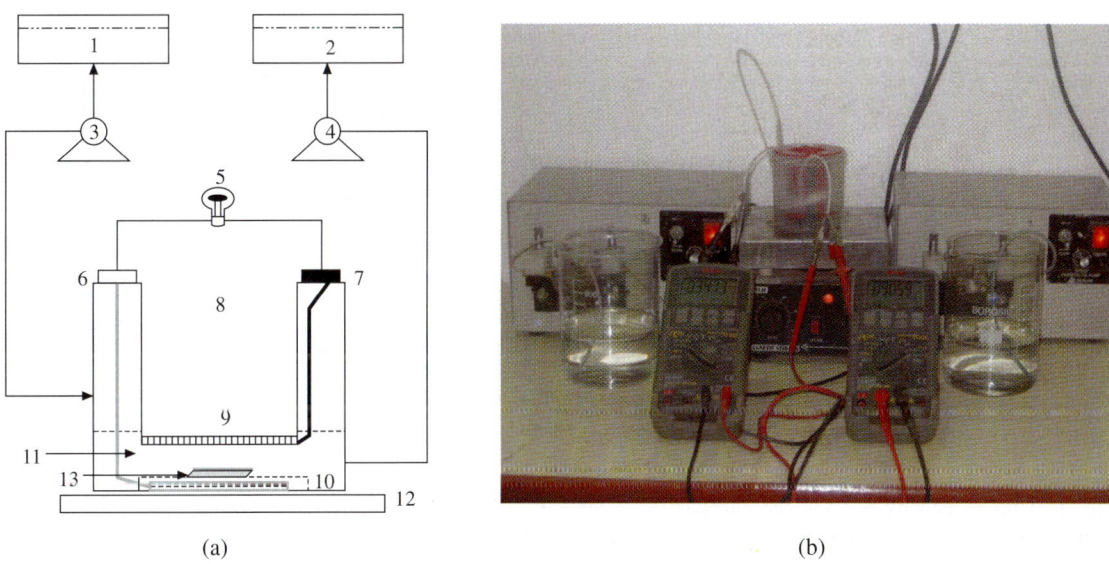

(a)　　　　　　　　　　　　　(b)

Fig. 14 (a) Schematic diagram of direct alcohol or borohydride alkaline fuel cell. 1. Fuel-electrolyte mixture storage; 2. Exhausted-fuel-electrolyte mixture storage; 3, 4. Peristaltic pump; 5. Load; 6. Anode terminal; 7. Cathode terminal; 8. Air; 9. Cathode electrode; 10. Anode electrode; 11. Fuel and electrolyte mixture; 12. Magnetic stirrer; 13. Anode shield. (b) Experimental set-up for direct alchol or sodium borohydride alkaline fuel cell.

Further, the mobility of hydroxyl ions reduces significantly with the increase in KOH electrolyte. The cyclic voltammogram results pointed out the formation of PtO layer at high KOH concentration. The formation of PtO layer deactivates the catalyst. The maximum power densities of 15, 10.5 and 22.5 mW cm^{-2} were obtained for methanol, ethanol and sodium borohydride, respectively, with 3 M KOH concentration at 25°C. The catalyst loadings used at anode and cathode were 1 mg cm^{-2} (Pt-black) and 3 mg cm^{-2} (MnO$_2$), respectively.

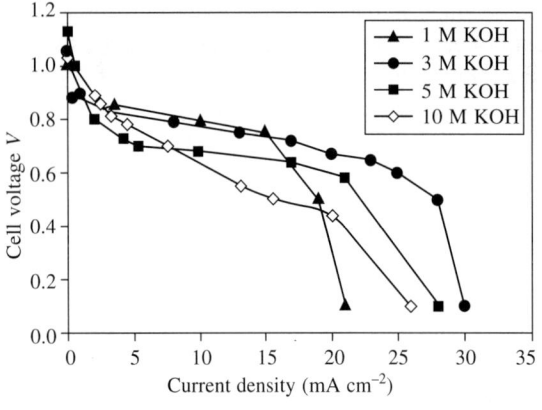

Fig. 15 Current density-cell voltage characteristics for methanol in different electrolyte concentration in direct alcohol fuel cell at 25°C, Anode; Pt-black; Cathode: MnO$_2$.

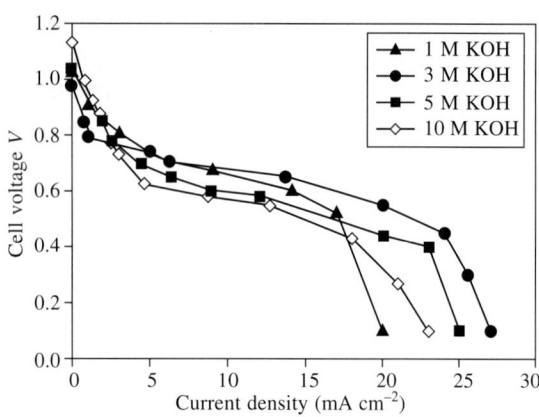

Fig. 16 Current density-cell voltage characteristics for ethanol in different electrolyte concentration in direct alcohol alkaline fuel cell at 25°C, Anode: Pt-black; Cathode: MnO$_2$.

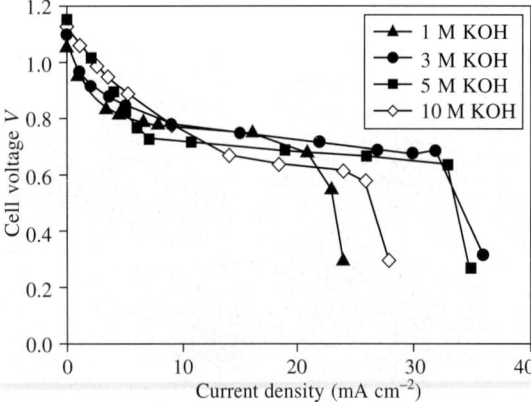

Fig. 17 Current density-cell voltage characteristics for sodium borohydride in different electrolyte concentration in direct borohydride alkaline fuel cell at 25°C, Anode: Pt-black; Cathode: MnO$_2$.

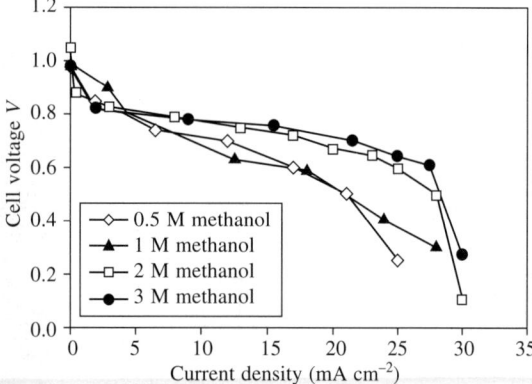

Fig. 18 Current density-cell voltage characteristics for different concentration of methanol in 3 M KOH electrolyte in direct alcohol alkaline fuel cell at 25°C, Anode: Pt-black; Cathode: MnO$_2$.

Figures 18 to 20 show that the equilibrium cell voltage increases with the increase in fuel concentration. Although the cell performance increases initially but it does not increase proportionally with further increase in fuel concentration. This is because the increase in fuel concentration leads to the decrease in hydroxyl ion mobility. The hydrolysis reaction dominates with the increase in sodium borohydride concentration and thus the performance increases rather slowly. Further at higher concentration of NaBH$_4$, viscosity of the fuel-electrolyte mixture increases leading to the rapid increase in concentration polarization at higher current densities and the performance decreases (Fig. 20). The maximum power density of 16.2 and 13.8 mW cm^{-2} were obtained for 3 M methanol and ethanol concentrations while 22.5 mW cm^{-2} for 2 M sodium borohydride. The fuel cell was operated at 25°C, 3 M KOH concentration and with 1 mg cm^{-2} of anode catalyst (Pt-black) loading catalyst and 3 mg cm^{-2} of cathode (MnO$_2$) loading, respectively.

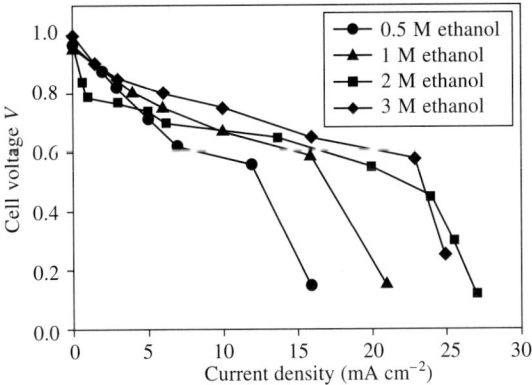

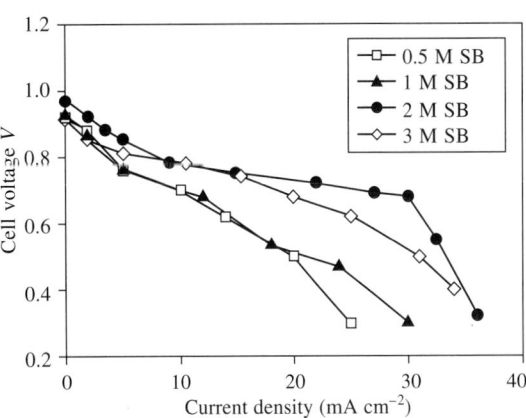

Fig. 19 Current density-cell voltage characteristics for different concentration of ethanol in 3 M KOH electrolyte in direct alcohol fuel cell at 25°C, Anode: Pt-black; Cathode: MnC.

Fig. 20 Current density-cell voltage characteristics for different concentration of sodium borohydride (SB) in 3 M KOH electrolyte in direct borohydride alkaline fuel cell at 25°C, Anode: Pt-black; Cathode: MnO_2.

Figure 21 shows that the cell performance increases with the increase in cathode catalyst (MnO_2) loading up to a certain level when methanol was used as fuel. The cell performance decreases with the further increase in cathode catalyst loading. The reason may be attributed to the saturation of the active site per unit area and the increase in flow resistance as large amount of catalyst compacted to the same given area. Also, the performance decreases because the MnO_2 is poor electrical conductor. Similar trend in cathode catalyst loading was observed for ethanol and sodium borohydride. To be specific, the performance of the fuel cell increases with the increase in MnO_2 loading at cathode from 1 to 3 mg cm^{-2} and then decreases with further increase in catalyst loading. The catalyst loading at anode was varied from 0.5 to 1.5 mg cm^{-2}. The optimum value of anode catalyst loading was 1 mg/cm^2. Beyond this loading no appreciable change in current density was observed at a particular voltage for three different catalysts (Pt/C, Pt-black, Pt/Ru) and fuels. The effect of catalyst loading is presented in Figs. 22 and 23 for Pt-black and Pt/Ru (40%:20% by wt.)/C when methanol was used as fuel.

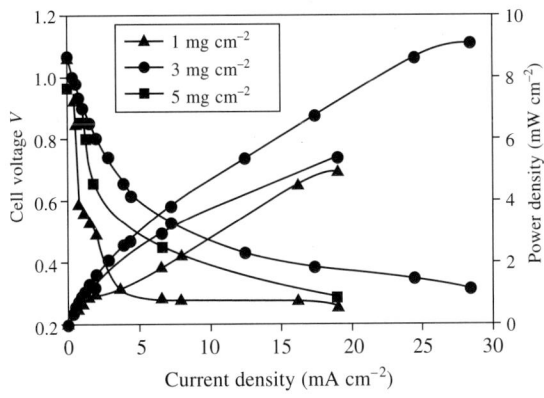

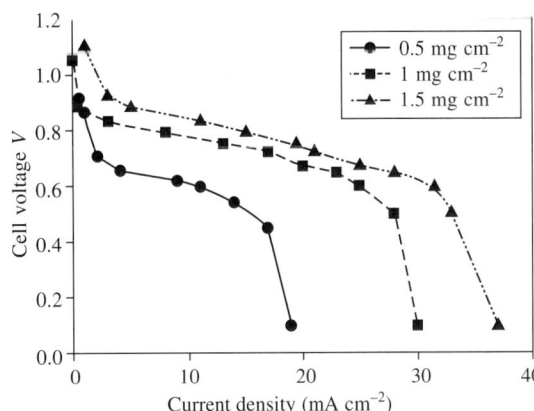

Fig. 21 Current density-cell voltage and power density characteristic for different concentration of MnO_2 catalyst at cathode in 2 M methanol/3 M KOH solution in direct alcohol alkaline fuel cell at 25°C, Anode: Pt-black.

Fig. 22 Current density-cell voltage characteristics for 2 M methanol and 3 M KOH solution at different loading of anode catalyst in direct alcohol alkaline fuel cell at 25°C, Anode: Pt-black; Cathode: MnO_2.

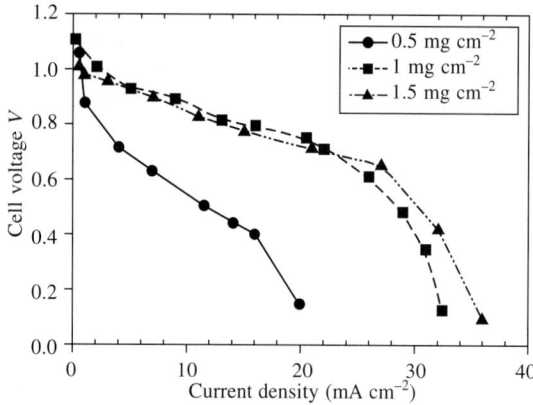

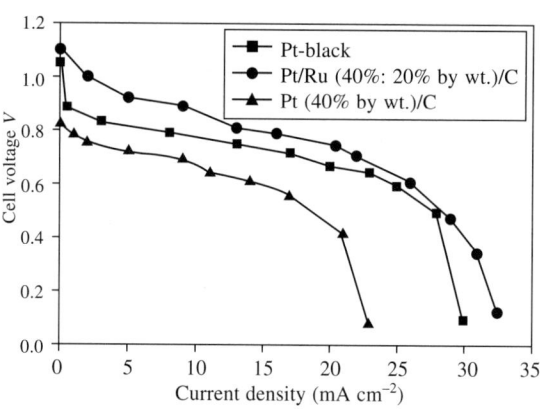

Fig. 23 Current density-cell voltage characteristic for 2 M methanol and 3 M KOH solution at different loading of anode catalyst in direct alcohol alkaline fuel cell at 25°C, Anode: Pt/Ru (40%: 20% by wt.)/C; Cathode: MnO₂.

Fig. 24 Current density-cell voltage characteristics for different catalyst in 2 M methanol/3 M KOH solution in direct alcohol alkaline fuel cell at 25°C, Anode: Different Pt-based catalyst; Cathode: MnO₂.

The different catalysts are tested at the anode, e.g. Pt-black, Pt/Ru (40%: 20% by wt.)/C and Pt (40% by wt.)/C. The performance of direct methanol and ethanol fuel cell is best with Pt/Ru (40%: 20% by wt.) electro-catalyst at anode (Figs. 24 and 25). The difference in performance between Pt-black and Pt/Ru is minimal. The performance of $NaBH_4$ is hardly affected irrespective of the catalyst used (Fig. 26).

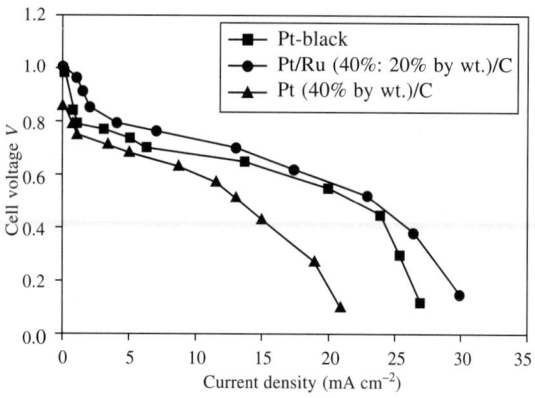

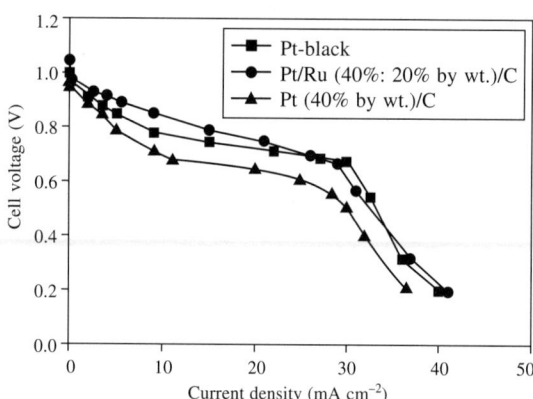

Fig. 25 Current density-cell voltage characteristics for different catalyst in 2 M ethanol/3 M KOH solution in direct alcohol alkaline fuel cell at 25°C, Anode: Different Pt-based catalyst; Cathode: MnO₂.

Fig. 26 Current density-cell voltage characteristics for different catalyst in 2 M $NaBH_4$/3 M KOH solution in direct borohydride alkaline fuel cell at 25°C, Anode: Different Pt-based catalyst; Cathode: MnO₂.

The current density-cell voltage characteristics for methanol, ethanol and sodium borohydride fuels were shown in Figs. 27 to 29 at three different temperatures e.g., 25, 45, and 65°C. The cell performance increases with the increase in temperature because of decrease in activation polarization, concentration polarization and increase in ionic conductivity and mobility at higher temperature. The performance of direct sodium borohydride alkaline fuel cell does not increase appreciably with the increase in temperature and in fact shows decreasing trend at 65°C (Verma et al. 2005d). The reason for this decrease may be because of hydrogen gas liberation from sodium borohydride and loss of fuel at higher temperature.

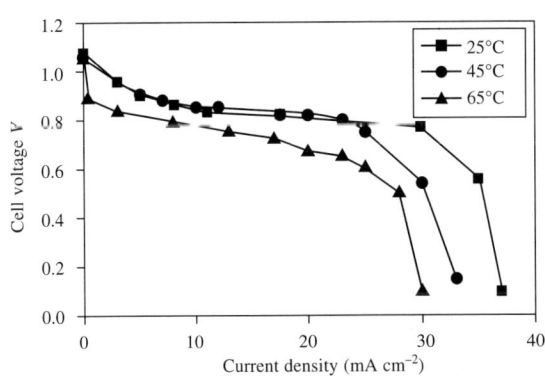

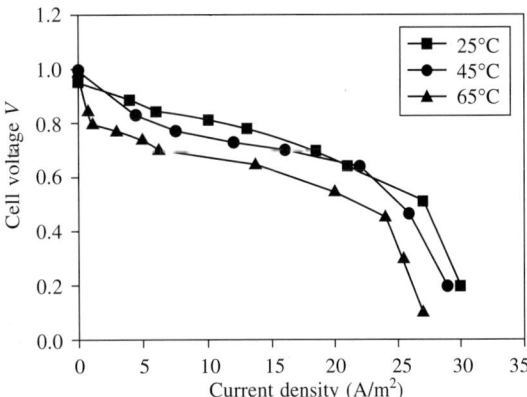

Fig. 27 Current density-cell voltage characteristics at different temperatures in 2 M methanol/3 M KOH solution in direct alcohol alkaline fuel cell. Anode: Pt Black; Cathode: MnO_2.

Fig. 28 Current density-cell voltage characteristics at different temperatures in 2 M ethanol/3 M KOH solution in direct alcohol alkaline fuel cell. Anode: Pt-black; Cathode: MnO_2.

The maximum power densities of 24.3 and 14.5 mW cm^{-2} were obtained for methanol and ethanol at 65°C while that for NaBH$_4$ obtained was 22.5 mW cm^{-2} at 25°C. The KOH concentration used was 3 M and fuel was 2 M. The catalyst loadings at anode (Pt-black) and cathode (MnO$_2$) were 1 and 3 mg cm^{-2}, respectively.

The lifetime test of direct alcohol and sodium borohydride alkaline fuel cell was conducted and the results pertaining to this is shown in Fig. 30. The useful operating lifetime of 380, 400, 510 h was found for methanol, ethanol and sodium borohydride fuels, respectively. The deterioration of performance of fuel cell may be because of the carbonate precipitate, oxide layer formation and adsorbed intermediate species on catalyst surface. The used up electrodes could be regenerated by treating the electrode with hydrochloric acid. The acid treatment might have removed the carbonates and other species from the electrode. The treated electrodes could regain more than 80% of the catalytic activity. The maximum power density with 2 M fuel and 3 M KOH obtained was 21.5 mW cm^2 at 33 mA cm^{-2} of current density for sodium borohydride at 60°C, whereas, methanol and ethanol produce 15 and 16 mW cm^{-2} of maximum power

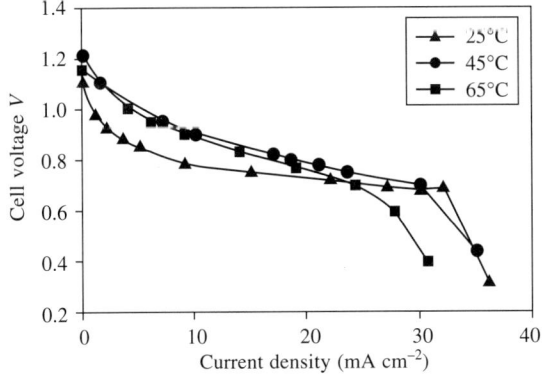

Fig. 29 Current density-cell voltage characteristics at different temperatures in 2 M NaBH$_4$/3 M KOH solution in direct borohydride alkaline fuel cell. Anode: Pt-black; Cathode: MnO_2.

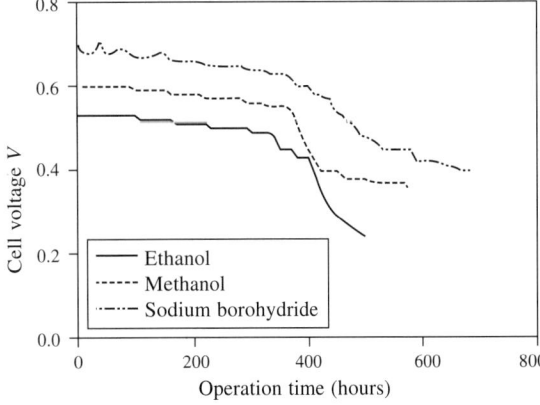

Fig. 30 Lifetime of the direct alcohol and sodium borohydride fuel cell at constant load.

density at 20 and 30 mA cm^{-2} of current density, respectively. A photograph of two cell stack lighting a 1.5 watt bulb by direct sodium borohydride alkaline fuel cell is shown in Fig. 31.

Fig. 31   A two cell stack of direct sodium borohydride alkaline fuel cell used for lighting a bulb.

Yu et al. (2004) carried out direct methanol fuel cell tests with anion exchange membrane electrolyte. Platinised Ti mesh was used as anode and it is compared with conventional Pt/C catalysts. The platinized Ti mesh anode showed maximum power density of 7.8 mW cm^{-2} as compared to 7 mW cm^{-2} for conventional Pt/C electrode when 2 M methanol and 1 M NaOH was used at 60°C. Prior to Yu et al. (2004), Amendola et al. (1999) used anion exchange membrane, Au/Pt as anode and borohydride with NaOH solution mixture as fuel (prevents hydrolysis reaction) to study borohydride alkaline fuel cell. However, the anion exchange based borohydride alkaline fuel cell suffered from the BH_4^- crossover. Seven electrons were generated instead of theoretically 8 electrons, as some portion of feed was lost due to hydrolysis reaction. The maximum power density of 20 mW cm^{-2} at 25°C and 60 mW cm^{-2} at 70°C were generated with the 97% Au/3% Pt on a conductive silk anode, cathode and anion exchange membrane. Gyenge (2004) studied the effect of additives (thio-urea and quaternary ammonium ions) on the fuel utilization with the aim to minimize the hydrogen evolution on Pt. They have found that the thio-urea (TU) minimizes the hydrolysis reaction due to the inhibiting effect of TU on the recombination of surface adsorbed species. The direct oxidation of BH_4^- is not affected by the presence of TU thus it could improve the BH_4^- utilization efficiency. Choudhury et al. (2005) reported an alkaline borohydride fuel cell with hydrogen peroxide as oxidant for use in underwater application. They have used treated Nafion membrane, LaCeNdPrNiAlMnCo or ZrTiVMnCrCoNi alloys as anode and Pt/C as cathode. The maximum power density of about 150 mW cm^{-2} was generated at 70°C. The reported studies only point out that there is a trend in surge of research activities on direct alcohols and borohydride based alkaline fuel cell.

5. Future Directions

The future of direct alcohol or borohydride alkaline fuel cell is promising in nature for powering future electronic equipments as the single cell results are quite impressive in terms of power density. The other advantages being its simplicity, possible operation close to ambient temperature and lower cost. The main hurdle to be overcome is the increase in life span to 5000 h without any interruption of power supply. To make the direct alcohol alkaline fuel cell competitive with other low temperature fuel cell technologies, certain issues need to be addressed. These are CO_2 poisoning, by-product identification and its effect on catalyst (if any) especially when alcohols are used as fuel. These issues may be dealt through half-cell analyses of different electrode systems in different fuel-electrolyte combinations and concentrations, understanding the reaction mechanism, the effect of by-product formation and catalyst deactivation. The fuel oxidation at cathode should be minimal as fuel-alkaline electrolyte mixture is in direct contact with both anode and cathode. This is partially achieved by using non-noble metal catalyst at cathode. However, development of anion exchange membrane and its use instead of liquid alkaline electrolyte would stop fuel oxidation at cathode and the loss of potential can be avoided. The specially designed anion exchange membrane should not allow crossover of fuel through the membrane. Finally, stack design, fabrication and stack testing is required for this particular type of fuel cell. Since direct alcohol alkaline fuel cell (DAAFC) may have potential application in portable electronic gadget, it worth to fabricate micro DAAFC stack for testing.

References

Ahuja, V. and Green, R. "Carbon dioxide removal from air for alkaline fuel cells operating with liquid hydrogen- Heat exchanger development", *Int. J. Hydrogen Energy*, **21** (1996) 415-421.

Ahuja, V. and Green, R. "Carbon dioxide removal from air for alkaline fuel cells operating with liquid hydrogen- A synergistic advantage", *Int. J. Hydrogen Energy*, **23** (1998) 131-137.

Al-Saleh, M.A., Gultekin, S., Al-Zakri, A.S. and Celiker, H. "Effect of Carbon Dioxide on the Performance of Ni/PTFE and Ag/PTFE Electrodes in an Alkaline Fuel Cell" *J. Appl. Electrochem.*, **24** (1994) 575-80.

Amendola, S.C., Onnerud, P., Kelly, M.T., Petillo, P.J., Sharp-Goldman, S.L. and Binder, M. "A novel high power density borohydride-air cell", *J. Power Sources*, **84** (1999) 130-133.

Appleby, A.J. and Foulkes, F.R. "Fuel cell handbook", Van Nostrand Reinhold (1989), 261-312.

Bard, A.J. and Faulkner, L.R. "Electrochemical methods: Fundamentals and Applications" John Wiley & Sons, 2nd ed. (2001) 226-260.

Burchardt, T., Gouérec, P., Sanchez-Cortezon, E., Karichev, Z. and Miners, J. H. "Alkaline fuel cells: contemporary advancement and limitations", *Fuel*, **81** (2002) 2151-2155.

Bezdička, P., Grygar, T., Klápště, B. and Vondrák, J., "MnO$_x$/C composites as electrode materials. I. Synthesis, XRD and cyclic voltammetric investigation", *Electrochim. Acta*, **45** (1999) 913-920.

Carrette, L., Friedrich, K.A. and Stimming, U. "Fuel cells—fundamentals and applications", *Fuel Cells*, **1** (2001) 5 39.

Choudhury, N.A., Raman, R.K., Sampath, S. and Shukla, A.K. "An alkaline direct borohydride fuel cell with hydrogen peroxide as oxidant, *J. Power Sources*, **143** (2005) 1-8.

Cifrain, M. and Kordesch, K. "Hydrogen/oxygen (air) fuel cells with alkaline electrolytes", in: "Handbook of fuel cells—fundamentals, technology and applications", Vielstich, W., Gasteiger, H.A., Lamm, A. (Eds.), John Wiley, Vol. 1 (2003a) 267-280.

Cifrain, M. and Kordesch, K.V. "Advances, aging mechanism and lifetime in AFCs with circulating electrolyes", *J. Power Sources*, **127** (2003b) 234-242.

Demarconnay, L., Coutanceau, C. and Léger, J.-M. "Electroreduction of dioxygen (ORR) in alkaline medium on Ag/ C and Pt/C nanostructured catalysts-effect of the presence of methanol", *Electrochim. Acta*, **49** (2005) 4513-4521.

Gojković, S.Lj., Gupta, S. and Savinell, R.F. "Heat-treated iron(III) tetramethoxyphenyl porphyrin chloride supported on high-area carbon as an electrocatalyst for oxygen reduction. Part II. Kinetics of oxygen reduction", *J. Electroanal. Chem.*, **462** (1999) 63-72.

Golikand, A.N., Shahrokhian, S., Asgari, M., Maragheh, M.G., Irannejad, L. and Khanchi, A. "Electrocatalytic oxidation of methanol on a nickel electrode modified by nickel dimethylglyoxime complex in alkaline medium", *J. Power Sources*, **144** (2005) 21-27.

Gouérec, P., Poletto, L., Denizot, J., Sanchez-Cortenzon, E. and Miners, J.H. "The evolution of the performance of alkaline fuel cells with circulating electrolyte", *J. Power Sources*, **129** (2004) 193-204.

Gupta, S.S., Mahapatra, S.S. and Datta, J. "A potential anode material for the direct alcohol fuel cell", *J. Power Sources*, **131** (2004) 169-174.

Gülzow, E. and Schulze, M. "Long-term operation of AFC electrodes with CO_2 containing gases", *J. Power Sources*, **127**, (2004) 243-251.

Gyenge, E. "Electrooxidation of borohydride on platinum and gold electrodes: implications for direct borohydride fuel cells", *Electrochim. Acta*, **49** (2004) 965-978.

Heller-Ling, N., Prestst, M., Gautier, J.-L., Koenig, J.-F., Poillerat, G. and Chartier, P. "Oxygen electroreduction mechanism at thin $Ni_xCo_{3-x}O_4$ spinel film in a double channel electrode flow cell (DCEFC)", *Electrochim. Acta*, **42** (1997) 197-202.

Klápště, B., Vondrák, J. and Velická, J. "MnO_x/C composites as electrode materials II. Reduction of oxygen on bifunctional catalysts based on manganese oxides", *Electrochem. Acta*, **47** (2002) 2365-2369.

Khalil, M.W., Rahim, M.A.A., Zimmer, A., Hassan, H.B. and Hameed, R.M.A. "Nickel impregnated silicalite-1 as an electro-catalyst for methanol oxidation", *J. Power Sources*", **144** (2005) 35-41.

Kordesch, K., Gsellmann, J., Cifrain, M., Voss, S., Hacker, V., Aronson, R., Fabjan, C., Hejze, T. and Daniel-Ivad, J. "Intermittent use of a low-cost alkaline fuel cell-hybrid system for electric vehicles", *J. Power Sources*, **80** (1999) 190-197.

Kordesch, K., Cifrain, M., Koscher, G., Hejze, T. and Hacker, V. "A survey of fuel cell systems with circulating electrolytes", Power Sources Conference 2004, Philadelphia, June 14-17.

Koscher, G.A. and Kordesch, K. "Alkaline methanol/air power devices", in: "Handbook of fuel cells—fundamentals, technology and applications", Vielstich, W., Gasteiger, H.A., Lamm, A. (Eds.), John Wiley, Vol. 4 (2003) 1125-1129.

Lamy, C., Belgsir, E.M. and Léger, J-M. "Electrocatalytic oxidation of aliphatic alcohols: Application to the direct alcohol fuel cell (DAFC)", *J. Appl. Electrochem.*, **31** (2001) 799-809.

Lee, S., Kim, J., Lee, H., Lee, P. and Lee, J. "The characterization of an alkaline fuel cell that uses hydrogen storage alloys", *J. Electrochem. Soc.*, **149** (2002) A603-A606.

Li, Z.P., Liu, B.H., Arai, K. and Suda, S. "A fuel cell development for using borohydrides as the fuel", *J. Electrochem. Soc.*, **150** (2003a) A868-A872.

Li, Z.P., Liu, B.H., Arai, K., Asaba, K. and Suda, S. "Evaluation of alkaline borohydride solutions as the fuel for fuel cell", *J. Power Sources*, **126** (2003b) 28-33.

Liu, B.H., Li, Z.P. and Suda, S. "Anodic oxidation of alkali borohydrides catalyzed by nickel", *J. Electrochem. Soc.*, **150** (2003) A398-A402.

Manoharan, R. and Prabhuram, J. "Possibilities of prevention of formation of poisoning species on direct methanol fuel cell anodes", *J. Power Sources*, **96** (2001) 220-235.

Mao, L., Sotomuro, T., Nakatsu, K., Koshiba, N., Zhang, D. and Ohsaka, T. "Electrochemical characterization of catalytic activities of manganese oxide to oxygen reduction in alkaline aqueous solution", *J. Electrochem. Soc.*, **149** (2002) A504-A507.

McLean, G.F., Niet, T., Prince-Richard, S. and Djilali, N. "An assessment of alkaline fuel cell technology", *Int. J. Hydrogen Energy*, **27** (2002) 507-526.

Morallón, E., Rodes, A., Vázquez, J.L. and Pérez, J.M. "Voltammetric and *in-situ* FTIR spectroscopic study of the oxidation of methanol on Pt(hlk) in alkaline media", *J. Electroanal. Chem.*, **391** (1995) 149-157.

Morris, J.H., Gysling, H.J. and Reed, D. "Electrochemistry of boron compounds", *Chem. Rev.*, **85** (1985) 51-76.

Ortiz, J., Puelma, M. and Gautier, J.L. "Indirect oxidation of phenol on graphite on $Ni_{0.3}Co_{2.7}O_4$ spinel electrodes in alkaline medium", *J. Chil. Chem. Soc.*, **48** (2003) 67-71.

Parsons, R. and VanderNott, T. "The oxidation of small organic molecules", *J. Electroanal. Chem.*, **257** (1988) 9-45.

Prabhuram, J. and Manoharan, R. "Investigation of methanol oxidation on unsupported platinum electrodes in strong alkali and strong acid", *J. Power Sources*, **74** (1998) 54-61.

Ponce, J., Rehspringer, J.-L., Poillerat, G. and Gautier, J.L. "Electrochemical study of nickel-aluminium-manganese spinel $Ni_xAl_{1-x}Mn_2O_4$. Electrocatalytic properties for the oxygen evolution reaction and oxygen reduction reaction in alkaline media", *Electrochim. Acta*, **46** (2001) 3373-3380.

Rashkova, V., Kitova, S., Konstantinov, I. and Vitanov, T. "Vacuum evaporated thin films of mixed cobalt and nickel oxides as electrocatalyst for oxygen evolution and reduction", *Electrochim. Acta*, **47** (2002) 1555-1560.

Rahim, M.A.A., Hameed, R.M.A. and Khalil, M.W. "Nickel as a catalyst for the electro-oxidation of methanol in alkaline medium" *J. Power Sources*, **134** (2004) 160-169.

Schulze, M. and Gülzow, E. "Degradation of nickel anodes in alkaline fuel cells", *J. Power sources*, **127** (2004) 252-263.

Shobba, T., Mayanna, S.M. and Sequeira, C.A.C. "Preparation and characterization of Co-W alloys as anode materials for methanol fuel cells", *J. Power Sources*, **108** (2002) 261-264.

Tarasevich, M.R., Karichev, Z.R., Bogdanovskaya, V.A., Lubnin, E.N. and Kapustin, A.V. "Kinetics of ethanol electrooxidation at RuNi catalyst", *Electrochemistry Communications*, **7** (2005) 141-146.

Tripković, A.V., Popović, K.Dj., Momčilović, J.D. and Dražić, D.M. "Kinetic and mechanistic study of methanol oxidation on a Pt(111) surface in alkaline media", *J. Electroanal. Chem.*, **418** (1996) 9-20.

Tripković, A.V., Popović, K.Dj., Momčilović, J.D. and Dražić, D.M. "Kinetic and mechanistic study of methanol oxidation on a Pt(110) surface in alkaline media", *Electrochimica Acta*, **44** (1998a) 1135-1145.

Tripković, A.V., Popović, K.Dj., Momčilović, J.D. and Dražić, D.M. "Kinetic and mechanistic study of methanol oxidation on a Pt(100) surface in alkaline media", *J. Electroanal. Chem.*, **448** (1998b) 173-181.

Tripković, A.V., Popović, K.Dj. and Lović, J.D. "The influence of oxygen-containing species on the electrooxidation of the C_1-C_4 alcohols at some platinum single crystal surfaces in alkaline solution", *Electrochim. Acta*, **46** (2001) 3163-3173.

Torresi, R.M. and Wasmus, S. "Product analysis", in: "Handbook of fuel cells – fundamentals, technology and applications", Vielstich, W., Gasteiger, H.A., Lamm, A. (Eds.), John Wiley, Vol. 2 (2003) p-163-190.

Verma, L.K. "Studies on methanol fuel cell", *J. Power Sources*, **86** (2000) 464-468.

Verma, A., Jha, A.K. and Basu, S. "Manganese dioxide as a cathode catalyst for a direct alcohol or sodium borohydride fuel cell with a flowing alkaline electrolyte", *J. Power Sources*, **141** (2005a) 30-34.

Verma, A. and Basu, S. "Direct use of alcohols and sodium borohydride as fuel in an alkaline fuel cell", *J. Power Sources*, **145** (2005b) 282-285.

Verma, A., Jha, A.K. and Basu, S. "Evaluation of an alkaline fuel cell for multi-fuel system", *J. Fuel Cell Science and Technology*, **2** (2005c) 234-237.

Verma, A. and Basu, S. "Power from hydrogen via fuel cell technology," *Chemical Weekly* July 12 (2005d) 177-181.

Verma, A., Sharma, S. and Basu, S. "Electrooxidation study of methanol and ethanol in alkaline medium" (2005e), manuscript submitted.

Wagner, N., Schulze, M. and Gülzow, E. "Long term investigations of silver cathodes for alkaline fuel cells", *J. Power Sources*, **127** (2004) 264-272.

Wang, Y., Li, L., Hu, L., Zhuang, L., Lu, J. and Xu, B. "A feasibility analysis for alkaline membrane direct methanol fuel cell: thermodynamic disadvantages versus kinetic advantages", *Electrochem. Commun.*, **5** (2003) 662-666.

Yang, J. and Xu, J.J. "Nanoporous amorphous manganese oxide as electrocatalyst for oxygen reduction in alkaline solutions", *Electrochem. Commun.*, **5** (2003) 306-311.

Yu, E.H. and Scott, K. "Development of direct methanol alkaline fuel cells using anion exchange membranes", *J. Power Sources*, **137** (2004) 248-256.

Recent Trends in Fuel Cell Science and Technology
Edited by S. Basu
Anamaya Publishers, New Delhi, India

8. Phosphoric Acid Fuel Cell Technology

Suman Roy Choudhury

Naval Materials Research Laboratory, DRDO, Shil-Badlapur Road, Ambernath-421 506, India

1. Introduction

Of the hydrogen-oxygen fuel cell systems the most mature is the phosphoric acid fuel cell (PAFC). It operates at 150-190°C and pressure ranging from ambient to 5 atm. PAFC systems use primarily Pt as catalyst both for hydrogen and oxygen electrodes. The operating temperature range of PAFC allows it to take up hydrogen directly from hydrogen sources like reformer gases. Less than one percent of CO present in the reformer gases are not adsorbed on Pt sites owing to high operating temperature. The other components used in PAFC are mainly made of graphite and carbon. All these factors make PAFC a versatile member of the hydrogen-oxygen fuel cell family.

Concentrated phosphoric acid (90-100% based on ortho phosphoric acid) is used as electrolyte in this fuel cell, that operates at 150 to 190°C. Some of the pressurized systems are reported to work upto 220°C. At lower temperatures, phosphoric acid is a poor ionic conductor (Der-tau and Chang, 1989), and CO poisoning of the Pt electrocatalyst in the anode becomes severe. The relative stability of concentrated phosphoric acid is high compared to other common acids; consequently the PAFC is capable of operating at the high end of the acid temperature range (100 to 220°C). In addition, the use of concentrated acid minimizes the water vapor pressure and hence water management in the cell is not as difficult as it is for polymer electrolyte fuel cell (PEMFC). The matrix, that is universally used to retain the acid, is silicon carbide. The electrocatalyst typically used in both the anode and cathode is Pt loaded on carbon.

2. Phosphoric Acid Fuel Cell (PAFC): System Definition and Principle of Operation

A phosphoric acid fuel cell (PAFC) is composed of two porous gas diffusion electrodes, namely, the anode and cathode (Fig. 1) juxtaposed against a porous electrolyte matrix. The gas diffusion electrodes are porous substrates that face the gaseous feed. The substrate is a porous carbon paper or cloth. On the other side of this substrate, which faces the electrolyte (phosphoric acid), platinized fine carbon powder electrocatalyst is roll coated with polytetrafluroethylene (PTFE) as a binder. PTFE also acts as a hydrophobic agent to prevent flooding of pores so that reactant gas can diffuse to the reaction site easily. At anode, hydrogen ionizes to H^+ and migrates towards cathode to combine with oxygen, forming water. The product water then diffuses out to the oxygen stream and comes out of the system as steam. An emf is generated between the two electrodes through conversion of reaction free energy to electricity and on connecting an external load, electrical power can be extracted.

The reactions at anode and cathode are as follows:-

At anode: $\qquad H_2 \rightleftharpoons 2H^+ + 2e^- \qquad \Delta E_{orev} = 0 \qquad$ (1)

At cathode: $\qquad \frac{1}{2}O_2 + 2H^+ + 2e^- \rightleftharpoons H_2O \quad \Delta E_{orev} = 1.229 \qquad$ (2)

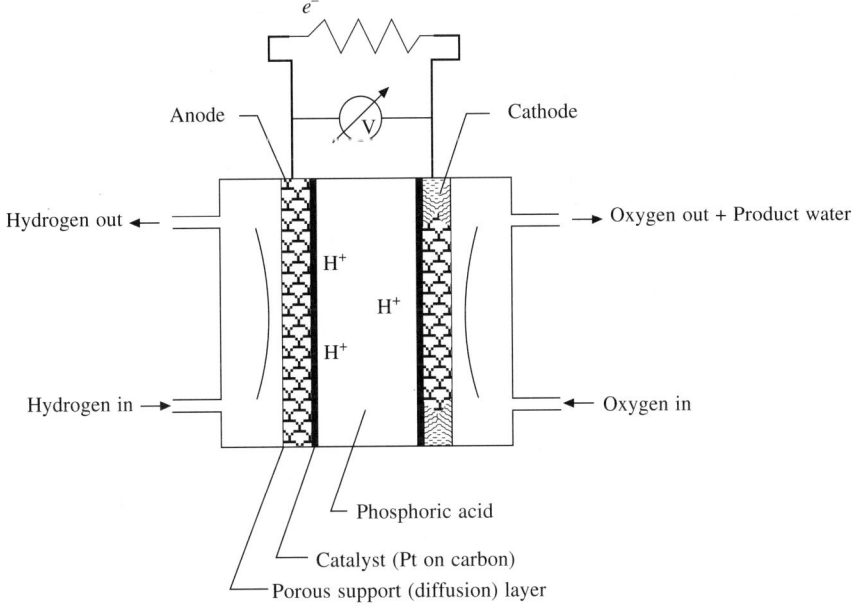

Fig. 1 Phosphoric acid fuel cell: Principle of operation.

$$O_2 + 2H^+ + 2e^- \rightleftharpoons H_2O_2 \quad \Delta E_{orev} = 0.67 \qquad (3)$$

$$H_2O_2 + 2H^+ + 2e^- \rightleftharpoons 2H_2O \quad \Delta E_{orev} = 1.7 \qquad (4)$$

$$H_2O_2 \rightleftharpoons H_2O + \frac{1}{2}O_2 \qquad (5)$$

Under reversible condition, maximum electrical work available is as follows (Appelby and Foulkes, 1989):

$$W_{max} = -\Delta G_{water\ formation} \qquad (6)$$

where ΔE_{orev} is standard electrode potential against standard hydrogen electrode (SHE).

The ideal open circuit potential is given by the Nernst equation (Appelby and Foulkes, 1989)

$$E = -\Delta G/nF = E_0 + (RT/nF) \ln [(a_{hydrogen} \cdot a_{oxygen}^{0.5})/a_{water}] \qquad (7)$$

PAFC electrodes are basically multilayered planar structures as shown in Fig. 2. Bulk gas, i.e. oxygen flows through the ribbed support over the electrode substrate layer. As the gas flows, part of the oxygen diffuses through the porous substrate of the electrode. The gas after passing through porous substrate (diffusion layer) reaches the catalyst layer also termed as the reaction layer. The catalyst layer is a porous layer partly filled with electrolyte. The gas diffuses through the dry pores, dissolves in the electrolyte and then diffuses to the catalytic site. Oxygen reduction takes place on the site and the product water diffuses back through the diffusion layer to escape in the oxygen stream.

Fig. 2 shows an exploded view of a practical stack (Appelby and Foulkes, 1989), where syrupy phosphoric acid is immobilized inside a SiC matrix. The high density graphite plates (Fig. 2) have grooves through which the reactant gases are provided to the electrodes (anode and cathode). The porous substrate of the electrodes face the grooved side of the graphite plates, whereas the catalyst is coated on the other side. In between the two electrodes, phosphoric acid is provided in the silicon carbide matrix.

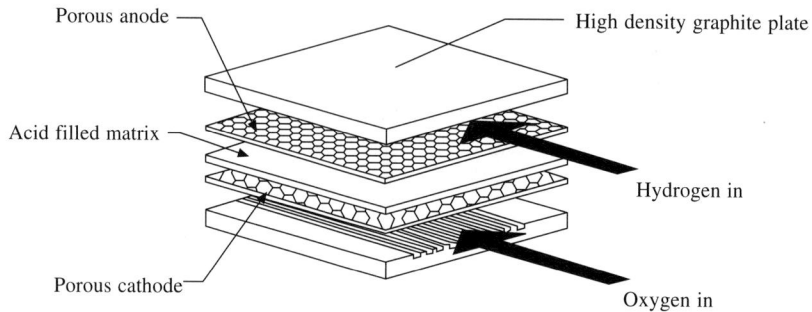

Fig. 2 Exploded view of PAFC.

The whole assembly including the grooved graphite plates is then sandwiched between two insulator plates (not shown in Fig. 2) and bolted tightly. The two graphite plates act as the terminals for connecting the electrical load. Such cells are stacked in series/parallel combinations for getting realistic power. In big stacks, the graphite plates are bipolar in nature where one side of a plate provides hydrogen to one cell and the other side provides oxygen to the adjacent cell in series.

2.1 Advantages and Disadvantages

A convenient hydrogen source for fuel cell is hydrogen rich reformed gas. The CO_2 in the reformed fuel gas stream and air does not react with the electrolyte in a phosphoric acid electrolyte cell, but acts as a diluent. This attribute, and the relatively low temperature of the PAFC make it a prime, early candidate for application. The cell performance is somewhat lower than the alkaline cell because of the slow oxygen reduction rate. However the PAFC system efficiency is good because of its higher temperature operation and less complex fuel conversion. The need for scrubbing CO_2 from the process air is also eliminated. The heat rejected from the cell is at a temperature high enough to heat water or air. Some steam is available from PAFC, a key point in expanding co-generation applications. PAFC systems achieve about 37 to 42% electrical efficiency (based on the lower heating value of natural gas). PAFC use high cost precious metal catalysts such as platinum. Also, the cell requires an external reformer for converting hydrocarbon fuels to H_2-rich gas and CO (formed as a byproduct) has to be shifted by a water gas reaction to below 3 to 5 vol%. This is required to avoid poisoning of the fuel cell anode catalyst. Following sections discuss various aspects of the PAFC technology in detail.

3. Catalysts

3.1 Cathode Catalyst

For low temperature acid electrolyte based fuel cells like PAFC, the primary cathode catalyst is Pt. At the onset of PAFC development, experimental data was only available for oxygen reduction on polished Pt surfaces. From these data, high surface area Pt black electrodes were developed. The initial electrodes used to have Pt black of about 25 m^2/g surface area, 20 mg/cm^2 metal loading and was bonded with PTFE (Teflon). The electrode was operated at around 160°C, at 96% H_3PO_4 and showed a Tafel slope of around 90 mV per decade (Vielstich et al., 2003a).

Such catalysts, though they performed satisfactorily, had certain limitations. Finely divided Pt blacks have limited surface area and thus Pt utilization is quite poor. Further, Pt black has a tendency to sinter at lower potential under PAFC environment owing to a process involving surface atom migration (Vielstich et al., 2003a). This reduces the electrode porosity as well. Reduction of surface area leads to lowering of current density at a particular potential, while reduction of electrode porosity increases reactant diffusion resistance thus leading to loss of current. At higher potential, Pt black also looses surface area by a combination of three processes viz, sintering, dissolution-precipitation, and crystallite migration. The dissolution and precipitation of Pt at high potential (Ostwald ripening) is a phenomenon that destabilizes very small Pt particles and finally forms agglomerated particles in the range of 80-120 Å.

In order to enhance Pt utilization per unit weight of the metal, the use of supported catalyst was attempted. The basic requirement of such catalyst supports are quite straight forward. They should be high in surface area, electronically conducting and thermally as well as electrochemically stable under phosphoric acid environment.

Based on these guidelines, it was soon observed that various forms of carbon only pass the requirement. Detailed studies employing various kinds of carbon blacks as support material were undertaken and it is such a versatile field that still newer forms of carbons like carbon nano tubes and mesoporous carbon are being studied all over the world.

Various types of carbon black that were initially tried include furnace blacks, pyrolytic carbon, charcoals etc. All the basic studies were guided by the fact that carbon tends to undergo oxidation at a potential of about 1.0 V with respect to SHE. Thus if carbon support is to be used, one has to keep the operating PAFC cell voltage below 0.8 V. Thus a support resistive to carbon corrosion is the ideal one. This allows the use of more graphitized forms of carbon supports as they are more resistant to corrosion. However, this leads to some problem for catalyst developers. With increased degree of graphitization of the carbon support, the surface area of the support decreases. Highly graphitized carbon support offers resistance to Pt particle deposition on the surface and the chances of Pt detachment from the support surface is higher.

Such problems forced fuel cell developers to pursue a trade-off between the surface area and the degree of graphitization of the support material. In view of this, Vulcan XC-72 from M/s Cabot Corporation is arguably the most popular catalyst support material. Vulcan XC-72 is basically a high surface area furnace black with reasonably good graphite content. This allows the support material to be corrosion resistant at PAFC cathode environment. Various properties of Vulcan XC-72 are provided in Table 1.

However, it was observed that at higher temperatures (>180°C) even Vulcan XC-72 faces significant corrosion, thus reducing the PAFC cathode life significantly. It was observed that, if the carbon support is heat treated prior to catalyst loading, then its corrosion resistance is greatly enhanced. A heat treatment temperature of above 2000°C is found to be effective in this regard (Vielstich et al., 2003b).

Pure Pt based supported catalysts have limited performance. Very fine Pt particles do not enhance the reaction rate due to the oxygen atoms inability to get adsorbed on very small particles (Kinoshita, 1992; Vielstich et al., 2003c). Additionally, smaller particles tend to dissolve and get precipitated again as bigger particles as well. Thus, various Pt based alloy catalysts were tried particularly for better performance with air.

Table 1. Properties of Vulcan XC-72 carbon black

Properties	Values
Surface area (m^2/g)	254
Particle size (nm)	30
Volatile content (%)	2.0
pH	5.0
App. density (kg/m^3)	96

Various studies on Pt alloy catalysts have been carried out world over. Wieckowski et al. (2003) provides a comprehensive analysis of these studies. The final conclusion that is drawn is as follows. Alloying of Pt with transition metal elements increases the Pt d-band/atom vacancy and decreases the Pt-Pt bond distance. The extent of the change depends upon the electronegativity of the transition element. X-ray absorption near edge spectroscopy (XANES) data indicates that the outer surface of the oxygen reduction reaction active alloy crystals are almost pure Pt. The alloying elements were relegated to the subsurface layers. After the formation of the initial Pt "Skin", the dissolution of the non-noble alloy elements in the acid are found to be insignificant even at higher potentials (Wieckowski et al., 2003).

Roques et al., 2005, reported experimental coupled with theoretical explanations on oxygen reduction activity on Pt and Pt_3Co alloy. It was concluded that the oxygen reduction rate depends on the metal surface composition and structure. Alloying with transition metals like Co and Cr clearly leads to better activity. Kim et al., 1993, reported a study on alloying with iron. Various binary alloy catalysts like Pt/Co, Pt/Cr, Pt/V, PT/Ni etc., and ternary catalysts like Pt/Co/Cr are developed for higher oxygen reduction activity (Vielstich et al., 2003c). A DC polarization comparative plot of Pt/Co catalyst is shown in Fig. 3. It can be seen that the potential gain for the cobalt alloy catalyst is much higher when air is used in the cathode. This is expected as the presence of cobalt increases the oxygen adsorption on the catalyst surface.

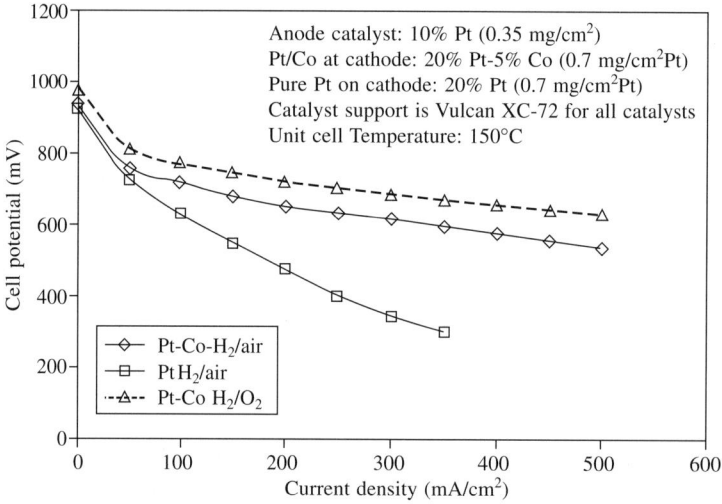

Fig. 3   DC polarization of pure Pt and Pt/Co supported cathode catalyst in unit cell (Courtesy: Naval Materials Research Laboratory, India).

3.2 Anode Catalyst

Oxidation of hydrogen to proton at anode has several candidates at acidic environment. For example various noble metals that are stable in hot phosphoric acid can be considered as alternatives. Table 2 gives exchange current densities of various possible anode catalysts. However, on close examination, it is obvious that Pt turns out to be the most active catalyst in PAFC environment. Note that the hydrogen reduction reaction in acid media contains following two steps:

$$M + H_2 \rightleftharpoons (M)2H_{ads} \qquad (8)$$

$$(M)2H_{ads} \rightleftharpoons M + 2H^+ + 2e^- \qquad (9)$$

Different metals shows different rate limiting steps. Thus, considering the overall kinetics, Pt seems to be the best choice for PAFC anode as well.

Table 2.   Exchange current densities (i_0) for hydrogen evolution reaction

Metal	$-\log(i_0)$ (in Amp/sq cm)
Pd	3.0
Pt	3.1
Rh	3.6
Ir	3.7
Au	5.4

Although hydrogen oxidation over platinum is around 10^5 times faster than oxygen reduction in acidic medium, still enough emphasis has been placed on the anode catalyst development. For pure hydrogen not much work is needed on the anode catalyst with Pt as active metal due to the faster kinetics. However, most of the hydrogen sources are impure and contain various contaminants like S and CO. The impurity like S comes from the basic feedstock like hydrocarbons that are used for generating the hydrogen. S tolerance is extremely poor for Pt based catalyst and to make the matters worse the poisoning of the catalyst is permanent.

The impurity for which the Pt based anode catalyst systems could be tailored is CO. CO originates mainly by reforming/cracking of carbon based feedstock like coal, hydrocarbons, alcohols etc. The typical CO level after low temperature/high temperature shift reactors are in the range of 0.5-2%. Various studies were conducted to understand the effect of CO on poisoning the Pt catalyst active sites, and the enhancement of the tolerance level by alloying Pt with other metals (Auer et al., 2000, Gasteiger, 1994, Kinoshita, 1992). CO blocks the active Pt sites by getting preferentially adsorbed on them. The percent coverage of CO on the active sites depends upon the CO concentration in the gaseous phase and system temperature. A graph showing the effect of CO poisoning of Pt is shown in Fig. 4. The acceptable limit of CO concentration in the reformer gas thus depends upon the PAFC system temperature. The CO poisoning is reversible in nature and on flushing the catalyst with pure hydrogen, the catalyst revives reversibly. A detailed discussion on various

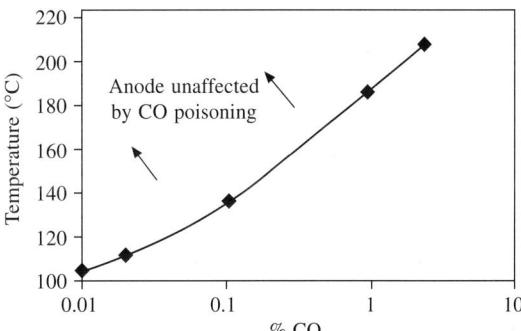

Fig. 4 Effect of CO concentration in the hydrogen stream for Pt anode.

mechanism of CO tolerance of PT/Ru catalysts is provided by various groups Kinoshita, 1992, Vielstich et al., 2003c.

The electrochemical oxidation of CO to CO_2 vide the reaction 10 takes place at a potential of +0.45 V w.r.t. SHE.

$$CO + H_2O \rightleftharpoons CO_2 + 2H^+ + 2e^- \tag{10}$$

Thus, the Pt catalyst based fuel cell anode gets polarized to 0.45V w.r.t. SHE potential before it stabilizes. Depending on the temperature and the CO concentration, the CO oxidation effect may or may not stabilize the anode even after polarizing to the CO oxidation potential.

The most effective CO tolerance is found to be with Pt/Ru alloys. Pt/Ru alloy catalysts, supported on carbon were found to be effective upto 2% CO without a significant drop in performance. However, the other promising metals are Co and Ni. It may be noted that the Ru attracts the CO molecules, thus keeping the Pt free for hydrogen oxidation. Additionally on Ru, the oxidation of CO to CO_2 occurs at a lesser potential of about 0.35 w.r.t. (SHE) at 50 atom% Ru and at 0.2V (SHE) for 90 atom% Ru. Thus, at a lesser polarization the electrochemical oxidation of CO starts for Pt/Ru based catalysts. These two effects make Pt/Ru based anode catalyst as the most accepted CO tolerant PAFC catalyst. As Ru percent increases, CO tolerance is enhanced at a cost of activity due to less Pt availability. In view of this, a trade-off is required, and the optimal composition that most of the developers claim is 1:1 atomic ratio of Pt:Ru.

For high temperature PAFCs, i.e. for 190°C and above, it was found that pure Pt is the best to use as at such high temperatures CO adsorption is negligible (Vielstich et al., 2003a).

3.3 Methods of Producing Supported Catalyst for PAFC

There are various ways to produce supported catalysts. These methods are similar for all acid fuel cells. Most of them can be categorized into two types, viz. (i) impregnation methods and (ii) colloidal sol methods.

Impregnation followed by reduction is the oldest method. The basic process involves loading of the support powder with a solution of the catalyst metal(s) salt followed by drying and reduction. Various reducing agents like direct hydrogen gas, formaldehyde, methanol, sodium-boro-hydride etc. can be used. The major parameters in this process are pH, solution concentrations, soaking time and reduction conditions. A major problem associated with this process is the presence of unreduced salt crystals adsorbed in the micro pores of the catalysts. These crystals may hinder the mass transfer process and require thorough washing for better performance. A pioneering patent by Kemp and George 1974, claims around 20 Å Pt particle size leads to successful large scale production of PAFC catalysts. Pt crystals in the range of 50-70 Å that remain highly stable under PAFC environment is reported by Choudhury et al., 2004.

The current state-of-the-art carbon supported catalysts are made by using variants of the colloidal "Sol" approach. The patent most frequently referred in this field is from United Technologies (Petrow and Allen, 1976). The overall process is based on the neutralization of H_2PtCl_6 with Na_2CO_3 and $NaHSO_3$ to change the pH to around 4. The pH is then adjusted to 7 by adding Na_2CO_3 resulting in a white precipitate $(Na_2Pt(SO_3)_4OH)$. This compound is heated in air for about an hour to convert it into a glassy substance and when dissolved in water forms a stable Pt colloidal sol. This colloid is then adsorbed on the porous carbon support like Vulcan XC-72.

Alloy catalysts are prepared either by co-precipitation with Pt or by sequential deposition. One widely used method is sequential deposition of alloying elements by carbo thermal process on Pt loaded carbon support. However, carbo thermal method is suitable for non-noble metals and does not yield high purity alloys. This is due to the presence of large quantity of unalloyed metal and the need to be "Cleaned" by soaking in acid. The other method involves formation of various types of complexes of Pt and the alloy metal to form a mixed colloid "Sol". This mixed colloid when adsorbed on the carbon support yields highly uniform alloy distribution.

4. Electrode

PAFC electrodes are made out of planar, porous substrate on which the catalyst is coated along with a binder. The reaction takes place at the catalyst surface, and the electrons are transferred from the catalyst particles to the electrode substrate. Further, the electrons are passed to the conducting separator gas plate and transmitted to the external circuit.

4.1 Electrode Substrate

PAFC electrode substrate material for anode or cathode are similar in nature, and perform similar tasks. It provides mechanical support to the catalyst layer. The substrate allows transport of gaseous reactants like hydrogen or oxygen (as it is for anode or cathode), by diffusion, to the catalyst layer from the bulk stream. Similarly, product water formed inside the catalyst layer, needs to be diffused out to the bulk stream though the substrate. Thus, the substrate has to be porus in nature. The electrons generated in the catalyst layer coated on the substrate, transfers the electron through the substrate, to the external circuit and hence, should be electronically conducting. The heat generated in the catalyst layer is also transferred out by conduction through the substrate and hence, the substrate has to be thermally conducting. A uniform contact pressure is provided between the catalyst layer and the matrix through the substrate, hence, it should not get crushed at that pressure. Obviously, the substrate material should be stable and oxidation resistant in PAFC environment.

The first generation PAFC electrodes were made of Tantalum 100 mesh, woven screens with gold plating, to avoid oxide layer on Tantalum in PAFC environment. However, the high cost of Tantalum and other problems like hydrogen embrittlement at anode and oxidation at cathode, made it unsuitable for PAFC electrode substrate.

Union carbide developed a carbon substrate by depositing pyrolytic carbon by chemical vapor deposition onto a polyacrylonitrile (PAN) precursor based carbon paper. This has about 70% porosity and was about 25 μ thick. Though this proved to be a very good electrode substrate, it was difficult to make in bigger size, and was not cost effective as well.

Kureha Chemical Company developed a paper made from pitch based carbon fibre and further made a prepeg by loading phenolic resin to it. This was carbonized and graphitized upto 2000-2500°C to imbibe the electrical properties. The porosity and stiffness were controlled by the amount of phenolic resin. Phenolic resin contains very high amount of carbon and after carbonization leaves no other impurities and is thus suitable for making electrode substrate. Stackpole Carbon company also developed a similar product, but based on rayon fibre instead of carbon fibre. This substrate has a porosity of about 80% and was better for PAFC air cathode than Kureha, which has a porosity of about 65%. However, the Stackpole paper was brittle, generated lot of scrap and required higher investments.

A ribbed structure substrate was made by United Technologies Corporation (UTC). Pitch based carbon fibre about 0.12-0.25 mm long was blended with phenolic resin powder and laid on a steel belt for curing in a belt press. This was subsequently graphitized. This thick ribbed substrate (around 2-3 mm) was successfully used in UTCs 40 kw PC-18 power plants. The main purpose for such high thickness was to hold sufficient phosphoric acid and to have the flow fields (the ribbed structure) embedded on the substrate itself. Thus, costly high density graphite plates that were otherwise required for flow field and separator plate can be reduced in thickness for cost efficacy.

The fifth generation PAFC electrode substrates are basically refinements of the Kureha's carbon substrate as mentioned before. These substrates manufactured by Toray have a mean pore size of about 20-30 μ, about 0.3-0.4 mm thick and are very successfully used in UTCs 200 kw PC-25 power plants, being one of the most sold fuel cell unit.

4.2 PAFC Electrode Making

The first part of the electrode manufacturing is the edge sealing of the electrode substrate. This is to prevent gas leakage through the thin porous edge of the catalyst substrate. Various kinds of sealing methodologies are tried, like wet sealing, using SiC particles as fillers (Schroll and Hartford, 1974). In this method the hydrophobised substrate is coated with the catalyst leaving a frame like portion blank near the edge. Further, while casting the matrix, this blank area is filled with the matrix material and the thickness is adjusted to match the total thickness of the matrix and the catalyst layer. The sealing is done by keeping the matrix and this extended thick matrix portion wet with acid. Other methods, like changing the density of the substrate at the edge while manufacturing the substrate, is reported by DeCasperis et al., 1981. By this method the electrode edge is of same substrate material and thickness but of much lower pore size. This pores gets impregnated with phosphoric acid due to high capillary action and acts as an edge seal. The other designs like sealing the edge with a resin without graphitization of the same, providing acid filled reservoirs at the edge are some of the methodologies employed as well.

Subsequent to edge sealing, it is required to hydrophobise the electrode substrate. This is necessary to prevent the acid from filling the pores of the substrate. Hydrophobisation is typically done by impregnating poly-tetra-fluro-ethylene (PTFE) or fluro-ethylene-copolymer (FEP) colloidal suspension. Methods like screen printing, or dip coating are used to load the targeted amount of the polymer. FEP is preferred over PTFE due to its lower melt viscosity, resulting in lesser gas diffusion resistance. High viscosity of the impregnating "Ink" yield better result, as after impregnating the polymer, during drying of the water (suspension medium), the polymer does not migrate towards the surface. After drying the ink, a heat treatment cycle may be given to the substrate, so that the polymer sinters and the substrate becomes hydrophobic.

The next stage is to coat one side of the hydrophobic substrate with the catalyst. Typically PTFE colloidal suspension is used as a binder. This has several functions. Apart from binding the catalyst particles, it provides necessary hydrophobicity to the catalyst layer. Hydrophobicity needs to be controlled for the catalyst layer. Less hydrophobicity will flood the catalyst layer with electrolyte by capillary action, and gas diffusion will be affected. Too much hydrophobicity will make most of the pores dry, leading to drying of the active sites. Ideally the active sites should have a thin electrolyte coating and there should be an electrolyte continuum for the ion movement, from the bulk acid side to the catalyst. Thus the amount of PTFE in the catalyst layer is changed to achieve the desired result. Depending upon the design and operating parameters, around 35-45% PTFE is used for cathode and around 40-50% PTFE for anode.

After coating the catalyst layer, the electrode is given a series of heat treatments. The first treatment is to dry the solvents like wetting agents and water. If traces of solvent remains, then at higher temperature in presence of air and Pt, it may start burning, causing partial burning of the carbon powder of the catalyst as well. After solvent removal, the temperature is increased to around 280°C and maintained for at least half an hour to remove the surfactant that are present in the PTFE suspension. Further, the temperature is increased to near the glass transition temperature of PTFE that is around 330-350°C to sinter PTFE for effective binding and hydrophobicity. The exact temperature and time duration at each stage varies with fuel cell manufacturers. Fig. 5 shows a SEM photograph of a catalyst layer.

5. Acid Holder Matrix

The phosphoric acid is required to be "Immobilized" in between the two electrodes. For this, an acid holder matrix is created on one or both the electrodes. The matrix is a sponge type material that holds the syrupy

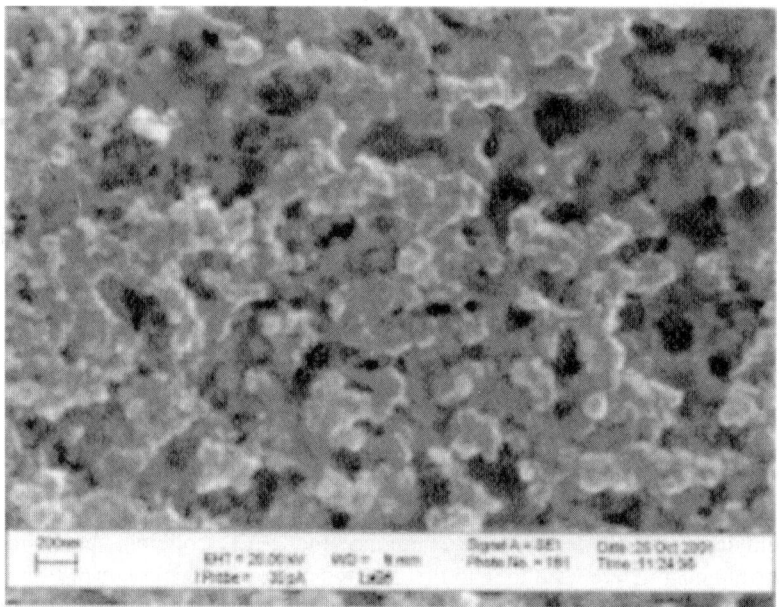

Fig. 5   SEM photograph of a PAFC cathode catalyst layer (Courtesy: Naval Materials Research Laboratory, India).

phosphoric acid similar to acid separator in VRLA batteries. After testing of various materials, SiC based matrix, as invented by Breault, 1977 is the basic material that is still in use, with various upgrades of the technology. The target is to reduce ionic resistance and chance of gas crossover through thinner matrix with high porosity and bubble pressure. Higher bubble pressure implies smaller pores and thus better acid retention capability. It may be noted that for acid holder matrix of about 150 μ or more, operating with current density of 200 mA/cm^2, the major potential loss occurs at the acid matrix in the form of ohmic loss. Song et al., 2002, has reported usage of a mixture of SiC powders to vary packing density of the matrix and its effect on the PAFC performance. The study suggests possible modifications of electrolyte matrix packing density using coarse and fine particles. However, the fine particles due to high surface area may face early corrosion and may lead to early loss of cell performance.

Most of the methods for coating the SiC matrix on the electrodes involve SiC based slurry handling. In order to maintain the slurry mixture without settling, a suitable vehicle is required. Aqueous polyethylene oxide base ink vehicle has certain advantages compared to other systems like glycols (Trocciola et al., 1977), and is used almost universally for making the matrix.

The SiC particles (1-5 μ) along with little PTFE suspension is mixed with or without a stabilizing agent like polyethylene oxide and are either screen printed or slurry casted. Stewart and Robert, 1979 developed a curtain coating method by which about 120 μ thick matrix can be casted. For further thinner matrix, a Grauvere coating process (Spearin, 1989) using 1 μ SiC particles was used. Such matrix is about 50% porous and is about 50 μ thick with a high bubble pressure of 70 kPa when wet. After coating the matrix on the catalyst layer of the electrode, it is dried and heat treated to sinter the PTFE binder. The PTFE binder used is generally 0.5-2%(by weight). However, it is found that for 1 μ SiC powder it may be possible to develop the matrix without PTFE binder.

6. Performance Evaluation of Catalyst, Electrode and Matrix

6.1 Catalyst Evaluation
Anode and cathode catalysts need to be evaluated through physical as well as electrochemical characterization techniques. Various types of non-electrochemical characterization techniques that are commonly used are as follows:

Techniques like X-ray diffraction (XRD) and Transmission electron microscopy (TEM) are used to understand the size distribution of the metal (Pt or Pt alloy) particles. TEM also helps to understand the shape factor of the particles. This is quite useful, as the surface roughness of the particles determines the local adsorption rate of reactants, particularly for oxygen at cathode.

BET analysis provides the total surface area of a catalyst (support and the metal particles). Chemisorption studies can be used to evaluate the active metal surface area. For example, hydrogen can be used to estimate the surface area available for Pt. Average particle size can be calculated from chemisorption surface area, and should be also matched with the XRD or TEM results, for higher confidence.

With temperature programmed reduction/oxidation method, it is possible to test the catalyst surface activity. For example, a gas phase reaction—typically cracking of a compound—is studied at various temperatures and the gas mixture coming out of the catalyst filled reactor is analyzed (generally using thermal conductivity detector, TCD) for determining the catalyst activity.

The catalyst need to be characterized further using electrochemical techniques. The most used ones are as follows:

Cyclic voltammetry (CV) or triangular pulse voltammetry is an important tool to measure quantitatively, the monolayer coverage of the catalyst active sites. Thus, it is possible to find out the electrochemically active surface area from the CV output (Pozio et al., 2002). The basic method is shown in Fig. 6. The CV

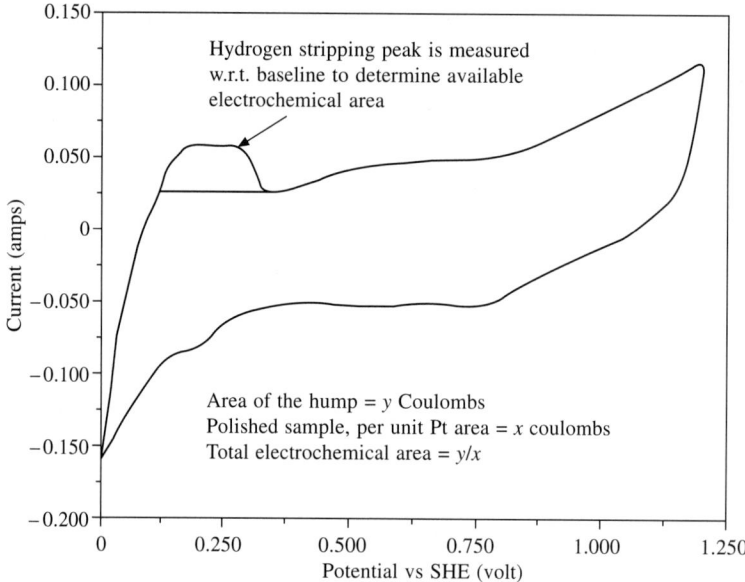

Fig. 6 Electrochemical surface area determination of Pt based PAFC electrode by cyclic voltammetry.

is generally done on a PAFC electrode using H_2SO_4 at ambient temperature or phosphoric acid at high temperature. An atmosphere of a neutral gas like Ar or nitrogen is maintained inside the electrochemical cell. The hydrogen desorption peak area is matched with that of a polished Pt standard (where geometric area equals electrochemical area), and the total electrochemical surface area of the catalyst is determined. Fig. 7 shows an actual CV plot of a PAFC cathode catalyst. CV can be done on the catalyst powder by holding the same in a catalyst holder or after forming a PAFC electrode. The electrochemically active area

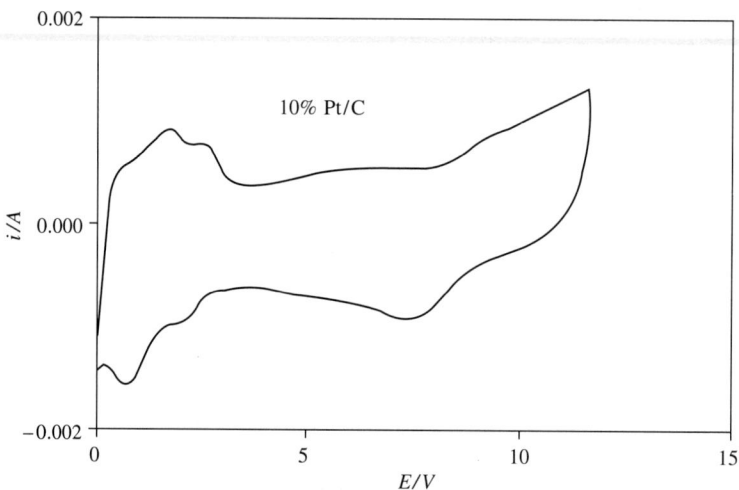

Fig. 7 A typical cyclic voltammery plot of cathode catalyst in phosphoric acid (Courtesy: Naval Materials Research Laboratory, India).

of a PAFC electrode is the actual area available for the reaction. This is generally lesser than the active area of the pure catalyst. The hydrophobic binder (PTFE) that prevents electrolyte from wetting part of the catalyst surface reduces the overall electrochemical area of the catalyst, in the PAFC electrode. Rice et al., 1998 has reported a combined method of Nuclear Magnetic Resonance (NMR) and CV to characterize commercial Pt electrocatalysts. The study also revealed Pt particles size dependency of CV output leading to particle size determination of the Pt particles.

The other method that is used for catalyst electrochemical performance, is called half cell testing. Half cell apparatus can be used for different kinds of electrochemical studies such as DC polarization, electrochemical impedance spectroscopy (EIS) and chrono-amperometry. Such a setup can be even extended for CV study of electrode samples. Fig. 8 shows the principle of a half cell apparatus, configured for DC polarization studies of a cathode. The whole setup is inserted in a thermostatic bath (not shown in the figure). The test electrode is placed in a holder typically made of PTFE and secured with a cap. The hole in the cap determines the exact area exposed and the current density is determined by dividing the operational current with the exposed area. A suitable metal tubular bush, typically gold plated, provides gas passage and electric contact to the test electrode sample.

The sample holder assembly, is then inserted in the acid pool (either Phosphoric acid or sulphuric acid). The temperature of the acid is maintained at

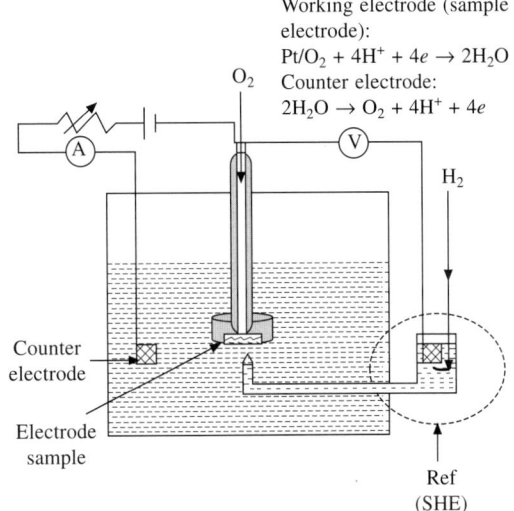

Working electrode (sample electrode):
$Pt/O_2 + 4H^+ + 4e \rightarrow 2H_2O$
Counter electrode:
$2H_2O \rightarrow O_2 + 4H^+ + 4e$

Fig. 8 Schematic half cell setup.

the required temperature by the thermostatic bath. The acid pool contains a counter electrode to complete the circuit and a reference electrode with a lug in capillary to sense the potential at the test electrode surface.

At high currents the potential gradient inside the acid pool is extremely steep near the electrode. Thus, distance between the lug in capillary end and the test electrode surface is highly sensitive to proper sensing of the potential. To maintain the acid concentration, a gaseous atmosphere with appropriate humidity is provided. For example, at around 150°C, for ortho-phosphoric acid of 90-95% H_3PO_4, around 20% humid air must be maintained over the head space of the cell. For this, a lid with proper gas feeding arrangement (not shown in the figure) is used. Fig. 9 shows a glass half cell apparatus, with teflon sample holder containing SS tube (gold-plated) gas feeder cum current collector, and Pt counter electrode suitable for DC polarization and EIS studies.

In a half cell, the electrode sample is very small and thus is free from diffusion related resistance to some extent. This allows, evaluation of the catalyst performance, after making it into an electrode form. It may be noted that, electrochemical methods like rotating disk electrode (RDE) system allow the catalyst performance to be studied under diffusion resistance free condition. The rotating ring disk electrode (RRDE) version allows to measure the intermediate products as well. However, unit cell performance is more realistic for estimating the electrode performance from the fuel cell point of view. Thus, RDE/RRDE evaluations are done mostly for scientific understanding, whereas half cell polarization and EIS indicate the highest performance of the electrode that is possible in actual conditions. Fig. 10 shows a typical PAFC cathode performance in half cell measured w.r.t. SHE. The flatness of the DC polarization curve at higher current density indicates very low resistance and the initial drop provides activation polarization information.

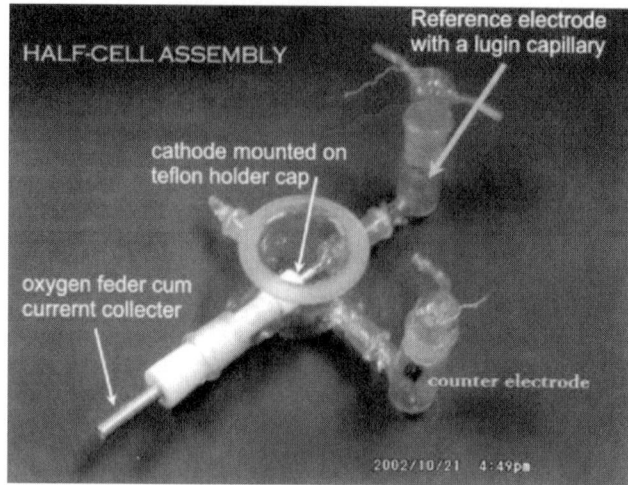

Fig. 9 Half cell apparatus (Courtesy: Naval Materials Research Laboratory, India).

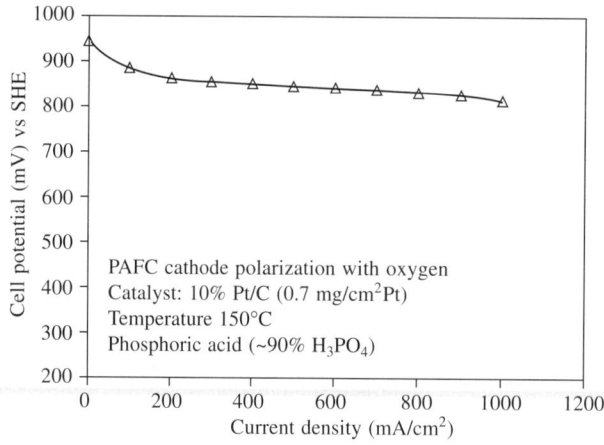

Fig. 10 Half cell polarization of PAFC cathode (Courtesy: Naval Materials Research Laboratory, India).

Combination of the above methods is useful to analyze change of catalyst structure and activity of PAFC electrode in the course of operation. Such an aging study helps to determine the various modes of catalyst decay. Maoka et al. (1996), reported various physical and electrochemical methods to determine PAFC cathode catalyst deactivation over time.

6.2 Electrode and Matrix Evaluation

Subsequent to satisfactory evaluation of catalyst and small electrode samples, it is required to test the matrix coated electrode assemblies. Several physical tests are carried out on the matrix, to evaluate its integrity, thickness, uniformity, porosity etc. Simple tests like bubble pressure check is done in a wet matrix (pre-wetted with water or acid). The wet matrix side is pressurized with a gas. The pressure at which the first bubble appears on the top of the other side (i.e. the electrode substrate side) kept under a pool of water, indicates the extent of differential pressure that the matrix can withstand. As the electrode layers are hydrophobised, they remain mostly dry during the bubble pressure test, and only the matrix holds the gas from crossing over.

Out of other physical methods used for porosity analysis, mercury intrusion porosimetry is worth mentioning. This method can be tried at all levels of electrode development, i.e. from catalyst to matrix stage. Testing of a matrix is done at low pressures (upto 400 kPa) so that the soft structure of the matrix is not deformed. Fig. 11 shows a typical result of PAFC cathode developed by the author's laboratory. The matrix is also characterized by other simple tests, like water loading, rate of water migration when dry matrix is dipped in one end etc. (Caires et al., 1997).

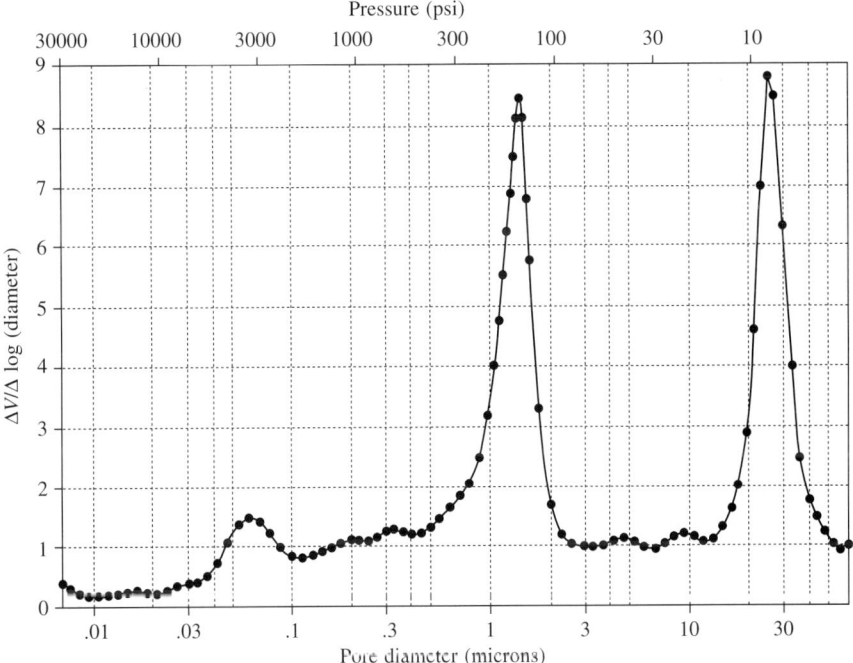

Fig. 11 Porosity data of PAFC cathode by mercury intrusion porosimetry, the peaks from the rightmost side pertains to average pore size of the electrode substrate, matrix and catalyst layer, respectively (Courtesy: Naval Materials Research Laboratory, India).

Subsequent to physical characterizations, the electrode coated with matrix needs to be evaluated electrochemically. The setup used for this purpose is called *unit cell setup*, as shown in Fig. 12. It consists of two planar graphite electrodes, viz, the cathode and anode, separated by a porous SiC matrix which holds the syrupy phosphoric acid electrolyte. The exposed area of the electrode generally remains small, around 20-100 sq. cm. However, depending upon the level of evaluation, even full sized electrodes for the stacks are tested in unit cells. The cathode-matrix-anode sandwich is held between two grooved graphite plates which feed oxygen to the cathode and hydrogen to the anode. Two stainless steel plates act as the current collector at the two ends. The entire assembly is press held by two end pusher plates which are electrically isolated. As an option, to test the effect of only cathode against a standard reference electrode, the matrix is connected to a standard hydrogen electrode (SHE) as a reference, through an ionic bridge (may be a glass filter paper, for e.g. Merck GF120 tried by NMRL), wet with dilute phosphoric acid. Another variation, where the reference electrode holder cylinder connects ionically to an extended SiC matrix, through a hole at the bottom of the cylinder, for the single electrode study, is reported by Aragane et al. (1994).

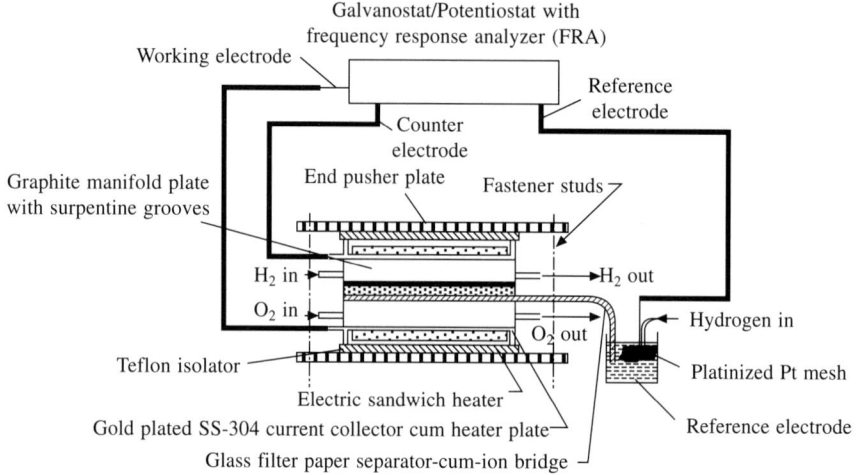

Fig. 12 Unit cell experimentation setup with third electrode option.

With such setups one can study the total cell performance and individual electrode performance as well with a third electrode, i.e., the reference electrode. The advantage of the reference electrode is that one can evaluate the electrode performance even with little gas crossover. The reference electrode can also be used for evaluating the matrix performance.

7. Separator/Bipolar Plates

The separator plates have various functionalities. Namely, it prevents the mixing of hydrogen at anode of one cell from oxidant (air) at cathode of the adjacent cell. Thus, it should be impermeable to the reactant gasses, at least along the "Through plane" direction of the plate. These plates transmit electrons and heat generated to the next cell and has to be electronically and thermally conducting. In many of the designs, the gas flow field for the reactants are provided in these separator plates making them bipolar gas manifold plates. For designs where ribbed electrode substrates are used, the flow field for the reactant gases are provided on the ribbed structures, and in that case the separator plates are flat sheets. Further, the open porosity of the separator plates should be minimum, so that little acid gets absorbed to it. This keeps the acid available in the acid reservoirs. The plates should be stable in phosphoric acid and resistant to corrosion.

All these factors make it difficult to develop a cost effective solution for the separator plates. Initial trials were with graphite polymer molded composites. The polymers that were tried includes polyvinylidenefluoride (PVDF), polyetheretherketone (PEEK), perfluoroalkoxy (PFA) etc. Although most of the graphite polymer composite passed the 200°C phosphoric acid test, few survived more than few thousand hours, when they are maintained in 200°C phosphoric acid with a potential of 1.0 V w.r.t. SHE. Of the various polymers, PFA was found to be the most suitable. However, it was observed that after long exposure to PAFC temperature, the polymers got relaxed, causing an increase in the plate electrical resistance and porosity. However, all these plates exhibited excellent gas crossover resistance at least in the initial phase.

In view of the above, several developments took place with carbon-carbon composite. The basic process is as follows. The base plate is molded with phenolic resin and graphite powder. Subsequently, it is carbonized in inert atmosphere with very slow temperature buildup. This prevents blister formation and creation of open porosity from the volatile components. Further, it is graphitized by raising the temperature. Several parameters like graphite powder size, phenolic resin type and content, moulding pressure, heat treatment cycle, plate dimensions etc. controls the final property of the plates. Generally, the flow fields are molded during the initial stage.

The latest separator plates developed by M/s Showa Denko uses a prepeg of laminated layers of felt containing cellulose fiber, carbon fiber, graphite powder and phenolic resin. These plates after compression molding are further carbonized and graphitized.

Non-carbon plates have also been tried. There are reports of using steel plates coated with conducting carbide materials but unfortunately are not cost effective and are quite heavy as well. In this connection it is safe to say that the polymer-carbon composite plates are cost effective if they are not considered for high capacity. Additionally, if the temperature of the PAFC is kept around 150°C, the life of these plates increases significantly. In view of this, it may be mentioned that the cost implication of the bipolar plates is very important for commercial success of PAFC based power plants.

8. Stack Designs and Assembling

8.1 Gas Feeder Manifold

A basic PAFC stack configuration is given in Fig. 13. The anode-acid holder matrix-cathode sandwich is held between two bipolar plates and this array is repeated to the end. At each end there is an one-side grooved plate—one with anode groove and the other end with cathode groove. Each of these plates have through holes, when assembled act as reactant gas flow headers (inlet and outlet). The inlet header of one gas has an opening in one side of each bipolar plate, and similarly the outlet header of the same gas collects excess unreacted reactants/products from the same side of the bipolar plates. The inlet and outlet headers of the other gas similarly feed and collect gas from the other side of the bipolar plates.

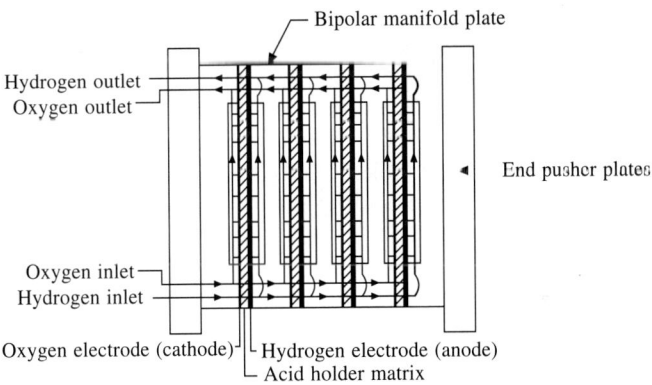

Fig. 13 Stack components and assembly scheme, Internal manifold design.

Various issues need to be understood for successful PAFC stack development. There are typically two types of gas feeding options to the cells. The first one is called an external manifold. In this case, both the grooves (at the anode and cathode side) open up on different ends of the plate as shown in Fig. 14. Further, the individual gases are fed to this open ended groove through a common manifold chamber. Similarly, the excess gas is collected through the other end of the grooves through another manifold chamber. This kind of design is useful, if the PAFC stack is of higher capacity. The main advantage of this configuration is the simple design and maximum utilization of the costly separator/bipolar plates for the active area.

However, various problems are associated with this type of external manifold design. First, is the high critical sealing area. For example, the entire edge of the cathode electrode is exposed to the anode feeder manifold chamber and vice versa. This means that the sealing of these exposed edges is critical. Other problem that is occasionally faced is due to the covering of the major portion of the stacks side by the manifold chambers. This often makes it difficult to take out the coolant/heating pipes. Further, after assembling, the side of the stacks where the manifold chamber should be fixed needs to be smoothened and gaskets have to be fitted for leak prevention. As there will be small dimensional variations of the length of the stack, the gas manifold chambers need to be oversized a bit and may require filler dummy plates. Another problem that is faced is related to

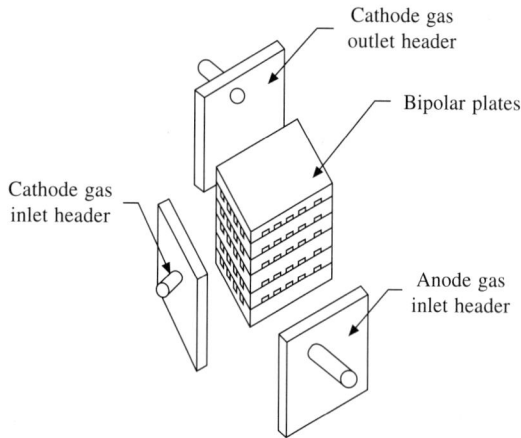

Fig. 14 External manifold design.

maintenance. Replacement of faulty cells becomes difficult as one has to maintain the gaskets and smoothing of the side of the stack at least for that module. This makes quick cell replacement difficult for external manifold design.

The other design option is known as the internal manifold design. In this case, part of the bipolar/separator plates have through holes, and when assembled act as the inlet and the outlet headers. Fig. 13, as mentioned before is an example of internal manifold. This design gives the stack a compact look and is quite useful for small capacity stack. The main advantage of this stack is that the critical edge area that has to be sealed to prevent gas mixing in the anode and cathode gas manifolds are small, unlike external manifold design. Further, as the side of the stacks are exposed, taking out of utility lines like heating and cooling pipes are quite easy. Another advantage of this design is that it allows quick replacement of faulty electrode sandwiches. However this design eats away part of the costly separator/bipolar plates area for providing the feeder holes and thus, effective utilization of these plates for the active area is less than that of the external manifold design. Another problem that is occasionally faced is at high fuel utilization condition. The inter cell distribution tends to be poor, due to relatively smaller header volume.

8.2 Importance of Gasket Materials and Design

Gasket materials and designing the gaskets for proper sealing, plays a crucial role in avoiding gas mixing and leakage. Even minor gas mixing or crossover may result in a loss of potential as well as stack overheating. The gaskets play various roles in the stack. Apart from sealing the gas leakage, it also helps to cover the machining/dimensional variations of the components. This prevents cracking of the separator/bipolar plates and the electrode substrate as well.

In some designs, it may be required to prevent the acid holder matrix from getting too much compressed. The matrix is soft and depending upon the tightening, gets compressed, causing reduction in acid holding capacity. The ion transport path becomes increasingly tortuous and provides increased resistance. Further in case of acid loss, the acid fill up from the acid reservoirs may be very slow. All these factors may reduce the cell performance after some time. To avoid such problems, gaskets are used like a soft "spacer frames" which prevent too much compression of the matrix.

The gaskets materials that are generally used consist of soft exfoliated graphite, fluro-carbons like Viton and PTFE. At fuel cell temperature, creep failure of the gaskets may reduce the tightening pressure, causing increased ohmic resistance of the stacks over a period of time and leakage related problems. Thin reinforced gaskets are preferred in case of PTFE or Viton like materials particularly if it has to be used as spacers as well.

Soft gasket materials are required for sealing gap between the external manifold chambers and the PAFC sack sides. For this purpose, one typical option is to use Viton *in-situ* cured (cross linked) sealing gaskets. Internal manifold designs do not require such systems.

8.3 Stack Assembling Experience

United Technologies Corporation (UTC) and their allied groups have done pioneering work of PAFC system development. Much of the PAFC development occurred under the US marines initiated industry partnership effort, "Team to Advance Research for Gas Energy Transformation, Inc. (TARGET), during mid 70s. Initial PAFC stacks suffered due to inadequate acid level in the cell resulting from poor acid management. Acid loss occurred through evaporation and mist carry over by the reactant streams, absorption of acid in the carbon-polymer bipolar plates etc.

Several designs are developed to provide enough acid inside the stack in reserved form. For example, in place of totally hydrophobised electrode substrate (diffusion layer) a partially hydrophilic substrate material that can hold reasonable volume of acid in some of its pores, and also does not pose diffusion resistance to the anode gas is developed (Breault, 1980, Bushnell and Kunz, 1977, Lamarine et al., 1977). Further this idea was extended to replace thin substrate with thick ribbed structure substrate that can hold more acid with controlled hydrophobicity. Also, the ribbed structures contain the reactant flow grooves, thus reducing the costly separator plate thickness.

The latest generation of PAFC units employ mixture of various technologies. Ribbed electrode design as well as Toray developed thin diffusion layer with separate acid reservoirs are employed depending upon the design, capacity and the availability of the fuel cell materials. For fully hydrophobised electrode substrate, one has to provide acid links from the reservoirs located in bipolar plates to the acid holder matrix. This is possible, by making part of the substrate hydrophilic, and is achieved by filling carbon or SiC particles inside the substrate to allow acid transport to the matrix by capillary action (DeCasperis et al., 1981, Schroll and Hartford, 1974).

In an effort, to reduce acid loss by absorption in separator plates, better materials with low open porosity are used. Additionally, acid loss can be reduced by low temperature operation of PAFC. Some groups including the author's laboratory is adopting a temperature range that is around 150-170°C. This reduces component corrosion and acid loss. However, for anode catalyst, Pt/Ru alloy catalyst has to be used for better CO tolerance with reduced current density. Additionally, the upstream reformer system has to be tuned to ensure CO level at < 1%. However, this design is found to be quite acceptable for small capacity power plants.

Another approach pursued by UTC is to add an "acid condensation zone" in the cell where no catalyst is present and have extra cooling. This keeps the condensation zone temperature to < 20-40°C than the core of the cell. This reduces acid loss significantly and is reported to be employed in various designs.

With all these efforts, the recent PAFC stacks perform anywhere between 200 and 300 mA/cm^2 current density at a cell potential of 500 to 700 mV during the initial phase of its life, at atmospheric pressure. Though pressurized cells are reported and can perform at a higher current density, they pose other problems (Bloomfield and Cohen, 2000). For example, the bulky components and added compressor power may overshadow the higher power density of the stack. The system that uses pressure canopy faces additional gas leakage problems and safety related issues. Hence, with present technology, atmospheric PAFCs are more realistic than their pressurized counterparts.

9. PAFC Stack: The Key Issues

9.1 The Thermal System

PAFC systems require heating to around 130°C before start-up. This is required, as at lower temperature the concentrated acid will mainly remain undissociated, and low availability of protons will offer high resistance. Further, the moisture generated, will not come out with the reactant stream due to the low vapor pressure of concentrated acid at low temperature. This will result in retention of moisture by the acid to get diluted and expand in volume. Part of this diluted acid will ooze out through the electrode and will be carried over as mist by the reactant streams causing heavy acid loss. Further as the system temperature increases to its normal operating value, the acid will again get back to its original concentration and shrink in volume. If further acid is not supplied to the matrix, this may lead to dry pores in the matrix resulting in gas crossover.

Stack heating can be achieved by various means, viz. recirculating hot inert gas like nitrogen at the anode and cathode flow field grooves, providing heaters at sides of the PAFC stacks, and so on. However, insertable heater system, that heats up the stacks by conduction is probably the best way to heat the system in a controlled manner. After preheating, during operation of the stack (depending upon the scale and load) it may be necessary to remove heat from the stack in order to maintain the temperature. In view of this, SS tube based inserted coolers after 5-10 cells in between two separator plates are found to be quite effective (DeCasperis et al., 1981, Grevstad and Gelting, 1976). Heat transfer medium could be water/steam, oil etc. Two types of gas cooling are also used. The first one is to use gas cooler plates using an independent stream of cooling air (Tajima et al., 1996), while the other type is to use excess air in the oxidant channel to have necessary cooling. For the second type of cooling additional cooler plates may be provided but there is no independent control on air for reaction oxidant and cooling air.

The waste heat from the PAFC stack can be utilized for other purposes. Miki and Shimizu, 1998, have reported a dynamic simulation study of water/steam based 50 kW PAFC cooling. The study uses lumped parameter models and suggested fuzzy logic type controls for better performance. The study considers an upstream reformer and the steam generated from the fuel cell waste heat is used for feeding the shift reactor.

The waste heat management of PAFC is greatly enhanced if there is a methanol reformer based hydrogen provider in the upstream. The vaporization heat load of the methanol and the reforming process steam demand can be taken care of by the fuel cell waste heat as the temperature of the PAFC stack is sufficiently high to allow such waste heat recovery.

For small stacks, the heating methodology may play a crucial role as it may be necessary to continue the heating of stack even under operation due to heat loss to the environment. Compact thermal systems employing steam or catalytic burner with high burner efficiency and low fuel loss are critical for economic operation of such systems.

9.2 Humidity/Acid Management

Humidity management is the basic requirement for long time operation of the stack. Water is generated at the cathode and dilutes the acid near the cathode. Thus, the oxidant steam (air/oxygen) carries more moisture out of the stack due to higher vapor pressure of the diluted acid. Further due to proton movement from anode to cathode, water migration occurs towards cathode as well. If the moisture content in anode gas is not sufficient, it will cause water evaporation into the anode side as well. All these phenomena lead to the development of acid concentration gradient from anode (concentrated) to cathode (diluted) resulting in low protonation at anode, and subsequent lowering of potential. However, if the stack operation is suspended for some time, the acid concentration gets uniform.

In view of this, there is a need to reduce water evaporation from the anode side. Thus the anode gas is sufficiently humidified so that the vapor pressure matches the target acid concentration at the system temperature. The cathode gas (air) is generally not humidified. Such arrangement maintains the acid concentration as the water migration from anode to cathode is replenished from the moisture present in the anode gas.

Another phenomena of "cathode weeping" is observed for very high current density. Owing to very high proton transfer, the acid (the phosphate ion) will get carried over to the cathode and may start oozing out or loading of too much acid at the catalyst layer and the electrode substrate interface. This aspect may not be of major concern, as at nominal current density this effect is less significant. Providing internal/external acid reservoirs and connecting these reservoirs with the acid holder matrix in a distributed format is the key to address uncontrollable acid imbalance and loss.

9.3 Cell Life and Mechanisms of Cell Decay

PAFC cells can last to anywhere between 10,000 and 50,000 hr. However, there are various issues that can reduce the lifetime. The most important is the one related to acid management. Effective humidity control is the key way to control the acid concentration inside that also helps to prevent loss of phosphoric acid. The other aspects of acid management is the uniform distribution of the acid in the matrix. Occurrence of any undesirable distribution of acid can be prevented through a well designed matrix-electrode assembly, along with good distributed contact with the internal/external acid reservoirs.

There are other reasons for acid loss in the form of absorption in the separator plates, mist carryover etc. This can weaken few of the cells in the stack. This may eventually reverse some or all of those cells, owing to high local polarization related problems and can reduce the stack performance seriously. Thus, the main technological challenge to keep PAFCs working is to ensure uniform performance of the cells. This has to be attained by close control of fabrication parameters and automation. A system that is operating with a variable load, should have a system controller for effective humidity adjustments and some form of fault diagnosis methodology by which the acid concentration/level can be monitored in the electrodes and the matrix.

Corrosion of matrix and the cell components in the presence of impurities also cuts short the cell life. Prevention of dust is an absolute necessity, as apart from choking the fine electrode structure it may react with the acid as well (most of the dust particles contain silica).

PAFC needs to be maintained at temperatures above 60°C to prevent acid freezing leading to frozen acid crystals from getting inside the hydrophobic pores. Under operation, these crystals melt and remain inside the hydrophobic pore. Such blockages, can offer very high gas diffusion resistance. Major damage occurs to the PAFC stacks during the shutdown and startup only, and hence it should happen only a minimum number of times.

High operating temperature and cell potential (> 0.8 V) corrodes the cell irreversibly. The controllers and the power conditioners should be so designed that under no load/low load condition, the build up of potential should be avoided and under high current temperature rise should be controlled as well.

Mitsuda and Murahashi, 1991, have carried out a detail study on the effects of air and hydrogen starvation on PAFC electrodes by employing multiple reference electrodes. The reference electrodes are positioned at the periphery of the unit cell with ion contact through extended SiC matrix. The reference electrode holder system is similar to that reported by Aragane et al. 1994. The study concluded that air starvation has little effect on cathode, but hydrogen starvation (around 92% fuel utilization and above) cause severe anode corrosion. Thus, local blocking of gas grooves in the anode is highly detrimental as it may allow local starvation of the order as reported. Hydrogen starvation in the cell causes proton and CO_2 generation by reacting with carbon and moisture in the affected anode, destroying the catalyst, electrode substrate and even the surface of the separator plates. Song et al. (2000), have conducted another study on gas starvation and reported loss of performance that is higher for anode gas (hydrogen) starvation compared to that of the cathode.

10. Accessories and Control Systems

10.1 Control Systems

Effective operation of PAFC power plants require control of process parameters. Several stack related parameters like temperature control, adjustment of flowrates based on electric load, humidity etc. are to be controlled effectively. Control of some of the parameters may look trivial, but operating the PAFC plant with moderate turn down while maintaining high overall efficiency may be tricky. This is due to the coupled effects of various parameters on fuel cell performance. Control of humidity under varied load may be considered, for example, to understand such issues.

The moisture adjustment at a single current density may not help, as the current density may change during course of operation. As current density changes, the moisture generation rate is also changed. Considering the reactant gasses flowrate remains unchanged, the excess volume of gases coming out of the system will change. These two phenomena will affect the total humidity balance.

If the current density is increased without changing the flowrates, the total amount of hydrogen and air that comes out of the system will decrease. Thus, the anode gas that is near saturated at the inlet, will come out with less moisture. This is due to lowering of the net gas coming out of the stack with same inlet concentration. The humidity level of exit air will tend to remain same as before due to unchanged contact time. However as the total air coming out is reduced, it will be able to take out less moisture. Thus the additional moisture that is generated gets added with the moisture not removed and starts diluting the acid. This dilution will increase water vapor pressure of the acid and thus will allow the air to take up more moisture after some time. Although the PAFC is able to adjust to such fluctuations upto a certain level, for large increase in current density, the moisture will start building up, diluting the acid too much. This will increase the acid volume as discussed in Section 9 and will cause extra acid loss.

The scenario is more complicated if change in hydrogen and air flowrates are considered as well as the change in current density. This should be considered as one has to use hydrogen economically and the air flow device at minimum power. Thus, online control of humidity of the inlet anode gas is necessary. In the same time, the local temperature gradient may also change. This is owing to different heat generation rate as change of current density is associated with change of cell polarization losses. Change of reactant flowrate will affect the heat removal feature as well. The systems with independent cooling systems, the temperature maintenance is decoupled, and can be done easily. However for cells, that cools the system by passing excess air, the control mechanism should consider electric load and humidity management factors as well.

Cell potential is directly proportional to the power conversion efficiency of the stack. If the current density increases the cell potential decreases. The extent of potential drop also depends upon the flow rates (Choudhury et al., 2002). Thus, there is an optimal setting of parameters for every current density. This is to be determined, keeping in view the cell design for minimum excess gas flowrates requirement, anode gas recirculation option and incremental increase of cell potential with respect to gas flowrates. To operate the system successfully for onsite power generation, a supervisory controller to operate the system optimally may be a good choice. The most simple way to reach optimal control is to provide a sort of look up table in which the humidity and flow rates are provided for various currents to the supervisory controller, that updates the set point of the relevant control system.

10.2 Power Conditioners

PAFC like other fuel cells, dynamically changes potential with current. There is thus a requirement to stabilize the output voltage. This is done with wide input variation DC-AC inverters or through DC-DC stabilizers. The efficiency of the converter is of critical importance. The minimum operating input potential range while maintaining high efficiency dictates the number of cells in a stack. In a low potential high current device like PAFC, the target efficiency should be greater than 85%.

There are various techniques available to stabilize the output potential from PAFC stack. A typical method is to convert unstabilized low potential DC into a high voltage AC. This is achieved by using switching devices in series with the primary of the transformer. The output from the fuel cell is connected to the primary of the transformer through the switching device. As the device switches ON and OFF at a pre-determined frequency, an AC potential is generated at the secondary of the transformer. The AC frequency is dependent on the switching frequency. The wave form of the AC potential is square wave and that for the current is rounded square to near triangular wave form depending upon the circuit impedance. The potential of the AC generated is dependent upon the transformer turnings. The overall stabilization of the high frequency AC voltage is carried out by dynamically changing the ON and OFF time of the switching device.

The high frequency AC is then either converted into a stabilized DC voltage through a rectifier circuit or is converted into a lower voltage AC as per requirement. The major losses that occur during the overall conversion is at the section where variable DC input is converted to high frequency AC. The switching devices and the switching control mechanisms are of primary concern to achieve high efficiency. Other aspects that are to be considered involve output voltage regulation and the variable DC input window. All these aspects are interrelated and have to be considered in a holistic manner.

The transformer weight (for smaller power plants) needs to be low for easy transportation. To achieve this one may go for higher frequency AC to allow smaller transformer, but this may also lead to more power loss in the switching device.

11. Experience of PAFC System Development

11.1 Developments at Naval Materials Research Laboratory: A Case Study

Naval Materials Research Laboratory (NMRL) is one of the oldest of the Defence Research and Development Organization (DRDO) laboratories in India and is a multidisciplinary materials laboratory. The laboratory pursues research on ceramics, polymer, metallurgy, marine microbiology, protective coatings, environmental sciences, corrosion and electrochemistry within one campus.

NMRL started an ambitious program on development of fuel cells for various defence and non-defence applications in early nineties. The primary emphasis of this program is indigenous development of all materials, system and accessories of fuel cell resulting in technology demonstration and deliverables. Under this program, all major components and materials related to PAFC have been developed and transferred successfully to the industries. These components are assembled to develop complete fuel cell based power packs from sub-Watt to 10 kW complete with accessories and various kinds of fuel processors. The program is described briefly in the following paragraphs.

11.2 Components Developed

Catalyst

Over years NMRL has developed various carbon supported Pt and Pt alloy based electrocatalysts suitable for PAFC and PEMFC. Some of the key issues that have been addressed during the catalyst development phase are controlled positioning of active sites on the porous carbon support, active metal particle size distribution and stabilization. Such expertise has allowed NMRL catalyst to pass 2000 hr of continuous operation without appreciable loss of activity in PAFC unit cells.

Catalyst developed in NMRL are high activity Pt on carbon for PAFC and PEMFC, CO resistant Pt-Ru on carbon for PAFC anodes, Pt-Co and Pt-Ni on carbon for PAFC/PEMFC cathode.

Porous Carbon Paper

NMRL is working primarily on development of PAFC, that requires carbon paper as the catalyst layer support cum current collector, and gas diffuser. In order to achieve totally indigenous fuel cell ideology, NMRL through subprojects to other R&D institutions has successfully developed the paper upto 10 cm^2 size. The quality of the paper is tested and proven at NMRL. Scaling up of the technology for commercial production is being pursued.

Graphite Bipolar Plate

NMRL from the initiation of the development of fuel cell technology program has co-ordinated with various Indian graphite component manufacturers. This has resulted in successful development of graphite material, that passes the stringent requirement of low porosity (< 1:5%) and apparent density (> 1:8 gm/cc) required form for PAFC. Right now, NMRL is using machined graphite manifold for the PAFC program due to lack of volume. However, to produce bipolar plates at a cheaper cost, NMRL has developed and standardized technology for 3-4 mm thick bipolar, molded, grooved, graphite composite plates for 1-5 kW PAFC stack fabrication.

Acid Holder Matrix

NMRL has developed a slurry casting technique to cast flawless matrix of thickness 100 μ with excellent acid holding capability and high bubble pressure. Average particle size of the SiC powder that is used is around 5 μ. Moderate matrix thickness along with not-so-fine SiC particles allows rugged handling of the electrode during assembling and longer life of the stack. The casting technique is partly automated, yields similar result to that of curtain coating methods. An automated, customized film applicator with draw speed controller is used for casting the matrix on the cathode. Vacuum compaction after casting is done mainly for the large size electrodes to provide uniform thickness. Electrode coated with matrix is left in open air (with controlled humidity) for drying up. Moisture in the ambient air changes drying rate and if not controlled may result in PTFE binder segregation and shrinkage related problems.

Fuel Cell Power Packs

NMRL has mastered the technology of PAFC stacks, ranging from 100 W to 10 kW. These compact systems have all necessary subsystems, namely, built-in acid management systems, catalytic pre-heaters with controlled gas diversion, sandwich type humidifiers (for certain applications) based on waste heat and tail gas catalytic burners.

NMRL developed systems are highly compact units suitable for small domestic, commercial and military applications. These power packs are thoroughly tested and last well beyond 5000 h of continuous operation. These systems can be operated intermittently, however, the acid addition to the built-in "fill port" needs to be more frequent. Fig. 15 shows a typical NMRL PAFC 1 kW stack and Fig. 16 shows the performance of the stack under DC polarization run. The stacks are air cooled systems, and the top mounted fan (Paul and Getting, 1976) provides air both for oxidation and cooling.

Power Conditioners

Fuel cell produces DC potential which varies dynamically as the load changes. This requires a post

Capacity: 1 kW
Overall stack dimensions
Length 300 mm, Width 320 mm
Height 250 mm

Fig. 15 The 1 kW PAFC stack (Courtesy: Naval Materials Research Laboratory, India).

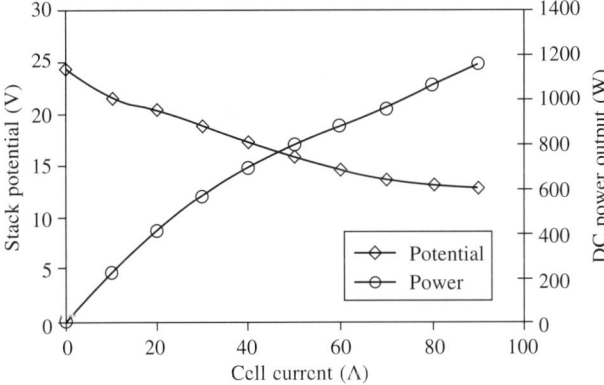

Active area = 250 cm^2; Number of cells = 30
Anode catalyst: 20% Pt, 10% Ru
    (0.35 mg Pt/cm^2 loading)
Cathode catalyst: 10% Pt (0.7 mg Pt/cm^2 loading)
Anode: Reformer gas (~0.8% CO)
Cathode: Air stack temperature = 150°C

Fig. 16 The 1 kW PAFC stack performance with air and methanol reformer (Courtesy: Naval Materials Research Laboratory, India).

processor, which will generate a stable DC/AC output as required. NMRL has developed through various Indian companies a wide array of DC-DC converters and DC-AC inverters which are economical, compact, light weight and highly efficient (> 90% at full load).

Online Hydrogen Generators

NMRL has developed various kinds of hydrogen generation devices, and has successfully integrated them with PAFC stacks for several applications. For continuous operation of fuel cells, NMRL has an array of compact, light-weight, rugged, methanol reformers. This includes a compact reformer for 2-5 kW PAFC reformer with methanol based catalytic burners. A natural convection flameless catalytic burner based portable reformer for camp powering is also developed. This system works with 100-300 W PAFC power packs and is free of instrumentation or any powered device. In the higher side, a 60 kW reformer has also been tested successfully.

For sub-watt to few watts systems, a reformer may not be a pragmatic choice. NMRL has developed several types of once use cartridge based systems, for online generation of hydrogen. Hydrogen generated from these systems is highly pure and can be used in PEM fuel cells as well, apart from PAFC. Typically the system consists of a powder filled cartridge which, on demand can be brought in contact with water at a controlled rate that hydrolyzes the powder to generate hydrogen. The powders used are alkali hydrides, borohydrides, a special super-corroding alloy etc. These disposable, snap-fit cartridges are very light in weight, for example a 100 g cartridge can generate power @100 watt for 1-1.5 hr.

These devices are suitable for remote area battery chargers of various capacity, operation of hand-held electronic devices like laptop, telecommunication equipments etc.

Fig. 17 The 8 kW PAFC power plant for area lighting (Courtesy: Naval Materials Research Laboratory, India).

Fuel Cell Controllers
Optimum usage of fuel with load following characteristics is possible only with smart controllers. A mini chip based controller card which controls the hydrogen flow optimally to maximize the efficiency is also developed. This chip based card is universal, i.e. can be used for smaller as well as larger stacks. It has self-learning ability for automatic customization and can be integrated with the online reformer system.

The basic principle of this hydrogen economizer device is to measure the current load and to determine the hydrogen flow that is decided by a characteristic function programmed into the chip. The control is done in feed forward mode and for better stability it also checks hydrogen flow at the PAFC outlet. Based on this feedback, a correction on hydrogen flow is made.

12. Future Challenges
PAFC stacks are being evaluated world wide for various purposes. The distributed onsite power generation is perhaps the most successful one. These power plants are operated with online fuel converters like reformers based on feedstock like methanol, CNG, propane, LPG, desulphurized diesel etc. The overall report show stable performance with very high (> 85%) availability.

However, there are few issues that are holding PAFC from penetrating deeper into the power sector. The primary reason is the overall cost of the PAFC power plants. Although it is one of the cheapest member of the fuel cell family, the cost is still prohibitive for onsite power generation.

Components like the separator or the bipolar plate are the more costly components. After the finding that graphite composite materials are the only ones successful for long time operation, a major ongoing research in this field is to develop cheaper precursor materials for producing these plates. For example, the latest trend is to mold a "Green" plate using alternate layers of graphite powder, phenolic resin and cellulose material. This plate is then taken for carbonization and graphitization. However, for making plates with large dimension (about 1 m characteristic length), the technology developed should be suitable for mass production. Shrinkage of the plate and loss of surface planarity are the major problems that must be addressed during mass production. Thus the development of low cost separator plates, suitable for mass production, is one of the R&D areas that must be considered very seriously for further development of PAFC technology.

Reducing the Pt metal in the catalyst is the other challenge for cost reduction. Although there is a significant reduction in Pt content of PAFC electrodes (from 1 mg/cm^2 to around 0:1-0:2 mg/cm^2) still further reduction is required. R&D efforts are on to develop catalysts free of Pt and in these regard metal phthalocyanines and similar molecules have shown some promise. However, the major problem associated with such compounds is the loss of stability primarily due to destruction of active sites by peroxide generated during oxygen reduction (Vielstich et al., 2003c). At present, a major research aim is to use trace amounts of Pt alloyed with suitable metals like Co, V, Ni, etc.

Development of technologies suitable for mass production of the matrix coating on the electrode, assembling of the stack is also quite important for cost reduction while maintaining uniform product quality. In this regard, coating of the matrix by curtain coating method or more recently by Grauvere coating method is the right step towards mass production. Reduction of matrix thickness has already reached the minimum and performance enhancement by further reduction of ohmic loss related to proton transfer may not be feasible. Unfortunately, the thinnest matrix uses very small (1 μ) SiC particles that is more susceptible to corrosion. Very small SiC particles are found to have pure Si on the surface and when employed for ultra thin matrix preparation show higher corrosion in phosphoric acid. Thus, the major emphasis should be to enhance current density by reducing the matrix pores tortuosity and thinner catalyst layers, rather than trying to reduce the matrix thickness too much.

At present, most of the PAFC designs cannot handle wide turn down of output power. In the case of high load variation, acid concentration inside the stacks change due to inadequate humidity and acid management. This may lead to excessive acid loss particularly from the cathode. Additionally, the effect of potential cycling may cause cumulative decay in the catalyst layer. For example, when a large stack is started, it may take days to reach its peak potential at a given current. During this time, wetting of the catalyst layers occur which means that the acid enters at many of the less accessible regions. As the acid fills up the dry portions of the electrode, on the surface of the carbon support of the catalyst traces of oxide gets reduced at anode, and the oxygenated groups fill the cathode. This phenomena makes the catalyst layer of the anode more hydrophobic and that of the cathode more hydrophilic. Ryu et al. (1998) reported formation of carboxyl group by reaction of H$_2$O$_2$ with carbon at lower oxygen partial pressure (of around 0.2 atm, similar to air) at PAFC operating temperatures. Formation of other oxygenated groups like quinone have also been reported. Thus, frequent potential swing changes the level of oxide reduction and formation in the catalyst support thus weakening the bond between the Pt crystals and the support. This causes Pt sintering and the effect is further magnified by changing oxide layer of Pt.

However, for a distributed power generation, it is necessary to accept high turndown of load. There is a need to study and develop electrodes, that are resistant to such sintering. One way is to use a very low ash content carbon support with very high graphitic structure. Supports like carbon nanotubes may prove to be a good replacement of the existing support in this regard. Suitable development of sinter resistant Pt alloy(s), is necessary as well.

As many of the parameters like inlet humidity, cell temperature, gas flow rates and catalyst state of health are interrelated, there is a need to develop comprehensive online fault diagnosis system. There is a big gap in the knowledge base of online techniques for understanding problems of PAFC electrodes. For example, for loss of power in an operating PAFC stack, there is a need to understand the state of health of the electrodes viz. the wetting level, Pt activity and diffusion resistance. Further, the acid level in the matrix and its concentration are other issues that have to be analyzed quickly as well. Thus, there is a need to develop suitable non-invasive methodologies that can be used to understand the state of the electrodes and the acid matrix of an operational stack.

In this regard perturbation analysis like step analysis and electrochemical impedance spectroscopy (EIS) can show the way (Choudhury et al., 2005, Jenseit et al., 1993). The popular trend of using transmission line model for EIS analysis may not be sufficient to diagnose the dynamic mechanisms. Thus to understand and decouple the effects, there is a need to develop and validate comprehensive transient models based on first principles. With the availability of increased computational power it may be possible to develop online fault diagnosis analyzer systems for the actual field units.

Similarly, there is a need to develop control philosophies that can be used with different types of fuel cells with little configurational change. At present, some of the power plants have load following controllers that economize hydrogen to a certain level under turn down conditions. However, such savings could be greatly enhanced with better control methodologies. To achieve that, more transient data of PAFC operation and the effect of parameters need to be generated and made available in the open domain.

PAFC systems use various other accessories as well. Sufficient effort to reduce the cost of these accessories like power conditioners, gas handling systems and down stream reformers/hydrogen generators also has to be undertaken. Small foot-print, low temperature reformers with coke resistant catalyst is one of the major challenges for wider acceptance of PAFCs in future.

Development of the power electronics for stabilizing and using the PAFC power is gathering momentum. Fortunately, other fuel cells like PEMFC also have similar electrical characteristics and requirement for power conditioners. However, fuel cell developers are still using customized power electronics. To reduce the cost, thus there is a need to develop a standard design for fuel cell operational range. This will lead to standardization of power electronic devices for output voltage stabilization of PAFCs.

13. Conclusions

PAFCs have shown limited commercial prospect. Their efficacy in onsite power generation is proven thoroughly. However, the main issue that prevents deeper penetration is the high production cost. The future of PAFC depends on how the component cost will be brought down as there are multiple high priced components that needs attention. The costly components are the catalyst, separator plate and the fabrication technology of the electrode. The accessories also add up significantly and this includes, among others, the thermal system, the hydrogen generator and power electronics.

Apart from onsite distributed power generation, PAFCs may find other applications as well, particularly in developing countries such as India. It is clear that the usage of PAFC is not very suitable for small vehicular applications. However, for countries like India, it is worth exploring usage of PAFC for bigger vehicles with routine movements like long distance busses. Usage of PAFC as auxiliary power for locomotives like AC coach powering, compartment lighting etc. may prove highly efficient. One of the main problems in Indian railways is high transmission losses. Keeping PAFC distributed power plants very near the railway track may prove to be useful.

PAFC, unlike PEMFC does not enjoy high R&D support globally. PAFC still can still be an attractive alternative, particularly when there is no suitable answer for small distributed onsite power generation. Long life membranes, along with CO tolerant catalyst for PEMFC is still not in sight. This has pushed many groups to develop further improved low cost PAFCs.

Acknowledgements

The author thanks the Director, Naval Materials Research Laboratory and the members of Energy Science and Technology division for their kind and wholehearted support.

References

Anthony, A. DeCasperis, Richard, Roethlein, J. and Richard Breault, D. Electrode reservoir for a fuel cell. US Patent No. 4,269,642, May 26, 1981.

Appelby, A.J. and Foulkes, F.R. Fuel Cell Handbook. Van Nostrand, New York, 1989.

Aragane, J., Urushibata, H. and Murahashi, T. *Journal of Electrochemical Society*, **141**(7): 1804-1808, July 1994.

Emmanuel Auer, Gerhard Heinz, Thomas Lehmann, Robert Schwarz and Karl-Anton Starz. Pt/Rh/Fe alloy catalyst for fuel cells and a process for producing the same. US Patent No. 6, 165, 635, December 26, 2000.

David, P. Bloomfield and Ronald Cohen. Pressurized fuel cell power plant. US Patent No. 3,972,731, August 3, 2000.

Breault, Richard, D. Silicon carbide electrolyte retaining matrix. US Patent No. 4,017,664, April 12, 1977.

Breault, Richard, D. Fuel cell electrolyte reservoir layer and method for making. US Patent No. 4,185,145, January 22, 1980.

Calvin L. Bushnell and Harold Russel Kunz. Electrolyte reservoir for a fuel cell. US Patent No. 4,064,322, December 20, 1977.

Caires, M.I., Buzzo, M.L., Ticianelli, E.A. and Gonzalez, E.R. *Journal of applied electrochemistry*, **27**: 19-24, 1997.

Choudhury, S. Roy, Deshmukh, M. B. and Rengaswamy, R. *Journal of power source*, **112**: 137-152, 2002.

Choudhury, Suhasini Roy, K.V. Nair, and Rangarajan, J. A process for depositing platinum on to carbon black for fuel cell. Indian patent application No. 1491/del/2004, 2004.

Suman Roy Choudhury, Suhasini Roy Choudhury, J. Rangarajan, and Rengaswamy, R. *Journal of power sources*, **140**: 274-279, 2005.

Der-tau and Howard Chang, H. *Journal of applied electrochemistry*, **19**: 95-99, 1989.

Gasteiger. H.A. *Journal of physical chemistry*, **98**: 617, 1994.

Grevstad, E. Paul and Raymond Gelting, L. Fuel cell electrode cooling system using a non-dielectric coolant. US Patent No. 3,969,145, July 13, 1976.

Jenseit, W., Bohme, O., Leidich, F.U. and Wendt, H. *Electrochimica Acta*, **38**(14): 2115-2120, 1993.

Kemp, S. Fred and Michael George, A. Sequential catalyzation of fuel cell supported platinum catalyst. US patent No. 3,857,737, December 31, 1974.

Kyong Tae Kim, Jung Tae Hwang, Young Gul Kim, and Jong Shik Chung. *Journal of Electrochemical Society*, **140**(1), January 1993.

Kim Kinoshita. Electrochemical Oxygen Technology. John Wiley & Sons Ltd., 1992.

Lamarine, H. John Jr. Robert Stewart, C. and Raymond Vine, W. Electrode reservoir for a fuel cell. US Patent No. 4,038,463, July 26, 1977.

Maoka, T., Kitai, T., Segawa, N. and Ueno, M. *Journal of applied electrochemisrty*, **26**(12): 1267-1272, 1996.

Miki, H. and Shimizu, A. *Applied Energy*, **61**: 41-56, 1998.

Mitsuda, K. and Murahashi, T. *Journal of applied electrochemistry*, **21**: 524-530, 1991.

Petrow, G. Henry and Robert, J. Allen. Finely particulated colloidal platinum compound and sol for producing the same, and method of preparation. US Patent No. 3,992,512, November 16, 1976.

Pozio, A., De Francesco, M., Cemmi, A., Cardellini, F. and Giorgi, L. *Journal of power sources,* **105**: 13-19, 2002.

Rice, Cynthia, Yuye Tong, Eric Oldfield, and Andrzej Wieckowski. *Electrochimica Acta*, **43**(19-20): 2825-2830, 1998.

Roques, J., Alfred Anderson, B., Vivek Murthi, S. and Sanjeev Mukerjee. *Journal of the Electrochemical Society*, **6**: E193-E199, 2005.

Ryu, Young-Gyoon, Su-Il Pyun, Chang-Soo Kim, and Dong-Ryl Shin. *Carbon*, **36**(3): 293-298, 1998.

Schroll, Craig R. and West Hartford. Liquid electrolyte fuel cell with gas seal. US Patent No. 3,855,002, December 17, 1974.

Song, Rak-Hyun, Dheenadayalan, S. and Dong-Ryl Shin. *Journal of power sources*, **106**: 167-172, 2002.

Song, Rak-Syn, Chang-Soo Kim, and Dong Ryul Shin. *Journal of power sources*, **86**: 289-293, 2000.

Spearin, W. Process for forming a fuel cell matrix. European patent No. 0,344,089, November 29, 1989.

Stewart and Robert, C. Process for forming a fuel cell matrix. US Patent No. 4,173,662, November 6, 1979.

Tajima, Osamu, Akira Hamada, Junji Tanaka, Yasunorj Yoshimoto, Keiogo Miyal, Nobuyoshi Nishizawa, Masaru Tsutsumi, Tomotoshi Ikenaga ans Kunihiro Nakato, and Kiyoshi Hori. Fuel cell using a separate gas cooling method. US Patent No. 5,541,015, July 30, 1996.

Trocciola, John, C. Dan Elmore, E. and Ronald, J. Stosak. Screen printing fuel cell electrolyte matrices using polyethylene oxide as the inking vehicle. US Patent No. 4,001,042, January 4, 1977.

Wolf Vielstich, Arnold Lamm and Hubert A. Gasteiger editors. Handbook of Fuel Cells, Fundamentals, Technology and applications, volume 1. John Wiley & Sons Ltd., 2003a.

Wolf Vielstich, Arnold Lamm and Hubert A. Gasteiger editors. Handbook of Fuel Cells, Fundamentals, Technology and applications, volume 4. John Wiley & Sons Ltd., 2003b.

Wolf Vielstich, Arnold Lamm and Hubert A. Gasteiger editors. Handbook of Fuel Cells, Fundamentals, Technology and applications, volume 2. John Wiley & Sons Ltd., 2003c.

Wieckowski, Andrej Elena Savionova, Constantinos, R. and Vayens, G. editors. Catalysis and electrocatalysis at nanoparticle surfaces. Marcel Dekker Inc., 2003. 49.

Recent Trends in Fuel Cell Science and Technology
Edited by S. Basu
Anamaya Publishers, New Delhi, India

9. Carbonate Fuel Cell: Principles and Applications

Hossein Ghezel-Ayagh, Mohammad Farooque and Hansraj C. Maru
FuelCell Energy Inc., 3 Great Pasture Road, Danbury, CT 06813 USA

1. Introduction

In recent years, carbonate fuel cell technology has enjoyed a steady improvement both in cost reduction and performance. Large-scale manufacturing combined with a multitude of power plant installations has resulted in the establishment of the carbonate fuel cell technology as a leading candidate for efficient and reliable distributed generation.

The carbonate fuel cell principle of operation is based on transfer of the oxygen in the form of carbonate ions from the cathode to the anode. In many ways, this is similar to Solid Oxide Fuel Cell (SOFC) technology, with the main differences being that the medium of the ionic transfer is a molten carbonate immobilized in a ceramic matrix. Various eutectic mixtures of lithium, potassium, and sodium carbonates have been used as electrolyte with the most prevalent ones being 38Li/62K or 60Li/40Na carbonate mixtures.

One of the most important characteristics of the carbonate fuel cell is its ability to generate electricity directly from a hydrocarbon fuel, such as natural gas, by reforming the fuel inside the fuel cell to produce hydrogen. This "one-step" internal reforming process results in a simpler, more efficient and cost-effective energy conversion system compared with external reforming fuel cells. External reforming fuel cells, such as proton exchange membrane (PEM) and phosphoric acid, generally use complex, external fuel processing equipment to convert the fuel into hydrogen. This external equipment increases capital cost and reduces electrical efficiency.

The carbonate fuel cell has been demonstrated using a variety of hydrocarbon fuels, including natural gas, methanol, diesel, biogas, coal gas, coal mine methane and propane. The commercial power plant products are expected to achieve electrical efficiencies exceeding 45%. Depending on location, application and load size, it is expected that a co-generation configuration will reach an overall energy efficiency between 70 and 80%.

2. Carbonate Fuel Cell Technology

The carbonate fuel cell power plant is an emerging high efficiency, ultra-clean power generator utilizing a variety of gaseous, liquid, and solid carbonaceous fuels for commercial and industrial applications. The carbonate fuel cell uses alkali metal carbonate mixtures as electrolyte and operates at ~650°C. Corrosion of the cell hardware and stability of the ceramic components have been important design considerations in the early stages of development. The material and electrolyte choices are founded on extensive fundamental research carried out around the world in the sixties and early seventies. The cell components were developed in the late 1970s and early 1980s. The present day carbonate fuel cell construction employs commonly available stainless steels in a majority of cell and stack components. The electrodes are based on nickel and well-established manufacturing processes. Manufacturing process development, scale-up, stack tests, and pilot system tests dominated throughout the 1990s. Commercial product development efforts started in late

1990s, leading to prototype field tests beginning in the current decade, leading to commercial customer applications. Cost reduction has been an integral part of the product effort. Cost-competitive product designs have evolved as a result. Approximately half a dozen teams around the world are pursuing carbonate fuel cell product development. The power plant development efforts to date have mainly focused on several hundred kW (submegawatt) to megawatt-class plants. Over 50 submegawatt units have been operating at customer sites in the U.S., Europe, and Asia. Several of these units are operating on renewable bio-fuels. A 1 MW unit is operating on the digester gas from a municipal wastewater treatment plant in Seattle, Washington (US). Presently, there are a total of approximately 13 MW carbonate fuel cell power plants installed around the world. Carbonate fuel cell products are also being developed to operate on coal-derived gases, diesel, and other logistic fuels. Innovative carbonate fuel cell/turbine hybrid power plant designs promising record energy conversion efficiencies approaching 75% have also emerged. This article reviews the historical development of the carbonate fuel cell and discusses the recent advances in this unique technology.

2.1 Fundamental Principles

The carbonate fuel cell utilizes a mixture of alkali carbonates as the electrolyte and operates at 550-650°C. The basic electrochemistry of carbonate fuel cells (Fig. 1) involves the formation of carbonate (CO_3^{2-}) at the cathode by the combination of oxygen, carbon dioxide, and two electrons (Eq. 1); transport of the carbonate ions to the anode through carbonate electrolyte; and finally, reaction of the carbonate ion with hydrogen at the anode, producing water, carbon dioxide, and two electrons (Eq. 2):

Cathode reaction:
$$CO_2 + \frac{1}{2}O_2 + 2e^- \rightarrow CO_3^{2-}$$
(1)

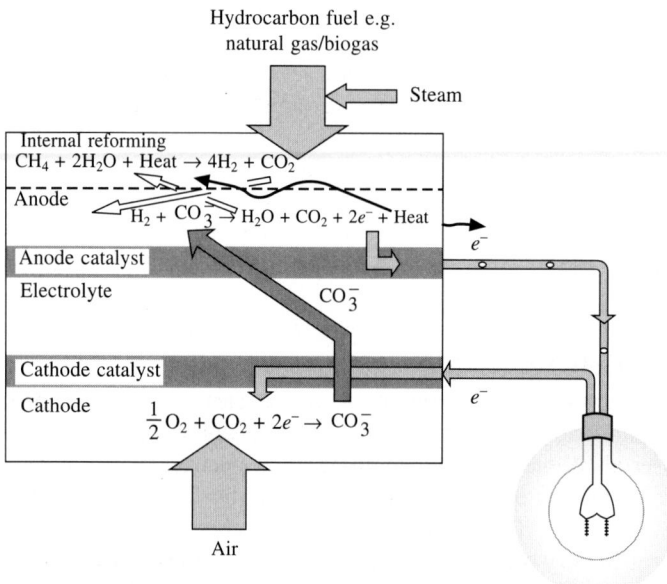

Fig. 1 Schematic showing carbonate fuel cell operating principles: CO_2 required at the cathode is supplied from anode exhaust.

Anode reaction: $$H_2 + CO_3^{2-} \rightarrow H_2O + CO_2 + 2e^- \tag{2}$$

Overall: $$H_2 + \frac{1}{2} O_2 \rightarrow H_2O \tag{3}$$

For each mole of hydrogen consumed in the anode compartment, one mole of carbon dioxide and one mole of water are produced. Hydrogen is made available to the anode in a carbonate fuel cell by extracting it from a common fuel (such as by steam-reforming natural gas). The fuel cell reactions use both H_2 and CO in the anode. The commonly available carbonaceous fuels need to be converted to a usable form, either inside or outside the cell (further discussed in Section 2.3). Oxygen is supplied from air, and carbon dioxide is made available by recycling it from the anode exhaust. For this purpose, the anode exhaust is oxidized with the feed air in an oxidizer prior to its introduction into the cathode. Details on carbonate fuel cell chemistry are avilable in the *Fuel Cell Handbook* (Fuel Cell Handbook 2002).

2.2 Historical Development Perspective

A brief development history of carbonate fuel cell technology is shown in Fig. 2. A concise review of early work was presented *Fuel Cells and Fuel Batteries* (Liebhafsky 1968). Broers (Broers 1958) started fuel cell testing in the late 1950s. The basic cell design was defined in the 1960s while the high performance components were identified in the 1970s. The chemical and electrochemical understanding developed during this period is reviewed in *Physical Chemistry and Electrochemistry of Alkali Carbonate Melts in Advances in Molten-Salt Chemistry* (Selman 1981). Fundamental pore equilibrium and capillary control

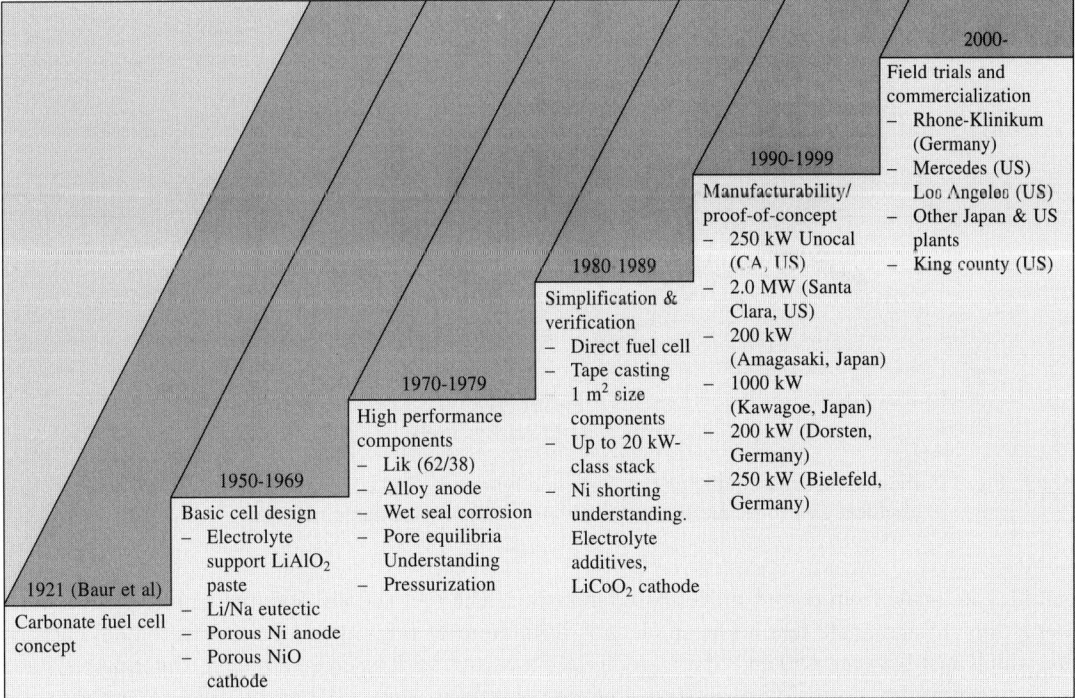

Fig. 2 Development history of carbonate fuel cell: Intensive global R&D during the past three decades advanced the technology to commercial level.

model (Maru 1976) provided the invaluable guidance to optimize cell components design. This understanding helped to optimize anode, cathode, and matrix pore sizes to keep matrix filled with electrolyte all the time while anode and cathode remain partially filled, providing optimum performance. Between 1980 and 2000, component and stack technologies were further improved, simplified and verified in large-area stacks. The internal reforming concept (named "Direct FuelCell") (Baker 1980) was a breakthrough, leading to an efficient and simple system (Farooque 1990). Manufacturability and proof-of-concept power plants were demonstrated in greater than 200 kW field-testing. Since 2000, the development focused on field trials and commercialization. Over 50 units ranging from 200 kW to 1 MW have been successfully operated, the majority of them by FuelCell Energy (Connecticut, USA).

The carbonate fuel cell operates at an optimal temperature of approximately 600-650°C that avoids the use of noble metal electrodes required by lower temperature fuel cells, such as PEM and phosphoric acid, and the more expensive metals and advanced ceramic materials required by higher temperature solid oxide fuel cells. Less expensive electrocatalysts and readily available commercial metals are used in the carbonate fuel cell design. The carbonate fuel cell uses a simple construction as illustrated in Fig. 3. Today's state-of-the-art carbonate fuel cell materials technology is a result of intensive worldwide research carried out over the last three decades with pioneering contributions from the laboratories of FuelCell Energy (formerly Energy Research Corporation), United Technology, General Electric and M-C Power in the US; MELCO, Hitachi, Toshiba, IHI and KIST in Asia-Pacific and ECN, Ansaldo Fuel Cells and MTU-CFC Solutions in Europe. The bipolar plate and the corrugated current collectors are made of 300-series stainless steel. The

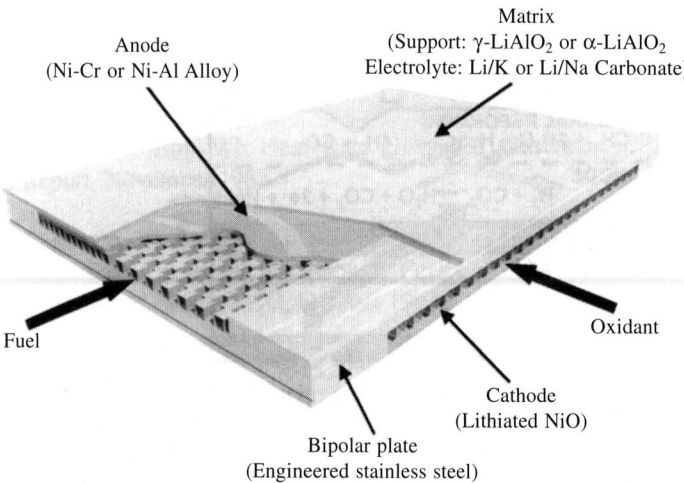

Fig. 3 Carbonate fuel cell bipolar cell package configuration: Cell construction employs 300 series stainless steel and electrodes are nickel-based; all commonly available materials.

electrodes are made from porous nickel-based materials. The fuel and air isolating membrane is a porous ceramic lithium aluminate that holds an electrolyte mixture of lithium and potassium/sodium carbonate salts, which melt between 450 and 510°C. Material properties such as creep strength, sintering resistance, low compaction, hot corrosion resistance, low carbonization, etc. will impact cell life and durability. Stable, long-term anode and cathode electrochemical activity is also required (materials technology is further discussed in Section 2.4).

2.3 Fuel Processing for Carbonate Fuel Cells

The carbonate fuel cell utilizes both H_2 and CO at the anode. The commonly available fuels being considered for carbonate fuel cells include natural gas, gaseous fuels derived from biomass and coal, landfill gas, biomass-derived ethanol, propane, diesel, and bio-diesel. Important properties of these fuels from fuel cells viewpoint are listed in Table 1. Sulfur and halogens present in some of these fuels are considered harmful to fuel processing catalysts and the fuel cell. Therefore, these undesired components need to be removed from the fuel stream. Other potential contaminants such as siloxane in the renewable fuels (biomass digester gas and landfill gas) also should be removed from the fuel prior to fuel processing for fuel cells.

The carbonaceous fuels need to be converted to fuel cell useable hydrogen and CO form to support fuel cell anode reactions. In the conventional carbonate fuel cell system concept (called "indirect" fuel cell), this conversion is carried out in an external fuel processor such as a catalytic steam reformer for natural gas. Steam reforming of a light hydrocarbon fuel in an external reformer is a well-established industrial process for hydrogen production for fuel cells. The natural gas reformers usually operate at temperatures 750 to 850°C, which is significantly higher than the 650°C carbonate fuel cell operating temperature. The steam reforming reaction is highly endothermic and the fuel cell anode reaction is exothermic. While the product of the reforming reaction (hydrogen) is a reactant in the fuel anode reaction, the fuel cell reaction product (water) is a reactant in the reforming reaction. Therefore, the thermal and chemical features of the fuel cell and reforming reactions are uniquely complementary for efficient integration of both of these reactions inside the anode compartment.

Nearly all fuels used to operate carbonate fuel cell power plants contain sulfur. Sulfur compounds deactivate nickel-based catalysts used in the carbonate fuel cell anode. Sulfur has a tendency to be chemisorbed on active nickel, forming nickel sulfide (as shown in Reactions 4-6). The catalyst deactivation causes loss of reforming activity and hence limits the catalyst life. For stable long-term carbonate fuel cell operation, the sulfur concentration in fuel needs to be reduced to a lower level prior to introduction to the anode by utilizing an efficient fuel desulfurization system. As a rule-of-thumb, sulfur should be removed to bring the concentration down to the sub-ppm level.

$$Ni + H_2S \rightarrow NiS + H_2 \tag{4}$$

$$Ni + COS \rightarrow NiS + CO \tag{5}$$

$$Ni + RSH \rightarrow NiS + RH \tag{6}$$

In natural gas fuel, some of the sulfur compounds are present naturally (from the wellhead) while other compounds are added as an odorant for leak detection. In other fuels such as anaerobic digester gas or coal bed methane, all sulfur present is naturally occurring. In either case, many different types of inorganic and organic sulfur-containing molecules may be present and sulfur levels must be reduced to sub-ppm level. A desulfurization system for natural gas and other fuels depends on the concentration as well as the nature of the sulfur compounds. Desulfurization can be accomplished by any of the following processes:

(a) Single-stage sulfur removal using sorbents at low temperatures,
(b) Hydrodesulfurization (HDS) using hydrogen at high temperatures and high pressures, followed by ZnO bed, or
(c) Selective partial oxidation of sulfur compounds to sulfur oxides, followed by adsorption of sulfur oxides in K_2CO_3 bed.

To-date, the focus of fuel desulfurization research has been on low temperature absorption systems such as activated carbon beds and molecular sieves (zeolite). This process has been selected for initial commercial

Table 1.   Typical fuels composition: A variety of compositions and impurities must be handled

Composition	Natural gas	Coal bed methane	Digester gas	Typical landfill gas	Coal gas entrained-flow O$_2$-blown gasifier	Propane	Diesel (Marine)	Ethanol (Fermentation)
Methane, vol%	80-100	42-90	50-70	40-55	2	–	–	–
Hydrogen	–	–	–	0.1	34	–	–	–
Carbon dioxide, vol%	0-3	6-3	50-30	35-50	16	–	–	–
Carbon monoxide, vol%	–	–	–	–	45	–	–	–
Nitrogen, vol%	0.3	50-5.0	–	–	2	–	–	–
Oxygen, vol%	0-0.2	0.05-0.1	<1.0	0-20	2	–	–	–
Higher hydrocarbons, vol%	0-10	0.5-1.0	–	0.5-1.0	–	0-20 (Propylene, butane)	C$_{15}$H$_{27}$ (Avg. Mol. wt) 100%	C$_2$H$_5$OH 9-10% dilute; 1-3% other oxygenated Hydrocarbons
Propane, vol%						80-95		
Typical impurities (ppm)								
Sulfur	2-12	0.5	50-500	200	70	100-200	5000-10,000	1000-2000
Chloride	–	0.25	1-5	5-20	–	–	5-25	–
Siloxanes	–	–	1-10	>5	–	–	–	~10,000
Others	–	–	–	Metals Particulates	Metals Particulates	–	Metals	(Inorganic materials)
Heating value	High	Medium-High	Medium	Medium	High	High	High	High

introduction over other processes because of the simplicity of design and deployment. In this process, fuel is passed through vessel(s) containing the desulfurization sorbent. The sulfur concentration is reduced to an acceptable level. Compared to other systems, the room temperature absorption system offers relatively low sulfur loading at bed breakthrough. Also, the sorbents tend to be selective to some sulfur species, allowing other species to pass through. In addition to the sorbents, parallel research in selective oxidation and hydrodesulfurization areas are currently under way. In selective oxidation, nearly all sulfur species are oxidized selectively (without oxidizing the fuel) to sulfur dioxide and then trapped on a potassium carbonate or sodium carbonate sorbent. In hydrodesulfurization, the sulfur species are converted to hydrogen sulfide and then trapped in a zinc oxide bed. With development and improvement in the room temperature absorption system, it remains as the preferred system for small sub-megawatt power plants.

The carbonate fuel cell system operates at high enough temperatures to allow practical reforming reaction kinetics for the natural gas and other light hydrocarbons inside the fuel cell. The in-situ fuel conversion concept is termed as the "Direct" Fuel Cell (DFC®). A review of the DFC technology and system was published (Farooque 1999). The internal reforming DFC® concept is illustrated in Fig. 1. The hydrocarbon fuel is directly introduced to the anode compartment. The fuel cell and reforming reactions are carried out in close proximity of the fuel cell anode to allow thermal and mass exchanges between the fuel cell and reforming reaction. Water produced by the fuel cell reaction is consumed by the reforming reaction. Hydrogen produced by the reforming reaction is continuously consumed in the fuel cell. A significant portion of the fuel cell reaction heat released at the anode is used up by the endothermic reforming reaction.

One advantage of the internal reforming design is its ability to drive the reforming reaction to completion, thus improving the system efficiency. In systems with external reformers, the extent of the hydrocarbon conversion is limited by the reforming reaction equilibrium. In these systems, the unconverted hydrocarbon is not utilized in the fuel cell and is often combusted in an after-oxidizer to provide heat for the reforming reactor. In order to increase the conversion of the hydrocarbon fuel, and hence to minimize its impact on system efficiency, the conventional external reformers are often designed to operate at high steam-to-carbon ratios (>3) and elevated temperatures of 750 to 850°C. Both of these aproaches have implications in increasing the cost and complexity and lowering efficiency of the system.

In an internal reforming carbonate fuel cell, hydrogen produced by the reforming reaction is utilized in the fuel cell electrochemical reaction and water produced by the fuel cell reaction is consumed by the reforming reactions. This mechanism results in an increased extent of reforming reaction, even though the fuel cell operates at ~650°C which is below the conventional reformer operating temperature of 750-850°C. In practice, a complete conversion (~100%) of the hydrocarbon fuel to hydrogen is achieved in DFC stacks at much lower temperatures than the external reformer (Fig. 4). With a steam-to-carbon ratio of 2.0 at 650°C, only 80% methane conversion can be obtained in an external reformer, whereas, 100% conversion has been achieved in the internal reforming fuel cell at fuel utilization of greater than 70%. Steam usage is dictated by carbon deposition consideration only. This is quite unlike the external reforming case, where a greater amount of water is used for achieving meaningful fuel conversion. As was reported previously by Mitsubishi Electric Company (MELCO) (Tanaka 1990), less steam in the fuel stream improves the cell conversion efficiency by reducing fuel dilution. Steam-to-carbon ratio also affects the oxidant side gas composition. These combined benefits of lower steam-to-carbon ratio were also shown in a 200 kW power plant test (Farooque 1999).

Another imporant feature of the internal reforming type fuel cell is its cooling contribution. Fuel cells usually process 20 to 35% excess fuel. In a direct fuel cell, approximately 60% of the by-product heat is used up by the reforming reaction. This 60% reduction of cooling load and the freedom to overlap the reforming zone with the heat producing section of the cell has the potential to produce uniform fuel cell

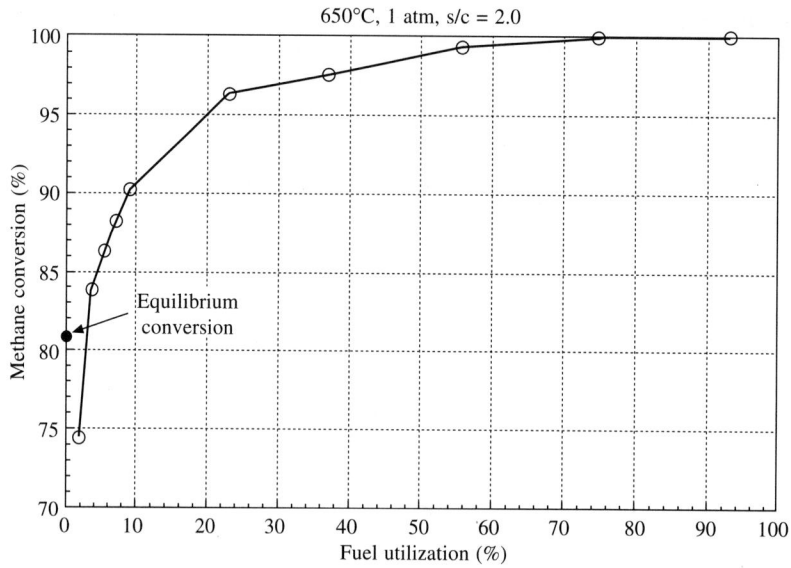

Fig. 4 CH$_4$ conversion as a function of fuel cell fuel utilization in an internally reformed carbonate fuel cell: ~100% equilibrium is achieved at greater than 76% fuel utilization.

temperature distribution. Reduced cooling load also requires less cooling gas, i.e. less dilution of the reactants, hence, higher conversion efficiency in the fuel cell.

The comparison of a conventional fuel cell power plant with an internal reforming fuel cell power plant is conceptually depicted in Fig. 5. The internally reformed fuel cell concept eliminates the external reformer and associated heat exchange equipment, resulting in the most simple and efficient power plant system. The single cycle internal reforming power plants can be expected to lead to a system efficiency of 50 to 60% on a lower heating value (LHV) basis (Patel 1983). The DFC combined cycle systems are expected to have greater than 70% (LHV) efficiency (Ghezel-Ayagh 2002) in large power plant configurations. Table 2 shows the selection of fuel processing approaches of currently active fuel cell developers.

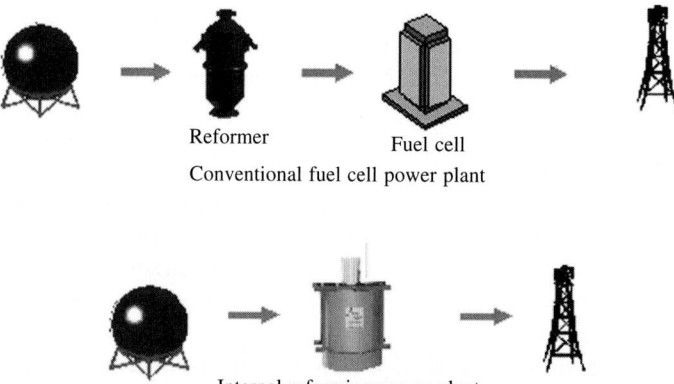

Fig. 5 Comparison of a conventional fuel cell power plant with a direct fuel cell power plant: Internal reforming eliminates equipment and results in a simpler system.

Table 2. Reforming approach of carbonate fuel cell developers: Both internal and external reforming has been selected by developers

Developer	Fuel processing approach	
	Internal reforming	External reforming
FCE (US)	×	
GenCell (US)	×	
IHI (Japan)		×
KIST/KEPRI (Korea)		×
MTU-CFC Solutions (Germany)	×	
Ansaldo Fuel Cell (Italy)		×

While the internal reforming fuel cells allow a simple overall system for low cost and enhanced reliability, the fuel cell anode needs to be integrated with the internal reforming catalyst. In one design approach called DIR (direct internal reforming), the catalyst (e.g. supported nickel catalyst) is located in the anode compartment where it gets exposed to the electrolyte-containing environment. The internal reforming catalyst is required to remain sufficiently active throughout the life of the fuel cell in the presence of the carbonate vapors. An alternate approach, named IIR (indirect internal reforming), has also been developed. In IIR concept, the reforming catalyst is placed in between cell groups in order to achieve much longer catalyst life as well as improved thermal management. A hybrid concept called IIR-DIR that combines both approaches has emerged as the design choice for fuel processing by FCE and MTU-CFC Solutions from performance, thermal management and life considerations. Fig. 6 shows the hybrid arrangement in which the reforming catalyst is placed in the anode compartment of each cell. Additionally, reforming units (RU) are placed in between fuel cell groups, typically every 8-10 cells.

The carbonate fuel cell also has a great potential for applications where liquid fuels such as diesel and propane are used. Liquid fuel operation capabilities are considered important for fuel flexibility as well as secure power production in the event the fuel gas supply is interrupted. Efforts are underway to adapt the

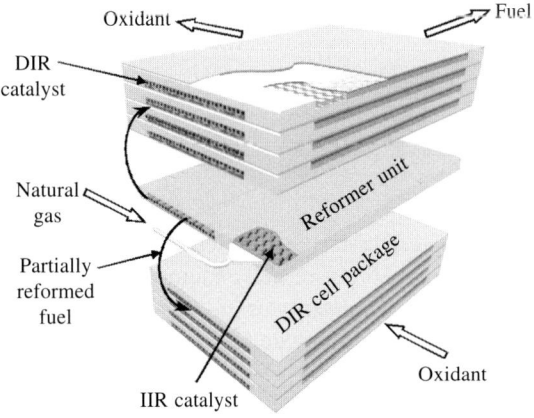

Fig. 6 Internal reforming stack concept: The hybrid reforming allows longer catalyst life and uniform cell temperature.

natural gas plant technology to relatively more difficult-to-process fuels such as propane, diesel, ethanol and bio-diesel. A unified fuel-preprocessing concept has evolved to take advantage of the internal reforming capability of the carbonate fuel cells for operation with liquid fuels. This is shown in Fig. 7, where the raw fuel is first cleaned up of undesired components and subsequently converted to methane-rich gas in an adiabatic converter. This methane-rich gas constitutes the fuel cell anode gas. Some example output compositions of the "adiabatic conversion" process with different fuels are reported in Table 3. As shown in the table, the product gas composition of the fuel converter is similar to the digester gas and not very different from natural gas. As a result, the front end of the plant may be used to process various types of fuels. Therefore, with appropriate fuel preprocessing equipment, the plant can acquire multi-fuel operation capability.

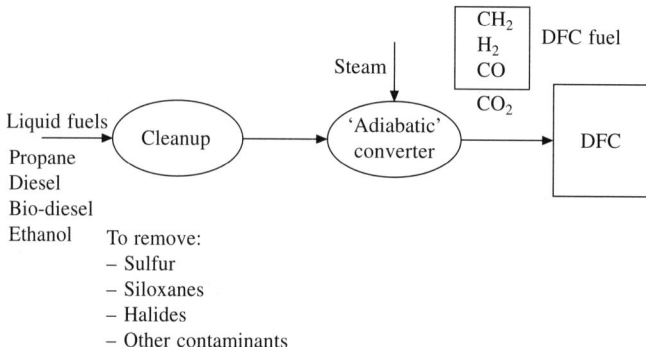

Fig. 7 Adaptation of transportable liquid fuels to DFC™ (internal reforming stack): The liquid fuels are converted to a methane-rich gas in an external "adiabatic" process.

Table 3. **Hydrocarbon composition with different fuels after preprocessing: Methane and CO₂ content varies with different fuels**

	Hydrocarbon composition (dry basis), vol%		
	Natural gas	Propane	Diesel
Methane	71	53	52
Hydrogen	21	27	23
Carbon dioxide	8	19	24
Carbon monoxide	–	1	1

2.4 Carbonate Fuel Cell Materials

The present day carbonate fuel cell materials design is based on intensive materials research carried out during the last three decades including in-cell/out-of-cell endurance test results and cost considerations. A discussion of various carbonate fuel cell component designs and improvement opportunities is presented below and additional details can be found in carbonate fuel cell literature (Hoffman 2003).

Anode: Unalloyed porous nickel anodes used by early developers were found to shrink during operation under the stack compressive load, resulting in undesired dimensional change, reduced surface area, and lower electrochemical performance. Alloying with chromium and/or aluminum provides oxide dispersion strengthening, resulting in adequate creep strength. Excellent mechanical and chemical stability of the baseline Ni-Al anode is verified in an 18,000 h field operation. Fig. 8 shows that the structure of the

Fig. 8 Anode after 18,000 h service: Excellent structural integrity is observed with no change in morphology.

Ni-Al anode in carbonate fuel cell is similar for both fresh and used anodes. The anode shows adequate structural integrity and electrochemical activity that can be projected to be greater than 5-year life.

Cathode: The carbonate fuel cell cathode material has been lithiated NiO from the beginning of development. This component is known to have a small but finite solubility in the electrolyte. The extent of its dissolution is controlled mainly by electrolyte composition, applied gas atmosphere, operation pressure and temperature. Some developers have selected an atmospheric pressure system to assure minimal dissolution and adequate long-term life for the cathode. Long-term field operation has shown no issues relating to particle coarsening, indicating a stable structure (Fig. 9).

Fig. 9 Cathode after operation: Only very slight particle coarsening occurred during long-term operation.

Opportunities exist in further extending the cathode material stability by modifying the electrolyte to increase the basicity and/or modifying the cathode materials. Developers are actively investigating both of these options.

Matrix: The matrix having micropores is sandwiched between the anode and cathode (Fig. 10) electrodes with larger pores. The electrolyte matrix provides ionic transport, reactant gas separation and perimeter seal. It is a layer of tightly packed ceramic powder bed impregnated by alkali carbonate electrolyte to form a composite paste-like structure at the operating temperature. The stability of the matrix support materials and matrix robustness to withstand thermo-mechanical stress, are important considerations that impact the fuel cell performance and endurance. A comprehensive review of the matrix considerations, issues, and status is provided by Yuh, Farooque and Maru (Yuh 1999).

The ceramic $LiAlO_2$ matrix support material, a product of the reaction between Al_2O_3 and lithium carbonate, has three allotropic phases (α, β and γ). High surface area submicron α-LiAlO$_2$ matrix has shown little change in stability after 18,000 h of operation. During fuel cell stack operation, the matrix experiences both mechanical and thermal stresses. Strong and tough matrices capable of withstanding such stress buildup to maintain good gas sealing capability are desired. Without sufficient strength, the matrix may crack and result in increased gas cross leakage. A cost-effective, strong matrix using an innovative strengthening approach has been described by Yuh (1995). This strong matrix has resolved the potential concerns for thermal cycleability of carbonate fuel cells.

Anode   Matrix   Cathode

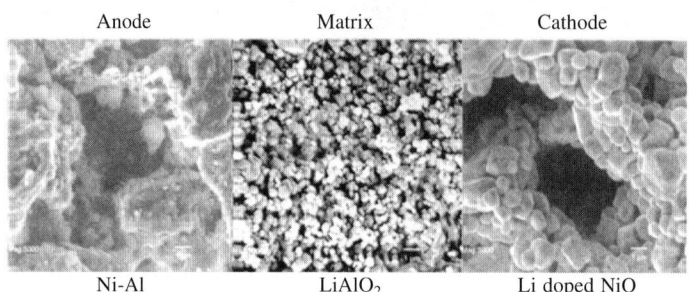

Ni-Al   LiAlO$_2$ Li doped NiO

Fig. 10   Morphology of carbonate fuel cell active components: The matrix (center image) with micropores is sandwiched between the fuel (Anode: Left margin) and oxidant (Cathode: Right image) electrodes.

Prevention or at least minimization of fuel and/or oxidant leakage or intermixing of the fuel and oxidant is the most important attribute of an operating fuel cell. An intermixing of fuel and oxidant not only reduces efficiency but can also lead to premature component failure due to excessive local heating. The carbonate fuel cell has successfully adapted a simple wet seal concept. Liquid electrolyte held up in the micro-pores of the matrix and on the metal surface where the metal to matrix interface seal is involved, provide adequate barrier to gas leakage.

Cell Hardware Materials: In the early years of development, hot corrosion of the metallic hardware in the carbonate environment was an important consideration. The metallic components include bipolar separator plates and corrugated current collectors, which provide the gas flow passages. Hot corrosion of the bipolar plate and current collector components in the presence of liquid alkali carbonate electrolyte is considered a challenge for material selections. This is especially important in the presence of two very different environments, i.e., the reducing fuel environment on the anode side and the oxidizing oxidant on the cathode side. Contact electrical resistance could increase due to oxide scale buildup, lowering output voltage. Also, electrolyte loss to the bipolar current collector due to corrosion and electrolyte creepage could further contribute to stack power decay. In general, the anode-side environment (particularly the fuel exit) is more corrosive than the cathode-side, except for pure nickel or high-nickel, nickel-base alloys. The exit condition is generally more corrosive than the inlet due to a higher operating temperature and possibly higher moisture content.

Endurance results at FCE and other laboratories have confirmed that a properly selected stainless steel provides adequate corrosion protection for this application. Currently, stainless steels, particularly the 300-series austenitic stainless steels, are the primary hardware materials of the cell stack and BOP (balance-of-plant). Ferritic Al-containing stainless steels have an adequate corrosion rate due to the formation of a dense thin protective inner Cr-Al oxide layer. However, the extremely high electrical resistivity of the alumina-containing scale prevents them from cell active hardware use. For the anode-side application, surface protection of stainless steels by Ni-cladding has generally been adopted. With the protection provided by a nickel-clad coating, the anode-side bipolar plate has shown virtually no corrosion attack during an 18,000 h field operation. Consequently, no significant ohmic loss due to anode-side contact was observed. Although a small amount of chromium-rich oxide forms at the grain boundaries of the nickel-clad layer, no delecterious effect on the corrosion protection is observed. The observed diffusion of iron and chromium from the substrate into the coating appears harmless. The cathode current collector materials with a promise to reduce the corrosion rate, contact resistance increase rate, and electrolyte loss rate are being pursured.

Wet-Seal Material: The wet seal simultaneously experiences reducing and oxidizing environments. The chromia-forming alloys experience high corrosion in the wet seal environment. Cost-effective corrosion protection of the wet seal surfaces has been a major focus of materials development efforts. Understanding the corrosion mechanism (Donado 1996) and finding aluminization to eliminate the observed wet seal surface corrosion are considered important steps in carbonate fuel cell development. Aluminizing methods used so far include slurry painting, vacuum deposition, thermal spraying, etc. The resultant diffused coating on stainless-steel surfaces generally consists of a MAl-M$_3$Al structure. The coating has been shown to provide sufficient protection for the substrate stainless steels (Fig. 11).

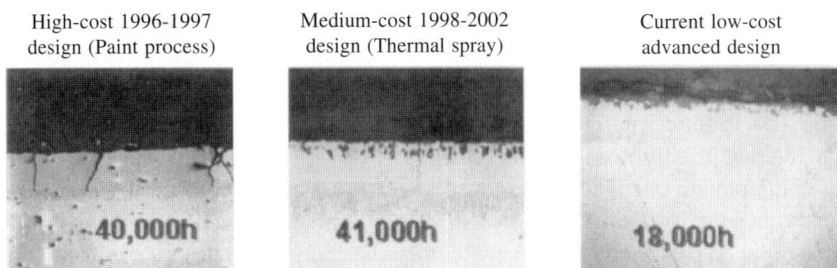

High-cost 1996-1997 design (Paint process) — 40,000h

Medium-cost 1998-2002 design (Thermal spray) — 41,000h

Current low-cost advanced design — 18,000h

Fig. 11 Wet-seal aluminum protection: The coating process reduced the cost significantly while maintaining excellent corrosion resistance.

3. Stack Development

Carbonate fuel cells are typically planar in shape. Major cell, stack and module design considerations include: uniform compressive load to ensure good cell-to-cell contact, thermal uniformity, and provision of uniform reactant gas flow. The uniformity of gas flows is a multifaceted design challenge including both the fuel and oxidant flows into the individual cells in a stack (cell-to-cell), between stacks in a module (stack-to-stack), and between modules in a large power plant. Over the past decade, several unique engineering tools have been utilized for design development, engineering analysis and verification that have greatly accelerated the pace for power plant development and performance improvements. The following are some of the tools utilized:

(a) Computer Aided Design (CAD) 3-D parametric solid modeling.
(b) Scale models and testing of key functional components.
(c) Finite Element Analysis (FEA) of material strengths and stresses.
(d) Water flow visualization.
(e) Hot Wire Anemometer flow analysis.
(f) 3-D Computational Fluid Dynamics (CFD) simulation models.
(g) Full scale testing of key functional components.

Techniques such as water flow visualization, hot-wire anemometer flow measurement and CFD modeling have been used to enhance reactant gas cell flow uniformity and performance. Simple water visualization equipment can be inexpensively purchased or fabricated to qualitatively identify flow deficiencies. Using water as the medium, the flow can be matched to a gas flow according to appropriate Reynolds number and luminescent dyes can be injected in the fluid path to elucidate flow characteristics. In this way, dead zone regions can be quickly and easily identified enabling engineers immediate feedback for corrective action. A more quantitative technique for cell flow characterization is the hot wire anemometer. Fig. 12 shows such an apparatus analyzing cell flow uniformity within a stack. The anemometer is set up to scan across each individual cell package quantitatively measuring flow and temperature. The integrated data identifies cell-to-cell flow variations with a high degree of precision and accuracy. Using the hot wire anemometer as a tool, improved cell-to-cell reactant gas flow uniformity can be achieved.

Operating power density of large size fuel cells is mainly dependent on thermal management and to a lesser extent on the kinetic and gas diffusion polarizations and internal resistive loss. The fuel cell in-plane temperature gradient needs to be managed

Fig. 12 Hot wire anemometer flow analysis set-up: Quantifies cell-to-cell reactant gas flow uniformity within a stack.

by employing an efficient cell cooling management system. At present, the power density-limiting process in direct carbonate fuel cells is the cell in-plane temperature uniformity.

Model predicted cell temperature profile

Measured cell temperature profile

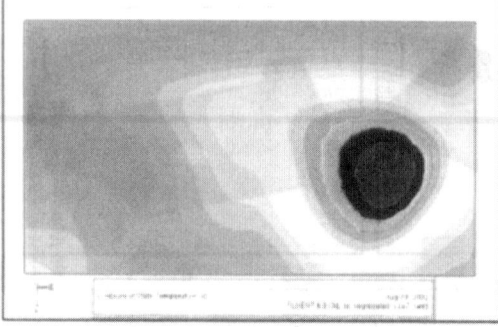
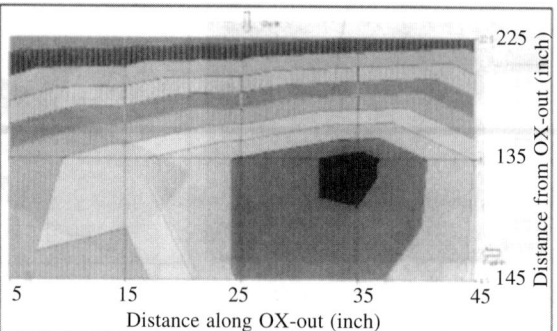

Fig. 13 Model predicted and experimental thermal profile comparisons: Good correlation of CFD model with test data enabled engineers to modify design for ~30% improvement in cell thermal profile within a stack.

The internal reforming carbonate fuel cell design employs the most efficient thermal management system where the endothermic internal reforming reaction is used to remove heat generated by the fuel cell. Furthermore, cooling uniformity is achieved by distributing the cell-cooling load between an indirect reformer plate and the direct internal reformer. Comprehensive stack models that simulate the hydrodynamics, kinetics, electrochemical, and heat transfer processes have been used to guide the development of an efficient internal reformer cooling design (Ma 2004). As a result, the stack temperature distribution has been improved recently, allowing 20% higher power density operation without penalty in cell temperature gradient. Fig. 14 compares the fraction of the cell area below a given temperature at the same current density operation for three different consecutive cell design iterations. The experimental data shows the gradual cell temperature distribution improvement by optimizing the internally reformed carbonate fuel cell design.

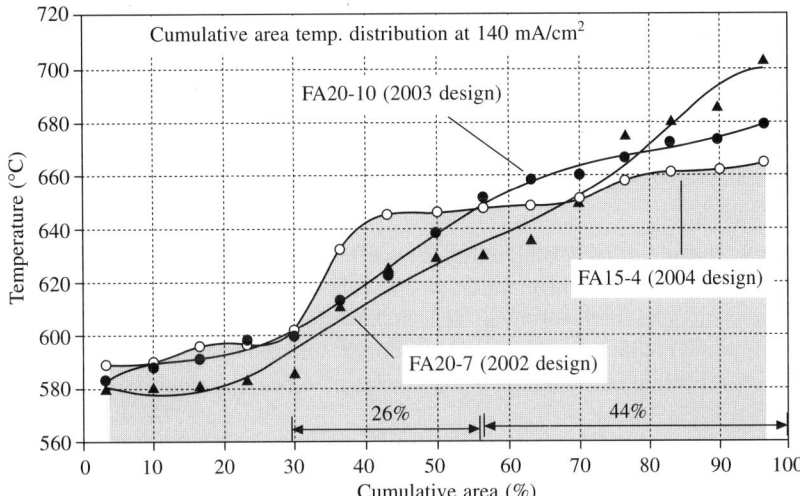

Fig. 14 Area based cumulative thermal distribution in stacks: New internal reformer unit design significantly reduces cell area above 660°C.

The electrochemical reactions at the carbonate electrodes are sufficiently fast and account for low polarization losses. Electrode pores have been sufficiently optimized over the years to cause low gas diffusion polarizations at the current level of power densities. There are further opportunities for electrode structure optimization for reduction of cathode gas diffusion polarization as well as reduction of internal resistance by electrolyte modification. These efforts are ongoing concurrent with improving temperature uniformity of the large area fuel cells of up to 9000 cm^2.

3.1 Stack and Module Engineering

Computer graphics are used to construct state-of-the-art CAD 3-D models enabling design verification in virtual reality to provide interference free fit during assembly. CAD solid models allow use of analysis tools for strength and stress analysis to enable quick, efficient design and materials verification. Finite Element Analysis (FEA) modeling of material strength and stress ensures correct design and material selection to prevent failure. Fig. 15(a) shows that 3-D computational FEA has been used to model stack movement through various operating modes. Fig. 15(b, c) present FEA models of end plates and compression hardware ensuring that engineered designs meet their functional requirments for flatness, uniform stack compressive load and strength under stack operating conditions (high temperature and load). Utilization of FEA has resulted in achieving 2-fold improvement in compressive load uniformity to the stacked cells.

The state-of-the-art carbonate fuel cell technology includes a stack module rated as high as one megawatt. Fig. 16 shows a picture of a 1-megawatt module being assembled on the factory floor. The 1 megawatt module consists of an arrangement of four stacks each rated at 250 kW. Special considerations were given to the flow distribution between the stacks during the design of the internal manifold system. Computer Aided Design (CAD) 3-D parametic solid modeling has been used (Doyon 2003) to construct individual

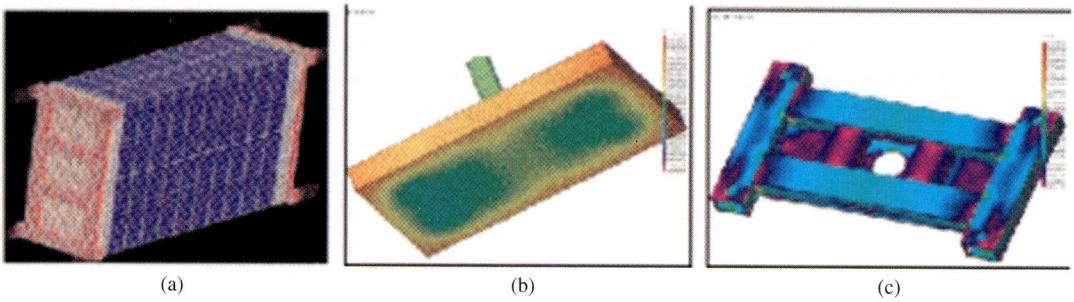

(a)   (b)   (c)

Fig. 15 Examples of computational FEA applications: FEA has proven a useful tool in developing and verifying new component designs for the carbonate fuel cell.

components and stack assemblies in virtual reality. This efficient technique identifies design oversights and interferences in advance of component manufacture saving costs and lost assembly time at the factory. Rapid design development of the fuel cell stacks and stack modules were made possible through construction of a CAD 3-D solid model containing all stack components and the assembled module. Full scale testing of key functional non-repeating components is also performed prior to stack integration to ensure design functionality.

Fig. 16   Four stacks installed in 1-megawatt fuel cell module: Carbonate fuel cell stack modules have been developed for multi-MW power generation applications.

Fig. 17 shows the historical improvement of subscale stack performance. Continuous efforts to improve cell components and module design have resulted in more than 40 mV higher cell voltage at 140 mA/cm^2 versus the 1997 stack design at comparable temperature conditions. Contributions to the performance

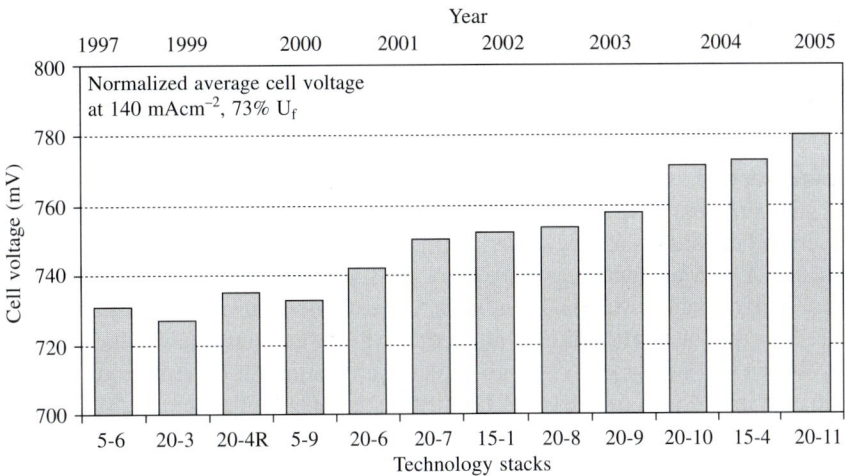

Fig. 17 Full-area carbonate fuel cell stack performance: Progression of performance improvements has been validated.

improvement are mainly due to an innovative electrolyte composition and optimized thermal management using an advanced internal reforming configuration. The new highly conductive electrolyte provided not only lower internal stack resistance, but also higher electrode efficiency. The improved internal reforming design configuration utilizes fuel cell reaction heat more effectively across the cell area. Fig. 18 shows an

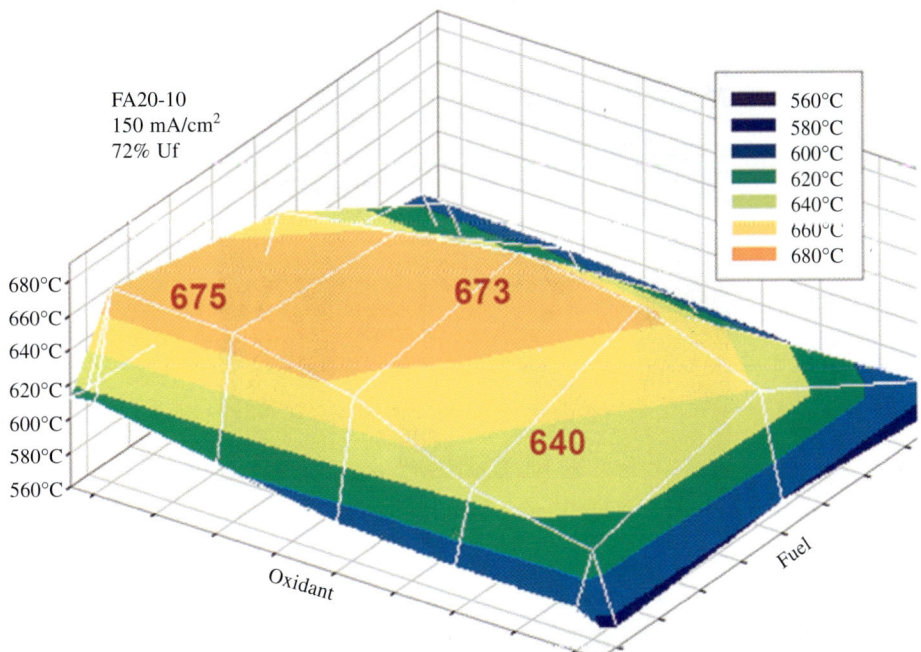

Fig. 18 Temperature distribution in a carbonate fuel cell stack during 150 mA/cm^2 operation: Excellent temperature uniformity is achieved.

example of in-cell temperature distribution during high power operation. As shown in Fig. 18, a steady improvement in the temperature distribution has been achieved leading to a reduction in hot-area temperature. Because of improved thermal management, the fuel cell stack durability is increased and operational life is extended.

3.2 Stack Manufacturing

In recent years, great progress has been made both in the scale-up of the cell area and in the stacking of the carbonate fuel cells. There has been a two orders of magnitude area increase from the early 3 cm^2 cell size, nearly three decades ago. Manufacturing of the fuel cells with the area in ~1m^2 (10 ft^2) range has been achieved. The full-height stack design is different for different developers with respect to gas manifolding and number of cells in the stack. The major characteristics of the carbonate fuel cells are: (1) large cell area (largest among various fuel cell types), (2) high cell area utilization for current production (less than 10% of the geometric area is seal area), (3) ease of fabrication, and (4) low material cost. FuelCell energy and MTU use approximately 350-400 cells in a full-height 250 kW stack. These stacks are built with the direct-indirect internal reforming concept described by Farooque, Katikaneni and Maru (Farooque 1999). A photograph of such a stack in Fig. 19 shows that ~400 cells are stacked in a bipolar configuration. The Japanese developers have pursued a building block approach. In this approach, a fewer number of cells, typically 80-120 are enclosed within the terminating plates forming a building-block unit. Several of the building-block units are stacked to build a truck-transportable full-height stack.

Fig. 19   Full-size stack completed at the FCE manufacturing facility: ~ 400 cells per stack can be fabricated.

The carbonate fuel cell employs well-known manufacturing processes. Standard sheet metal forming, bending and welding operations are involved in the manufacture of cell hardware components. The anodes and cathodes are manufactured by standard processing techniques such as tape casting or powder doctoring followed by sintering. Matrix manufacturing by tape casting is also well developed. Most of the developers have demonstrated full-size components manufacture capability. Five developers have already tested full-area cell stacks. At least one developer, FuelCell Energy, currently has 50 MW/year manufacturing capability.

4. Carbonate Fuel Cell Systems

Carbonate fuel cell systems suitable for operation with various types of fuels have been developed and demonstrated. In general, the operating temperature of the fuel cell stacks (~920 °K) is low enough to avoid NO_x production exhibited by conventional technologies, but high enough to promote rapid electrochemical reactions—providing high conversion efficiency while avoiding the need for the noble-metal catalysts used in lower temperature fuel cell systems. This also avoids the problem of carbon monoxide catalyst poisoning, providing additional flexibility in configuring low-cost balance-of-plant systems to support the operation of carbonate fuel cell stacks.

The operating temperature of the carbonate fuel cell confers another advantage on the technology—the ability to reform hydrocarbon fuels to hydrogen-rich gas (which is required for the fuel cell) within the fuel cell stack. Other fuel cell technologies utilize external reformers, in which the hydrocarbon fuel is mixed with steam, heated to a high temperature, reformed, cooled, and then often processed further (with shift reactors or carbon monoxide polishers) before being delivered to the fuel cell stack. A simplified flow schematic of an internally reformed carbonate fuel cell power plant, which can operate on natural gas or syngas is shown in Fig. 20. The fuel, such as natural gas, is fed to the fuel cell power plant where methane is internally reformed and CO is shifted to CO_2 and H_2. Spent fuel exits the anode and is consumed in the anode exhaust oxidizer to supply oxygen and CO_2 to the cathode. The resulting electrochemical reactions in the fuel cell anode and cathode produce DC power, which is inverted to AC. The cathode exhaust supplies heat to the fuel clean-up and fuel moisturizer as it is vented from the plant. The exhaust gas is available for cogeneration to provide heating or cooling

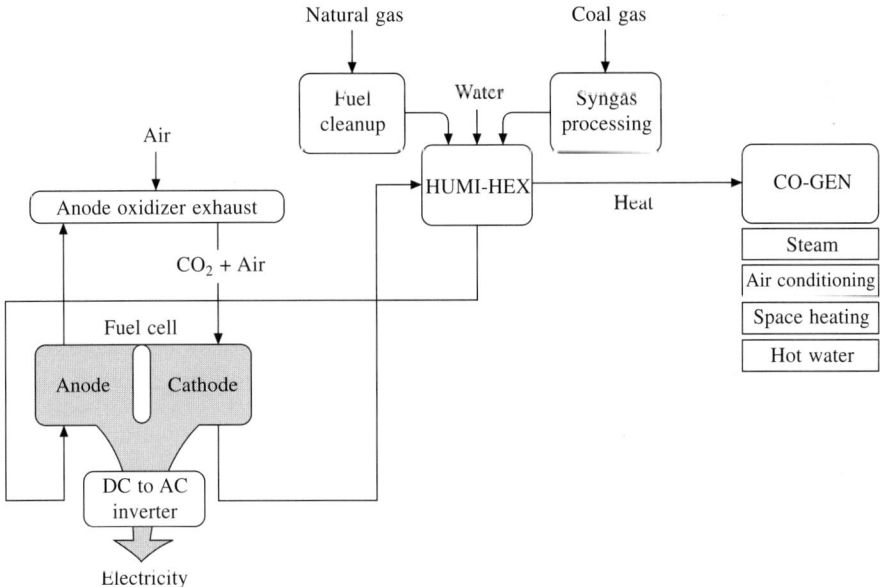

Fig. 20 Carbonate fuel cell power plant simplified process schematics: Waste heat in the power plant exhaust can be used for variety of co-generation applications.

4.1 Natural Gas Systems

The Santa Clara Demonstration Project (SCDP) was the first megawatt scale demonstration of the carbonate fuel cell, in which a 1.8 MW DFC® power plant was successfully built and operated in Santa Clara, California in mid-1990s. The SCDP plant is shown in Fig. 21. The plant utilized 16 internally reformed carbonate fuel cell stacks; each rated at 125 kW DC output, configured into four 4-stack truck-transportable modules. The plant was operated for 4000 h of grid-connect time, producing up to 1.9 MW net AC power into the utility grid. In addition to meeting and exceeding its rated power criteria and ramping criteria, the power plant operations met many other key project objectives. Specific project criteria, which were successfully demonstrated, included rated output, peak operation, voltage harmonic power quality, low NO_x and SO_x emissions, and operation within noise limits. The SCDP balance-of-plant proved to be exceptionally reliable. The system rode through minor grid disturbances and responded to major grid problems exactly as intended. The overall availability of the balance-of-plant during the test program was 99%. These would be excellent results for any plant, but are particularly impressive in this first-of-a kind "proof-of-concept" demonstration plant.

Fig. 21 Santa Clara Demonstration Power Plant: Technology capabilities were demonstrated by operation of this plant above its rated power output.

The decade after the SCDP demonstration has witnessed steady progress in performance and durability of the carbonate fuel cells. A new generation of stack hardware has been developed with cells that are 50% larger in area, 40% lighter per unit area, and 30% thinner than the SCDP design. These improvements result in a doubling of power output from a full-height stack. A low-cost and high-strength matrix has also been developed for improving product ruggedness. The low-cost advanced cell design incorporating these improvements has been refined through subscale stack tests in which the new cells have been shown to be more tolerant of load and thermal changes, and have shown no loss of performance in up to 10 thermal cycles.

4.2 Systems for Petroleum Distillates

Carbonate fuel cell systems are being developed for operation on liquid fuels in general, and on marine distillates, specifically (Abens 2002, Abens 2003). The immediate interest is the 500 kW hotel power for ship service application. The United States Navy has taken the lead in supporting the development of the ship service fuel cell (SSFC) for the next generation of the naval ship (Nickens 2004). A principal goal of the Navy's SSFC program is to demonstrate that commercially developed fuel cell technology can utilize naval logistic fuels and operate in a marine environment. However, the eventual development of the SSFC system will spillover to many other applications where liquid petroleum distillates such as diesel or jet fuels are the fuel of choice. The applications may include generation of power for remote locations, islands, cruise and marine ships.

Fig. 22 shows the principal elements of the SSFC power plant process flow diagram. The logistic fuel processor section of the power plant has been designed for operation with NATO F-76 marine diesel fuel, which may contain as much as 1 weight percent sulfur. The fuel processing system removes the sulfur from the fuel by passing it through a hydrodesulfurization (HDS) reactor in the presence of hydrogen. An on-board electrolyzer is used to generate the required make-up hydrogen gas. The hydrogen sulfide produced in this process is absorbed in dual regenerable zinc oxide beds. The desulfurized fuel is mixed with steam and converted to a methane-rich gaseous fuel in a pressurized adiabatic reactor for use in the internally reforming carbonate fuel cell stacks. The fuel gas pressure is reduced to near one atmosphere through a turbo expander before the gas enters the internally reforming fuel cell stacks.

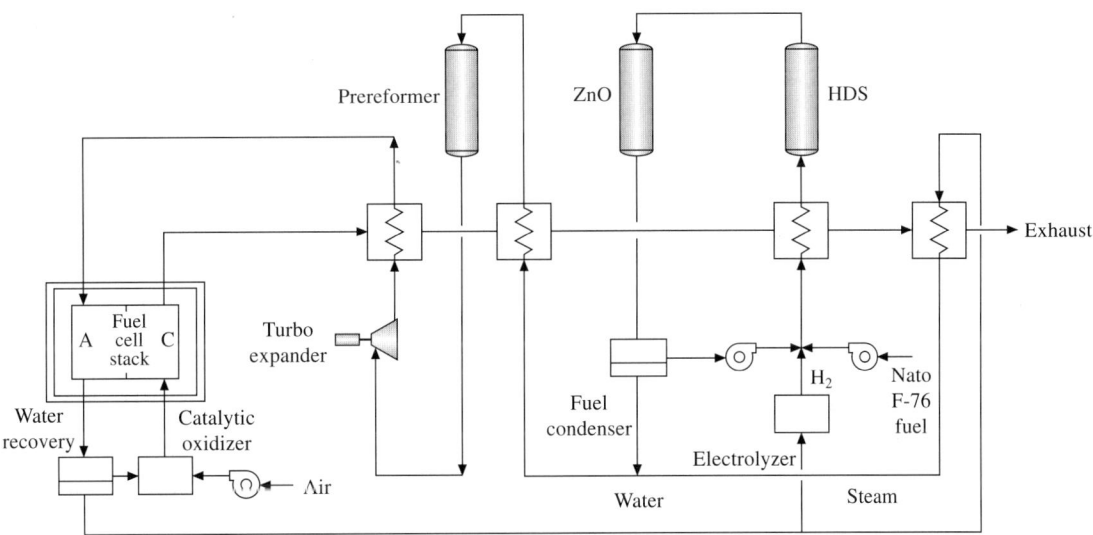

Fig. 22 Ship service fuel cell power plant process flow diagram: Logistic fuel is desulfurized and converted to methane before entering the fuel cells.

The general arrangement of the overall SSFC equipment within the assembly is shown in Fig. 23. The balance-of-plant skid includes the fuel processing equipment and the thermal management and water recovery systems. The process equipment is located within an enclosure sealed with acoustical panels to provide environmental safety and attenuation of noise from the ancillary system rotating equipment. The DC output from the two fuel cell stacks provides input to an inverter producing 450 V AC power output to the load. The inverter also conditions the output of the turbo expander generator. The packaging of equipment is compact to minimize the volume of the module. Controls, filters, and other equipment requiring periodic maintenance are located on the periphery of the assembly. Reactor process beds and large pipe runs occupy the less accessible inner regions of the skid. Compact plate-fin heat exchangers were used for the thermal management system.

4.3 Coal Gas Carbonate Fuel Cell
Fuel cell systems operating on coal have been studied extensively in past years. Fig. 24 shows a simplified

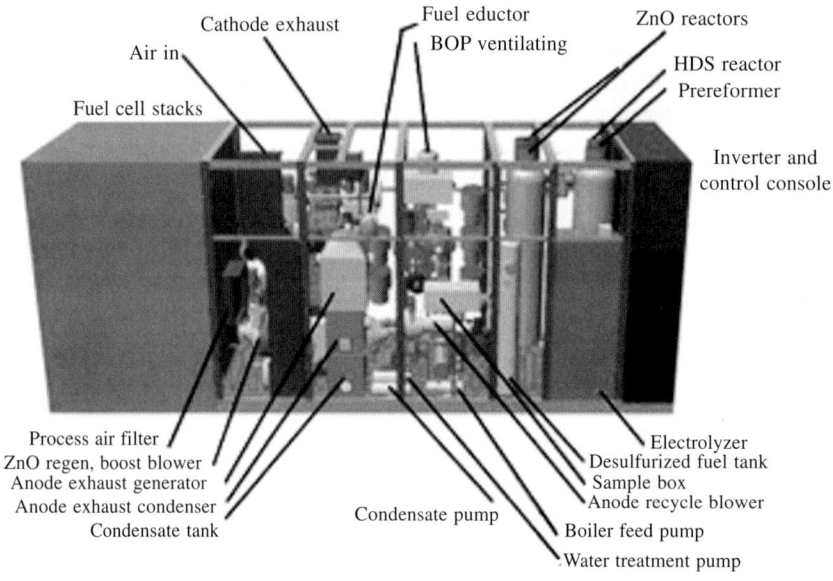

Looking starboard

Fig. 23 Ship service fuel cell equipment arrangement: Skid is designed with sulfur removal and water recovery equipment.

process flow diagram for an Integrated Gasification Fuel Cell (IGFC) system. Gasification is used to convert the solid fuel to a gas, which is processed to remove sulfur compounds, tars, particulates, and trace contaminants. The cleaned fuel gas is converted to electricity in the carbonate fuel cell. Waste heat from the carbonate fuel cell is used to generate steam required for the gasification process and to generate additional power in a bottoming cycle.

Design studies for development of a 200 MW scale power plant (Farooque 1990, Sander 1992, Sandler 1992) indicated that using conventional gasification and clean-up technologies, a heat rate of 7186 Btu/k

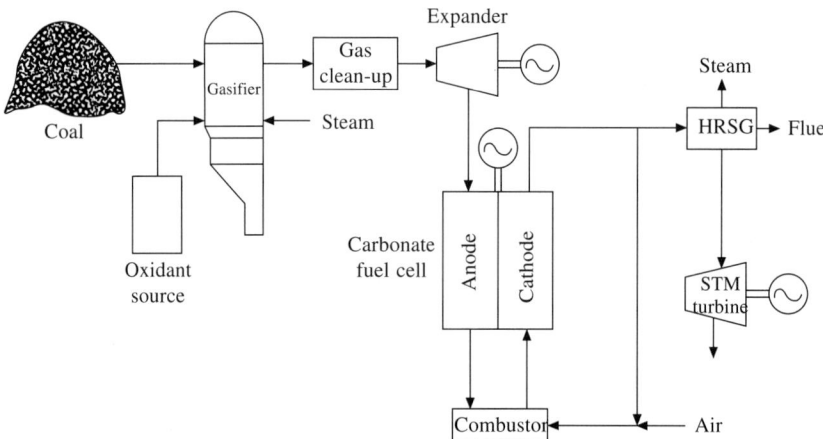

Fig. 24 Simplified process flow diagram for integrated gasification fuel cell (IGFC): Carbonate fuel cell operation on coal gas can be combined with a steam turbine bottoming cycle.

Wh (47.5% HHV efficiency) can be achieved with an IGFC power plant. This plant would require about 2140 tons/day coal and generate a net output of 250 MW. Later studies (Steinfield 1993) have indicated that higher efficiencies, 51.7-53.5%, could be achieved with higher methane producing gasifiers and by using hot gas clean-up. Fig. 25 shows a schematic of the E-GAS™ coal gasifier (Clean Coal Technology 2000) plant with a slipstream feeding the 2-MW fuel cell. The design of the oxygen blown, continuous-slagging, two stage, slurry-fed, and entrained flow E-gas gasifier is based on the Dow/Destec Louisiana Gasification Technology, Inc. (LGTI) gasifier, originally tested at LGTI between 1987 and 1995. Particulate cleanup is by a hot metallic candle filter system. Sulfur removal is by COS hydrolysis, an MDEA acid gas removal system, and a Claus sulfur recovery unit. Prior to the COS hydrolysis unit, the syngas is scrubbed to remove chlorides to protect the COS hydrolysis catalyst. High quality sulfur (greater than 99.99% pure) is recovered from the Claus unit in liquid form and sold for agricultural applications. A typical analysis of the syngas produced is shown in Table 4. Syngas from the E-Gas entrained flow gasification plant clean-up system is required to be further processed to make it suitable for the fuel cell power plant (~ 0.1 ppm sulfur). The treated syngas is fed to the fuel cell to generate power.

4.4 Carbonate Fuel Cell-Gas Turbine Combined Cycle

Integration of a high temperature fuel cell with a gas turbine has recently been the focus of development by various organizations (Williams 2004). A variety of system configurations including pressurized and non-pressurized (atmospheric pressure) carbonate fuel cells have been developed (Lunghi 2003, Grillo 2003, Ghezel-Ayagh 2002). The DFC/T® hybrid system concept is based on integration of an atmospheric pressure internally reforming carbonate fuel cell with a gas turbine. The power plant design utilizes a heat recovery approach (Ghezel-Ayagh 2002) for extraction of heat from the balance-of-plant. The fuel cell plays the key role by producing the larger share of the power (greater than 80%). The gas turbine is utilized for generation of additional power by recovering the fuel cell byproduct heat in a Brayton cycle, as well as for providing the air for the fuel cell operation (Leo 2000).

The DFC/T system concept is schematically shown in Fig. 26. The feed water humidifies natural gas in a waste heat recovery unit (HRU). The mixed fuel and steam are then preheated to about 550°C prior to entering the fuel cell anode. The methane in the natural gas is reformed in the fuel cell and its chemical potential is converted to electrical energy. The anode exhaust, containing some unreacted fuel, is mixed with air and then oxidized completely in a catalytic oxidizer. In the turbine cycle, air is compressed to the operating pressure of the gas turbine and heated in a recuperator using waste heat from the fuel cell. The compressed air is then heated further to the operating temperature of the gas turbine expander by a high temperature recuperator (HTR) located between the oxidizer and fuel cell cathode. The hot compressed air is then expanded in the turbine providing additional electricity. The expanded air flows into the oxidizer, into the HTR, and subsequently into the fuel cell cathode. At the cathode, the oxygen in the air and the CO_2 from the anode are reacted to complete the electrochemical fuel cell reaction. The cathode exhaust provides the heat for preheating the air and fuel, and for generation of steam in the HRU before exiting the power plant.

The key features of the DFC/T system include:

(a) Net electrical efficiencies approaching 70% (LHV) based on natural gas and 60% (HHV) based on coal gas

(b) Minimal emissions including ultra-low NOx and reduced carbon dioxide release to the environment

(c) Simplicity in design

(d) Direct reforming internal to the fuel cell

(e) Potential cost competitiveness with existing combined cycle power plants

(f) Uncoupled fuel cell operating pressure and gas turbine compression ratio

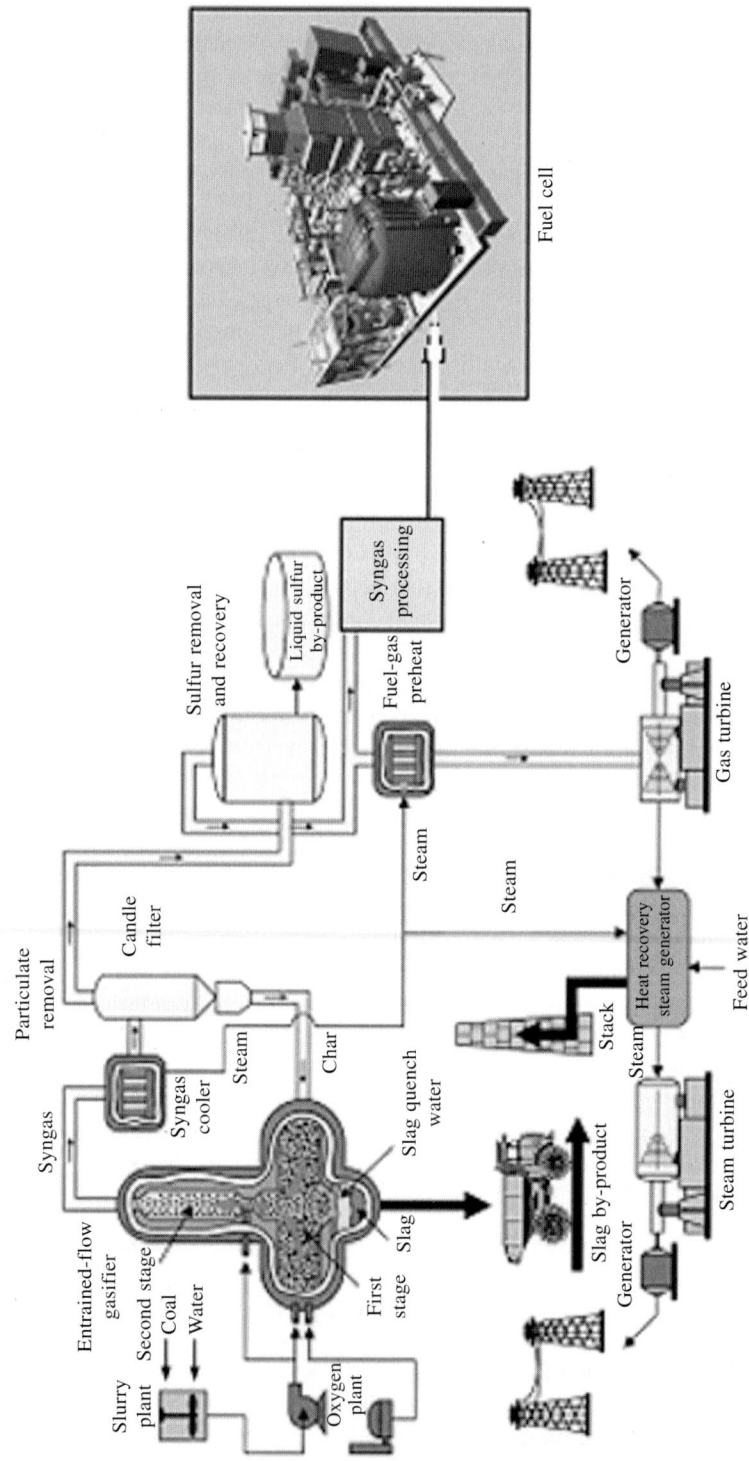

Fig. 25 Simplified process flow diagram of the IGCC with fuel cell: 2MW carbonate fuel cell power plant operating on coal gas from an E-gas gasification plant.

Table 4. Typical product syngas analysis (Clean Coal Technology 2000):
Sulfur content is greater than natural gas

Analysis	Typical coal	Petroleum coke
Nitrogen, vol%	1.9	1.9
Argon, vol%	0.6	0.6
CO_2, vol%	15.8	15.4
CO, vol%	45.3	48.6
H_2, vol%	34.4	33.2
CH_4, vol%	1.9	0.5
Total Sulfur, ppmv	68	69
HHV, Btu/SCF	277	268

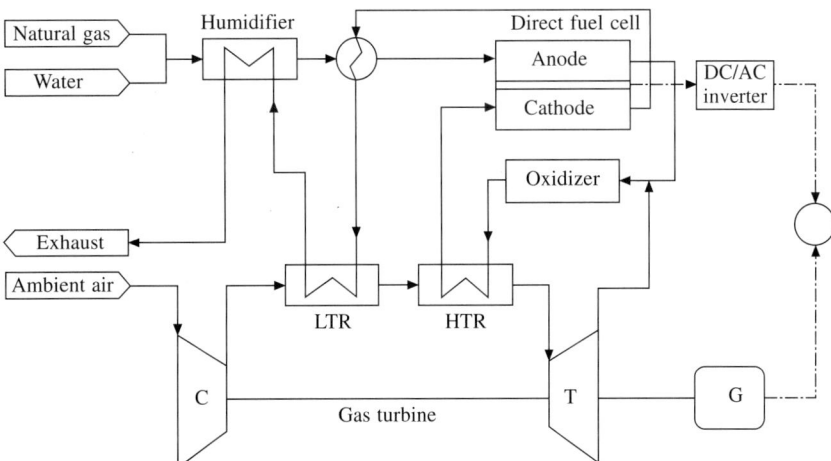

Fig. 26 DFC/T® ultra high efficiency system concept: The system utilizes a gas turbine to supplement power from the fuel cell.

The DFC/T system design allows the turbine compression ratio to be independent of fuel cell pressure. Typically, microturbines use a low compression ratio (3 to 5), while MW-size units are designed for high compression ratios (9 to 15). This characteristic extends the application of the DFC/T hybrid cycle to both the small sub MW units for distributed generation as well as the large multi-MW power plants for base-load utility power production and grid support.

The DFC/T system concept was implemented in a power plant test facility by integration of a 250 kW carbonate fuel cell stack and a Capstone MicroTurbine™. The focus of the power plant tests was on the verification of the DFC/T concept, the developmental testing of critical system components, and acquiring design information for development of power plant products (Ghezel-Ayagh 2003). The proof-of-concept tests demonstrated that a substantial gain in efficiency is feasible in small size power plants by integration of the microturbine with the fuel cell. The operation of the power plant benchmarked a near 5% increase in efficiency due to the microturbine integration.

4.5 Operations and Control of Carbonate Power Plants
Electrical and control subsystems are two key components in a carbonate fuel cell power plant. The electric

subsystem performs the tasks of inverting the fuel cell direct current (DC) power to alternating current (AC) power, and providing electricity to auxiliary units such as pumps and blowers. The control subsystem performs very important supervisory and regulatory tasks of maintaining process parameters such as temperatures and pressures, while responding to customer needs for power. Therefore, design of the control system consists of very challenging, nevertheless rewarding, opportunities for improving plant's performance and for enhancing its reliability.

In power dispatch mode, the plant's controller responds to a power demand ensuring adequate supply of reactants (fuel, water and air) to the fuel cell. If the control system is not designed properly, it may lead to undesirable consequences such as fuel starvation or process overtemperature. Based on the controller's regulatory response to a power demand, the fuel cell produces a DC power corresponding to the demand. Essentially, as the power demand increases (or its set point is raised), the fuel cell stack produces higher current, and vice versa. A typical current, voltage, and power relationship for a 250 kW carbonate fuel cell stack operating on natural gas is shown in Fig. 27. The gross DC power is inverted to AC power using a semiconductor based power-conditioning unit. A DC-to-AC power-conditioning unit has the important function of regulating output voltage, synchronizing both current and voltage waveforms with the utility grid, and preventing harmonic distortions. Performance, efficiency, and cost of DC-to-AC inversion are important considerations for power-conditioning system design. Carbonate fuel cell power generators are available for both grid-connected (grid-parallel), as well as grid-connected islanding, schematically shown in Fig. 28. The islanding mode is considered a high value feature by allowing power delivery to a local critical bus during grid outages.

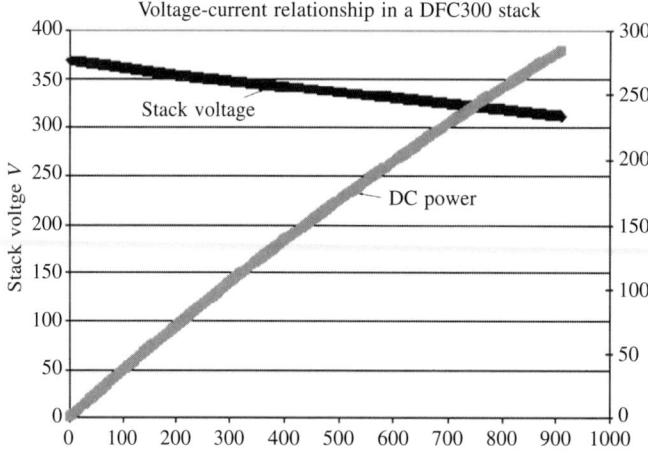

Fig. 27 Current-voltage-power relationship of a 250 kW carbonate power plant.

Grid-parallel power generation mode requires that AC power be supplied at the grid distribution voltage and frequency, meeting grid power quality for voltage and current waveform harmonic distortions as specified by IEEE-1547 (IEEE Standard for Interconnecting Distributed Resources with Electric Power Systems). The DC-to-AC inverter output current is varied to meet a desired power output. The inverter also provides the ability to regulate reactive power production, both in magnitude and direction, independent of real power. The imaginary power dispatch ability for grid VAR (volt-ampere reactive) control is considered an added feature of the fuel cell system. Safety considerations, particularly for linemen who may be

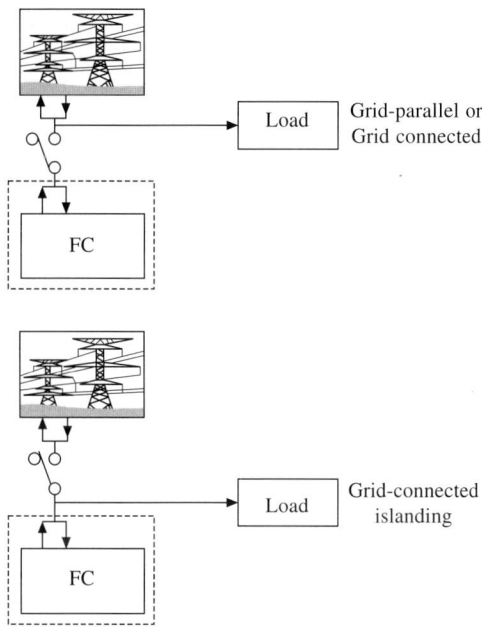

Fig. 28 Carbonate fuel cell system power electronic configuration alternatives.

working on the system during a grid outage, require that appropriate contactor and sensors for disconnecting from grid be implemented. In grid-connected operation, the power plant stops power production and is switched to either idle or shutdown when grid outage occurs. In islanding operation, the fuel cell source is disconnected from utility grid at the time of grid outage, but continues to provide power to its dedicated local load.

The power electronics topology depends on external load requirements as well as the fuel cell DC voltage. In addition to an inverter, depending on the fuel cell voltage and the external load voltage, the power electronics system may also include a DC-to-DC converter and/or a step-up transformer to boost output to match grid voltage. The inverter, DC converter, transformer, and parasitic power consumption by auxiliary components such as inverter power electronics cooling system and filters, affect the overall conversion efficiency. A typical power plant inverter has an overall DC to 480 V 3-phase AC conversion efficiency of about 95%. The best way to achieve higher overall efficiency as well as lower cost is to maximize fuel cell output DC voltage. An overall DC-to-AC inverter efficiency approaching 98% has been achieved for delivery of 480 V 3-phase AC power from an ~800 V DC fuel cell source. However, attainment of similar overall efficiencies with lower voltage fuel cell sources (< 500 V) remains a goal for future system developers.

Although Distributed Control System (DCS) architecture is making inroads in control system of larger power plants, control systems based on Programmable Logic Controller (PLC) still offer an inexpensive alternative for smaller power plants. The PLC gathers information from the power plant subsystem instruments and directs the operation of various components in the plant. The PLC scheme uses a Human Machine Interface (HMI) computer to acquire and store operational data. The remote access software allows the off site specialist to access plant data and controls via phone modem or through a high-speed Internet line. Extensive remote monitoring of key parameters, diagnostics, and process control via HMI allow off-site specialists to interface with the plant when necessary and/or instruct the site operators for necessary action.

An operator and/or a specialist can access the monitoring and control functions locally at the power plant via a screen. Typically, carbonate fuel cell products are designed for unattended and remote operation capabilities. Also for cost reasons, the operators who infrequently interface with power plants are intended to be generalists and not dedicated fuel cell technologists. Their interactions are limited to initiating startup, starting power generation, changing power level, taking plant off line, and commencing cool down. These activities are single button functions. The unattended plant operation requires incorporation of protection logic against all possible credible events without relying on operator intervention. The protective actions include ramp holds, power back ramp, total load shed, and complete plant shut down. Warning alarms use more conservative set points than protective functions to notify operation personnel early about potential problems. Notification is carried out through pager and/or phones. When the pager alarm arrives, operation/ or specialist personnel can either log in remotely or go to the plant HMI to address the situation. A first-out function identifies the initiating condition for any plant trip to assist in diagnosis, correction for plant recovery, and restart. The modern digital controllers and telecommunication technology perform this job inexpensively and within a small footprint.

5. Carbonate Fuel Cell Products

Due to its simple system design and high efficiency, several developers around the world are pursuing carbonate fuel cell products for stationary power generation applications in the near term and large (10 MW or larger) power plants in the long term. IHI (Ishikawajima Harima Heavy Industries Co. Ltd.) of Japan has focused on the development of coal-fueled large hybrid power plants based on its pressurized carbonate fuel cell. In the near-term, the company has launched field-testing of 300 kW natural gas units. The first prototype unit was tested in 2003 at Kawagoe, Japan. Additional unit tests on coal gas and biogas have also been planned. A Korean team consisting of KIST (Korean Institute of Science and Technology) and KEPRI (Korean Electric Power Research Institute) is in an early stage of large multi-megawatt systems for coal gas use. This team is testing a 100 kW system and has a plan to test a 250 kW system in the near-term.

The Italian-Spanish team led by Ansaldo Fuel Cell (AFCO) is focusing on large systems based on its pressurized fuel cell technology. In the near-term, the team is focusing on a 500 kW system. MTU CFC Solutions GmbH of Ottobrunn, Germany, in collaboration with FuelCell Energy has developed a 250 kW stationary fuel cell plant (the HotModule® brand) for combined heat and power (CHP) applications. The MTU CFC Solutions plants operating at Bad Berka and Munich in Germany, and at Cartagena in Spain are being used for CHP and air conditioning (trigen). MTU has already placed ten hot module units at customer sites. Fig. 29 shows an indoor hot module operating at the Michelin Karlsruhe, Germany factory. MTU has reported 47% electrical and greater than 80% overall thermal efficiency for their cogeneration system. The sub-megawatt fuel cell power plant is a collaborative effort using the Direct FuelCell® technology of FCE and the HotModule® balance-of-plant design of MTU CFC Solutions GmbH (a subsidiary of DaimlerChrysler).

FCE is currently commercializing carbonate fuel cell products for commercial and industrial customers and continuing to develop the next generation of large size carbonate fuel cell products. FCE's current products, the DFC300A (to the replaced by enhanced modular design DFC300MA), DFC1500 and DFC3000, are rated in capacity at 250 kW, 1 MW and 2 MW, respectively, and are scalable for distributed applications up to 10 MW or larger. These products are designed to meet the base load power requirements of a wide range of commercial and industrial customers including wastewater treatment plants (municipal, such as sewage treatment facilities, and industrial, such as breweries and food processors), telecommunications/ data centers, manufacturing facilities, office buildings, hospitals, universities, prisons, mail processing facilities, hotels and government facilities, as well as in grid support applications for utility customers. Through October 2005, over 78 million kWh of electricity has been generated from power plants incorporating the DFC technology at customer sites throughout the world.

Fig. 29   MTU's HotModule® in operation: An indoor 250 kW unit operating in an industrial environment.

The DFC plants have achieved electrical efficiencies of 45-48% in single-cycle applications and have the potential to reach an electrical efficiency 57% at product maturity. In addition, power plants can achieve overall energy efficiency of 70-80% for combined heat and power applications. This is greater than the fuel efficiency of competing fuel cell and combustion-based technologies of similar size and potentially results in a lower cost per kWh over the life of the power plant. Carbonate power plants have significantly lower emissions of greenhouse gases and particulate matter than conventional combustion-based power plants. They emit virtually no NO_x and SO_x and have been designated "ultra-clean" by the California Air Resources Board (CARB).

Biogas offers a unique opportunity for carbonate fuel cell products. Industrial wastewater treatment facilities represent a promising market for carbonate fuel cells. The methane produced from the anaerobic digester process is the fuel to generate electricity to power the wastewater treatment plant. The fuel cell heat can be used to heat the sludge to facilitate the anaerobic digestion. Moreover, wastewater treatment gas is a renewable fuel eligible for incentive funding for project installations throughout the world. FCE has fielded several units (two in japan and four in the U.S.A.) to operate on digester gas including a 1 MW DFC1500 at King County, W.A. The King County unit (Fig. 30) operates seamlessly on pipeline natural gas or on the digester gas.

6. Conclusion

While various fuel cell systems are emerging as alternatives for power generation, the carbonate fuel cell plants are attractive for applications where the superior efficiency or combined high quality heat and power are desired. Over fifty units ranging in 250 kW to 1 MW size have been in field operation worldwide. These units have shown 45 to 48% LHV electrical conversion efficiencies and overall thermal efficiency approaching 80% in combined heat and power (CHP) applications. The plant emissions are ultra clean. These attributes

Fig. 30. King County fuel cell power plant: Plant operates on digester gas and on natural gas when digester gas is not available.

and the various incentives available for high efficiency, ultra clean renewable fuel technologies are helping market entry of the product in stationary applications. As the technology matures and the cost is lowered through cost-out efforts, the product is expected to capture broader commercial acceptance paving the way for larger multi-megawatt systems.

References

Abens, S., Ghezel-Ayagh, H., Lukas, M., Sanderson, R. and Steinfeld, G. "Ship Service Fuel Cell Power plant Development", Fuel Cell Seminar Abstracts, November 2002.

Abens, S., Steinfeld, G., Sanderson, R. and Lukas, M. "Ship Service Fuel Cell Power Plant Development", Fuel Cell Seminar Abstracts, November 2003.

Baker, B.S. and Dharia, D.J. "Fuel Cell Thermal Control and Reforming of Process Gas", U.S. Patent 4,182,795 (1980).

Broers, G.H.J. "High Temperature Galvanic Fuel Cells", Vol. 1, Dissertation, University of Amsterdam, 1958.

Clean Coal Technology, Topical Report Number 20, The Wabash River Coal Gasification Repowering Project, a 262 MWe Commercial Scale Integrated Gasification Combined Cycle Power Plant: An Update, September 2000.

Donado, R.A., Marianowski, L.G. and Maru, H.C. "Corrosion of the Wet-Seal Area in Molten Carbonate Fuel Cells-Part I, Analysis", *J. Electrochem. Soc.*, **131** (11), 2535-2540, 1984 and Part II, Experimental Results, 2541-2544, 1996.

Doyon, J., Farooque, M. and Maru, H. "The Direct FuelCell™ Stack Engineering", *Journal of Power Sources*, **5195** (2003) 1-6.

Farooque, M. Internal Reforming Fuel Cell System Requiring No Recirculated Cooling and Providing a High Fuel Process Gas Utilization, US Patent 4,917,971, April 17, (1990).

Farooque, M., Katikaneni, S. and Maru, H.C. "The Direct Carbonate Fuel Cell Technology and Products Review", Carbonate Fuel Cell Technology V, Electrochemical Society Proceedings, Vol. 99-20, pp. 47-65 (1999).

Farooque, M., Steinfeld, G., McCleary, G. and Kremenik, S. "Assessment of Coal Gasification/Carbonate Fuel Cell Power Plants", Topical Report to DOE/METC, June 1990, DOC/MC/23274-2911. NTIS/DE90015579.

Fuel Cell Handbook, EG&G Technical Services Inc., November 2002, Science Applications International Corporation, 6th edition, published by DOE/NETL, US.

Ghezel-ayagh, H., Leo, A.J. and Sanderson, R. "High-Efficiency Fuel Cell System", US Patent 6,365,290, April 2002.

Ghezel-Ayagh, H. and Maru, H. "Direct Fuel Cell/Turbine System for Ultra High Efficiency Power Generation", 2002 Fuel Cell Seminar, Nov. 18-21, 2002, Palm Springs, CA.

Ghezel-Ayagh, H., Leo, A.J. and Sanderson, R. "High-Efficiency Fuel Cell System", U.S. Patent No. 6,365,290, April 2002.

Ghezel-Ayagh, H., Daly, J.M. and Wang, Z. "Advances in Direct Fuel Cell/Gas Turbine Power Plants", Proceedings of ASME/IGTI Turbo Expo 2003, ASME paper GT 2003-38941.

Grillo, O., Magistri, L. and Massardo, A.F. "Hybrid Systems for Distributed Power Generation Based on Pressurization and Heat Recovering of an Existing 100 kW Molten Carbonate Fuel Cell", *Journal of Power Sources*, **115**, 2003, 252-267.

Hoffmann, J., Yuh, C. and Godula Jepek, A. "Electrolyte and Material Challenges", Chapter 67 in Handbook of Fuel Cell, Wiley, Fundamentals, Technology and Applications, Vol. 4 Fuel Cell Technology and Applications, John Wiley & Sons, 2003.

Leo, A.J., Ghezel-Ayagh, H. and Sanderson, R. "Ultra High Efficiency Hybrid Direct Fuel Cell/Turbine Power Plant", Proceedings of ASME TURBOEXP 2000, ASME paper 2000-GT-0552.

Liebhafsky, H.A. and Cairns, E.J. 1968, Fuel Cells and Fuel Batteries, Wiley, New York, Chapters 2 and 12.

Lunghi, P., Bove, R. and Desideri, U. "Analysis and Optimization of Hybrid MCFC Gas Turbines Plants", *Journal of Power Sources*, **118**, 2003, 108-117.

Ma, Z., Venkataraman, R. and Farooque, M. "Study of the Gas Flow Distribution and Heat Transfer for Externally Manifolded Fuel Cell Stack Using Computational Fluid Dynamics Method", *Journal of Fuel Cell Science and Technology*, **1**, No. 1, 2004, 49-55.

Maru, H.C. and Marianowski, L.G. "Composite Model of Electrode-Electrolyte Pore Structures", Extended Abstracts, 76-2, Electrochemistry Society, Pennington, NJ (1976).

Nickens, A., Cervi, M., Abens, A. and Hoffman, D. "US Navy Ship Service Fuel Cell Program", Fuel Cell Seminar Abstracts, November 2004.

Patel, P., "Assessment of a 6500-Btu/kWh Heat Rate Dispersed Generator", EPRI EM-3307, Project 1041-12 Final Report, November 1983.

Sander, M.T. and Steinfeld, G. "Cost and Performance Analysis for a 220 MW Phased Construction Carbonate Fuel Cell Power Plant", 11th Annual Conference on Gasification Power Plants, EPRI, October 1992.

Sandler, H.S. and Meyers, S.J. "Integrated Coal Gasification in Carbonate Fuel Cell Power Plants" 11th Annual Conference on Gasification Power Plants, EPRI, October 1992.

Selman, J.R. and Maru, H.C. "Physical Chemistry and Electrochemistry of Alkali Carbonate Melts", Advances in Molten-Salt Chemistry, Volume 4, Plenum Press, New York, p. 159 (1981).

Steinfeld, G. and Willson, W. "Advanced Power System Featuring a Closely Coupled Catalytic Gasification Carbonate Fuel Cell Plant" Presented at the 17th Biennial Low-Rank Fuels Symposium, May 10-13, 1993, St. Louis, Missouri.

Tanaka, T., Matsumura, M., Gonjo, Y., Hirai, C., Okada, T. and Miyazaki, M. "Development of Internal Reforming Molten Carbonate Fuel Cell Technology", Proceedings of the 25th Intersoc. Energy Conv. Eng. Conf., Vol. 3, pp. 201, Reno, NV, 1990.

Williams, M.C., Strakey, J.P. and Singhal, S.C. "U.S. distributed generation fuel cell program", *Journal of Power Sources*, **131**, 2004, 79-85.

Yuh, C., Farooque, M. and Maru, H. "Advances in Carbonate Matrix and Electrolyte, Electrochemical Society Proceedings", Vol. 99-20, pp. 189 (1999).

Yuh, C., Johnsen, R., Farooque, M. and Maru, H. "Status of Carbonate Fuel Cell Materials", *Journal of Power Sources*, **56**, pp. 1-10 (1995).

Recent Trends in Fuel Cell Science and Technology
Edited by S. Basu
Anamaya Publishers, New Delhi, India

10. Direct Conversion of Coal Derived Carbon in Fuel Cells

John F. Cooper

Energy Systems, Materials Science and Technology Division, Chemistry and Materials Science Directorate,
Lawrence Livermore National Laboratory, L-352, Livermore CA 94550, USA

1. Introduction

A long-held dream of early energy science and technology has been a fuel cell that would directly convert coal char (if not raw coal) into electric power—bypassing the emission problems of combustion and the efficiency limitations of thermal cycles. Such a fuel cell would generate electric power from an electrochemical reaction similar to the combustion reaction of carbon:

$$C + O_2 = CO_2 \ (\Delta H^0_{298 \ K} = -94.05 \ \text{kcal/mole}, \ E^0 = 1.02 \ \text{V at } 750°C) \qquad (1)$$

As an example, the fuel cell (Fig. 1) might use carbon plates or particulates wetted with a molten alkali carbonate electrolyte, a melt saturated porous ceramic separator, and an air electrode catalyzed by lithiated NiO, such as used in the molten carbonate hydrogen fuel cell (Cherepy et al, 2005; Vutetakis, 1985; Weaver et al, 1979).

The advantages of a carbon fuel cell in efficiency is derived from the thermodynamics of reaction (1). The entropy change is very small ($\Delta S^0 = 0.67$ cal/K-mol), resulting in a theoretical efficiency ($\Delta G^0_T / \Delta H^0_{298 \ K}$) of nearly 100% for temperatures as high as 1000 °C. Because both carbon and the product CO_2 exist as pure substances in separate phases, the chemical potentials are constant and independent of extent of fuel conversion. This invariant activity allows all of the fuel to be converted in a single pass through the cell. The polarization associated with anode reaction has been found to be small for atomically disordered "turbostratic" carbon at $T > 700°C$ (Cherepy et al, 2005). The efficiency

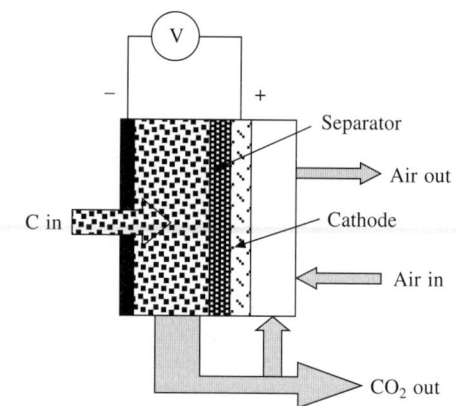

Fig. 1 Direct carbon fuel cell makes use of a rigid or particulate carbon anode, a molten salt electrolyte saturating the anode and porous matrix separator, and an oxygen-depolarized cathode supplied with air and CO_2.

of the carbon/air cell has been demonstrated at 80% referenced to the higher heating value of the fuel (32.8 MJ/kg-C) at rates up to about 100 mA/cm^2 that are practical for many fuel cell applications. This efficiency is over twice that achieved with combustion of carbon or coke in conventional steam power generation cycles.

There are further advantages of using carbon in a fuel cell rather than as a combustion fuel. Air is excluded from the anode chamber, which generates the CO_2 reaction product. Therefore the volume of CO_2 product gas that has to be treated for removal of contaminants is reduced by a factor of 10 compared with stack gas of conventional combustion plants—twice because of the greater efficiency, and five-fold by elimination of the nitrogen component of air. Most (but not all) of the net carbon dioxide production may be sequestered or used in enhanced oil or gas recovery without additional costs of collection and separation. The temperature is sufficiently low that NO_x is not produced. The higher efficiency makes pre-cleaning of the fuel cost-effective, removing many of the contaminants that would foul the energy conversion system as slag or be entrained in stack gas at low levels.

2. Thermodynamic Basis

The net reaction (1) is generally written as the sum of two half-cell reactions, both of which involve the carbonate ion (Weaver et al., 1979).

$$O_2 + 2CO_2 + 4e^- = 2CO_3^{2-} \text{ (cathode reaction)} \tag{2}$$

$$C + 2CO_3^{2-} = 3CO_2 + 4e^- \text{ (anode reaction)} \tag{3}$$

the carbon anode may also partially oxidize to CO in a competitive reaction:

$$C + CO_3^{2-} = CO + CO_2 + 2e^- \text{ (anode reaction)} \tag{4}$$

Reactions (2-4) are net reactions, and do not indicate the mechanisms.

Weaver calculated the open circuit potentials of these and other possible reactions that might occur under open circuit conditions, finding agreement between measured potentials and the potentials calculated from thermodynamic tables (Weaver et al, 1979). Hemmes and Cassir (2004) recalculated the cell open circuit potentials. They determined the equilibrium concentrations and electrode potentials in a system comprised of carbon, carbonate, CO_2, CO, O^{2-}, and electrons, using the phase rule modified for electrochemical systems by Coleman and White (1996). Hemmes expressed the half-cell potentials of the anode reactions (3) and (4) referenced to an idealized cathode reaction (unit oxygen and CO_2 partial pressures):

$$E = E^0_{C/CO_2} - RT/4F \ln [CO_2]^3 \tag{5}$$

$$E = E^0_{C/CO} - RT/2F \ln [CO][CO_2] \tag{6}$$

Figure 2 shows calculated open circuit potentials for this idealized carbon fuel cell, calculated from thermodynamic data. (These are our calculations, after Hemmes and Cassir, 2004). This figure shows that at practical temperatures for carbon fuel cell operation ($T > 650°C$), the open circuit potential will be in excess of the standard potential of reaction (1). The concentration of CO_2 is suppressed because of the reaction of CO_2 with carbon to produce CO according to the Boudouard reaction:

$$C + CO_2 = 2CO \tag{7}$$

To achieve the full utilization of carbon (4 electrons per atom) requires an electrochemical reaction producing only CO_2. For this to occur, it is necessary to polarize the carbon anode by at least the difference between the open circuit potential and the carbon/carbon dioxide equilibrium potential (1.02 V), i.e. by about 0.10 V at 750°C. This lowers the electrode potential into the range where evolution of CO_2 is possible.

3. Historical Basis of Carbon Fuel Cells and Fuel Batteries

Early experimenters in carbon or coal fuel cells encountered difficult problems many of which still occupy our attention today. Coal is not electronically conductive and must be decomposed to a conductive char to

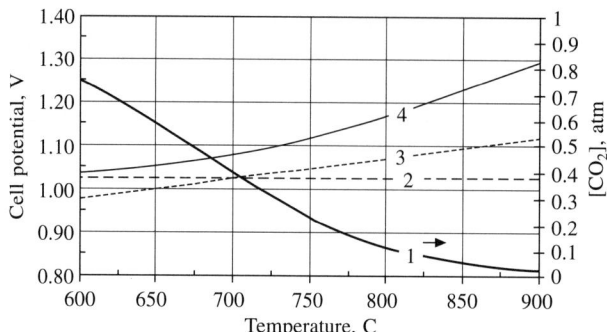

Fig. 2 Calculations of (1) equilibrium carbon dioxide partial pressure in the C/CO/CO₂ system according to the Boudouard reaction (right); (2, 3) standard potential of the C/CO₂ and C/CO cells, respectively; and (4) the open circuit potential calculated for CO and CO₂ assuming Boudouard equilibrium. For simplicity, unit partial pressures of O_2 and CO_2 are assumed for the cathode half-cell for purposes of calculation of cell potentials.

react efficiently as an anode. The reactions of the carbon anode are sluggish, and high temperatures (> 650°C) are required for practical rates in a chemically stable electrolyte such as carbonate. Boudouard corrosion of the carbon expected at such temperatures would nearly halve the energy yield by liberating only two of the four valence electrons per atom of carbon. The entrainment of impurities from natural carbon resources (i.e., coal, petroleum coke, charcoal, etc.) into the melt would soon exhaust the electrolyte by reaction with the melt or occlusion of current flow by "ash." Sulfur entrained as pyrite or chemically bound into the coal "molecule" leads to corrosion of most metals used for containment, electrodes or conductors. The distribution of carbon plates to millions of cells comprising even a small power plant presents a formidable logistics problem.

Early work from the 19th and 20th century is reviewed by Howard (1945). Liebhafsky and Cairns (1968) give a more thorough critical review of solid carbon fuel cells and fuel batteries. In addition to his work in powdered coal anode slurries in molten salt, Vutetakis (1985) cataloged the carbon material/electrolyte combinations that have been examined and collected useful materials data on many electrolytes used in such cells.

In the late 19th century, William Jacques (1896a) demonstrated large (multi-kilowatt scale) fuel batteries that reacted rods of processed coke with molten caustic soda at temperatures of 400-500°C to produce electric power according to the net reaction,

$$C + 2NaOH + O_2 = Na_2CO_3 + H_2O \ (E^0 = 1.4 \ V) \tag{8}$$

Jacques (1896a, 1896b) noted that the caustic soda was gradually converted to sodium carbonate. Cairns observed that Jacques' cell was not a fuel cell, but rather a *fuel battery*—reacting carbon, NaOH (or KOH) and atmospheric oxygen in an energy intensive molten salt electrolyte that was consumed to form the carbonate in proportion to the electric energy generated. It failed to achieve practical or economic significance because (1) the cost of caustic was prohibitive (an order of magnitude greater than the stoichiometric equivalent of coal-based carbon), (2) the efficiency was poor compared to alternatives when the energy cost of caustic production from carbonate ($\Delta H° = +40.3$ kcal/mol-C) was weighed against the heat of combustion of carbon (–94.05 kcal/mol-C), and (3) the caustic electrolyte was unstable and was progressively converted into higher melting carbonate (Na₂CO₃, m.p. = 851°C) preventing *sustained* operation at the lower temperatures of molten caustic (m.p. = 318°C). Later, the reaction was attributed to the formation *in situ* of electro-active hydrogen by anode reduction of the electrolyte, at a low overall efficiency (Haber, 1904). Still, this work demonstrated high anode rates and a simple cathode comprised of an iron container sparged with air in a simple configuration envisioned for domestic use.

4. Relationship of Direct Carbon Fuel Cell to Molten Carbonate Fuel Cells

The molten carbonate fuel cell (MCFC) shares some aspects in common with the direct carbon fuel cell (DCFC) as envisioned in this article, but there are important differences as well. The MCFC uses hydrogen or steam reformates at a catalyzed anode surface separated from the cathode by (typically) a porous lithium aluminate ceramic tile saturated with a mixture of molten carbonate salts. Current status is reviewed by Larminie and Dicks (2000). The DCFC may make use of a similar cathode, but the higher temperature of operation (750°C vs 650°C for MCFC) allows other catalysts to be considered. The DCFC operates with an excess and variable amount of electrolyte, while the MCFC has a fixed amount requiring control over losses and composition changes. The MCFC uses an anode catalyst sensitive to sulfur poisoning; the DCFC uses no anode catalyst (other than the carbon surface) and is not sensitive to poisoning. Finally, if hydrogen is used as fuel, the system is subject to steam corrosion, while the DCFC is an anhydrous system. Since melt-wetted carbon is non-explosive, the reliance on the separator to rigorously isolate fuel from air is relaxed.

5. Recent Research in Carbon Fuel Cells and Fuel Batteries

5.1 Alkaline Systems

Work continues on the Jacques cell in various configurations by SARA, Inc. (Cypress, CA) (Zecevic, 2004; Pesaventeo, 2001). Current work features a cell with a porous separator, with a hydroxide electrolyte in the air-sparged cathode half-cell, and a mixed hydroxide and carbonate electrolyte on the anode side. (Patton, 2005). Two parallel net anode reactions are posited, although no mechanism is proposed:

$$C + 6OH^- = CO_3^{2-} + 3H_2O + 4e^-$$

$$C + 2CO_3^{2-} = 3CO_2 + 4e^-$$

These are balanced for invariant CO_3^{2-} concentration, leading to a hypothesized steady state anode reaction, $C + 4OH^- = CO_2 + 2H_2O + 4e^-$. If there is no interfacial or bulk reaction of ambient NaOH with the evolved carbon dioxide, then the net reaction would be the same as (1). To the extent that the CO_2 product reacts with ambient hydroxide in the electrolyte on or near to the anode, the net reaction approaches that of the Jacques cell (reaction (8)). In this case, the alkaline reactant will be consumed, and cost and energy efficiency might more accurately be estimated on the basis of (8). At this writing, a rigorous experimental mass balance would help resolve these issues.

5.2 Fluidized Beds

Vutetakis investigated the behavior of carbon (or coal) slurries in mixed alkali carbonate melts at temperatures of 700-800°C. Slurries of carbon, coal, graphite and charcoal were polarized using a gold wire current collector in a stirred melt (up to 25% carbon). Stable polarization was reached only after minutes of application of current, and polarograms showed hysteresis. He measured the CO_2 content of the offgas and found it proportional to the polarizing current (1 mole per 4 equivalents of charge) at current densities of 110 mA/cm².

Agarwal and Kornhauser (2004) proposed and analyzed carbon/salt slurry cells in fluidized configurations with flow-through cathodes, which should adapt well to a large-scale utility plant comprised of very large cells. Ash would be removed continuously from the anode stream, and CO generated by the anode bed would be burned to produce the CO_2 necessary to sustain the cathode half reaction. The concept might make use of the observed delay in the recovery of open circuit potentials for polarized carbon, once the circuit is interrupted. This phenomenon gives rise to the hysteresis observed in cyclic polarization and may be expect to protect the carbon from Boudouard corrosion when out of contact with the anode current collector.

5.3 Ionic Conductors
Cocks (2004) is pursuing a carbon ion conductor, analogous to the oxide conductor defect zirconia. Success here would revolutionize the possibilities of carbon fuel cells by providing a carbon-ion conducting solid electrolyte analogous to oxide-conducting defect zirconia.

5.4 Solid Oxide Electrolyte Cells
Gur experimented with concentration cells for production of electrical energy (~ 1-3 mA/cm^2 at 0.8 V for 725-755°C) associated with the EMF that develops across a platinum-catalyzed zirconia membrane. (Gur and Huggins, 1992). The cathode side reduces oxygen at atmospheric pressure to oxide while the anode is depolarized by CO oxidation in a mixed CO/CO_2 atmosphere. The CO is regenerated by the Boudouard reaction of carbon and CO_2 at points out of electrical contact with the electrolyte. Nakagawa and Ishida (1988) examined this system further and presented a graphical analysis of exergy flow indicating that the energy efficiency was jeopardized by the mixed product gas (CO + CO_2). Ihara (2003) experimented with the direct conversion of carbon formed by pyrolysis of dry methane on the anode surfaces of Ni/YSZ and nickel/gadolinium-doped ceria. In all four studies, the cells were found to evolve mixtures of CO and CO_2, the rate of discharge of CO being low at the oxide/anode interface.

Chuang (2005) is investigating coal oxidation in a solid oxide fuel cell-producing voltages of 0.4-0.9 V at open circuit. He reports CO_2 as the major product and CO as a minor product at 950°C. An important issue is whether the reaction proceeds through an electrochemically active CO intermediate generated by the Boudouard reaction of carbon and CO_2, or (as Chuang concludes) at substantial amounts of electrons are withdrawn directly from elemental carbon.

5.5 Partial Carbon Conversion Fuel Cell with Cogeneration of Hydrogen
Hemmes (2003) and coworkers Au (1999) and Peelen (2000) have examined the exergy flow in systems using a carbon fuel cell to oxidize coal only to CO at low current densities. The product gas is then used for its thermal value in combustion or shifted to H_2. This approach takes advantage of the large entropy increase of the electrochemical reaction $C + \frac{1}{2} O_2 = CO$ ($\Delta S^0_{298\ K} = 21.43$ cal/K-mol) and resultant high theoretical voltage ($E^0_{1500} = 1.27$ V) (Fig. 2). Inexpensive sources of heat and the waste cell heat production balance the heat input required by the entropy increase. The cell delivers the maximum electrical energy from the expensive fuel cell component and yields (after shift reactions) a hydrogen-rich byproduct.

6. Carbon Anode in Molten Salt Electrolytes
Research throughout the last century focused on reactions of the carbon anode in various molten salts. Since gas-diffusion type air electrodes were not readily available, this research was conducted almost exclusively in half-cell configurations. Some technical achievements of this anode research are important in the potential application to fossil fuel chars.

6.1 Polarization
The minimum temperature for initiation of oxidation of carbon in air depends on various aspects of crystallographic disorder and characteristic scale of microcrystallinity (Kinoshita, 1988). Anodes made of

atomically disordered carbon are used in the Hall-Heroult process for smelting of aluminum. The disordered or "turbostratic" carbon show far greater reactivity than the highly ordered, more graphitized carbon. This is demonstrated in the comparative polarization curves in cryolyte melts at 1100°C (Thonstad, 1970). Weaver et al. (1979) were perhaps the first to recognize that the polarization of the carbon anode in the carbonate melt depended not on purity or specific surface area but rather on disorder. The anodic current at a fixed electrode potential of 0.8 V (vs. Au/0.28 CO_2, 0.14 O_2, 0.58 N_2) varied from 5×10^{-5} A/cm^2 for natural graphite to > 0.1 A/cm^2 for charred bituminous coal. Rates obtained at LLNL from Xerogel or aerogel composite electrodes are still greater. Comparative polarization curves are shown in Fig. 3.

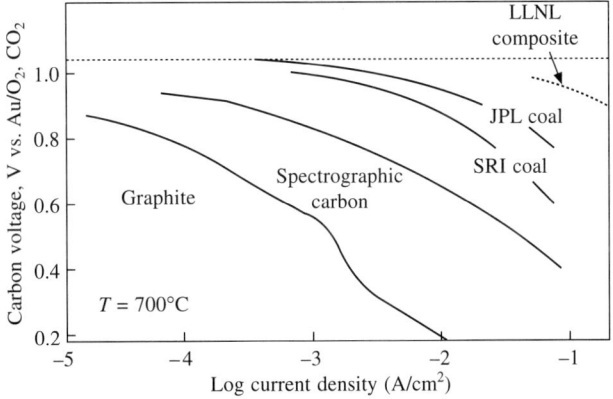

Fig. 3 Comparative anode half-cell polarization of various carbon materials as reported by Weaver (1979), showing increase in rate at fixed polarization from graphite to devolatilized bituminous coal. The rate at 0.8 V spans four orders of magnitude.

Table 1. Anode rates obtained at fixed polarization for diverse carbon materials in full cell configuration

Samples tested[a]	Current density (mA/cm^2) at $E_{cell} = 0.8$ V
Desulco graphite particles	58
Calcined petroleum coke[b]	58
Acetylene black	77
Furnace black	110
Coal-derived activated carbon	65
Coconut activated carbon	102
Peach pit char	124
Aerogel carbon	87

[a]In order of increasing crystallinity parameter (increasing 'disorder') and
[b]Needle coke, calcined at 1400°C.

Cherepy et al. (2005) conducted a more thorough examination of the dependence of anode rate at fixed polarization. The carbon materials included a variety of charred vegetable material, furnace and thermal blacks, calcined petroleum and charred coal-derived pitch, and fine graphite particles. The structure of these materials was characterized with use of a 'crystallinity index' after Fugimoto (1994). This index is a product of factors measuring the actual lattice dimensions relative to those of perfect graphite, and the characteristic crystallite dimensions normal and parallel to the basal plane. The current achieved at a fixed polarization (0.8 V or 0.2 V below the standard cell voltage) was found to correlate with this index. No correlation of rate with specific area was noted in the range 0.4-1200 m^2/g.

Electronic conductivity of the anode material is also important because it decreases the ohmic component of polarization. For example, results for coal in Fig. 3 were obtained for baked coal plates and rods having density of 0.8-1.2 g/cm^3 and electrical resistivity of 0.008-0.036 Ω-cm (Weaver, 1979).

6.2 Current Efficiency

At practical rates of 0.1 A/cm^2, the carbon/carbonate anode was found to yield predominately CO_2 resulting in high anode utilization and current efficiency. Current efficiency is defined in reference to reaction (1) by $e = 4[CO_2]/(4[CO_2] + 2[CO])$. Tamaru and Kamada (1935) first noted efficient production of CO_2 and reported the net cell reaction to be the same as that of coal combustion. Hauser (1964) analyzed the gas evolved from graphitized carbon anodes and found current efficiencies for reaction (1) to exceed 99% for applied current densities of 20-120 mA/cm^2 over the range, $T = 650$-$800°C$. At $870°C$, current efficiency was below 95%, and dropped to less than 75% as the current density was reduced to below 10 mA/cm^2. Weaver (1979) measured current efficiency as a function of current density and temperature, from gas analysis and by comparing weight loss with equivalents of charge passed. He concluded that CO_2 was the reaction product (100%, to within experimental error) at $700°C$ for various samples of thermally decomposed coal and spectroscopic carbon. For samples of granular coal and coke suspended in molten salt slurry, Vutetakis (1985, 1987) measured the dependence of CO_2 evolution rate on current density at a gold wire anode suspended in the slurry to establish that the anodic product was carbon dioxide, although portions of the slurry out of contact with the anode underwent Boudouard corrosion. These experiments support the conclusion that the carbon anode is substantially oxidized to CO_2 in molten carbonate electrolyte at polarizations greater than about 0.10-0.15 V, depending on temperature.

The predominance of carbon dioxide in the anode product gas is significant in that the equilibrium calculations for the Boudouard reaction would indicate that carbon monoxide would be favored

$$C + CO_2 = 2CO \ (K_{eq} = 2.59 \text{ at } 750°C)$$

The mechanism for carbon anode oxidation to CO_2 in carbonate electrolytes may be similar to that of the carbon anode in the cryolite/alumina electrolyte used in the smelting of aluminum by the Hall-Heroult process (reviewed by Grjotheim, 1982). The carbon evolves nearly pure CO_2 even at temperatures as high as $1100°C$. The accepted reaction sequence is initiated by the equilibrium thermal decomposition of an alumino-fluoride species to form low concentrations of O^{2-} that adsorbs on the surface of the carbon electrode. This ion is discharged to form strongly bound –CO functional groups. A second adsorption of O^{2-} on this layer and subsequent discharge requires polarization and results in —C_2O_3 groups that readily decompose into free CO_2 and —CO groups. (Frank and Haupin, 1985).

7. Current Approaches and Results

7.1 Test Configurations

Cell configurations (Fig. 4) used in typical laboratory tests consist of a cathode made of nickel oxidized in the presence of lithium ion; a porous separator, saturated with molten salt; an anode current collector; and the anode material, also saturated with molten salt. The cell assembly is generally held at an angle to horizontal to allow drainage of excess electrolyte liberated as the salt-saturated anode is consumed. Fig. 5 shows typical assemblies used in rapid generation of polarization data. Scale up from 2 cm^2 to 60 cm^2 cell

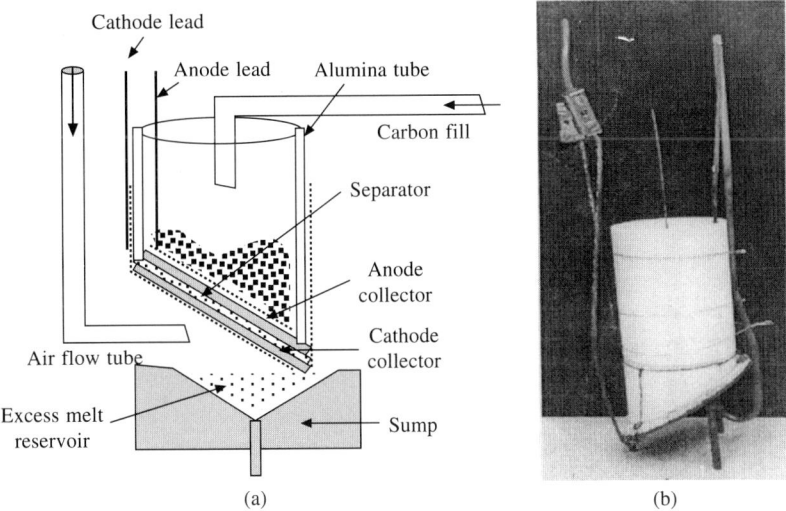

Fig. 4 (a) Schematic of a tilted cell allowing exchange of electrolyte between cell components and an underlying sump. and (b) a 60 cm^2 refuelable cell.

Fig. 5 Small cells (2.5 cm diameter) provide a convenient and inexpensive method of surveying carbon behavior. The reference electrode is a gold wire separated from the anode by a porous ceramic plug and flooded with an overflow of 0.28 CO_2, 0.14 O_2, 0.58 N_2.

area was achieved without significant loss of polarization, but the resistance of the electrode current leads is controlling at elevated temperatures.

The polarization curve shown in Fig. 7 is of particular interest in the extension of direct carbon conversion to coal char conversion. The sample of coal was chemically cleaned by a process developed to manufacture carbon particulate fuel for gas turbines. (Ultra Clean Coal Pty. Ltd.; Langley, 2004). Ash may be removed by a modified Bayer caustic digestion down to the level of 0.17%. At this level, the accumulation of ash necessary to reach 10%-volume criterion found by Weaver would require 1.5 yrs of operation at 0.1 A cm^2. The particulate coal (10 µm size) was charred in situ at 750°C in the presence of molten carbonate. The coal showed no tendency to agglomerate and flowed freely following pyrolysis as melt-wetted slurry.

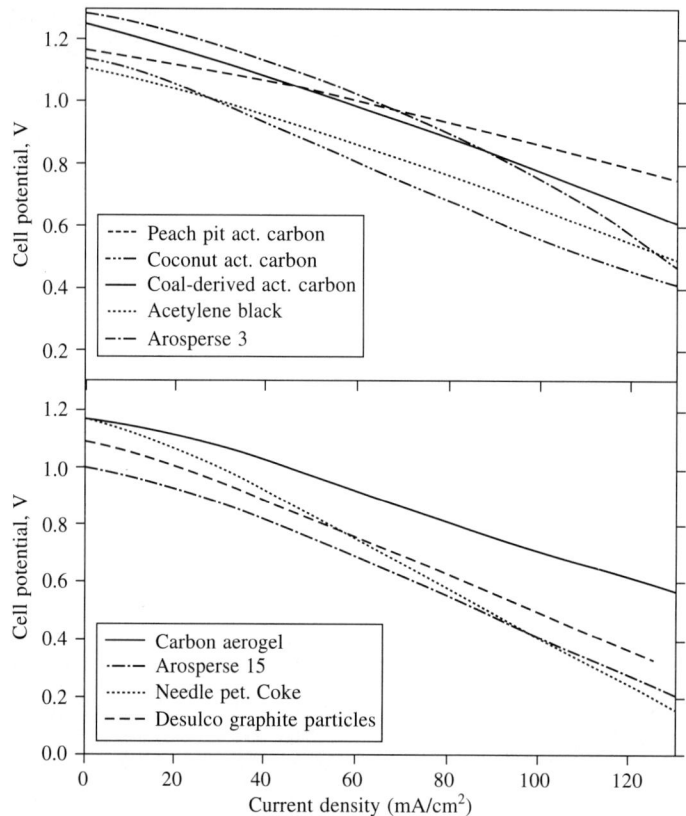

Fig. 6 Polarization curves of diverse carbon materials are shown (after Cherepy et al., 2005) at 800°C.

7.2 Decomposition of Cell Voltage

The measured cell voltage of 0.8 V, which serves as a performance benchmark, can be resolved into the various contributions of anode and cathode overpotential, Nernst corrections for gaseous reactants under non-standard conditions, and IR losses in the separator, anode and bipolar plate (Table 2). The open circuit potential (Fig. 2) is calculated to be 1.12 V at the reduced CO_2 partial pressure resulting from Boudouard corrosion at 750°C. From this we subtract first the 100 mV polarization required to bring the anode into the potential regime of the C/CO_2 electrode.

In current experimental work, carbon dioxide

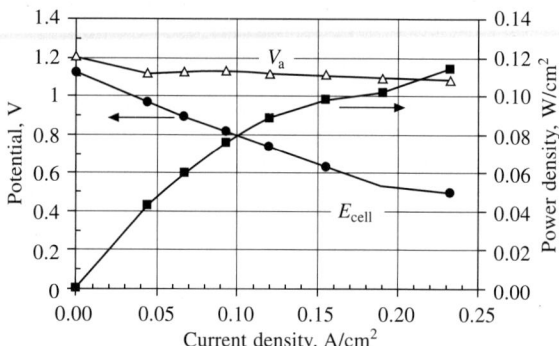

Fig. 7 Polarization of a char from *in situ* decomposition of chemically cleaned coal (Ultra Clean Coal Limited, Sydney, Australia).

mixed with air $\left(\frac{1}{7}O_2, \frac{2}{7}CO_2, \frac{4}{7}N_2\right)$ is allowed to flow past the cathode at very high multiples of the stoichiometric rate of consumption. This overstates the efficiency of a practical cell that must conserve on the use of carbon dioxide while producing an acceptable Nernst potential. The Nernst potential correction

Table 2. **Decomposition of experimental cell voltage into thermodynamic and kinetic contributions at 750°C for a current density of 100 mA/cm^2**

Contribution	Current experimental (mV)[a]	With improved cathode (mV)[b]
Open circuit potential: Boudouard anode composition and unit cathode CO_2 and O_2 partial pressures	1120 mV	1120 mV
Losses:		
Minimum polarization for CO_2 product	−100	−100
Nernst Correction (cathode)	−100	−145
Nernst Correction (anode)	−0	−0
Cathode overpotential (measured, and limiting)	−80	−30
Anode polarization (estimate from Fig. 7)	−10	−10
Electrolyte IR drop (2 mm separator, 50% porosity)	−20	−20
IR drop in anode and bipolar plate		−4
Net cell voltage	800 mV	801 mV

[a]Cell open circuit potential and potential at 0.1 A/cm^2 taken from data for ultra-clean coal char (Fig. 8).
[b]This estimate assumes reduced cathode flooding and a realistic Nernst loss associate with airflow at 2 stoics. IR drop in anode and bipolar plate assume properties of typical carbon rigid anodes and graphite respectively.

for uniform gas composition is 100 mV. For a finite flow rate of twice the stoichiometric rate of consumption, the Nernst loss will be 145 mV. There is no Nernst potential loss associated with the anode reaction.

The greatest loss at present is associated with the cathode. This loss results from failure to control the wetting of the cathode. Unlike the molten carbonate fuel cell (which has a fixed quantity of salt entrapped in the matrix or porous ceramic separator), carbon paste and porous plate fuel cells first absorb and then liberate molten salt during the process of anodic reaction. This tends to flood the fine porosity of the sintered nickel cathodes used in MCFC applications. A high-porosity, flow through structure is required for the carbon fuel cell in our configuration. The cathode losses are estimated at 30 mV for a well-engineered cathode at 100 mA/cm^2. This is slightly larger than the 25 mV losses for air electrodes measured at 650°C in the MCFC (Bregoli and Kunz, 1982).

In summary, the experimental data fits well into a picture of the carbon fuel cell reaction that indicates an open circuit potential (calculated from Boudouard equilibrium in the anode chamber), a minimal polarization of the carbon anode at 0.1 A/cm^2, and a significant loss associated with the cathode. Improvement of the cathode and realistic assumptions about Nernst losses in the cathode gas flow indicate that 0.8 V (80% efficiency) should be readily achieved in practical industrial cells. The greater efficiency of the cathode offsets the greater Nernst potential loss at practical air cathode flow rates.

8. Problems in Application to Coal

The use of charred coal or petroleum coke requires solutions to three significant problems: (1) the extraction of the mineral material phase including metal oxides, pyrite, and alumino silicates (collectively called 'ash'), (2) management of the sulfur content, and (3) use of the kinetic heat of the fuel cell to complete thermal decomposition of the fuel (bake out). The latter, not required for cell operation, improves overall efficiency and makes use of the high thermal diffusivity and conductivity of elemental carbon. Finally, some strategy must be devised for recovery of most of the carbon dioxide evolved from the anode.

8.1 Contamination of the Electrolyte with Ash Entrained with Coal

Weaver (1975) reports that addition of 10 wt-% of fly ash to the melt did not measurably change the polarization curves. It formed a separate phase of fine inclusions. Assuming a mass of salt per unit area (W_{el} ~ 2.3 g/cm^2), the current density of operation ($i = 100$ mA/cm^2) and mass fraction of ash in the carbon feed f and the time t_c to reach this critical concentration ($f_c = 0.10$) is given by

$$t_c = \frac{f_c \, n F \, W_{el}}{i \, M_C \, f} \tag{9}$$

Here M_C is the atomic weight of carbon and F is the Faraday constant. This criterion suggests a useful life of the melt of at least t_c (days) ~ 0.86/f. Thus for the mechanically cleaned coal granules with 0.17 or 1% ash, the critical time T_c is 506 or 586 days, respectively. For the solvent-extracted sources of fuel having < 0.05% ash, the time to exhaust the melt approaches 5 years—the life expected for any high temperature cell. We have not tested 'Weaver's criterion.' Nor does this test predict the rate of fouling of the anode current collector metal surface by ash impurities, as noted by Vutetakis (1985, 1987). The ash will likely be swept from the interface and the cell by the influx of carbon, which moves at a very high relative velocity compared with the accumulation of ash in the melt.

The salt may be separated from ash by aqueous techniques, and pure carbonate may be recovered by dissolution and recrystallization. At 750°C, the sodium-potassium carbonate eutectic (MP 710°C) can be used, which (unlike lithium carbonate) is quite water soluble. Electrical resistivity of the ternary eutectic $(Li_{0.43}Na_{0.32}K_{0.25})_2CO_3$ and of the binary eutectic $(Na_{0.58}K_{0.42})_2CO_3$ are 0.53- and 0.55 Ω-cm at 700 and 730°C, respectively (Weaver, Leach and Nanis, 1981). If lithiated NiO catalysts are used, one must take into consideration the rate of leaching of the lithium and resultant loss of activity, which is expected but not yet measured. Also important in the choice of electrolyte is the solubility of the NiO catalyst; Doyon found that the solubility is depressed by 1% additions of SrO (Doyon, 1987).

8.2 Removal of Ash from Coal

Several new processes are available for the pre-cleaning of coal for reduction or removal of ash, summarized by this author in Table 3 (Cooper, 2004a). The UCC process is illustrative of adaptation of Bayer digestion to remove silica and alumino-silicates from particulate raw coal. The process was developed and evaluated for production of particulate fuel for gas turbines and uses base, acid and hydrothermal treatment of coal fines (Langly, 2004). The product contains ash at 0.17-0.26% and sulfur at 0.4%. Fuel cost is reportedly $3.00-3.30/GJ. The cleaning technique was designed for Australian coals containing considerable amounts of clay, and uses an expensive hydrothermal extraction process following the base digestion. With Eastern US coal, for which silica is a predominate impurity, the hydrothermal step may be eliminated. The coal is a non-agglomerating (non-tarring), and the pyrolysis of the material *in situ* (750°C, 30 min.) produced slurry that flowed freely from an upturned beaker.

A simple cleaning process developed by University of Kentucky (Parekh, 2003) begins with the pulverization of coal in a rotating attritor down to 10 µm dimensions. Surfactants and oils are used to separate the purified coal particles from the heavier solid inclusions of silicates, alumino-silicates and iron pyrite. The process produces an acceptable level of ash at a total energy degradation of 1%-primarily associated with the electric attritor that consumes 0.33 MJ-th/kg-C (65 kWh/ton). The cost (raw coal plus cleaning) is estimated at $60/ton.

Solvent extraction of the soluble fraction of the coal has been pursued in Japan, for the application as a feed for pulverized coal-fired power plants as well as for production of fuel for turbines (Yasumuro, 2004). The process combines solvent extraction and ion exchange to reduce ash below 0.02% and Na and K content to < 0.5 ppm levels (far below our needs) at a cost estimated at $2/GJ (about $60/ton). The yields are 60% of the raw coal, the balance being available for combustion.

Zondlo et al. (2003) used the benign NMP solvent to extract pitch at low temperature (202°C). The solvent loss was 0.7% per cycle. The unconverted 40-50% fraction is reported to be especially suitable for gasification. Sulfur and ash content in the pitch is 1% and 0.04-0.3%, respectively. The cost of pitch was estimated (Mitre Corporation) at $174/ton of calcined carbon extract (5.8 $/GJ)

Berkovich (2003) uses the anthracene oil derived from coal as a solvent for pitch extraction under moderate temperature (425°C) and pressure. Sulfur and ash are very low. The cost of the process as a means of production of electrochemical fuel has not been estimated.

These processes were developed for producing high value products such as carbon fibers, under more rigorous materials constraints than those of a consumable electrochemical fuel. The processes need to be re-evaluated from the perspective of low cost requirements and relaxed constraints needed for electrochemical fuels.

Steinberg (2002, 2003) proposed an electric arc decomposition of coal or methane to form carbon black anode fuel. The carbon is collected by a molten salt stream and pumped through the cell stack using technologies developed for transporting the coolants in molten salt nuclear reactors. While the use of electric arc to produce fuel for a fuel cell may be counter-intuitive, the process could be economically attractive because of the low equipment cost, low heat of decomposition of dry coal and the near-perfect coupling of electrical energy to the decomposition reaction.

Table 3.   Costs of carbon anode fuel extracted from coal

Process/Developer	Yield (%)	Ash (%)	S (%)	Fuel cost
University of Kentucky, Mechanical separation	90	<1	1-2	$60/ton
Ultra clean Coal Energy Pty. Ltd. Leaching	–	0.17-0.27	0.42	$3.0-3.3 /GJ 1.4 ¢/kWh
NEDO[a], Japan	60%	<0.02	–	~ $2/GJ
University of Kentucky, anthracene oil	40-70	0.01-0.06	0.5	–
West Virginia University, solvent	40-50	0.04-0.3	1.0	$140/ton[b], 2.7 ¢/kWh

[a]National Energy Technology Development Organization, Japan.
[b]Calcined extract.

8.3 Sulfur: Form and Transport

Inexpensive coal or petroleum coke fuel will contain sulfur. The hydraulic cleaning processes involving pulverization to 0.1 mm particle size and flotation separation of pyrite (FeS_2) particles from the coal, will remove roughly half of the initial sulfur. Sulfur chemically bound to the coal molecule will be entrained with the char to some extent.

We examined the thermodynamics of the system carbon, carbon dioxide, and sodium carbonate (with assumed Boudouard equilibrium) alone and with additions of sulfur in various oxidation states. Table 4 gives the composition resulting from additions of small (1 mole quantities) of various sulfur compounds to an initial composition of 50 moles carbon, 50 moles sodium carbonate and 10 moles of carbon dioxide. Regardless of the initial oxidation state of sulfur, the product consists of carbonyl sulfide gas (COS) in equilibrium with a condensed phase sulfide (Na_2S or FeS), and in roughly the sample concentration (0.02 atm).

The anodic reaction of one mole of carbon produces three moles of CO_2 at ambient pressure (reaction 3). For equilibrium between COS, condensed phase sulfide, C and CO_2 (Case 1, Table 4), 0.66 moles of

Table 4. Equilibrium composition of an initial composition 50 mol C + 50 mol Na$_2$CO$_3$ +10 mol CO$_2$ with addition of 1 mole of a minor component containing sulfur; at 750°C (ref. FACT code (Bale et al. 1996))

Initial conditions 750°C (1023 K): 50 mol C + 50 mol Na$_2$CO$_3$+ 10 mol CO$_2$ + minor component

Case	Minor component (mol)	Gas products, moles								Solid products, moles			
		CO	CO$_2$	COS	CS$_2$	H$_2$O	H$_2$S	H$_2$	CH$_4$	Na$_2$CO$_3$	C	Na$_2$S	FeS
1	1 mole COS	15.197	4.358	0.021715	2.74E-05					49.02	42.4	0.978	
2	1 mole FeS$_2$	14.56	4.176	0.02081	2.62E-05					49.02	42.2	0.979	1
3	1 mole CH$_3$SCH$_3$	14.76	3.744	0.014596	1.44E-05	0.4629	7.56E-2	2.419	2.08E-02	49.09	44.4	0.91	
4	1 mole S	14.56	4.176	0.02081	2.62E-05					49.02	42.2	0.979	
5	1 mole Na$_2$SO$_4$	15.197	4.358	0.021715	2.74E-05					50.02	40.4	0.978	
6	1 mole Na$_2$SO$_3$	14.56	4.177	0.02081	2.63E-05					50.02	41.2	0.979	

sulfur may be removed as COS per mole of carbon, corresponding to the removal of all sulfur upto 18%-wt. Table 4 is overly simplified in listing the possible reactions that might occur with bound sulfur in a non-equilibrium electrochemical or pyrolysis reaction. Experimentation is needed to corroborate these predictions for a fuel cell.

Carbonyl sulfide is found in the vicinity of the carbon/electrolyte interface in Hall-Heroult cells. It is oxidized on contact with air to form CO_2 and SO_2, or hydrolyzed to form H_2S. The acidic gas SO_2 may also be removed by, for example, precipitation with CaO.

8.4 Pyrolysis

Considerations of process simplicity as well as economy suggest that the waste heat produced by the fuel cell (about 0.25 kW per kW of net electrical DC output) should be fed back to thermally decompose the raw coal feed. This requires that the cell produce sufficient heat at a sufficiently high temperature to effect thermal decomposition within a time span that is short compared with that of electrochemical conversion. Fig. 8 (after Howard, 1981) reproduces data underlying Dryden's correlation (Dryden, 1957) for many British and American coal seams and shows the extent of decomposition (relative to that at prolonged pyrolysis at 1000°C) as a function of temperature and time. The reference to 1000°C is useful, as the yields observed at this temperature approach those of higher temperature asymptote. Also in Fig. 8 is the data from Anthony et al. (1975) taken after various exposure intervals between 0.1 and 14400 s, showing that devolatilization is 90% complete in the range of 5-20 s.

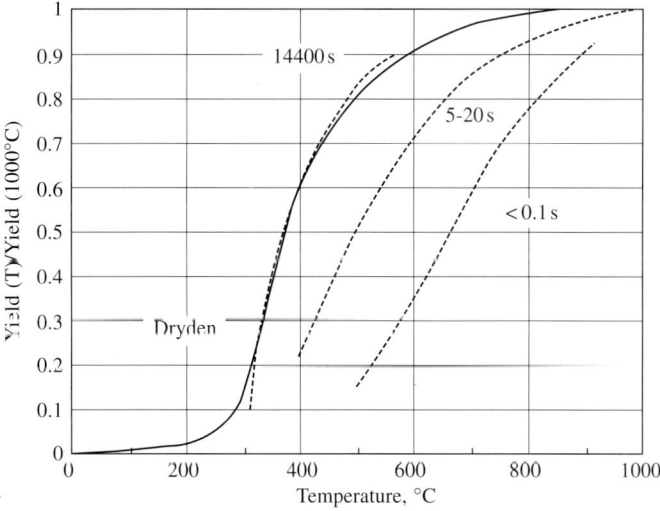

Fig. 8 Extent of thermal reduction to char (steady state 1000°C) as a function of temperature, for various treatment times (s). Solid line: correlation of Dryden (1957), Broken lines: data from Anthony et al. (1975) from a compilation by Howard (1981).

The heat of decomposition of coal is roughly 3-5% of the HHV, excluding volatilization of adsorbed moisture. The direct transfer of heat near the operating temperature of 750°C to the fuel, counter current to feed to the cell, would effect decomposition with a time constant that is small compared with the time required to displace the content of the hopper—roughly 8 days (7×10^5 s). Thus the counter-current flow of heat from the cell to the coal feed has sufficient time, temperature, and heat to result in asymptotic high temperature pyrolysis char.

Subbituminous and bituminous coal have C/T atomic ratios of 0.5-0.9. Thus the pyrolysis offgas, consisting of low molecular weight hydrocarbons, hydrogen, and tars, will have substantial thermal value. It is beyond the scope of the work to suggest how or for what the devolatilization products might best be used. Depending on the market demand for fuels, the products could be burned for process heat, cracked (perhaps on the surface of particulate fuel chars) to produce carbon and a hydrogen-rich product gas, or reformed and shifted and purified to form a stream of hydrogen for energy or chemical applications. At most, only 10% of the carbon is delivered to the offgas; 70% of the sulfur is devolatilized. One can perhaps calculate a 'formal' maximum energy efficiency of the use of coal assuming all the carbon content is used in a fuel cell at 80% efficiency HHV and all the hydrogen content is used in a fuel cell at 55%; for C/H ratio of 0.8, the weighted efficiency is 75% HHV.

8.5 Recovery of Carbon Dioxide

Part of the carbon dioxide evolved from the anode chamber must be recycled through the cathode to complete reaction (3). The requirement of mixing the pure carbon dioxide product with the incoming air stream leads to losses at the exit stream, as not all of the CO_2 will be consumed by the cathode reaction. By reducing the air flow rate through the cathode, or by placing cells in series airflow connection, can minimize the ultimate loss of CO_2 to the atmosphere (or into more expensive recovery equipment). Fig. 9 shows the dependence of cathode potential on the ratio of $CO_2:O_2$ in the airflow as a function of current density (United Technologies, 1983). Low CO_2 system losses are balanced against reduced cell voltage.

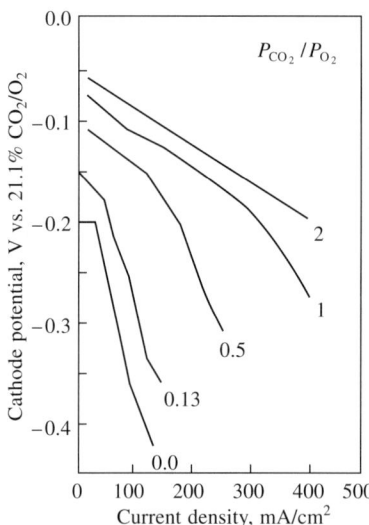

Fig. 9 Polarization of the oxygen cathode at specified partial pressure ratios of $CO_2 : O_2$, for oxygen partial pressure of 0.15 atm. (after United Technologies, 1983).

9. A Possible Configuration for Pyrolysis and Conversion of Coal

Of the many configuration concepts that could support direct conversion of fossil chars, the one depicted schematically in Fig. 10 offers advantages of simplicity and low cost. The wedge-shaped cell configuration has been used for many years in various ambient temperature cells. At this laboratory, such cells have been successfully tested on slabs of aluminum anode in aluminum/air fuel cells (Cooper, 1984) and on pellet-fed zinc/air fuel cells on scales of up to 1.4 kW. In these systems, the anodes are introduced into the cells at the rate of consumption, and conform to the shape of the cell by dissolution adjacent to the separator.

Carbon particles or plates are gravity fed from an overlying hopper into the electrochemical cell, where the anodic reaction near the anode/separator interface removes material from the anode. Whether the anode is introduced as a rigid slab or as a free-flowing paste of particles in electrolyte, the anode is expected to conform to the shape of the cell during prolonged operation. The electrochemical cell is comprised of two materials (dense alumina, and highly graphitized carbon plate) in addition to the porous alumina separator. The graphite bipolar plate, protected against oxidation by metal cladding on the cathode side, connects with the cathode.

Heat is conducted upwards from the cell to the overlying compartment, where thermal decomposition takes place. Because of the short time constants for thermal decomposition relative to electrochemical

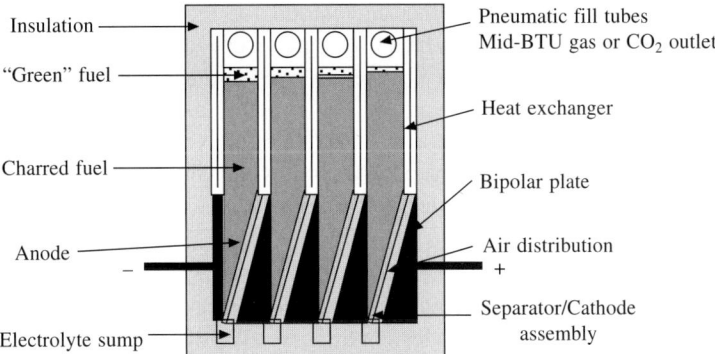

Fig. 10 Schematic shows a possible fuel cell design with overlying hopper charged with green coal particles. The hopper collects heat from the cell for prolonged bake-out, although pyrolysis is substantially complete within 1000 s at 750 °C. The costs of the components have been estimated (from those of similar mass produced articles of dense alumina, porous alumina and graphite) to be about $500/kW at 1 kW/m^2 (Cooper, 2004b).

conversion, the mid BTU gas is withdrawn immediately after charging the cell. During normal operation, carbon dioxide is vented at the top of the cell, without flowing through the carbon particle bed.

The process of pyrolysis in the presence of alkaline molten salts may be altogether different from the pyrolysis that occurs in a coker. In the presence of molten salt, we have pyrolyzed both non-agglomerating and agglomerating coal samples; the process results in a free-flowing melt or paste. The non-agglomeration of particulates of carbon treated with alkali has been observed for various ranks of coal, and results from chemical oxidation of the surface of the coal, as reviewed by Habermehl (1981). The oxidation in the presence of alkali hydroxide induces changes in the surface groups leading to decaking at temperatures above 300°C (Crew et al. 1975) With Pittsburgh Seam coal (a caking coal), treatment with brief exposure to oxygen/nitrogen mixtures at 375°C resulted in a loss of tendency toward caking (Johnson, 1975). Caking or agglomeration is not normally observed in the molten salt gasification of coal in molten carbonate electrolytes. The chemistry involved needs further study if this process is to rely on a free-flow of melt wetted chars.

10. Conclusions: Future of 'Electricity Direct from Coal'

Today, over half of the US electric energy production is based on the combustion of coal at net efficiencies of about 35%. Sixty percent of the world's fossil reserves are coal, and of this amount 80% is located within China, former Soviet Union and North America. Since electric power production from coal is expanding world wide, it is advisable to approach the problem of CO_2 emissions control by pushing the efficiency of power production from coal as close as possible to the theoretical natural limit.

The direct carbon conversion fuel cell (in its many variants) is one method of achieving this goal. Its advantages lie in: (1) power production resulting from a single unit process and (2) minimization of the entropy increase in the intermediate production of fuel from raw resources. With efficient use of the volatile fraction of the coal, the conversion efficiency is over twice that of today's combustion steam generators, allowing (even without CO_2 capture and sequestration) a doubling of today's electricity generation from coal without increasing CO_2 emissions. In a well-designed fuel cell, most (but not all) of the carbon dioxide product gas can be recovered as pure substance without additional separation steps.

Beyond the mechanics of scale-up and refueling, important areas of research concern anode mechanisms at open circuit and under moderate loads (0.05-0.4 A/cm^2). The balance of CO- and CO_2-producing reactions and their mechanisms must be better understood for rational design of large-scale systems. Cathode structures and catalysts used today in DCFC research were adapted from the molten carbonate fuel cell technology perfected for operation at 650°C, where a fixed amount of electrolyte is maintained within a porous lithium aluminate tile. The cathode catalyst and structure must be redeveloped for an excess and variable amount of electrolyte, and for a range of catalysts effective at the higher operating temperature (750-800°C). With chemical pre-cleaning of the coal, the salt will have an extended but finite life. Strategies for eventual removal and recycle of the salt need development.

Acknowledgements

This work was performed under the auspices of the U.S. Department of Energy by University of California Lawrence Livermore National Laboratory under contract No. W-7405-Eng-48. I gratefully acknowledge the support of the National Energy Technology Laboratory for support of this work.

References

Agarwal, Ritesh and Alan A. Kornhauser (2004) "Energy Balance for a Direct Carbon Molten Carbonate Fuel Cell," Proceedings of the 2004 ASME Heat Transfer Fluids Engineering Summer Conference, Charlotte, NC, July 11-15 2004.

Anthony, D. B., Howard, J. B., Hottel, H. B. and Meissner, H. P. (1975) Rapid devolatilization of pulverized coal (15th Symp. (Int.) Combustion, Combustion Institute, Pittsburgh, p. 1303).

Au, S. F., Peelen, W. H. A., Hemmes, K. and Woudstra, N. (1999) "Fuel cells, the next step: energy and exergy analysis of partial oxidation, direct carbon fuel cell and internal direct-oxidation carbon fuel cell," (Proc. 3rd International Fuel Cell Conference, Nagoya Congress Center November 30-Dec 3; co-organized by the New Energy and Industrial Technology Development Organization (NEDO) and Fuel Cell Development Center (FCDIC).

Bale, C. W., Pelton, A. D. and Thompson, W. T. (1966) Facility for the analysis of Chemical Thermodynamics (FACT 2.1), Ecole Polytechnique de Montreal, July 1996.

Berkovich, Adam, J. (2003) Low Severity Extraction of Coal for Production of Carbon Fuel for Direct Carbon Fuel Cells, DOE Direct Carbon Fuel Cell Workshop, NETL, Pittsburgh, PA; July 30; proceedings online, http://www.netl.doe.gov/.

Bregoli, L. J. and Kunz, H. R. (1982) *J. Electrochem. Soc.*, **129**, p. 2711.

Cherepy, N. J., Krueger, R. Fiet, K. J. Jankowski, A. F. and Cooper, J. F. (2005) Direct conversion of carbon fuels in a molten carbonate fuel cell, *J. Electrochem Soc.* **152**(1), A80.

Chuang, Steven S. C. (2004) Carbon Based Fuel Cell, Report to DOE/NETL (Manager Travis Shultz).

Clark, Keith, John Langley, Shigeki Sasaharra, Mitsuru Inada, Toru Yamashita and Yukitoshi Kozai (2003) "Ultra Clean Coal as a Gas Turbine Fuel: A Report on the Collaborative Study to Evaluate the Production and Utilisation of UCC as a Direct Fired Fuel in Gas Turbine Combined Cycle Power Plants".

Cocks, F. H., Klenk, P.A. and Simmons, W. N, (2005) Carbon Ionic Conductors for use in Novel Carbon-Ion Fuel Cells, Presented at the University Coal Research Contractors Review Conference, June 7-8 2005, Pittsburgh, PA.

Coleman, C. H. and White, R. E. (1996) *J. Electrochem. Soc.*, **143**, p. 1781-1783.

Cooper, J. F., Cherepy, N. Berry, G. Pasternak, A. Surles, T. and Meyer Steinberg (2001) Direct Carbon Conversion: Application to the Efficient Conversion of Fossil Fuels to Electricity Proc. Global Warming Conference, PV 20-2000, The Electrochemical Society; see also paper No. 50, Fall Meeting of the Electrochemical Society, Phoenix AZ Oct. 2000.

Cooper, John F. (1984) Aluminum-Air Power Cell Research and Development, UCRL-53536, December 1984.

Cooper, John F. (2004a) Direct Conversion of Coal or Coal-derived Carbon in Fuel Cells, Paper Fuel Cell 2004-2495, Proc. 2nd International Conference on Fuel Cell Science, Engineering and Technology, June 14-16, 2004, ASME, Rochester, New York, USA (Keynote address).

Cooper, John F. (2004b) Electric power generation from coal chars and elemental carbon in fuel cells, extended abstract and visuals for Gordon Fuel Cell Conference presentation (July 2004), UCRL-ABS-205258.

Cooper, John F., Roger Krueger and Nerine Cherepy (2002) Fuel cell apparatus and method thereof, US Patent 2002/010654 9 A1 August 8, 2002. See also Cooper et al., "Tilted Fuel Cell Apparatus", US No. 6, 878, 479 B2, April 12, 2005.

Crewe, George F., Uri Gatr and Vijay K. Dhir (1975) Decaking of bituminous coals by alkaline solutions, *Fuel*, **54**, p. 20.

Doyon, Joel D., Thomas Gilbert, Geoffrey Davies and Lawrence Paetsch (1987) *J. Electrochem. Soc.*, **134**, p. 3035.

Dryden, I. G. C. (1957) Chemistry of coal and its relation to coal carbonization, *J. Inst. Fuel*, **30**, 193.

Frank, W. and Haupin, W. et al. (1985) Aluminum, in Ullmann's Encyclopedia of Industrial Chemistry, 5th Ed., Vol. A1, Aluminum; VCH, Weinheim, FRG.

Fujimoto, H., Tokumitsu, K., Mabuchi, A. and Kasuh, T. (1994) Carbon 32, 1249.

Grjotheim, K., Krohn, C., Malinovsky, M., Matiasovsky, K. and Thonstad, J. (1982) Aluminum Electrolysis: Fundamentals of the Hall-Heroult Process Aluminum-Verlag, Dusseldorf.

Gur, T. M. and Huggins, R. A. (1992) Direct electrochemical conversion of carbon to electrical energy in a high temperature fuel cell, *J. Electrochem. Soc.*, **139**, L-95.

Haber, F. and Bruner, L. (1904) "Das Kohlenelement, eine Knallgaskette," *Z. Elektrochem*, **10**, 697.

Habermehl, Diethard, Fred Orywal, and Hans-Dieter Beyer (1981) Plastic Properties of coal. In: Chemistry of Coal Utilization, John Wiley, NY.

Hauser, Victor Emerald (1964) A study of carbon anode polarization in fused carbonate fuel cells, Ph.D. Thesis, Oregon State University, June 1964.

Hemmes, Kas (2003) Using the Full Exergetic Quality of Solid Fuels, DOE Direct Carbon Fuel Cell Workshop, NETL, Pittsburgh, PA; July 30; Proceedings online, http://www.netl.doe.gov/, 2003.

Hemmes, Kas and Michel Cassir (2004) "A theoretical study of the carbon/carbonate/hydroxide (electro-) chemical system in a direct carbon fuel cell," Paper FUELCELL 2004-2497, Proc. 2nd International Conference on Fuel Cell Science, Engineering and Technology, ASME, Rochester, NY June 2004.

Howard, H. C. (1945) Direct Generation of Electricity from Coal and Gas (Fuel Cells), (Vol. II, Chapter 35 in Chemistry of Coal Utilization, ed. Lowry, John Wiley and Sons, New York.)

Howard, Jack B. (1981) Fundamentals of Coal Pyrolysis and Hydropyrolysis In: Chemistry of Coal Utilization, Martin A. Elliott (ed.), John Wiley, NY.

Ihara, Manabu; Keisuke Matsuda, Hikaru Sato, and Chiaki Yokoyama, (2003) Solid State Fuel Storage and Utilization through Reversible Carbon Deposition of a SOFC Anode, Paper 14th International Conference on Solid State Ionics, Monterey, USA, June 22-27, 2003.

Jacques, W. W. (1896) Method of Converting Potential Energy of Carbon into Electrical Energy, US Patent No. 555,511, March 3, 1896.

Jacques. W. W. (1896b) Harper's Magazine 94, p. 114.

Johnson, G. E. et al. (1975) Prepr. Am. Chem. Soc., Div. Fuel Chem. 20(3).

Kinoshita, K. (1988) Carbon: Electrochemical and Physicochemical Properties, John Wiley, New York.

Langley, John, March (2004) UCC Energy Pty Ltd., L14 213 Miller Street, North Sydney, NSW 20, Australia, Private Communication to J. F. Cooper.

Larminie, James; and Andrew Dicks (2000) Fuel Cell Systems Explained, John Wiley & Sons Ltd, Chichester.

Liebhafsky, H. A. and Cairns, E. J. (1968) Fuel Cells and Fuel Batteries: A Guide to their Research and Development, John Wiley, New York.

Nakagawa, N. and Ishida, M. (1988) Performance of an internal direct-oxidation carbon fuel cell and its evaluation by graphic exergy analysis, *Ind. Eng. Chem. Res.*, **27** (7).

Parekh, B. K. (2003) Beneficiation of Ultra Clean Coal: An Economical Approach for Producing Carbon for the Fuel Cell Application, DOE Direct Carbon Fuel Cell Workshop, NETL, Pittsburgh, PA; July 30; Proceedings online, http://www.netl.doe.gov/

Patton, E. and Zacevic S. (2005) Assessment of Direct Carbon Fuels, EPRI, Palo Alto, CA: 2005. 1011496; prepared by, SARA, Inc., Cypress CA 90630.

Peelen, W.H.A., Olivry M. and S. F. Au, J. D. (2000) Fehribach, and K. Hemmes, Electrochemical oxidation of carbon in a 62/38 mol% Li/K carbonate melt, Preprint, *J. Appl. Electrochem*, **30**.

Pesaventeo, Philip V. (2001) Carbon-Air Fuel Cell, US Patent 6,200,697 B1, March 13.

Steinberg, Meyer (2003) Future Gen—Tomorrow's Clean Energy Interactive and Combined Economic Drivers, DOE Direct Carbon Fuel Cell Workshop, NETL, Pittsburgh, PA; July 30; proceedings online, http://www.netl.doe.gov/.

Steinberg, Meyer, J. F. Cooper and Cherepy, N. (2002) "High Efficiency Direct Carbon and Hydrogen Fuel Cells for Fossil Fuel Power Generation," Proc. American Institute for Chemical Engineers Spring Meeting 2002, New Orleans; March 10-14.

Tamaru, Setsuro; and Minoru Kamada (1935) Brennstoffketten, deren Arbeitstemperataure Unterhalb 600°C Liegt, z. Elekrochem. Bd. 41, No (2).

Thonstad, J. (1970) The electrode reaction on the C, CO_2 electrode in cryolite-alumina Melts—I. Steady state Measurements, II: Impedance Measurements, *Electrochemica Acta*, **15**, p. 1569.

United Technologies (1983) "Development of improved molten carbonate fuel cell technology," Final Report, United Technologies, Inc., for Electric Power Research Institute, Palo Alto, CA; contract #RP1085-4, July 1983; as reproduced in Fuel Cell Handbook, 4th Edition (U.S. Department of energy, Office of Fossil Energy, Federal Energy Technology Center, Morgantown WV; DOE FETC-99/1076).

Vutetakis, D. G., Skidmore, D. R. and Byker, H. J. (1987) *J. Electrochem. Soc.*, **134** (12) 3027.

Vutetakis, D. J. (1985) Electrochemical oxidation of carbonaceous materials dispersed in molten salt, Ph.D. Dissertation, Ohio State University, Columbus OH.

Weaver, R. D. Leach, S. C. and Nanis, L. (1981) Electrolyte Management for the Coal Air Fuel Cell, Proc. 16th Intersoc. En. Conv. Eng. Conf., ASME. NY; Paper No. 891344, p. 717.

Weaver, Robert D., Steven C. Leach, Arthur E. Bayce and Leonard Nanis (1979) Direct Electrochemical Generation of Electricity from Coal, Report May 16, 1977-Feb. 15, 1979; SRI, International, Menlo Park, CA 94025; SAN-0115/105-1.

Weaver, R. D., Laura Tietz and Daniel Cubicciotti (1975) Direct Use of Coal in a Fuel Cell: Feasibility Investigation, EPA-650/2-75-040, June.

Yasumuro, Motoharu (2004) Development of Hyper Coal (ash-free coal) Production Technology, Report, Environment Technology Development Department of NEDO (New Energy and Industrial Technology Development Organization) MUZA KAWASAKI, 20F, 1310 Omiya-cho, Saiwai-ku, Kawasaki-shi, Kanagawa, 212-8554 Japan.

Zecevic, Strahinja, Edward M. Patton and Parviz Parhami (2004) Direct carbon fuel cell with molten hydroxide electrolyte, Proc. 2nd International Conference on Fuel Cell Science, Engineering and Technology, June 14-16, 2004, ASME, Rochester New York, USA.

Zondlo, John; Peter Stansberry, Alfred Stiller and Elliot Kennel (2003) Coal Processing via Solvent Extraction, DOE Direct Carbon Fuel Cell Workshop, NETL, Pittsburgh, PA; July 30; Proceedings online, http://www.netl.doe.gov/.

Recent Trends in Fuel Cell Science and Technology
Edited by S. Basu
Anamaya Publishers, New Delhi, India

11. Solid Oxide Fuel Cells: Principles, Designs and State-of-the-Art in Industries

Roberto Bove

D4-Joint Research Centre, Institute for Energy, P.O. Box 2, 1755 ZG Petten, The Netherlands

1. Solid Oxide Fuel Cell basics

1.1 Origin and Evolution of Solid Oxide Fuel Cells

A solid oxide fuel cell (SOFC) is composed of two porous ceramic electrodes and a solid state electrolyte, made of solid metal oxides. For this reason, the SOFC is also referred to as 'ceramic fuel cell'. The idea of using a stabilized zirconia material as an electrolyte is derived from the experiments conducted by Nernst in 1899. Further studies, including those of Bauer and Preis (1937), showed that the so-called "Nernst Mass" (85% zirconia and 15% yttria), and other zirconia-based materials, at high temperature (600-1000°C), present an ionic conduction that meets the SOFC requirements. The 'typical modern' SOFC is composed of an electrolyte made of yttria-stabilized zirconia (YSZ), a porous anode made of nickel and yttria stabilized zirconia (Ni/YSZ) cermet and a porous cathode composed of doped $LaMnO_3$ (LSM). In order to obtain acceptable ionic conductivity, YSZ needs to operate at a relatively high temperature, typically above 700°C. Historically, the first SOFC operated at 1000°C. At this temperature, the protonic conductivity of the YSZ is negligible, while the ionic conductivity is about 15 S/m (Bossel). Due to the poor mechanical characteristics of the electrodes manufactured until the late 90s, SOFC had to be electrolyte supported. This is the main factor that constrained the operating temperature at typically 1000°C. Because of the minimum required thickness of the supporting structure (electrolyte), in fact, a lower temperature would lead to unacceptable ohmic resistance. Alternative configurations, like for example the previous design of the tubular Westinghouse cell (Bessette and Wepfer 1996) were based on an external (porous) support. However, this solution limits the mass transport through the cell and makes the manufacturing process more complex. By the end of the 90s and the early 2000s, the good mechanical characteristics, mostly due to the development of nano-powders for the manufacturing process, enabled the realization of electrode supported cells. Although some developers, like for example Siemens-Westinghouse, based their design on a cathode supported structure (Minh and Takahashi, 1995) many SOFC developers are now focusing on anode supported cells (Patel et al. 2004, Minh et al. 2004, Patel-Maru et al. 2004, Minh, Amaha et al. 2004). As a consequence of the electrode supported structure, the electrolyte thickness can be reduced to several microns. Due to the reduced thickness and, the related reduction of ohmic losses, the performance obtainable at lower temperature (~800°C and lower) is comparable to that of the electrolyte supported cells operating at 1000°C. These types of SOFCs are also known as intermediate temperature solid oxide fuel cells (ITSOFC). The advantages of operating at reduced temperature can be summarized as follows:

(a) Possibility of using stainless steel as the current collectors, rather than expensive ceramic interconnects
(b) Reduced complexity and cost of the system

(c) Improved chemical stability of individual components
(d) Reduced mis-matching of the thermal expansions of single components, and, consequently, reduced possibility of crack formation
(e) Reduced start-up and shut-down time.

On the other hand, when realizing a hybrid SOFC-Gas Turbine power plant system, the reduced temperature leads to an efficiency reduction of the gas turbine, and consequently of the entire system. This is one of the reasons that drives some companies, like Rolls-Royce, to keep on developing SOFC operating at 1000°C (Agnew and Spangler 2004).

1.2 Basics

Although proton conductive materials are being investigated as possible SOFC electrolytes (Coors et al. 2004, Smirnova et al. 2004, Hassan et al. 2003, Shimada et al. 2004), in the 'typical' SOFC oxygen ions migrate from the cathode to the anode. Oxygen ions are formed at the cathode, according to the following overall reaction:

$$O_2 + 4e^- \rightarrow 2O^{2-} \tag{1}$$

At the operating temperature, the electrolyte presents very high electrical resistivity and ionic conductivity, thus only ions can flow to the anode. If hydrogen is considered as fuel, the following overall anode reaction occurs:

$$H_2 + O^{2-} \rightarrow H_2O + 2e^- \tag{2}$$

If carbon monoxide is present, the shift reaction occurs:

$$CO + H_2O \leftrightarrow H_2 + CO_2 \tag{3}$$

Since CO has no poisoning effect on the anode materials, carbon monoxide represents additional fuel for the fuel cell.

Hydrocarbon can also be internally reformed, according to the following reaction:

$$C_xH_y + xH_2O \rightarrow xCO + \left(x + \frac{y}{2} \right) H_2 \tag{4}$$

Reaction (4) is usually not at chemical equilibrium in the anode, thus the shift reaction (3) takes place, providing additional hydrogen.

The overall fuel cell reaction is

$$H_2 + \frac{1}{2} O_2 \rightarrow H_2O \tag{5}$$

Direct oxidation of carbon monoxide and hydrocarbons in the anode is also possible, however, depending on the catalysts, the shift reaction (3) and the reforming process (4) are much faster.

When no current flows (i.e. no electrical load is connected to the cell), the voltage difference between the two electrodes is provided by Nernst equation

$$E = E^0 + \frac{RT}{4F} \ln P_{O_2} + \frac{RT}{2F} \ln \frac{P_{H_2}}{P_{H_2O}} \tag{6}$$

where E^0 is the reversible voltage at standard pressure, R the universal gas constant, T the temperature, F the Faraday constant, and P_i the partial pressure of the ith species. The value provided by the Nernst equation, however, usually shows a small deviation from the experimental values of the open circuit voltage (OCV). In particular, Costamagna and Honnegger (1998) noticed a variation of the OCV when the fuel flow rate changes, while, according to relation (6), only the gas composition should influence the OCV.

The difference between the experimental OCV and that computed by Nernst equation is mainly due to the electronic conduction of the electrolyte. When the electrolyte is constructed with a mixed oxygen-ion and electronic conductor, the electric current can flow through the electrolyte even at open circuit condition. If the SOFC is schematically represented as the electrical circuit (Fig. 1), it is possible to write

$$V = E - \frac{L}{\sigma_i} J_i \tag{7}$$

$$V = \frac{L}{\sigma_e} J_e \tag{8}$$

where the subscripts i and e refer to the ionic and electronic current, respectively.

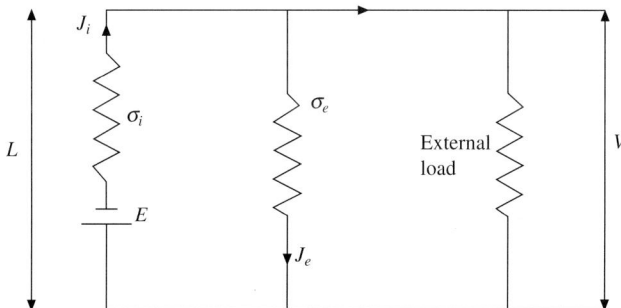

Fig. 1 Equivalent electric circuit of a solid oxide fuel cell.

Combining Eqs. (7) and (8), we obtain

$$E - \frac{L}{\sigma_i} J_i = \frac{L}{\sigma_e} J_e \tag{9}$$

At open circuit condition, no current passes through the external load and $J_e = J_i$, i.e.

$$J_e = \left(\frac{E\sigma_i\sigma_e}{\sigma_e + \sigma_i} \right) \frac{1}{L} \tag{10}$$

Substituting Eqs. (10) into (8), it is possible to obtain

$$V = \mathrm{OCV} = \frac{E\sigma_i}{\sigma_e + \sigma_i} \tag{11}$$

Another phenomenon leading to an OCV reduction is the so called cross-over that causes the passage of the cathodic gas into the anode and vice-versa. The result is precisely the same, i.e. a molar transfer of gas from an electrode to another without producing external electric current. Cross-over on SOFCs is usually due to micro-cracks in the electrolyte.

When an electric load is connected to the fuel cell, the cell voltage is lower than the open circuit voltage, because of three main losses, viz. ohmic, activation and concentration. The voltage can be expressed as

$$V = \mathrm{OCV} - J \cdot r - \eta_{\mathrm{act}} - \eta_{\mathrm{conc}} \tag{12}$$

where V is the cell voltage, OCV the open circuit voltage, J the current density, r the ohmic resistance expressed as [ohm m^2], η_{act} the activation loss and η_{conc} the concentration loss. The activation loss is due to the energy to be overcome for enabling reactions (1) and (2). The potential reduction is proportional to the current passing through the cell, and is dependent on the operating temperature. For SOFC, due to the high temperature, activation loss is relatively low, compared to low temperature fuel cells (Larminie and Dicks 2003). Concentration loss is due to the mass transport within the porous media that reduces the hydrogen concentration at the reaction zone, compared to the concentration at the electrode bulk. For a complete review of the mass transport phenomena in SOFC, the reader is referred to Beale (2004) and Suwanwarangkul et al. (2003).

1.2.1 Internal and External Reforming

If methane is considered as the primary fuel, the steam reforming reaction can be written as

$$CH_4 + H_2O \rightarrow CO + 3H_2 \tag{13}$$

Reactions (13) and (3) can be considered intermediate of the following reaction (14), even though CO is always present in the final gas mixture.

$$CH_4 + 2H_2O \leftrightarrow CO_2 + 4H_2 \tag{14}$$

The enthalpy of reaction (13) under standard conditions is 206 kJ/mol, while for (14) is 165 kJ/mol (Heinzl 2002). The enthalpy of the exothermic reaction (5), that occurs inside the fuel cell is 285.8 kJ/mol. Obviously, this energy is not totally released as heat, but a part is converted into electric energy and the rest into heat. Considering that for every CH_4 mole, a number between 3 and 4 moles of H_2 are produced (depending on the reaction conditions of (13) and (3)), it is reasonable to suppose that the heat produced by the fuel cell is enough for the steam reforming of the methane. However, heat transfer from the fuel cell section to the reformer must be ensured. For high temperature fuel cells, thermal energy can easily be transferred from the outlet anodic and cathodic gases to the reformer section (external reformer with thermal energy recovery).

Another possible solution for heat management is the use of the so-called internal reforming fuel cells. In this case, reactions (3), (13) and (14) occur inside the cell itself, thus solving the heat transfer problem. This is possible only for high temperature fuel cells, namely solid oxide fuel cells and molten carbonate fuel cells. Low temperature fuel cells, in fact, employ precious catalysts for the hydrogen reaction, thus the presence of hydrocarbons and carbon monoxide would poison the catalysts activity. Furthermore, due to the high temperature, nickel, that is the main constituent of the anode, represents an excellent catalyst for both anodic activity and reforming (Eguchi et al. 2002, Weber et al. 2002, Peters et al. 2002), however, Ni is prone to carbon deposition.

When internal reforming is performed, no external devices are needed and the heat transfer can take place with minimum losses. Moreover, reaction (14) removes H_2 and produces H_2O, thus promoting reaction (4).

However, internal reforming creates temperature gradients inside the fuel cell and carbon deposition problems. While carbon deposition can be avoided using an adequate steam to carbon ratio, temperature gradient reduces the efficiency and, in the case of SOFC, can cause cell cracking (Meusiger et al. 1998). In order to avoid these problems, and for reforming complex hydrocarbons, part of the fuel is usually subject to reaction (13) in an external reactor, usually called 'pre-reformer'.

2. Cell Components

2.1 Electrolyte

The electrolyte represents the media through which ions migrate from one electrode to the other, thus causing a voltage difference between anode and cathode, and, consequently, an electric current through an external load. For this reason, the electrolyte must meet the following requirements:

(a) High ionic conductivity
(b) High electronic resistivity (nearly zero electronic conductivity)
(c) Thermal expansion compatible with those of the other cell components
(d) Chemical stability in contact with the two electrodes
(e) Resistance to thermal cycling
(f) Low cost

Currently, most of the SOFC developers use ytrria stabilized zirconia (YSZ), with a variable percentage of yttria, as the electrolyte. Due to the reduced ionic conductivity at temperature below 800°C, alternative materials are also being investigated. One candidate is scandia stabilized zirconia (ScSZ) that presents high ionic conductivity at reduced temperature, as shown in Fig. 2. However, electrolytes manufactured with ScSZ based materials present high costs (Mori et al. 2004) and fast degradation (Müller et al. 2004). Gadolina-doped ceria materials also present high ionic conductivity at reduced temperature, however it is well known from the literature that Ce^{+4} is prone to reduce to Ce^{+3}, in a reducing atmosphere (Godickemeier and Gaucker, 1998). This phenomenon leads to an electronic conduction that allows electrons to pass through the electrolyte, with a relevant reduction of the cell voltage, even at open circuit condition. Other interesting alternative materials for ITSOFC are: $LaGaO_3$ (in particular the lanthanum gallate with the strontium doping on the A site of the perovskite and magnesium on the B-site (LSGM), proposed by Goodenough and Huang (1997), and ytterbia-stabilized zirconia (YbSZ). A comparison of the ionic conductivity of the mentioned materials is illustrated in Fig. 2.

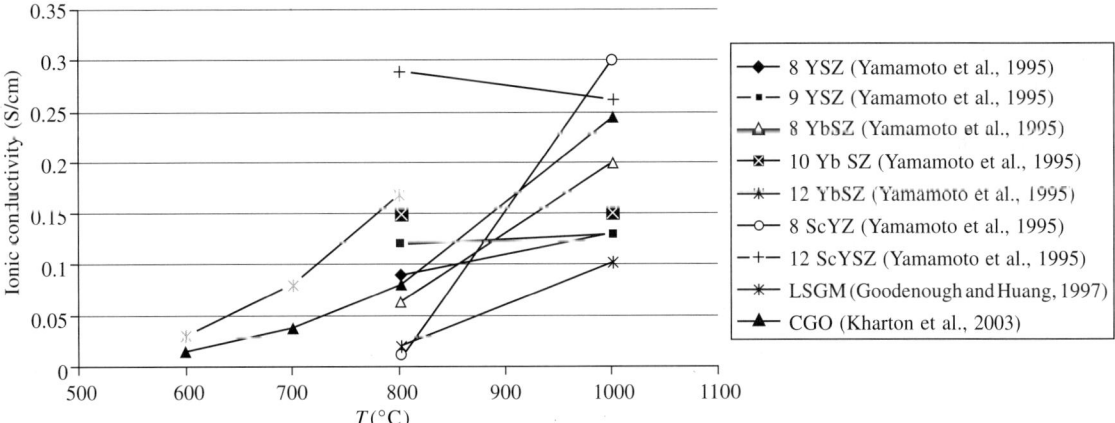

Fig. 2 Ionic conductivity of candidate materials for SOFC electrolyte. Data from Yamamoto et al. (1995) at 800°C are measured after aging at 1000°C for 1000 h.

Alternatively to the typical ionic conductive electrolytes, proton conductive electrolytes are being investigated (Iwahara et al., 1981, Iwahara, 1995, Bonano et al., 1991, Antoine et al., 2003, Coors et al., 2004, Hassan et al., 2003, Shimada et al., 2004, Smirnova et al., 2004).

2.2 Anode

The anode needs to be manufactured, so that reaction (2) takes place in the most efficient and effective way. The main requirements of the anode are:

(a) High electronic and ionic conductivity.
(b) Porous structure optimized for the mass transport of the gas species.
(c) Thermal expansion compatible with those of the other cell components.
(d) Chemical stability in contact with the two electrodes.
(e) Resistance to thermal cycling.
(f) High catalytic activity.

In addition to the above requirements, the choice of low cost materials and manufacturing processes are required for realizing a low cost fuel cell.

The reaction takes place at the so-called triple boundary zone (TPB), where electrons, ions and gas phase coexist.

The typical SOFC anode is composed of a mixture of Ni and small percentage of YSZ, that is added for reducing the anode sinterability and making the thermal expansion close to that of the electrolyte. Since the nickel structure is obtained from NiO powders, before operating, the nickel oxide needs to be reduced to Ni. This is usually achieved, running the cell with hydrogen at open circuit voltage, for a specific amount of time. Ni/YSZ provides the anode with high electrical conductivity, an adequate ionic conductivity, and a high activity for the electrochemical reactions and reforming process. Due to the low sulfur tolerance of Ni/YSZ anodes (Matsuzaki and Yasuda 2000), alternative materials are being considered. Sulfur, in fact, is present in most of the commercial fossil fuels, including natural gas and coal. Other drawbacks of the Ni/YSZ anodes are the tendency of Ni to oxidize to NiO if in contact with oxygen (Tikekar et al. 2003), and the reaction of NiO with LSGM during cell fabrication (Goodenough 2004). Examples of alternative materials are $Y_{0.2}Ti_{0.18}Zr_{0.62}O_{1.9\pm\delta}$ (YZT) (Pudmich et al. 2000), $La_{0.8}Sr_{0.2}Cr_{0.97}V_{0.03}O_3$, $La_xSr_{1-x}VO_{3-\delta}$ (Aguilar et al. 2004), ceria based oxide materials (Holtappels et al. 2001, Uchida et al. 2003, Rösch et al. 2003), and chromite/titanate based perovskites (Pudmich et al. 2000).

2.3 Cathode

The requirements of the cathode are analogous to those of the anode. The traditional material is the perovskite system $La_{1-x}Sr_xMnO_3$ (LSM). This material, however, can cause some problems when the temperature is lower than 1000°C. As mentioned above, in fact, one of the main advantages of operating at reduced temperature, is the possibility of using stainless steel as current collector, instead of expensive ceramic and Cr alloys. Stainless steel, however, usually contains Cr, that tends to evaporate and react with LSM, thus reducing cathode activity. Alternative materials are, e.g., $La_{1-x}Sr_xCoO_{3-\delta}$(LSCO), $SrCo_{1-x}Fe_xO_{3-\delta}$ (SCF), $La_{1-x}Sr_xFe_yNi_{1-y}O_{3-\delta}$(LSFN), $La_{1-x}Sr_xCo_yNi_{1-y}O_{3-\delta}$(LSCN), $La_{1-x}Sr_xCo_yFeO_{3-\delta}$(LSCF). The cathode overvoltages produced by cathodes produced using some of the previous materials are compared by Goodenough (2004).

2.4 Interconnects

When the operating temperature is about 1000°C, interconnects need to be made of ceramic materials. Although excellent performances of fuel cells based on ceramic interconnects have been proved, costs are still a relevant restriction. At reduced temperature, the use of less expensive materials, like for instance stainless steel, is possible. Several organizations are currently investigating the interaction of different metals with the electrodes (De Jonghe et al. 2004, Zahid et al. 2004, Pedersen et al. 2004). The two main issues to overcome are the oxidation of the metal when in contact with air at the cathode side and the interaction of the chromium vapors with the cathode.

3. Benefits and Open Issues

There are several benefits related to the use of SOFC, compared to traditional energy conversion systems. Compared to internal combustion engines (gas turbine, diesel engine and Otto engine) and external combustion engines (steam turbine, Stirling engine), SOFCs, as well as other fuel cell technologies, produce electricity through an electrochemical reaction, without any combustion. The main consequence is that NO_x, SO_x, and other combustion related pollutants are not produced. Furthermore, while the anodes of low temperature fuel cells, like proton exchange membrane fuel cells (PEM), alkaline fuel cells (AFC) and phosphoric acid fuel cells (PAFC), need precious metals like platinum as the catalyst, in SOFC the catalytic activity is conducted through the use of nickel. As a consequence, carbon monoxide does not have any poisoning effect on the cell. Contrarily, carbon monoxide can be oxidized at the anode or shifted to hydrogen, according to reaction (3). For this reason, SOFC presents a high flexibility to a vast variety of fuels, including, landfill gas (Staniforth and Kendall 2000, Pusz et al. 2005), biomass derived gas (Baron et al. 2004, Van herle et al. 2004) coal gas (Wotzak et al. 2003, Ziock et al. 2002).

For a distributed power generation scenario, where electricity is not produced in large size power plants and then distributed to the users (as happens today), but is locally produced, where and when needed, SOFC presents several attractive characteristics. First of all, due to the absence of combustion, the noise emissions are related only to the auxiliary components (mainly blowers or compressors). Secondly, the systems size does not influence the performance significantly, i.e., contrarily to internal combustion engines, the efficiency of a few kilowatts system is as high as that of a large size system. Moreover, the high efficiency (~40-50%) and high power density (~300 mW/cm^2 (Minh et al., 2004)) make the system very compact and characterized by a low fuel consumption.

Although SOFCs present very high potentials, on the other hand there are still some open issues that need to be solved, before their market introduction can take place. Due to the high temperature and the all-solid state, even small differences between the coefficients of thermal expansion of the components can cause a relevant difference of the expansions and, consequently, high internal stress. As a result, delamination and cracking can easily occur, especially under thermal cycling. In particular, the sealing layers need to be compatible with all the cell components, thus most of the cell failures are imputable to a sealing problem. Besides technical issues, costs still remain a limitation for the market introduction.

4. Overcoming Technology Obstacles

There are two main approaches for solving the main issues that affect SOFC development. One consists in re-designing the single cells, to avoid or reduce phenomena that cause cell failure. Another is to employ new materials and manufacturing processes to provide the cell with the needed characteristics. One example of the first approach is the tubular design that has been developed specifically for SOFC for avoiding the sealing and cracking problems, as explained in the following section.

4.1 Tubular Configuration

The most advanced tubular SOFC design is the sealless one, realized by Siemens Westinghouse (SW). The support of the cell is guaranteed by a supporting porous tube (cathode). As illustrated in Figure 3, the electrolyte and the anode cover the cathode, with

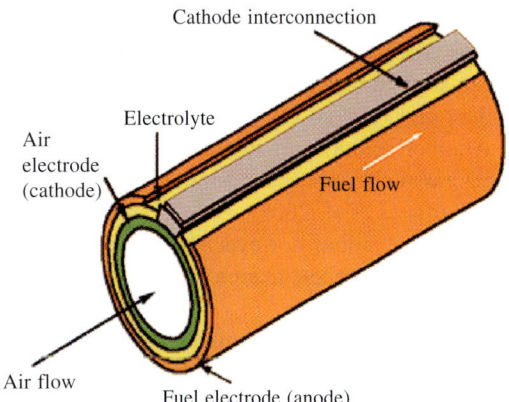

Fig. 3 Siemens-Westinghouse single cell design.

the exception of a small strip, where a current collector is placed (Minh and Takahashi 1995). The cell is closed at one end, thus air is provided through an injector tube.

In this configuration, there is no need for sealing; moreover, the tubular configuration allows for a better distribution of the mechanical stress. For these reasons, only SOFCs based on this design have been scaled-up to 220 kW. Systems based on the planar design, instead, are still in the range of 1-50 kW. The SW single cells are stacked together, according to the configuration of Fig. 4, i.e. a nickel felt connects the cathode of one cell with the anode of the next one.

However, as shown in Fig. 5, this configuration leads to a long current path around the tube, thus generating high ohmic resistance.

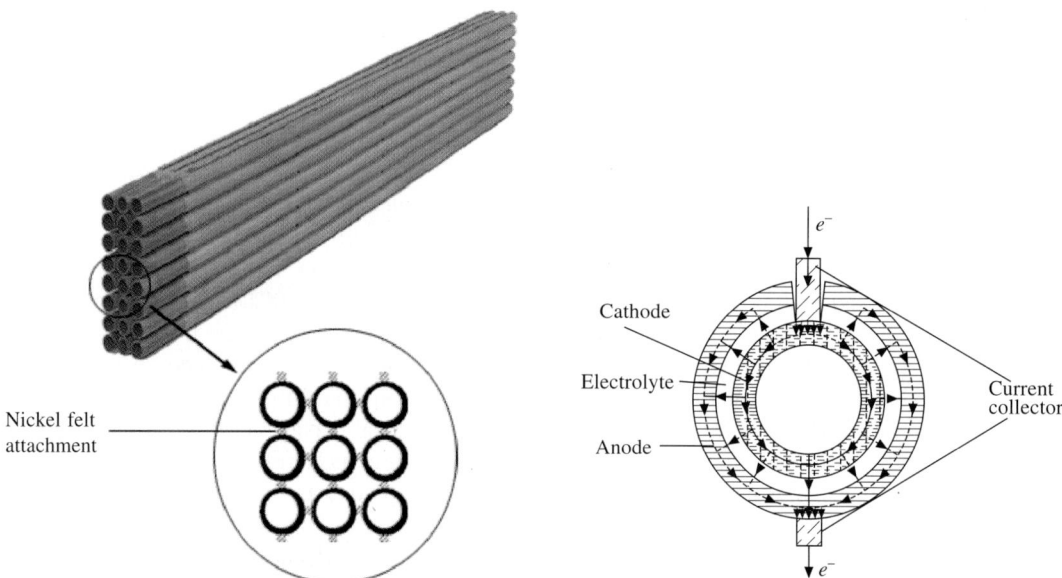

Fig. 4 Connection concept of Siemens-Westinghouse tubular cells.

Fig. 5 Current path in a Siemens-Westinghouse tubular SOFC.

In order to overcome this limitation, but, at the same time, to maintain the advantages of the sealess design, SW has developed another design, named High Power Density (HPD), where the current path is drastically reduced (Fig. 6). Currently, SW is considering to develop, a system based on the HPD design in the range of 5-10 kW (Vora 2004).

Another solution for tubular SOFC is to place current collectors at the two ends of the tube, one in contact with the anode, and another with the cathode. Fig. 7 represents this case for an anode supported fuel cell (Sammes et al. 2005).

In this situation, however, the current path is all along the tube surface, thus producing high ohmic resistance. This phenomenon can be reduced by wrapping a wire around and inside the tube (Bove and Sammes 2005).

Another approach for reducing the ohmic resistance is to fabricate small length tubes connected in gas flow and electric series configuration. This configuration is usually referred to as "segmented" cells, and an example is given in Fig. 8. Better performance, compared to that of traditional tubular SOFC, is due to the fact that the first fuel cell units work with a reduced fuel utilization, thus producing high voltage.

Fig. 6 Siemens-Westinghouse tubular and HPD configurations (Vora, 2004).

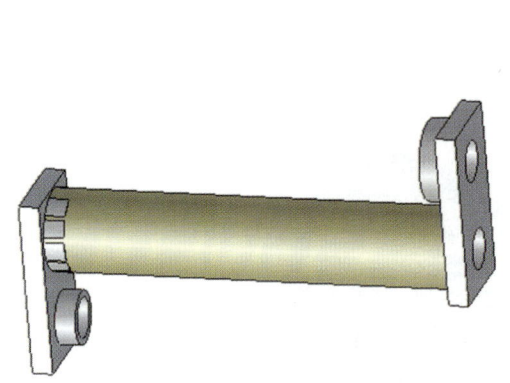

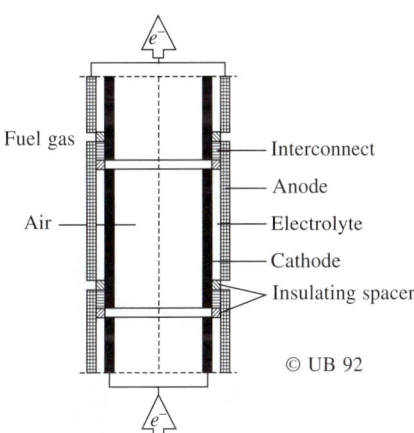

Fig. 7 Tubular SOFC with terminal current collectors (Sammes et al., 2005).

Fig. 8 Example of segmented SOFC (Bossel).

4.2 Other Designs

Other designs are based on planar configuration. Fig. 9 represents a cross section of the flat planar configuration, i.e. the configuration that is typically employed for all the fuel cell technologies.

The main advantage of this design is the low ohmic resistance, compared to the tubular SOFC. The current path from one electrode to the other, in fact, is fairly straight, thus minimizing the so-called 'in-plane' resistance. On the other hand, as stated before, sealing and cracking can represent a serious limitation, especially for large size cells. Another disadvantage is the poor contact between different layers, due to the all-solid state of the components.

An evolution of the flat planar design is the monolithic SOFC (Fig. 10). This consists of corrugated thin cell components. The result is that the volumetric power density is very high. On the other hand, the structure is more complex to realize, as well as the manifold system compared to the flat planar one. Based on this configuration, the Japanese Chubu Electric Power Company, INC. (CEPCO) and Mitsubishi Heavy Industries developed the so-called MOLB (Mono-block Layer Built) cells (Nakanishi et al. 2004).

Another variation of the flat planar is the disk shape. In this case, the layers are still flat planar, however, the cell presents a circular shape, rather than rectangular. The most evolved disk shaped SOFC is the

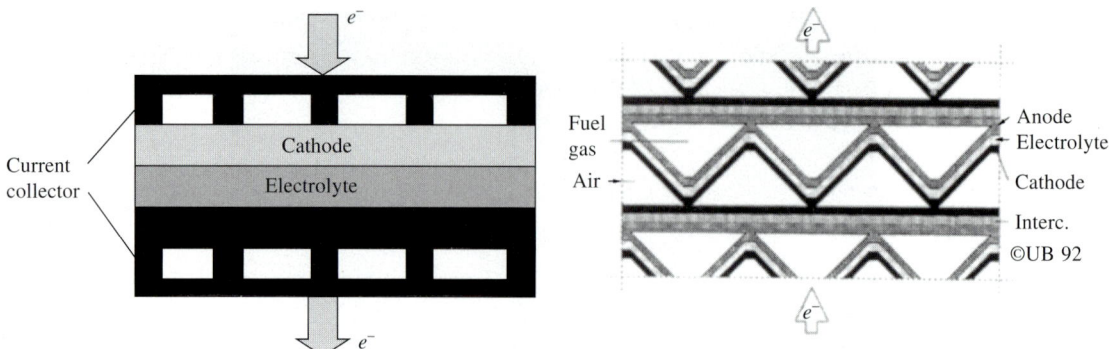

Fig. 9 Schematic cross-section of a flat planar SOFC.

Fig. 10 Schematic cross section of a monolithic SOFC (Bossel 1992).

Sulzer-Hexis, represented in Fig. 11. In this design, air and fuel move in a radial direction, in a co-flow configuration, i.e. proceeding from the center to the external side. An integrated heat exchanger allows the gas to be internally pre-heated.

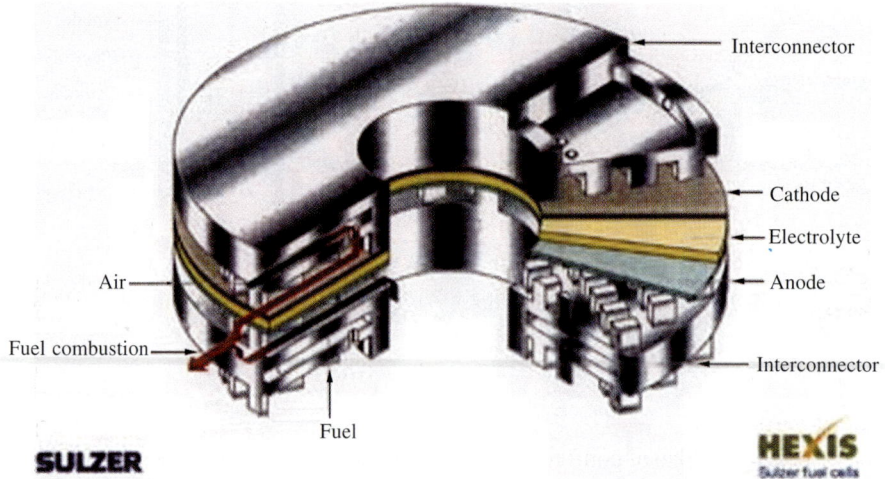

Fig. 11 Sulzer-Hexis design.

Table 1 summarizes the main advantages and disadvantages associated with the planar and tubular configurations.

Table 1. Main characteristics of tubular and planar SOFC

	Tubular	Planar
Power density	Low	High
Volumetric power density	Low	High
High temperature sealing	Not necessary	Required
Start-up cool down	Faster	Slower
Interconnect	Difficult	High cost
Manufacturing cost	High	Low

5. State-of-the-Art and Future Directions

Table 2 summarizes the major SOFC developers, the relative country, technology, as well as the results achieved. As for the other fuel cell technologies, most of the developers are located in US, Europe and Japan.

Table 2. Main manufacturers and technology status (US DOE 2004)

Manufacturer	Country	Achieved	Year	Technology
Acumentrics Corp.	USA	2 kW	2002	Microtubular
Adelan	UK	200 W	1997	Microtubular
Ceramic Fuel Cells Ltd.	Australia	5 kW	1998	Planar
		25 kW	2000	
Delphi/Battelle	USA	5 kW	2001	Planar
Fuel Cell Technologies	Canada/USA	5 kW		
		2 kW	2002	Flat Tubular/Tubular
General Electric (formerly Honeywell and Allied Signal)	USA	0.7 kW	1999	
		1 kW	2001	Planar
Fuel Cell Energy (formerly Global Thermoelectric)	USA	1 kW	2000	Planar
MHI/Chubu Electric	Japan	4 kW	1997	
		15 kW	2001	Planar
Rolls-Royce	UK	1 kW	2000	Planar
Siemens-Westinghouse	USA	25 kW	1995	
		110 kW	1998	Tubular
		220 kW	2000	
SOFCo (McDermott Technologies and Cummins Power Generation)	USA	0.7 kW	2000	Planar
Sulzer Hexis	Switzerland	1 kW	2002	Planar
Tokyo Gas	Japan	1.7 kW	1998	Planar
TOTO/Kyushu Electric Power Nippon Steel	Japan	2.5 kW	2000	Tubular

An overview of SOFC patents (Kozhukharov et al. 2001) shows that most of the recent inventions are focused on increasing power density and efficiency, enhancing the life-time, finding simple and low cost Balance of Plant (BoP) solutions.

5.1 United States: The SECA Program

There are several companies, national laboratories and universities that are currently developing SOFC technology in United States of America. In 1999, the most relevant formed an alliance, called the Solid State Energy Conversion Alliance (SECA). The main goal of SECA is to develop an SOFC module in the 3-10 kW range. Targets are: achieving a stack cost of 800 \$/kW by 2005 and 400 \$/kW by 2010. The stack life-time target is 40000 hours. These targets will allow to develop, by 2015, very high efficiency SOFC/ gas turbine systems (SOFC-GT). SECA is also a supporting activity of the so called FutureGen project, a 1 billion dollar project, that has the main goal of developing an integrated coal gasifier-SOFC system, with CO_2 sequestration. ***Siemens-Westinghouse***, whose technology is described in Section 4.1, is part of the SECA alliance.

Fuel Cell Energy, Inc. (FCE) started focusing on SOFC development, when, in 2003, it acquired the

Canadian company Global Thermoelectric, which started developing SOFC in 1997 (Borlum, 2003). The main goal of FCE is to develop an SOFC system, characterized by an electric power in the range of 1-10 kW, based on a flat planar design. Fig. 12 depicts a picture of the FCE stack. Each cell is composed of a Ni/YSZ anode, whose thickness is about 1 mm, and acts as the cell support. A thin electrolyte, made of YSZ, and a cathode of a conducting ceramic material complete the cell. The operating range is 700-750°C, which enables the use of inexpensive stainless steel interconnects.

Fig. 13 shows a 2 kW natural gas system, realized in 2001 at Global Thermoelectric.

Fig. 12 Fuel cell energy SOFC stack.

Fig. 13 The 2 kW natural gas SOFC system.

The Acumentrics technology is based on micro-tubular fuel cells. The small tube design, together with all the features of the tubular cells, permit a rapid fuel cell start, particularly useful in back-up power applications. Acumentrics has recently demonstrated stable cell performance for over 6000 h, with a degradation rate of 0.25% after 500 h. This achievement is close to the Phase III SECA goal of 0.1%/500 h (Bessette 2004).

Delphi Corporation, in partnership with Battelle, is developing a 5 kW SOFC system, capable of running on a variety of fuels, including gasoline, diesel, natural gas and coal derived gas. The system is designed to satisfy several applications, including small residential, auxiliary power unit (APU) for cars and heavy-duty trucks, as well as military applications. The fuel cell design is planar, the operating temperature is 750°C, and a power density of 330 mW/cm^2, when running on hydrogen, has been demonstrated (Shaffer 2004). Fig. 14 represents the compact APU realized in 2002; the volume is 44 litres and the weight 70 kg (Zizelman 2003).

General Electric (GE) after acquiring Honeywell SOFC technology, started the development of an SOFC system, based on an anode supported flat planar design. The system is targeted for residential applications and presents a net power of 5 kW; 3000 h has been demonstrated. The operating temperature is about 800°C. The fuel processor is a catalytic reactor that has the function of pre-reformer. Depending on the type of fuel and operating conditions, the reaction can be a Partial Oxidation (POX) or an Autothermal Reforming (ATR) (Minh 2004). GE is the only industrial member of the SECA alliance that demonstrated a fuel utilization as high as 95%, at 800°C. Together with the stack and balance of plant (BoP) development for residential applications, GE is also performing studies on hybrid SOFC-GT combined systems for power plant realizations.

Cummins Power Generation and SOFCo-EFS Holdings LL (formerly McDermott Technology), have

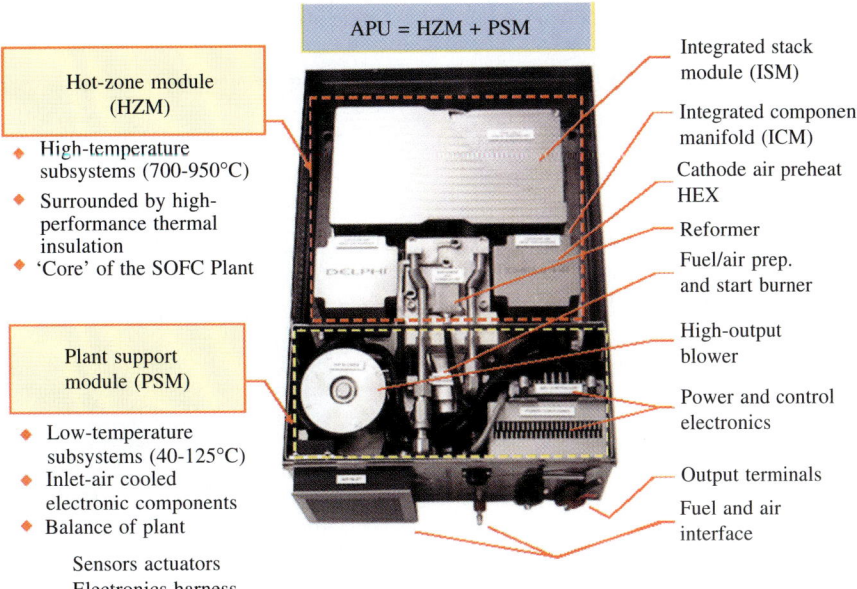

Fig. 14 APU Delphi system (Zizelman).

teamed together to develop a 10 kW power system. While Cummins has a strong background in designing, manufacturing and servicing of power generation solutions, SOFCo previously developed SOFC and fuel processing technologies. The stack currently under development is based on planar configuration, and is "all ceramic". Two stack prototypes realized are: (i) composed of 55 single cells 10×10 cm (C1 Prototype), and (ii) named C2 Prototype, composed of 55 cells, 15×15 cm. Fig. 15 depicts the systems developed at Cummins and SOFCo.

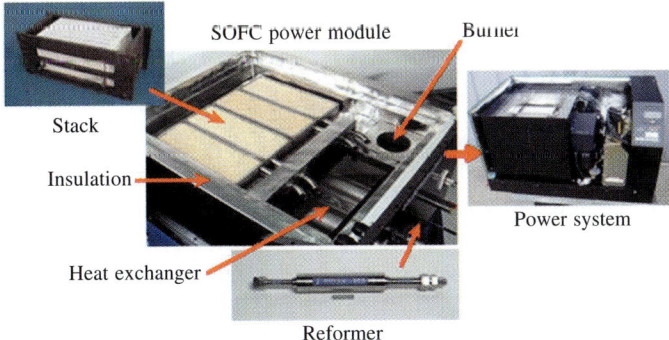

Fig. 15 Power unit developed at Cummins and SOFCo.

5.2 Europe

Sulzer Hexis was founded in 1997 as a division of the Sulzer Group. Although the theoretical Hexis "idea" was born in 1989, the first stack run during 1997-1998 for more than 12000 hours on hydrogen. The Hexis stack has a unique configuration, characterized by a disk shape and current collectors, that also act as heat exchangers and fuel distribution channels. A schematic representation is shown in Fig. 11, while a brief description is given in Section 4.2.

The Sulzer-Hexis system is the HXS 1000 PREMIER, illustrated in Fig. 16. The system has a nominal power of 1 kW$_e$ and 3 kW$_t$, and, consequently, the electric efficiency is about 33%. Although natural gas is the primary fuel, Sulzer demonstrated the use of bio-fuel, installing an HXS Premier on a farm in Switzerland (Van herle et al. 2004). The biogas production rate is 70 m^3 per day, corresponding to 0.55 TJ per year (Van herle et al. 2004). Other HXS 1000 PREMIER systems have been installed at several employees' homes, thus getting a relevant amount of data of 'real applications', before launching its product to the market.

In parallel with HEXIS 1000 Premier, Sulzer Hexis developed the more compact 1 kW system, called *Galileo*.

Fig. 16 HEXIS 1000 PREMIER 1 kW system.

Rolls Royce

Although some theoretical studies on PEM and SOFC were occasionally undertaken at Rolls-Royce between 1987 and 1992, a fuel cell program started in 1992 (Gardner et al. 2000). Within this program, Rolls-Royce has developed the Integrated Planar Solid Oxide Fuel Cells (IP-SOFC) design (Fig. 17). In analogy with the principles that lead to the development of the tubular segmented cells (Fig. 8), a single cell is composed of small cells, displayed in a series configuration. The development stage, that took place in 2003, has the main aim of producing small stacks in the range of 2-10 kW. Using these blocks, Rolls-Royce plans to realize a 10 kW system and later on 60 kW. The final step is to realize a 1 MW hybrid SOFC-GT system, that combines a 800 kW SOFC system and 200 kW Rolls-Royce Gas Turbine.

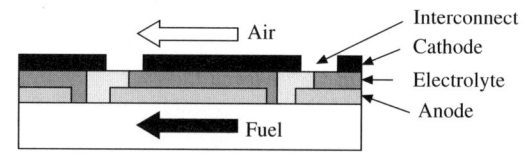

Fig. 17 Schematic representation of the integrated planar solid oxide fuel cells (IP-SOFC) design.

Energy Research Centre of The Netherlands (ECN)

ECN principally works on SOFC fuel cells development both on materials, cells, stack and systems design. During the past years, several SOFC solutions have been developed, including electrolyte supported cells (specifically developed for Sulzer-Hexis); anode supported cells, operating at intermediate temperature (680-800°C). An interconnect supported fuel cell with ferritic stainless steel is currently under development.

At the moment ECN is also working on stack development with two different designs, but the project is still in the early stages. In 1999 ECN created a spin off company called InDEC whose sole mission was commercialising their own technologies. In 2003, the German company H.C. Starck bought the majority of InDEC stocks from ECN.

Forschungszentrum Jülich GmbH (FZJ) is one of the largest research center in Europe, focusing, among others energy conversion technologies, on SOFC. Their main objectives are to develop reliable, low cost, durable and pollutant tolerant materials; to produce high power and mechanically strong cells; to develop highly efficient, low weight, low volume and thermal cycle resistant stacks; to manufacture whole system in a cost effective way. They have also developed numerical models for understanding SOFC thermodynamics and behavior. Their present research efforts are mainly on reducing steel corrosion, developing protective layers, enhancing resistance to thermal cycling and testing fuel cells under long term operation.

One of the main goal of the SOFC group at FZJ is to demonstrate a 20KW SOFC for industrial CHP, whose cost is lower than 1000 Euro/kW, and durability is >40000 h.

5.3 Japan

Japan, together with USA and Europe is a lead developer of SOFC technology. All the basic SOFC designs (planar, tubular and monolithic) are being developed. For example, *Mitsubishi Heavy Industy (MHI)* has realized a stack made of 414 tubular cells, each 1.5 cm in diameter and 70 cm active length, achieved a maximum output of 21 kW has been realized (Iritani and Kougami 2001). *Toto and Nippon Steel* (Nakayama and Suzuki 2001) have developed a 36-cell stack, made of single cells 2.2 cm in diameter and 90 cm in length. The system runs on natural gas, operates at 1 atm, 70% fuel utilization, and the electric power is 3 kW.

5.4 Australia

Ceramic Fuel Cell Ltd. (CFCL) has been developing SOFC technology in Australia since 1992. The CFCL technology is based on all ceramic stacks. A scheme of a so-called all ceramic "layer set" is shown in Fig. 18.

Stacks are designed and manufactured in a highly modular approach.
The CFCL stack, in fact, consists essentially of four main components:

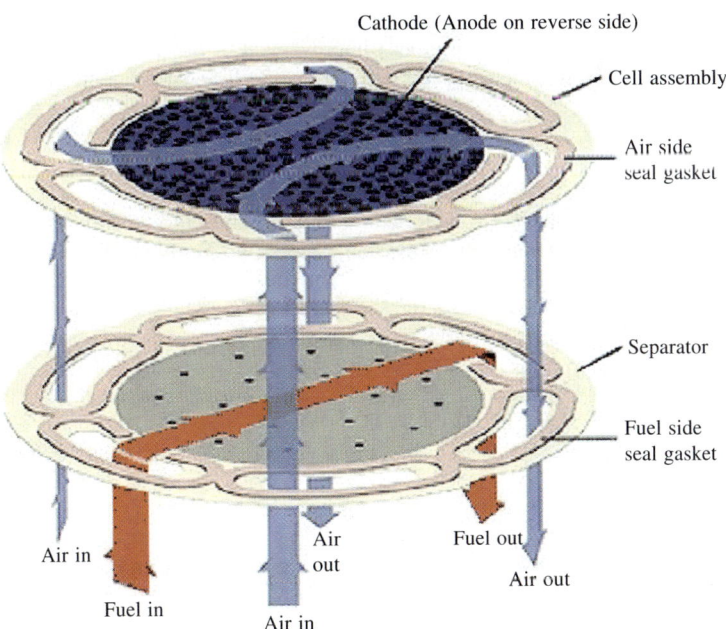

Fig. 18 Ceramic fuel cell single cell.

(a) Solid oxide cell plate with anode and cathode coatings.
(b) Solid oxide separator plate with conductive coatings (interconnect).
(c) Glass-ceramic air seal.
(d) Glass-ceramic fuel seal.

These four components form a layer set. The layer sets, assembled together, form a stack. First tests conducted on a stack made of a 4 × 2 array, made up of 50 layers, electrolyte supported, with metallic interconnects, showed an electric power of 5.5 kW (Godfrey et al. 2000). After that, a 10 layers 2 × 2 array stack was tested for 3500 h at 820°C, generating about 1.3 MWh, with a degradation rate of 1-2 % per 1000 h (Godfrey et al. 2000). Lately a 2 kW, 22 layer 2 × 2 anode supported stack operating at 760°C has been tested (Godfrey et al. 2000).

Tests conducted on a 25 kW system ended in June 2000, providing feedback to realize a 40 kW system designed to run on natural gas (Foger and Godfrey 2002). This stack is composed of a 10 YSZ electrolyte, that forms the cell support, a La-Sr Manganite cathode, and a Ni-10YSZ anode. The operating temperature is 850°C. The power section of the system is composed of 6 modules, each composed of 4 stacks with a nominal power of 2.1 kW. The total DC power is about 50 kW, while the AC power of the whole system is 40 kW. Important achievements of the 40 kW stack, compared to the 25 kW previously realized, are a substantial cost reduction, high reliability and low degradation (Foger and Godfrey 2002).

References

Agnew, G. and Spangler, A. (2004) Reducing fuel cell system cost without lowering operating temperature. Proceedings of the 2nd International Conference of Fuel Cell Engineering and Technology, 14-16 June 2004, Rochester, NY, keynote CD.

Aguilar, L., Zha, S., Cheng, Z., Winnick, J. and Meilin, L. (2004) A solid oxide fuel cell operating on hydrogen sulfide (H_2S) and sulfur-containing fuels. *J. Power Sources*, **135**: 17-24.

Amaha, A., Baba, Y., Yakabe, H. and Sakurai, T. (2004) Improvement of the performance for the anode supported SOFCs. Proceedings of the 2nd International Conference on Fuel Cell Science Engineering and Technology, June 14-16, Rochester, NY, pp. 45-48.

Antoine, O., Hatchwell, C., Mather, G. and McEvoy, J. (2003) Structure and conductivity of a Yb-doped $SrCeO_3$-$BaZrO_3$ solid solution. Proceedings of the eight International Symposium on Solid Oxide Fuel Cells (SOFC VIII), The Electrochemical Society, PV 2003-07, pp.379-387.

Baron, S., Brandon, N., Atkinson, A., Steele, B. and Rudkin, R. (2004) The impact of wood-derived gasification gases on Ni-CGO anodes in intermediate temperature solid oxide fuel cells. *Journal of Power Sources*, **125**: 58-66.

Baur, E. and Preis, H. (1937) Uber Brennsto.-Ketten mit Festleitern. *Z. Elektrochem.*, **43**(9): 727-732.

Beale, S.B. (2004) "Calculation procedure for mass transfer in fuel cells". *Journal of Power Sources*, **128**: 185-192.

Bessette, N. (2004) Development of a low cost 10 kW tubular SOFC power system. SECA Annual Workshop and Core Technology Program Peer Review, May 11, 2004, Boston, MA.

Bessette, N.M. and Wepfer, W. J. (1996) Prediction of on-design and off-design performance for a solid oxide fuel cell power module. *Energy Convers. Mgmt*, **37** (3): 281-293.

Bonano, N., Ellis, B. and Mahmood, M N. (1991) Construction and operation of fuel cells based on the solid electrolyte $BaCeO_3$:Gd. *Solid State Ionics*, **44**: 305-311.

Bossel, U.G. (1992) Facts and Figures. Report to the International Energy Agency, Swiss Federal Office of Energy, Bern/Switzerland.

Bove, R. and Sammes, N.M. (2005) The effect of current collectors configuration on the performance of a tubular SOFC, proceedings of the Ninth International Symposium on Solid Oxide Fuel Cells (SOFC IX), May 15-20, Quebec City, Canada.

Coors, W.G., Sidwell, R. and Anderson, F. (2004) Characterization of electrical efficiency in protonic ceramic fuel cells. Proceedings of the sixth European Solid Oxide Fuel Cell Forum, 28 June to 2 July 2004, Lucerne, Switzerland, Vol. 1, pp. 117-124.

Costamagna, P. and Honegger, K. (1998) Modeling of solid oxide heat exchanger integrated stacks and simulation at high fuel utilization. *Journal of Electrochemical Society*; **145** (11): 3995-4007.

De Jonghe, L.C., Jacobson, C.P. and Visco, S.C. (2004) Alloy supported thin-film SOFCs. Proceedings of the sixth European Solid Oxide Fuel Cell Forum, 28 June-2 July 2004, Lucerne, Switzerland, Vol. 1, pp. 91-96.

Eguchi, K., Kojo, H., Takeguchi, T., Kikuchi, R. and Sasaki, K. (2002) Fuel flexibility in power generation by solid oxide fuel cells. *Solid States Ionics*, **152-153**: 411-416.

Foger, K. and Godfrey (2002) SOFC product development at Ceramic Fuel Cells Ltd., Proceedings of the 5th European Solid Oxide Fuel Cell Forum, Lucerne, Switzerland, ed. J Huijsmans.

Gardner, F.J., Day, M.J., Brandon, N.P., Pashley, M.N. and Cassidy, M. (2000) SOFC technology development at Rolls-Royce. *Journal of Power Sources*, **86**: 122-129.

Godfrey, B., Foger, K., Gillespie, R., Bolden, R. and Badwal, S.P.S. (2000) Planar solid oxide fuel cells: the Australian experience and outlook. *Journal of Power Sources*, **86**: pp. 68-73.

Godickemeier, M. and Gaucker, L.J. (1998) Engineering of solid oxide fuel cells with ceria-based electrolytes. *J. Electrochem. Soc.*, **145**(2): 414-420.

Goodenough, J.B. (2004) Oxide components for solid oxide fuel cell. In: Mixed Ionic Electronic Conducting Perovskites for Advanced Energy Systems, edited by Nina Orlovskaya and Nigel Browning, NATO Science Series, Kluwer Academic Publishers.

Goodenough, J.B. and Huang, K. (1997). Lanthanum gallate as a new SOFC electrolyte. Proceedings of the Fuel Cells '97 review meeting.

Hassan, D., Janes, S. and Clasen, R. (2003) Proton-conducting ceramics as electrode/electrolyte materials for SOFC's-part 1: preparation, mechanical and thermal proprieties of sintered bodies. *J. Eur. Ceram. Soc.*, **23**: 221-228.

Heinzl, A., Vogel, B. and Hübner, P. (2002) Reforming of natural gas-hydrogen generation for small scale stationary fuel cell systems. *Journal of Power Sources*, **105**: 202-207.

Holtappels, P., Bradley, J., Irvine, J.T.S., Kaiser, A. and Mogensen, M. (2001) Electrochemical characterization of ceramic SOFC anodes. *J. Electrochem. Soc.*, **148**(8): A923-A929.

Iritani, J. and Kougami, K. (2001) "Pressurized 10 kW class module of SOFC". Proceedings of the Seventh International Symposium on Solid Oxide Fuel Cells (SOFC VII), Tsukuba, Japan.

Iwahara, H., Esaka, T., Uchida, H. and Maeda, N. (1981) Proton conduction in sintered oxides and its applications in steam electrolysis for hydrogen production. *Solid State Ionics*, **3-4**: 359-363.

Iwahara, H. (1995) Technological challenges in the application of proton conducting ceramics. *Solid State Ionics*, 77: 289-298.

Kozhukharov, V., Machkova, M., Ivanova, M. and Brashkova, N. (2001) Patents state of the art in SOFCs application. Proceedings of Seventh International Symposium on Solid Oxide Fuel Cells (SOFC VII), Tsukuba, Japan.

Larminie, J. and Dicks, A. (2003) Fuel Cell Systems Explained. John Wiley & Sons, Second Edition, UK.

Matsuzaki, Y. and Yasuda, I. (2000) The poisoning effect of sulfur-containing impurity gas on a SOFC anode: part I. Dependence on temperature, time and impurity concentration. *Solid State Ionics*, **132**: 261-269.

Minh, N. and Takahashi, T. (1995) Science and technology of ceramic fuel cells. Elsevier Science, The Netherlands.

Minh, N. (2004) SECA solid oxide fuel cell program-General Electric SECA industry team. Proceedings of the SECA Annual Workshop and Core Technology Program peer reviewed abstracts, May 11-13, 2004, Boston, MA, p. 2.

Minh, N., Andrews, R. and Campbell, T. (2004) General Electric Solid State Energy Conversion Alliance (SECA) solid oxide fuel cell program. Fuel Cell program Annual Report, Office of Fossil Energy, U.S. Department of Energy, pp. 27-32.

Mori, K., Miyamoto, H., Takenobu, K., Kishizaba, H. and Sakaki, Y. (2004) Characteristics and power generation test results of heavy rare earth stabilized zirconia. Proceedings of the sixth European Solid Oxide Fuel Cell Forum, 28 June to 2 July 2004, Lucerne, Switzerland, Vol. 3, pp. 1208-1213.

Müller, A.C., Weber, A. and Ivers-Tiffée, E. (2004) Degradation of zirconia electrolytes. Proceedings of the sixth European Solid Oxide Fuel Cell Forum, 28 June-2 July 2004, Lucerne, Switzerland, Vol. 3, pp. 1231-1238.

Nakanishi, A., Hattori, M., Sakaki, Y., Kimura, K., Miyamoto, H., Kanehira, S., Takenobu, K., Nishiura, M. and Oozawa, H. (2004) Development of MOLB type SOFC. Proceedings of Fuel Cell Seminar 2004, November 1-5, 2004, San Antonio, TX, pp. 105-108.

Nakayama, T. and Suzuki, M. (2001) Current status of SOFC R&D program at NEDO. Proceedings of the Seventh International Symposium on Solid Oxide Fuel Cells (SOFC VII), Tsukuba, Japan.

Nernst, W. (1899) Uber die elektrolytische Leitung fester K¨orper bei sehr hohen Temperaturen. *Z. Elektrochem.*, **6(2)**: 41-43.

Patel, P., Maru, H.C., Borglum, B., Stokes, R.A., Petri, R., Remick, R.J., Sishtla, C., Krist, K., Armstrong, T. and Virkar, A. (2004) Thermally integrated power systems (TIPS), high power density SOFC generator. Proceedings of the Fuel Cell Seminar 2004, November 1-5 2004, San Antonio, TX, pp. 132-135.

Patel, P., Borglum, B. and Huang, P. (2004) Thermally integrated high power density SOFC generator. Fuel Cell program Annual Report, Office of Fossil Energy, U.S. Department of Energy, 23-26.

Pedersen, T.F., Linderoth, S. and Laatsch, J. (2004) Oxidation behaviour of iron-chromium steels for solid oxide fuel cell interconnect. Proceedings of the sixth European Solid Oxide Fuel Cell Forum, 28 June-2 July 2004, Lucerne, Switzerland, Vol. 2, pp. 897-907.

Peters, R., Dahl, R., Klüttgen, U., Palm, C. and Stolten, D. (2002) Internal reforming of methane in solid oxide fuel cell systems. *Journal of Power Sources*, **106**: 238-244.

Primdahl, S., Hansen, J.R., Grahl-Madsen, L. and Larsen, P.H. (2001) Sr-Doped LaCrO$_3$ anode for solid oxide fuel cells. *J. Electrochem. Soc.*, **148(1)**: A74-A81.

Pudmich, G., Boukamp, B.A., Gonzalez-Cuenca, M., Jungen, W., Zipprich, W. and Tietz, F. (2000) Chromite/titanate based perovskites for application as anodes in solid oxide fuel cells. *Solid State Ionics*, **135**: 433-438.

Pusz, J., Bove, R. and Sammes, N.M. (2005) Landfill gas energy recovery based on micro-tubular solid oxide fuel cells. Proceedings of the Ninth International Symposium on Solid Oxide Fuel Cells (SOFC IX), May 15-20, Quebec City, Canada.

Rösch, B., Tu H., Stömer, A.O., Müller, A.C. and Stimming, U. (2003) Electrochemical behaviour of Ni-Ce$_{0.9}$Gd$_{0.1}$O$_{2-\delta}$ SOFC anodes in methane. Proceedings of the eight International Symposium on Solid Oxide Fuel Cells (SOFC VIII), The Electrochemical Society, PV 2003-07, pp. 737-744.

Sammes, N.M., Du, Y. and Bove, R. (2005) Design and Fabrication of a 100 W Anode Supported Micro-Tubular SOFC Stack. *Journal of Power Sources*, **145(2)**: 428-434.

Shaffer, S. (2004) Solid State Energy Conversion Alliance—Delphi SECA Industry. SECA Annual Workshop and Core Technology Program Peer Review, May 11, 2004, Boston, MA.

Shimada, T., Wen, C., Taniguchi, N., Otomo, J. and Takahashi, H. (2004) The high temperature proton conductor BaZr$_{0.4}$Ce$_{0.4}$In$_{0.2}$In$_{3-\alpha}$. *J. Power Sources*, **131**: 289-292.

Smirnova, A., Prakash, P., Phillips, R. and Sammes, N.M. (2004) Electrolyte proton-conductive materials for protonic ceramic fuel cells (PFCFs). Proceedings of the sixth European Solid Oxide Fuel Cell Forum, 28 June to 2 July 2004, Lucerne, Switzerland, Vol. 3, pp. 1029-1039.

Staniforth, J. and Kendall, K. (2000) Cannock landfill gas powering a small tubular solid oxide fuel cell: A case study. *J. Power Sources*, **86**: 401-403.

Suwanwarangkul, R., Croiset, E., Fowler, M.W., Douglas, P.L., Entchev, E., Douglas, M.A. (2003), "Performance comparison of Fick's, dusty-gas and Stefan–Maxwell models to predict the concentration overpotential of a SOFC anode". *Journal of Power Sources*, **122**: 9-18.

Tikekar, N.M., Armstrong, T.J. and Virkar, A.V. (2003) Reduction and re-oxidation kinetics of nickel-based solid oxide fuel cell anodes. Proceedings of the eight International Symposium on Solid Oxide Fuel Cells (SOFC VIII), The Electrochemical Society, PV 2003-07, pp. 670-679.

Uchida, H., Suzuki, S. and Watanabe, M. (2003) A high performance electrode for medium-temperature SOFC: mixed conducting ceria based anode with highly dispersed Ni electrocatalysts. Proceedings of the eight International Symposium on Solid Oxide Fuel Cells (SOFC VIII), The Electrochemical Society, PV 2003-07, pp. 728-736.

US Department of Energy (DOE). Fuel Cell Handbook, Seventh Edition. November 2004.

Van herle, J., Maréchal, F., Leuenberger, S., Membrez, Y., Bucheli, O. and Favrat, D. (2004) Process flow model of solid oxide fuel cell system supplied with sewage biogas. *Journal of Power Sources*, **131** (1-2): 127-141.

Van herle, J., Membrez, Y. and Bucheli, O. (2004) Biogas as fuel source for SOFC co-generators. *Journal of Power Sources*, **127**: 300-312.

Vora, S.V. (2004) Small-scale low-cost solid oxide fuel cell power systems. US Department of Energy, Office of Fossil Energy, 2004, Fuel Cell Program Annual Report, pp. 33-35.

Weber, A., Sauer, B., Müller, A.C., Herbstritt, D. and Ivers-Tiffée, E. (2002) Oxidation of H_2, CO, and methane in SOFCs with Ni/YSZ-cermet anodes. *Solid States Ionics*, **152-153**: 543-550.

Wotzak, G., Balan, C., Rahman, F. and Minh, N. (2003) Coal integrated gasification fuel cell system study. Pre-baseline topical report, prepared under the DOE/NETL Cooperative Agreement DE-FC26-01NT40779.

Yamamoto, O., Arati, Y., Takeda, Y., Imanishi, N., Mizutani, Y., Kawai, M. and Nakamura, Y. (1995) Electrical conductivity of stabilized zirconia with ytterbia and scandia. *Solid State Ionics*, **79**: 137-142.

Zahid, M., Tietz, F., Sebold, D. and Buchkremer, H.P. (2004) Reactive coatings against chromium evaporation in solid oxide fuel cells. Proceedings of the sixth European Solid Oxide Fuel Cell Forum, 28 June to 2 July 2004, Lucerne, Switzerland, Vol. 2, pp. 820-827.

Ziock, H.-J., Anthony, E.J., Brosha, E.L., Garzon, F.H., Guthrie, G.D., Johnson, A.A., Kramer, A., Lackner, K.S., Lau, F., Mukundan, R., Nawaz, M., Robison, T.W., Roop, B., Ruby, J., Smith, B.F. and Wang, J. (2002), Technical Progress in the Development of Zero Emission Coal Technologies. Proceedings of the 20th Annual International Pittsburgh Coal Conference, September 2002.

Zizelman, J. (2003) Development Update on Delphi's Solid Oxide Fuel Cell Systems: From gasoline to electric power. 4th SECA Workshop, April 15-16, Seattle, WA.

Recent Trends in Fuel Cell Science and Technology
Edited by S. Basu
Anamaya Publishers, New Delhi, India

12. Materials for Solid Oxide Fuel Cells

Rajendra N. Basu

Fuel Cell and Battery Section, Central Glass and Ceramic Research Institute, Kolkata-700 032, India

1. Introduction

Over the past one decade, several cell component materials and their combinations have been attempted to match with the appropriate requirements of solid oxide fuel cells (SOFCs). A large number of cell component materials with superior properties have been developed. The general observation is that most of the technological challenges associated with the development of SOFCs are related to materials science. For example, development of superior oxide-ion conductor electrolyte as well as cost-effective fabrication processes involves tremendous materials challenges. The improvements of the materials properties mostly include electrical conductivity, catalytic activity, stability and thermal expansion coefficient. Of late, significant improvements have also been made in the area of fast oxide-ion conductors. These oxide-ion conductors show extraordinarily high electrical conductivity compared to traditional zirconia-electrolyte. This helps SOFC not only to operate at lower temperature but also minimizes the polarization losses which is the key factor for a high performance cell (high power density or power per unit area). The differences between the operating cell voltage and the expected reversible voltage is termed as polarization or overpotential. More clear understanding of the fundamentals of these materials has been published by several groups through numerous articles (Steele 1993, 2000, Mogensen et al. 2000, Goodenough 2003, Singhal et al. 2003, Stöver et al. 2003, Kilner 2005). Reduction of electrolyte thickness also has tremendous advantages particularly from the technological point of view. Lower the electrolyte thickness lower is the internal resistance of the electrolyte, which in turn helps the cell to operate at a considerably lower temperature. The current research trend undoubtedly is more focused towards the development of high performance SOFC at low temperature (650°C and below). For making such high performance cell, interfacial contacts between two adjacent cell components is very critical. Therefore, an excellent compatibility (connectivity) between electrolyte and electrodes, and also with the interconnect (while stacking) is absolutely necessary. As all these materials are dissimilar in nature, so the improvement of the connectivity plays a crucial role in fabricating high performance SOFC. Continuous efforts have also been made to improve various deposition (fabrication) techniques. Normally, two types of fabrication techniques have been reported in the literature: (i) deposition at room temperature followed by high temperature firing, for example, tape casting, tape calendaring, screen printing, sol-gel and colloidal deposition such as slip casting, electrophoretic deposition and spray/dip coating and (ii) deposition at elevated temperature such as chemical vapor deposition (CVD), electrochemical vapor deposition (EVD), plasma spraying and spray pyrolysis (Minh et al. 1995, Singhal et al. 2003, Perednis 2004). Although there is a debate on the cost-effectiveness of the high temperature deposition processes (such as EVD) for mass production, yet this has delivered most robust and effective cell structure with a proven record of long term operation stability.

The various SOFC designs and their current status have been discussed in the earlier chapter. Among them, two SOFC designs—tubular and planar—are currently being pursued for the technology development. Although a large number of demonstration units are performing successfully, but the mass production of this important energy technology has not yet taken place. It is expected that it will be possible only when the synthesis and fabrication cost of the cell components are made inexpensive. A host of low cost fabrication processes such as tape casting, calendaring, colloidal techniques (such as electrophoretic deposition, slip casting), dip/spray coating, screen printing are considered as the extremely low cost and best suited for mass production. However, some degree of robustness in the fabricated cell components is required so that it serves the typical lifespan of SOFC (~ 40,000 h) with cell degradation rate less than 0.5% per 1000 h.

2. Materials Requirements and Related Issues

The state-of-the-art SOFC cell components materials are: 8mol% yttria stabilized zirconia (YSZ, electrolyte), Sr-substituted $LaMnO_3$ (LSM, cathode), Ni-YSZ cermet (anode), Ca-substituted $LaCrO_3$ (LCR) or metallic alloys (interconnect) and glass or glass-ceramic (sealants). Few more electrolytes, viz. Sc-doped ZrO_2 (SCZ), Gd or Sm-doped CeO_2 (CGO or SDC), and Sr- and Mg-substituted $LaGaO_3$ (LSGM) are also being used. The general requirements of the cell components are available in almost all the literatures and books dealing with the subject. Nevertheless, a comprehensive outline of the materials requirements for all the cell components except the sealant is given below:

1. Electrolyte and interconnect must be fully densed to prevent the mixing of oxidizing and reducing gases, whereas the electrodes must be porous to allow gas transport.

2. High electrical conductivity is necessary for both the electrodes (cathode and anode), the electrolyte, and the interconnect. While the electrolyte is pure ionic conductor, the electrodes could be either electronic- or mixed-conductor (electronic and ionic). For mixed conductor, the oxygen ion diffusivity is very high, which enhances the catalytic activity and as a result reduces the overpotential losses. Aging or degradation of conductivity at cell operating temperature is not desirable, particularly for electrolyte material.

3. Chemical compatibility is required with the adjoining cell components, both during fabrication as well as during cell operation. In addition, the cell components must also be chemically stable to withstand highly oxidizing and/or reducing atmospheres during operation at elevated temperature. For example, electrolyte and interconnect must be stable in both the atmospheres. Similarly, anode and cathode should be stable in reducing- and oxidizing-atmosphere, respectively.

4 Each component must have phase, morphological, chemical and dimensional stability. Changes are not desirable at cell operating temperature as well as during long term operation.

5. High thermal and mechanical shock resistance capabilities.

6. The supporting cell components—electrolyte for electrolyte-supported and electrode for electrode-supported—must have high enough mechanical strength and toughness at room temperature and during high temperature operation.

7. Thermal expansion coefficient values must be within a close proximity among all the cell components in order to avoid delamination, cracking and to reduce internal stresses during fabrication and cell operation.

8. Low cost and easily available technologies for fabrication of cell components and cell stack are important aspects, necessary to look into particularly when the technology is ready for mass production.

Although, YSZ is the most established and proven electrolyte material for SOFC, yet it has certain limitations. The main problem is its oxygen-ion carrying capability, which is highly temperature dependent. Therefore, operation with YSZ as the electrolyte (in thin film form) is limited at 700°C or above. Although, some attempts have been made to operate such cells at ~ 650°C, but could not achieve much success due to extraordinarily high polarization losses. Nevertheless, from practical point of view, a decrease in operating temperature (preferably ≤ 600°C) is beneficial particularly to increase the power density per unit volume, to use common metals or alloys as interconnect, to use less-complicated sealing method, to quick start up of the system and to withstand large number of thermal cycles during cell operation. Another important aspect is that the thermal expansion matching at around 600°C is also within a close regime for metals and ceramics. All these favorable attributes of low temperature application of ceramic fuel cells has led to an intensive research interest on alternative oxide-ion conducting electrolyte and compatible electrodes (Ishihara 2003). Like YSZ, doped CeO_2 is another electrolyte material for SOFC, which is addressed for low temperature operation (below 600°C). It is reported that at temperature above 600°C, Ce^{4+} ions are reduced to Ce^{3+} under the highly reducing condition in the anode side. Hence, the resulting electronic conductivity and deleterious lattice expansion produces an internal short-circuiting which significantly lowers the efficiency and performance of ceria-based fuel cells. Nevertheless, Steele (2000) and Goodenough (2003) observed that at temperatures below 600°C, the reduction of Ce^{4+} ions to Ce^{3+} at the anode side is negligibly less and can be neglected under typical cell operating conditions. Their observation paved the path for lower temperature SOFC operating in the range between 500 and 600°C (Bance et al. 2004).

Another alternate electrolyte material having extremely high oxide-ion conductivity is doped $LaGaO_3$. This electrolyte has been identified for operation within the temperature range 500-800°C (Ishihara 2003). Very recently, the advantages of thinning down the electrolyte to achieve high performance cell has been demonstrated by Yan et al. (2005). An overall temperature range of SOFC along with some selected materials normally used as cell components is listed in Table 1. In recent times even stainless steel has also been attempted as an interconnect (Ishihara 2003, Bance et al. 2004). It may be noted that the information given in the table is based on most commonly used materials. There could be several other choices of materials or combination of materials (Singhal et al. 2003). Table 2 summarizes typical SOFC cell performance data for several combinations of cell components used mostly in anode-supported thin film electrolyte structure. While more than 3 W/cm^2 has been reported to be highest power density at 700°C (Yan et al.

Table 1. Choice of SOFC materials and operation temperature

SOFC operating temperature (°C)	Material				Cell design	Electrolyte thickness (μm)
	Electrolyte	Cathode	Anode	Interconnect		
950-1000	YSZ	LSM	Ni-YSZ	LCR	Tubular	40
					Planar	150-300
650-800	YSZ, SCZ	LSM	Ni-YSZ	Ferritic steel	Planar	5-20
400-650*	LSGM, CGO	LSCF	Ni-SDC	Ferritic steel	Planar	500 or less
				Stainless steel	Planar	5 (LSGM)
				Stainless steel	Planar	10-30 (CGO)

*Several other materials options are available in Section 4.1.

Table 2. Intermediate temperature SOFCs—materials and performance

Operating temperature (°C)	Electrolyte/ thickness (μm)	Cathode	Anode	Fuel	Current density (A/cm²)	Peak power (W/cm²)	References
700	LSGM/5	SSC	Ni-Fe-SDC	H_2	5.0	3.3	Yan et al. 2005
800	LSGMN/200	Sr-Co-Fe-O	Ni-LDC	H_2	2.5	1.4	Wan et al. 2005
800	YSZ/5	LSM-YSZ	Ni-YSZ	H_2	1.0	—	Basu et al. 2005
800	YSZ-SDC/10	LSC-SDC	Ni-YSZ	H_2, CO, CO_2	3.5	2.0	Jiang et al. 2003
800	YSZ/10	LSM-YSZ	Ni-YSZ	Methanol	2.5	1.3	Jiang et al. 2001
800	YSZ/10	LSM-YSZ	Ni-YSZ	H_2	2.5	1.8-1.9	de Suza et al. 1997, Tsai et al. 1997, Kim et al. 1999

2005), around 2 W/cm^2 was obtained at 800°C (Jiang et al. 2003, de Souza et al. 1997, Tsai et al. 1997, Kim et al. 1999).

3. Existing and Alternative Cell Component Materials

The components of a single cell are electrolyte, cathode and anode. When single cells are stacked together to generate useful power, then it requires two more cell components, viz. interconnect and sealant. As mentioned in previous section, the electrodes (cathode and anode) must have mixed electronic/ionic conductivity and matching thermal expansion with electrolyte and interconnect. The electrolyte must have very high ionic conductivity and capable of withstanding both oxidizing and reducing environment down to 10^{-18} atm at high operating temperature. Interconnect must be an electronic conductor and have similar characteristics to that of electrolyte. A sealant needs to fulfill all the above mentioned criteria for all the above cell components. An overview of the existing and alternate cell component materials for electrolyte, cathode, anode, interconnect and sealant are given in this section. A brief description on the preparation of such cell components are also given.

3.1 Electrolytes

Electrolyte is considered as the main component of the ceramic fuel cell whose general requirements have already been mentioned in the previous section. Over the years many electrolytes have been developed. During this period, through detailed basic and fundamental studies on structure-property relationship of these electrolytes, a strong understanding has been developed. The electrolyte may carry either oxide-ion (O^{2-}) or proton (H$^+$). In this section both types of conductors are discussed. Detailed information on these electrolytes is available in several literatures/books (Subbarao 1984, Strickler et al. 1964, Minh et al. 1995, Boivin et al. 1998, Mogensen et al. 2000, Yamamoto, 2000, Singhal et al. 2003, Ishihara, 2003, Goodenough 2003, Joshi et al. 2004).

3.1.1 Oxide-ion Conductors

Oxide-ion conductors are best categorized in the following structural groups:
Fluorite-based (ZrO$_2$, CeO$_2$); δ-Bi$_2$O$_3$ and doped Bi$_2$O$_3$; Perovskites—doped LaGaO$_3$; Brownmillerites—Ba$_2$Ln$_2$O$_5$; Pyrochlores—Gd$_2$Zr$_2$O$_7$, Gd$_2$Ti$_2$O$_7$ and rare earth-based Apatites—La$_9$Sr$_1$Si$_6$O$_{26.5}$.

3.1.1.1 ZrO$_2$

Eight mol% of Y$_2$O$_3$ stabilized ZrO$_2$, commonly known as YSZ or 8YSZ, is the most widely used electrolyte for SOFC which fulfills almost all the criteria for an electrolyte. It is a well studied material and showed extremely good performance as SOFC electrolyte in several demonstration units of Westinghouse which are running for several years (Singhal et al. 2003). Choudhary et al. (1980) studied the phase diagram and defect structures of various fluorite structures including ZrO$_2$-Y$_2$O$_3$ system. Zirconia in its pure form has three polymorphs. From room temperature monoclinic structure it changes to a tetragonal phase above 1170°C and finally to a cubic fluorite structure above 2370°C which exists up to the melting point of 2680°C. These transformations are martensitic in nature and are reversible. There is also about 4% volume change reported during monoclinic to tetragonal phase transformation. However, the tetragonal or cubic phase can be stabilized at room temperature by direct substitution of small amount of divalent or trivalent cations at the host lattice (Zr^{4+}). The common dopants for zirconia are Y$_2$O$_3$, Yb$_2$O$_3$, Sc$_2$O$_3$, Gd$_2$O$_3$, Dy$_2$O$_3$, Nd$_2$O$_3$, Sm$_2$O$_3$, CaO, MgO etc. However, the stability limit depends on the type and the quantity of dopant used. The dopant metal ion substitutes Zr^{4+} site in the crystal lattice and in this process create vacancies in the oxygen sub-lattice because of lower valency (2+ or 3+). These vacancies in the oxygen sub-lattice are responsible for ionic conduction (Badwal et al. 1997). For example, the dissolution of yttria in fluorite phase of ZrO_2 creates large concentration of oxygen vacancies and can be expressed using Kröger-Vink notation. Such notation is used to describe the various defects where V stands for vacancy, the term within third bracket represents concentration, e' and h denote effective charges of electron and hole, and n and p for electron and hole concentration, respectively. Thus, the defect formation reaction for yttria doped ZrO_2 in Kröger-Vink notation can be expressed as

$$Y_2O_3 \xrightarrow{ZrO_2} 2Y'_{Zr} + V_O^{\cdot\cdot} + 3O_O^x \qquad (1)$$

Each additional yttria molecule creates one oxygen vacancy. The high oxygen vacancy concentration gives rise to high oxygen-ion mobility, and thus the oxygen-ion conduction takes place in stabilized zirconia by the movement of oxygen ions via vacancies (Ishihara et al. 2003). Hence, the conductivity can be written as

$$\sigma = e\mu[V_O^{\cdot\cdot}] \qquad (2)$$

where μ is the mobility, $[V_O^{\cdot\cdot}]$ the concentration of oxygen vacancies and e the effective charge. Under electrical neutrality condition $2[Y'_{Zr}] = [V_O^{\cdot\cdot}]$. The relationship is a clear indication that the vacancy concentration is linearly dependent on the doping level. Hence, the conductivity in the ZrO_2-M_2O_3 (or MO) system depends on the doping concentration and the dopant ionic radius. Further studies established that the variations in conductivity with different dopants of the same valency might be due to the defect interactions and lattice relaxation phenomena. It has also been reported that the binding energies of the defect clusters and the effective mobility of the vacancies depend on the dopant size and concentration (Meyer et al. 1997, Yamamuru et al. 1999, Zacate et al. 2000, Fleig et al. 2003).

The dependence of zirconia conductivity at 1000°C on the dopant concentration is shown in Fig. 1 (Arachi et al. 1999). Similar conductivity data at 800°C has also been reported by Badwal et al. (1997). The maximum conductivity occurs at a vacancy concentration of around 3.5-4% (about 8 mol% M_2O_3) for trivalent dopants and 6-7% (13% MO) vacancy concentration for divalent dopants. Although, the ionic conductivity of 3 mol% Y_2O_3-ZrO_2 having tetragonal phase (commonly known as 3YSZ) is lower by a

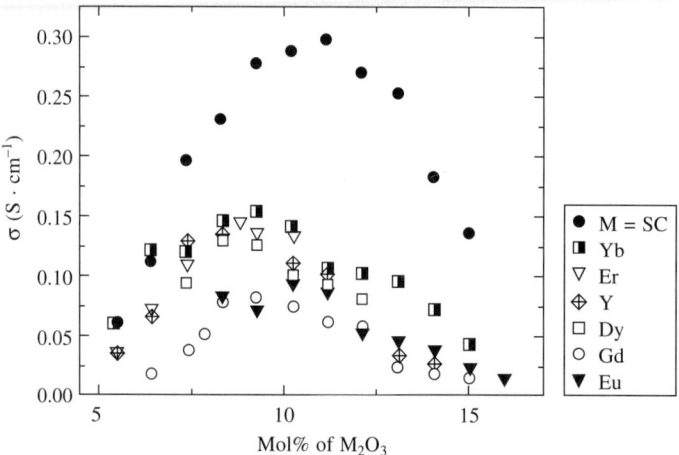

Fig. 1 Composition dependence of the electrical conductivityt at 1000°C for ZrO_2-M_2O_3 systems (M=Lanthanide) (Archi et al. 1999).

factor of about three as compared to the cubic 8 mol% Y_2O_3-ZrO_2 composition at 1000°C, but it has extremely good mechanical properties (Table 3). An in-depth study on the structure-property relationship of Y_2O_3-ZrO_2 electrolytes has been done by Badwal (1992). On the other hand, the maximum conductivity was obtained for Sc_2O_3-ZrO_2 system (at 1000°C, 0.3 S/cm) which is more than that of the commonly used 8 mol% Y_2O_3 doped ZrO_2. The possible reason for obtaining such high conductivity is because of the fact that the dopant Sc^{3+} has the closest ionic radius to the host Zr^{4+} ion. Some important properties pertaining to fuel cell application for Y_2O_3- and Sc_2O_3-ZrO_2 systems are given in Table 3. Despite of having such excellent electrical properties, Sc-doped ZrO_2 (SSZ) has not gained much attention by the SOFC developers.

Table 3. **Important properties of Y_2O_3 and Sc_2O_3 substituted ZrO_2 (compiled from Minh et al. 1995 and Ishihara et al. 2003)**

Dopant (M_2O_3)	Mol% of M_2O_3	Ionic radius (nm)		Conductivity (S/cm) at 1000°C	Bending strength (MPa)	TEC ($\times 10^{-6}$/K)
		Zr^{4+}	M^{3+}			
Y_2O_3	3	0.079	0.092	0.05	1200	10.8
Y_2O_3	8	0.079	0.092	0.13	230	10.5
Sc_2O_3	11	0.079	0.081	0.30	255	10.0

Three possible reasons often cited in the literature are: (i) availability, (ii) cost and (iii) aging effect (electrical conductivity deteriorates with long term operation). But, recently the long-term cell testing 10 mol% Sc_2O_3 (10 SSZ) electrolyte was demonstrated by Herbstritt et al. 2001 (Fig. 2). Also a sharp fall of the price of scandium oxide in the international market has brightened the possibilities of using Sc_2O_3-ZrO_2 electrolyte in a relatively larger user domain (Ralph 2001).

Influence of grain boundaries on the electrical and overall electrochemical performances is also another area of research interest for these fluorite-based electrolytes. It has been observed that SiO_2 impurities (glassy impurities) presence in the commercial grade powder segregates to the grain boundaries and consequently causes an increase in grain boundary resistances. Detailed study in this area has been done by Badwal et al. (1997). They confirmed this effect through impedance spectroscopy (Fig. 3). However, this effect is dominant in material only when substantial level of impurities is present, and at low temperature. It has been observed that with the increasing temperature, the contribution of

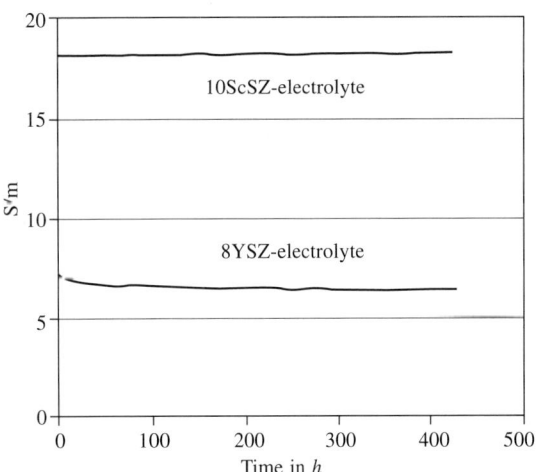

Fig. 2 Comparison of long term conductivity measurements of 8YSZ and 10 ScSZ electrolyte substrate at 900°C (Herbstritt et al. 2001).

grain boundary resistance diminishes due to high activation energy associated with the process. Glassy impurities have also been reported to be responsible for the aging behavior. However, the conductivity deterioration is reported to be lowest for compositions in the 9-11 mol% Y_2O_3-ZrO_2 system (Badwal et al. 2000).

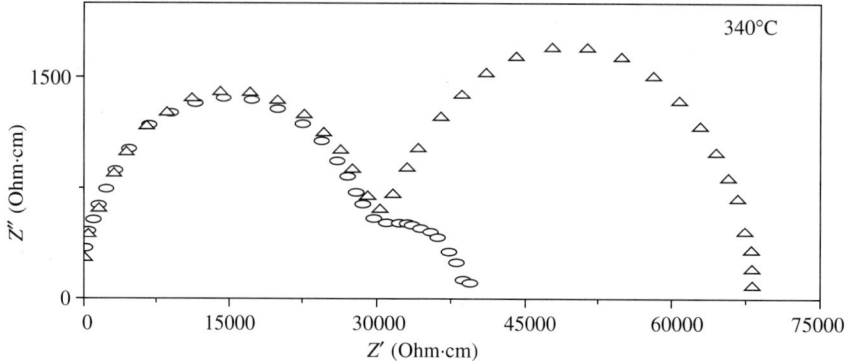

Fig. 3 Impedence spectra at 340°C of two 3 mol% Y₂O₃-ZrO₂ (3YSZ) compositions with different SiO₂ impurities (O: 20 ppm SiO₂, D: 800 ppm SiO₂) showing the effect of impurity segregation at grain boundaries on the grain boundary impedance (arc on the right hand side) (Badwal et al. 1997).

3.1.1.2 CeO₂

Ceria doped with alkaline earths or rare earths has received considerable attention and numerous research publications are available on this subject (Huang et al. 1998, Gödickemeier et al.1998, Steele 2000, Ralph et al. 2000, Mogensen et al. 2000). Pure stoichiometric CeO₂ has the fluorite structure having space group Fm3m over the entire temperature range from room temperature to the melting point (~ 2750 K). CeO₂ structure is known to tolerate a considerable reduction without phase change, especially at elevated temperatures. Therefore, CeO₂, unlike ZrO₂, does not need any stabilization. Depending upon the temperature and oxygen partial pressure (pO_2), the material exhibits large oxygen deficiencies in the formula $CeO_{2-\delta}$, where δ could be as large as 0.3. The doubly ionized oxygen vacancies are the principal ionic defects with compensating electrons when $\delta < 10^{-3}$. Some physical properties of CeO₂-based electrolyte are listed in Table 4.

Table 4. **Some physical properties of pure stoichiometric CeO₂ (Mogensen et al. 2000)**

Property	Value (Unit)
Density	7.22 g cm^{-1}
Melting point	2750 K
Specific heat	460 JKg^{-1}K^{-1}
Thermal conductivity	12 Wm^{-1}K^{-1}
Refractive index	2.1 visible
	2.2 infrared
Relative dielectric constant (0.5-50 MHz)	11
Young's modulus	165×10^9 Nm^{-2}
Poisson's ratio	0.3
Hardness	5-6

Gadolinium (Gd) and samarium (Sm) are the most commonly used dopants for CeO₂ (commonly termed as CGO and SDC, respectively). The dopant produces oxygen vacancies (analogous to that of Y₂O₃ doping with ZrO₂, Eq. (1)) and gives rise to the ionic conductivity. The creation of anion vacancy is governed by the following defect equation (Kröger-Vink notation):

$$M_2O_3 + CeO_2 \rightarrow 2\,M'_{Ce} + V_O^{\cdot\cdot} + 3O_O^x \qquad (3)$$

and in reducing environments, Ce^{4+} is easily reduced to Ce^{3+} leading to an electronic contribution to conductivity which could be expressed as

$$O_O^x \rightarrow V_O^{\cdot\cdot} + \frac{1}{2}O_2\,(g) + 2e' \qquad (4)$$

Gödickemeier et al. (1998) studied the electrical conductivity (ionic and electronic) of $Ce_{0.8}Sm_{0.2}O_{1.9}$ system at three different temperatures, viz. 600, 700 and 800°C as a function of oxygen partial pressure (pO_2). At high pO_2, these conductors show pure ionic conductivity, but at lower partial pressure (below 10^{-6} bar), they become electronically conducting. Due to this electronic conduction, electronic current flows through the electrolyte even at open circuit condition, as a result of which the open circuit voltage (OCV) shows lower value (0.89 V) than that of the theoretical value (Ishihara et al., 2003). Information on cell performance based on CeO_2 electrolyte has been reported by Ralph et al. (2001). Even having a lower OCV value, very recently Lu et al. (2003) have obtained a maximum power density of 130 mW/cm^2 at 650°C by introducing composite electrode (Goodenough 2003). Electrical conductivity data at 500 and 700°C are summarized in Table 5. The highest conductivity is reported in case of 10 mol% Sm substitution. The conductivity is strongly dependent on the ionic radii of the dopant. Therefore, the electronic conductivity of doped ceria plays an insignificant role in the temperature regime 500-550°C and is identified as the safe zone of operation for low temperature SOFC.

Table 5. Electrical conductivity data for CeO_2-Ln_2O_3 (Ishihara et al. 2003)

Ln_2O_3	Mol%	Conductivity (S/cm)		Activation energy (kJ/mol)
		700°C	500°C	
Sm_2O_3	10	3.5×10^{-2}	2.9×10^{-3}	68
	10	4.0×10^{-2}	5.0×10^{-3}	75
Gd_2O_3	10	3.6×10^{-2}	3.8×10^{-3}	70
Y_2O_3	10	1.0×10^{-2}	0.21×10^{-3}	95
CaO	5	2.0×10^{-2}	1.5×10^{-3}	80

Attempt has also been made to improve the electrical properties further by adding a small amount of additional dopant in CGO in addition to Gd or Sm (Yahiro et al. 1988, Ralph et al.1997). For this purpose, two compositions, e.g. $Ce_{0.9}Gd_{0.09}Ca_{0.01}O_x$ and $Ce_{0.8}Gd_{0.19}Ca_{0.01}O_x$ were studied and an enhancement of grain boundary conductivity vis-à-vis higher ionic conductivity phase at the grain boundary was found. It is believed that added calcia to CGO segregate to the grain boundary regions, where they provide good ionically conducting pathways.

3.1.1.3 Bi_2O_3
Compared to ZrO_2- and CeO_2-based ceramic electrolytes, δ-Bi_2O_3 electrolyte (fcc structure) exhibits the highest oxide-ion conductivity, 2.3 S/cm at around 800°C due to its extremely open crystal structure. Besides δ-Bi_2O_3, there also exist few other forms of Bi_2O_3 (Table 6). This electrolyte, however, is reported to be stable only within a very narrow temperature range between 730° and 830°C, the melting point of Bi_2O_3 (Boivin et al., 1998, Goodenough, 2003). At oxygen partial pressure below 10^{-14}atm, Bi_2O_3 is unstable and reduces from Bi^{3+} to Bi^0 (metallic bismuth) in presence of highly reducing gases such as H_2

Table 6. Temperature regions of the stable and metastable forms of Bi_2O_3 (Gao et al. 1999) (modified)

Phase	α	δ	β	γ
Phase stability range (K)	<1002	1002-1097	603-923	773-912
Temperature (K)	298	1047	916	298
Structure	monoclinic	fcc	tetragonal	bcc

or CH_4. However, in order to stabilize the δ-Bi_2O_3 phase at lower temperature, aliovalent metal oxides (smaller in size than Bi^{3+} ion) have been used. Several compositions of this system, $(Bi_2O_3)_{1-x}(Ln_2O_3)_x$ (where Ln represents Sm, Gd, Nd, Y, Er etc.), have been examined by Takahashi et al. (1977), Verkerk et al. (1981), Waschsman et al. (1992), Boivin et al. (1998) and Sammes et al. (1999). Among all the compositions, $(Bi_2O_3)_{0.75}(Y_2O_3)_{0.25}$ and $(Bi_2O_3)_{0.75}(Er_2O_3)_{0.25}$ are well studied and have attracted considerable attention. The problem of reduction in conductivity at the anode side of $Bi_{0.75}Y_{0.25}O_{1.5}$ electrolyte could be solved by applying a thin layer of Sm-doped ceria, but the overall thermal expansion coefficient is so high, 15×10^{-6}/K, that till now no working stack has been constructed. Another family of Bi_2O_3-based electrolyte which is stable at room temperature is bismuth metal (dopant) vanadium oxide, commonly known as BIMEVOX. The $Bi_2VO_{5.5}$ family had shown some ionic conductivity, but reported to encounter the similar redox problems as that of Bi_2O_3.

3.1.1.4 LaGaO₃

Although, over the last couple of decades the most extensively studied electrolytes for ceramic fuel cells are based on Bi_2O_3, CeO_2 and ZrO_2, the perovskite oxide (ABO_3)-based electrolytes have also received equally good attention because of their structural tolerance to various sizes of A and B cations and can also dissolve large concentration of aliovalent cations on both the cation sublattices. In addition, the perovskite-type oxide electrolytes are easy to tailor made because of their great geometrical and chemical flexibility. Thus they are considered as promising candidates for many important applications, including SOFC electrolyte. $La(Ca)AlO_3$ is the first perovskite-based electrolyte which shows a reasonably good ionic conductivity (~5 × 10⁻³ S/cm) at around 800°C and naturally shows a good performance in SOFCs. However, the landmark discovery of the new fast ion conductor, doped $LaGaO_3$, by the group of Ishihara (Japan) and Goodenough (USA) in 1994 brought tremendous enthusiasm and since then numerous research papers were published on this electrolyte (Ishihara et al. 1994, Feng et al. 1994).

Pure $LaGaO_3$ is a perovskite having orthorhombic structure at low temperature which transforms to rhombohedral phase at 900°C. Various dopants have been used and substituted either on La- or both on La- and Ga-site and the effect of these substitutions have been examined in detailed (Fig. 4). On the La and Ga sites, the best results among the alkaline-earth cations are obtained for Sr^{2+} and Mg^{2+} respectively. For instance, the composition, $La_{0.9}Sr_{0.1}Ga_{0.8}Mg_{0.2}O_{2.85}$ was shown to have oxide-ion conductivity > 10^{-2} S/cm at 600°C with a transport number (t_i) ≈ 1, over the oxygen partial pressure range 10^{-20} atm < pO_2 < 0.4 atm (Ishihara 1994). The transport number is defined as the ratio between the oxide-ion conductivity and the total conductivity (ionic plus electronic). Since in the case of solid electrolyte the ionic conductivity is always higher than that of the electronic conductivity, so the ideal electrolyte must have t_i close to unity. Incidentally, ionic conductivity value is significantly higher than that of the traditional YSZ electrolyte. Further studies on the effect of transition metal in the system, $La_{0.8}Sr_{0.2}Ga_{0.8}Mg_{0.2}O_3$ were carried out by the same group (Ishihara et al., 1999). Among the various transition metal cations (e.g., Co, Ni or Fe) used as dopant in B-site, the Co-doping (within 10 mol%) having composition, $La_{0.8}Sr_{0.2}Ga_{0.8}Mg_{0.115}Co_{0.085}O_3$ (LSGMC), further enhances the oxide ion conduction without introducing any additional electron conduction or lowering the transport number of the oxide ion.

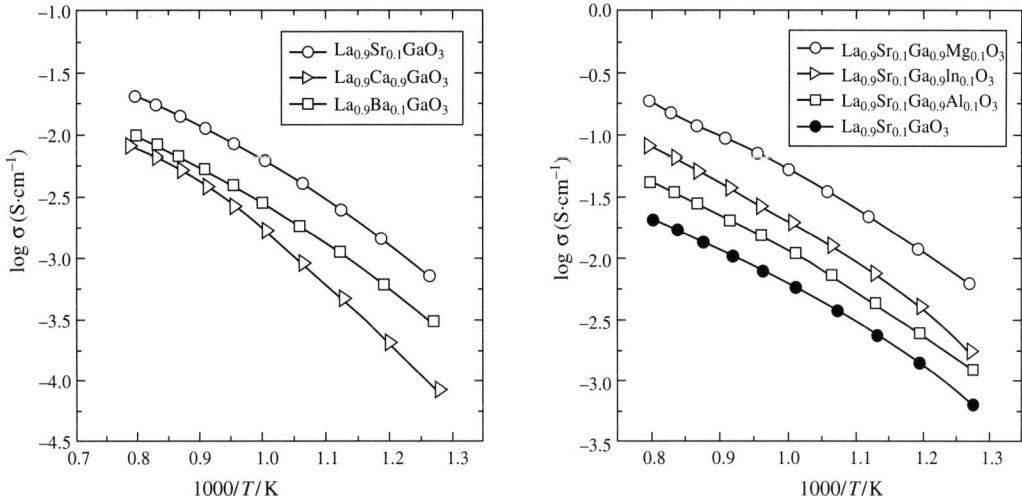

Fig. 4 Effect of various cations substitution for La and Ga sites in LaGaO$_3$ at $pO_2 = 10^{-5}$ atm on electrical conductivity (Ishihara et al. 1994).

So far, few laboratory sized stacks has been fabricated using LSGM electrolyte and very recently, a short stack has been fabricated (using LaGaO$_3$-based electrolyte) with stainless steel as a bipolar plate (Ishihara 2003). An output power of 65 W and volumetric power density of around 500 W/litre is obtained from this stack. The anode and cathode used in these stacks are Ni/Ce$_{0.8}$Sm$_{0.2}$O$_3$ and Sm$_{0.5}$Sr$_{0.5}$CoO$_3$ respectively. Fig. 5 is a typical performance curve of such an alternate electrolyte-based (LSGM) short stack operated at 650°C.

During this period, a large group of researchers also studied the crystal structure of LSGM which is

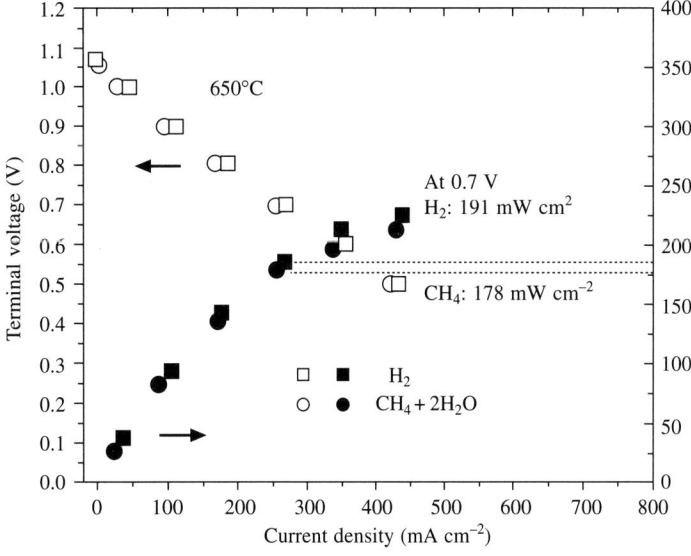

Fig. 5 Comparison of *I/V* curves of the LSGMC-5 cell using H$_2$ and CH$_4$/H$_2$O for fuels at 650°C (Ishihara et al. 2003).

basically an isolated rhombohedral domain. They are averaged to cubic on a larger scale and have higher order of symmetries (Slater et al. 1998, Skowron et al. 1999). Obtaining a single-phased LSGM is not an easy task—formation of secondary phases such as $LaSrGaO_4$ and $LaSrGa_3O_7$ have been often reported while synthesizing LSGM. Solid solution range of LSGM has been examined by Huang et al. in1996. Later on, the phase relations in quasi-ternary systems, e.g., $LaGaO_3$-$SrGaO_{2.5}$-$LaMgO_{2.5}$, and the phase equilibrium of the system Ga_2O_3-La_2O_3 were studied in detail (Majewski et al. 2001). In the literature, $La_{0.9}Sr_{0.1}Ga_{0.8}Mg_{0.2}O_{2.85}$ is a material of high research interest and study.

Some improvements of mechanical properties of doped $LaGaO_3$ have been reported by addition of extremely small amount of Al_2O_3 (Drennan et al. 1997, Wolfenstine 1999, Yasuda et al. 2000). Thermal expansion coefficient of LSGM is in the range of $11\text{-}12 \times 10^{-6}$/K which is well matched with the existing cathode (LSM). Besides thermal expansion and mechanical properties, it is very important to examine the chemical compatibility between the electrolyte and electrodes. Unlike YSZ, LSGM is chemically compatible with doped cobaltite ($LaCoO_3$), another well investigated cathode material. It is expected that the lattice mismatch at the electrolyte/cathode interface could be avoided to a large extent as both electrolyte (LSGM) and cathode material are preovskite. On the other hand, making of Ni/LSGM cermet anode using ceramic electrolyte (LSGM) and metal-phase (Ni) is not that straightforward like Ni/YSZ cermet. In fact, LSGM and NiO are prone to react at fabrication temperature (above 1400°C) as there is considerable solubility of NiO in doped $LaGaO_3$. During fabrication, $LaNiO_3$ is formed at the interface between the anode (Ni/LSGM) and the electrolyte (LSGM). This unwanted $LaNiO_3$ phase is metallic in nature and unable to carry O^{2-} ions (Huang et al. 1997). To alleviate this problem, a thin layer of $Sm_{0.2}Ce_{0.8}O_{1.9}$ (SDC) has been applied on the anode side of the LSGM electrolyte. The anode in this case is typically a mixture of NiO and SDC. However, a chemical reaction at elevated temperature between LSGM, CeO_2, and NiO causes alteration of composition in LSGM. During reaction La^{3+} is found to be fairly mobile in the LSGM electrolyte, leading to the formation of phases ($LaSrGaO_4$ and $LaSrGa_3O_7$) at the interface between the ceria and LSGM. These phases blocked the oxide-ion transport. However, a lower anode overpotential was obtained with $La_{0.4}Ce_{0.6}O_{1.8}$ (LDC) in place of SDC and the LDC was chosen in order to achieve an iso-La chemical activity across the interface (Huang et al. 2001). Nickel oxide does not react with LDC to form $LaNiO_3$. Similarly, the possible interactions between $LaGaO_3$-based electrolyte and CeO_2 have been examined and no ternary compound was found (Horvat et al. 1999). Among the several other attempts to stop the interactions at the anode/electrolyte interface, Deng et al. (1999) demonstrated a solution. In their experiments they used one additional porous LSGM layer in order to shield the anode-electrolyte interface contamination.

Electrolyte efficiency of fluorite- and LSGM-based ceramic fuel cells has been estimated and such efficiency has a strong relationship between fuel utilization and internal resistance of the electrolyte (Ishihara 2003). Thinner the electrolyte, the lower is the internal resistance. For example, in case of YSZ, the thickness limit is approximately 10 μm and cell operational temperature is between 700 and 750°C. But for LSGM, it can go down to 5 μm having operational even at 450°C. Therefore, it is believed that for lower temperature SOFC, the materials choice would be limited to only with the existing $LaGaO_3$- and CeO_2-based electrolytes. High performance and low temperature operated (< 650°C) anode-supported thin film LSGM has very recently been constructed and demonstrated by Yan et al. (2005).

For low temperature SOFC, some important features of the electrolyte which must be looked into are: (i) excellent ionic conductivity and variety of choices of possible compositions (ii) transport number close to unity (signature of a superior ionic conductivity), (iii) stable over the entire oxygen partial pressure range, (iv) not sensitive to moisture and (v) fulfills other essential characteristics of an electrolyte. Although $LaGaO_3$-based electrolyte has many favorable attributes, still there are certain concerns related to: (i) volatization of gallium oxide and formation of reaction phases while processing/co-firing with typical electrode materials, (ii) long-term stability with some of the candidate fuels and (iii) possible high creep rate compared to YSZ (Yamaji et al. 2000, Nguyen et al. 2000, Tao et al. 2000, Wolfenstine et al. 1999).

Even though variety of electrolytes have been investigated, but so far yttria stabilized zirconia is still considered as a potential solid electrolyte material ever since SOFC has been successfully demonstrated by Baur et al. (1937). Since then search is on for an oxide-ion conductor which can give superior property than YSZ. Numerous such oxide-ion electrolytes have been attempted and demonstrated. However, only a few could finally emerge with acceptable solid electrolyte properties. Fig. 6 demonstrates the temperature dependence of conductivity of some important solid electrolytes. The advantages and disadvantages of four most extensively studied oxide-ion conductors are given in Table 7.

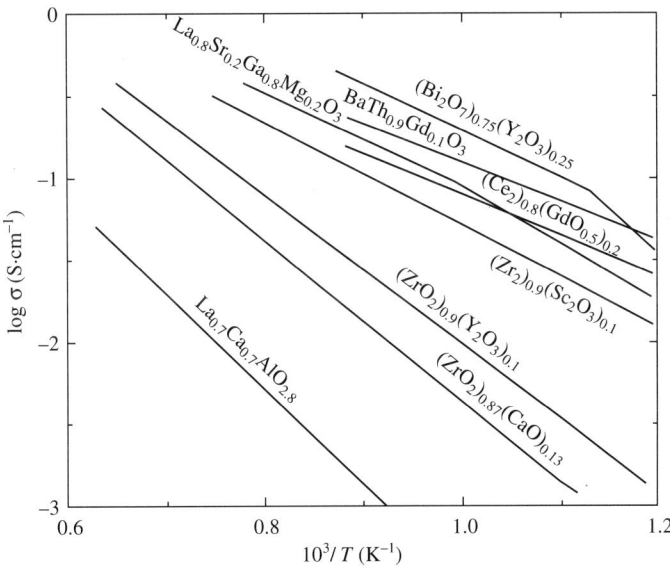

Fig. 6 Temperature dependence of electrical conductivity for selected oxide ion conductors (Ishihara et al. 2003).

Table 7. Advantages and disadvantages of possible electrolyte candidates for SOFC (Weber et al. 2004)

YSZ	CGO	LSGM	SSZ
Advantages			
Excellent stability in oxidizing and reducing environment	Mixed electronic and ionic conductor at low pO_2	Good compatibility with cathode materials	Excellent stability in in oxidizing and reducing environment
Excellent mechanical stability (particularly, for 3YSZ) > 40,000 h of fuel cell operation possible High quality raw materials available			Better long-term stability than 8YSZ
Disadvantages			
Low ionic conductivity (especially for 3YSZ) Incompatibility with some cathode materials	Electronic conduction at low pO_2 Low OCV, mechanical stability, availability and price of Gd	Phase stability, Ga-evaporation at low pO_2 Incompatible with NiO, mechanical stability, availability and price of Ga	Availability and price of Sc

3.1.1.5 Brownmillerites

Brownmillerites is another form of perovskite having general formula $A_2B_2O_5$. It is derived from the ABO_3 perovskite structure by an ordering of the oxygen vacancies into alternate BO_2 sheets to give layers of corner-shared $BO_{6/2}$ octahedra alternating with layers of corner-shared $BO_{4/2}$ tetrahedra—the tetrahedra and octahedra share corners along the c-axis. This structure is formed when the B cations are stable in both octahedral and tetrahedral symmetry, as is the case for the main group of ions Al(III), Ga(III), In(III) and Ge(III). In this structure, oxide-ion vacancies are introduced without aliovalent substitution on one of the cation subarrays (Goodenough 2003). Such vacancy ordering results in an increased unit cell relative to the perovskite. In some cases, the oxygen vacancies do not order, which results a perovskite structure with a statistical distribution of oxygen vacancies on the oxygen sites. Hence high oxide ion conductivity is also expected in this structure. The oxide ion conductivity of Brownmillerites is rather high compared to fluorite structure (like ZrO_2 or CeO_2). A large number of such compositions are available in the literature (Ishihara et al. 2003). Among them $Ba_2In_2O_5$ has been widely investigated; this material is interesting because of large size of the cations, which offered the greatest free volume for oxygen diffusion. Temperature dependence of conductivity of $Ba_2In_2O_5$ is shown in Fig. 7, and compared with traditional zirconia electrolyte. It shows a sharp discontinuity at ~ 930°C which is believed to be due to first-order order-disorder transition, and the change in slope near 650°C is due to temperature-dependent enthalpy to excite an oxide ion to an interstitial site (Goodenough 2003).

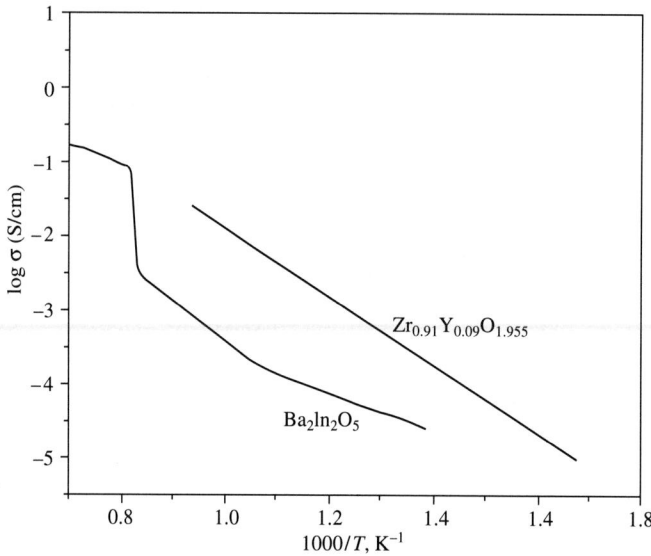

Fig. 7 Arrhenius plots of total conductivity in air of the brownmillerite $Ba_2In_2O_5$. For comparison the conductivity behaviour of commonly used YSZ electrolyte is shown (Goodenough 2003) (modified).

3.1.1.6 Pyrochlores

Another class of oxides, pyrochlores $(A_2B_2O_7)$ is a defect fluorite superstructure. It is derived from an oxygen-deficient fluorite structure by an ordering of both cations and oxygen vacancies suggests that an order-disorder transition may lead to a disordered phase exhibiting high oxide-ion conductivity. The key factor that determines the stability of cation ordering is the ratio of the A and B cations. The smaller is the value of the ratio, the lower is the expected order-disorder transition temperature (Goodenough 2003).

Tuller's group (Cambridge, MA) has done extensive studies on this class of materials, and initially obtained oxygen-ion conductivity value of $\sim 10^{-2}$ S/cm at 1000°C, which however was improved when they focused on some interesting compositions like $Gd_2Ti_2O_7$ and $Gd_2Zr_2O_7$ (Kramer et al. 1994, Tuller 1997). In particular, the electrolyte based on $Gd_2(Ti_{1-x}Zr_x)_2O_7$ solid solution is of great interest because at $x = 0$, the composition becomes an insulator whereas for $x = 1$, the composition turns to be a very good oxide-ion conductor and the conductivity value is close to that of YSZ electrolyte. Another composition, $Gd_2(Zr_{0.4}Ti_{0.6})O_7$ has been studied extensively and its ionic domain is found to be stable down to pO_2 of 10^{-21} atm (Boivin et al., 1998).

3.1.1.7 Rare Earth-based Apatites
In general it is believed that high symmetry in the lattice is an essential requirement for fast ion conduction. Recent interest in rare earth-based apatite has attracted considerable attention because of their high ionic conductivity and well known chemical stability of the apatite structure. It is well known that hydroxyapatite $Ca_{10}(PO_4)_6(OH)_2$, the most commonly available apatite-based material in nature, is a major component of teeth and bone. However, by changing the nature of cations it is possible to make these systems (e.g., $Ln_{10-y}Si_6O_{26-z}$), a high oxygen ion conductor. The structure of cation-deficient apatites consists of isolated SiO_4 tetrahedra and O^{2-} anions, in tunnels along the c-axis. Those anionic species in tunnels are responsible for the high ionic conductivities at low temperature. Conductivities as high as 0.01 S/cm have been reported for $La_{10}Si_6O_{27}$ at 700°C by Nakayama et al. (1998, 2001)—the value which is much higher than YSZ at 700-800°C (Nakayama et al. 1998). Very recently, conductivity values of large number of apatite-based rare earth oxides have been reported (Beaudet Savignat et al. 2003, Sansom et al., 2005). They found the oxide ion conductivities at 700°C for $La_{9.25}Ca_{0.75}Si_6O_{26.625}$, $La_9Sr_{1.0}Si_6O_{26.5}$ and $La_{9.6}Ge_6O_{26.4}$ are 9.2×10^{-3}, 1.2×10^{-2} and 1.4×10^{-2} S/cm respectively. Temperature dependence of conductivity of most promising material, lanthanum silicate ($La_9Sr_{1.0}Si_6O_{26.5}$) shows straight line behavior (Brisse et al. 2005). It shows a lower value of the activation energy (0.8 eV)—this value is even lower than that of the traditional YSZ electrolyte. The stability of the electrical conductivities down to 10^{-24} atm of oxygen partial pressure is a clear indication of pure oxygen ion conductivity in the system (Fig. 8). Though oxides of this category appear to be very promising but still lot more studies are necessary before their acceptance as SOFC electrolyte.

3.1.2 Proton Conductors
Similar to oxide ion conducting electrolytes (described in Section 3.1.1), some proton conducting perovskite oxides are also considered as electrolytes for SOFC. It is well known that proton is the smallest positive ion (H^+) having high mobility, thus a reasonably good electrical conductivity is expected from this group of perovskite oxides in the intermediate temperature range (600-850°C). The existence of such oxides and their proton conduction at high temperature was first discovered by Iwahara's group in Japan in early eighties (Iwahara et al. 1983). Although, $BaCeO_3$ and $SrCeO_3$ are considered to be the two most widely studied proton conducting oxides, but at a later stage some interesting properties were also observed in certain zirconates, viz., $CaZrO_3$, $BaZrO_3$ and $SaZrO_3$ (Iwahara et al. 1988, Bonanos et al. 1995, Ma et al. 1999, Higgins et al. 2005). Very recently, an excellent review article covering the entire proton conducting oxides been published by Kreuer et al. (2003). Pure cerates and zirconates do not show good electrical conductivity. However, similar to other perovskites, doping with suitable cations enhances their conductivity. In case of $BaCe_{1-x}M_xO_{3-\delta}$, the common dopants are: Gd_2O_3, Nd_2O_3, La_2O_3, Y_2O_3 etc. Similarly for $SrCeO_3$, the common dopants are Yb_2O_3, Y_2O_3 and Sc_2O_3 and typically Y-doped $BaZrO_3$ has been commonly used.

Proton incorporation into such materials in general occur through two step processes (Bonanos 1995).

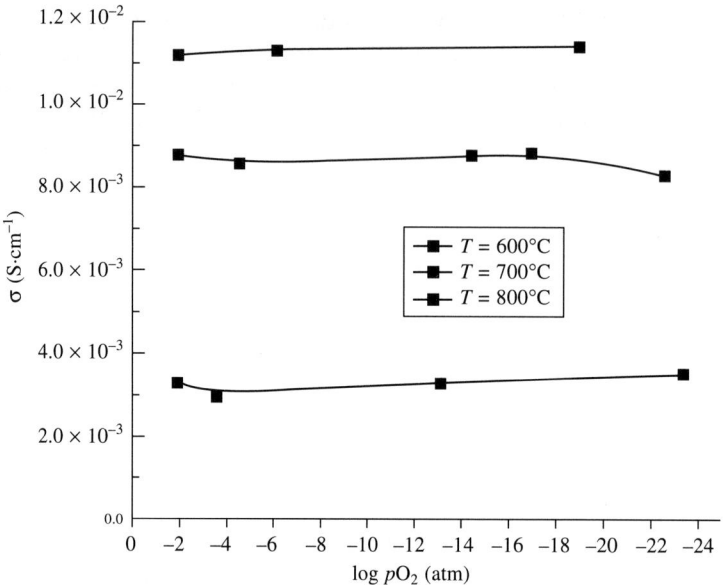

Fig. 8 Electrical conductivity of the apatite; a function of oxygen partial pressure at 600, 700 and 800°C (Brisse et al. 2005).

In the first step, creation of vacant oxygen sites or in oxidizing atmospheres creation of electronic holes. For example, replacement of Ce^{4+} in $BaCeO_3$ with Gd^{3+} can be written in Kroger-Vink notation as

$$2Ce^x_{Ce} + O^x_O + Gd_2O_3 \rightarrow 2Gd'_{Ce} + V^{..}_O + 2CeO_2 \qquad (5)$$

$$2Ce^x_{Ce} + Gd_2O_3 + \frac{1}{2}O_2(g) \rightarrow 2Gd'_{Ce} + 2h^. + 2CeO_2 \qquad (6)$$

In the second step, upon exposure to water vapor (H_2O) or H_2 containing atmosphere, the electron hole reacts with hydrogen or water to produce proton in the material,

$$\frac{1}{2}H_2 + h^. \rightarrow H^+ \qquad (7)$$

$$H_2O(g) + 2h^. \rightarrow 2H^+ + \frac{1}{2}O_2(g) \qquad (8)$$

The other possibility could be

$$H_2O(g) + V^{..}_O + O_O \rightarrow 2(OH^.)_O \qquad (9)$$

The protons are believed to be not bound to any particular oxygen ion, but are rather free to migrate from one ion to the next. This easy migration results in the high proton conductivity in the doped perovskite oxides.

The temperature dependence of conductivity of few proton conducting oxides are shown in Fig. 9. While the highest conductivity is obtained in case of Nd-doped $BaCeO_3$, the lowest conductivity was normally observed for zirconates. At 600°C in 5% H_2/N_2 atmosphere, $SrCe_{0.95}Yb_{0.05}O_{3-\delta}$ exhibits conductivity value

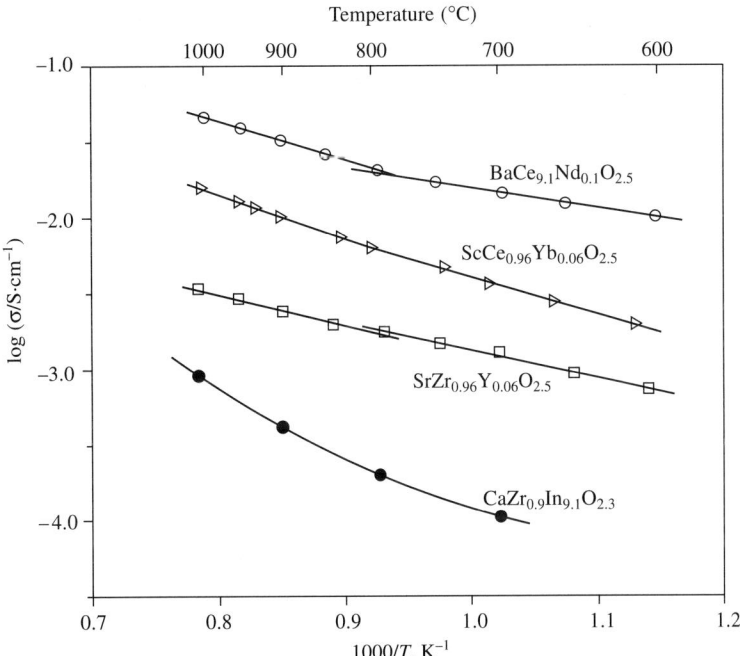

Fig. 9 Comparison of proton conduction in perovskite oxides (Ishihara et al. 2003).

of ~2 × 10^{-2} S/cm. Very recently Sammes's group, USA, examined the possibilities of using modified BaCeO$_3$ (Higgins et al. 2005).

Although, zirconates show poor electrical conductivity values, but their chemical stability is reported to be very strong compared to cerates. The cerates normally show poor stability in carbon dioxide containing atmosphere due to the formation of BaCO$_3$ (Shima et al., 1997). Nevertheless, considering few characteristics of the doped BaZrO$_3$, namely, high stability, reasonably good electrical conductivity and wide ionic domain, Veith et al. (2000) has fabricated proton conducting oxide thin films by sol-gel technique. Although, some encouraging results have being published on proton conducting oxides, yet their acceptance still lies on the complete elimination of the stability related problem which they suffer mostly.

3.1.3 Electrolyte Powder Synthesis
Electrolyte grade sub-micron size YSZ powders are now commercially available at a reasonable price which gives full densification when sintered between 1350 and 1400°C. Similar densification is possible to achieve at temperature as low as 1200°C using commercially available nanocrystalline size powders, but at relatively higher price. The powders could also be produced in several possible routes such as conventional solid state route, co-precipitation and soft chemistry route in the laboratory. One can also prepare YSZ by mixing the stoichiometric amount of ZrO$_2$ and Y$_2$O$_3$ followed by ball milling under ethanol and drying overnight at 60°C and finally calcining at 1200°C for 4-6 h. Such powder after compaction may require to sinter at 1650°C for 6 h to obtain full densification. Similar to YSZ, CeO$_2$- based (e.g. CGO and SDC) and LaGaO$_3$-based (e.g. LSGM) powders are also available commercially. However, in laboratory one can prepare these electrolytes following the above mentioned solid state route. To lower the sintering temperature below 1400°C, more reactive powders are required (nanocrystalline size) for which the soft chemistry route is so far the best choice. In this technique, a gel is formed from the mixed metal nitrate solutions in the

presence of fuel (e.g. citrate, glycine etc.). The mixture of nitrates and citrate are then finally ignited as a result of the thermally induced oxidation-reduction reaction. Upon calcination within 500-700°C for 3-4 h in air, the ash obtained from the ignition yields the desired powder. Subsequently, grinding in ball mill or planetary mill under ethanol or similar solvent produces ultrafine/nanocrystalline powder. Normally the powders prepared in this route are single-phased and homogeneous.

3.1.4 Processing of YSZ Electrolyte Films

Several processing techniques are involved for fabricating SOFC electrolytes. However, these electrolytes must be fully densed (gas-tight) and the choice of processing technique becomes critical when the electrolyte is deposited in the form of thin film on a porous supported electrode. It has already been mentioned that only two designs (tubular and planar) are being practiced by the SOFC developers for commercialization. Both the designs are on a porous electrode-supported—cathode or anode. Electrolyte in the thin film form gets support from this porous electrode. For example, in case of tubular, the porous cathode tube acts as the main support, whereas for planar, the main support is the porous anode. The electrodes are typically 1-2 mm thick.

The SOFC technology is approaching towards more matured stage with large number of demonstration units in operation in various parts of the world. Keeping the mass production in view, most of the developers are now trying to exploit the existing traditional cost-effective fabrication technique which has already been discussed in the beginning of this chapter. Table 8 represents a host of such processing techniques which are now being currently used by U.S. DOE's SECA industrial team and is expected to lead the mass production of SOFCs. Stöver et al. (2003) in their review article has described most of these processing techniques. Electrochemical vapor deposition (EVD) technique, developed by Siemens Westinghouse, is

Table 8.　U.S. DOE-SECA industrial team, design and manufacturing (Williams et al. 2005 (modified))

Company	SOFC design and features	Processing techniques
Cummins-SOFCo	Electrolyte-supported planar (825°C), thermally matched materials and seal-less stack	Tape casting and screen printing followed by co-firing
Delphi Battelle	Anode-supported planar (750°C), ultra compact having rapid transient capability	Tape casting and screen printing followed by two stage sintering
General Electric Company	Anode-supported planar (750°C), hybrid compatible with internal reforming	Tape casting and screen printing followed by two stage sintering
Siemens Westinghouse Power Corp.	Cathode-supported flatten oval (800°C) having seal-less stack	Extrusion and plasma spraying
Acumentrics Corporation	Anode supported-microtubular (750°C) with thermally matched materials robust structure with rapid start-up	Extrusion and plasma and spray deposition followed by co-sintering
FuelCell Energy, Inc.	Anode-supported planar (< 700°C) with low cost metals having thermal integration	Tape casting and screen printing followed by co-sintering having additional electrostatic deposition

DOE: Department of Energy, SECA: Solid State Energy Conversion Alliance.

the only technique which has the ability to deliver the most robust gas-tight zirconia electrolyte film and superior adherence with the porous LSM cathode tube (Pal et al. 1990). However, difficulties in high temperature (1000°C) deposition, extremely slow rate of deposition (3 μm h^{-1}) (Isenberg 1981), expensive precursors, equipment and maintenance pushed the company to switch over to relatively less expensive technologies like plasma spraying (Table 8).

Forchungszentrum Jülich, Germany has demonstrated the anode-supported (NiO/YSZ) planar design with YSZ electrolyte thickness < 25 μm and examined the feasibility of operating such cell as low as 750°C using an inexpensive vacuum slip casting technique and operating the stack at ~ 750°C (Buchkremer et al. 1996, 1997) with ferritic steel as the interconnect. Later on, it is reported that the zirconia electrolyte thickness could even be brought down to 5-10 μm (Stöver et al. 1999, Ghosh et al. 1999, Kim et al. 1999, De Jonghe et al. 2003, Minh 2004). Attempt has also been made to reduce the number of processing steps involved to make these thin-film-supported-structures (Basu et al. 2005). Typical cross-section of such cells and their performances are shown in Figs. 10 and 11, respectively. Current density of ~ 1 A/cm^2 at 800°C has been achieved for 5 μm electrolyte obtained by co-firing technique.

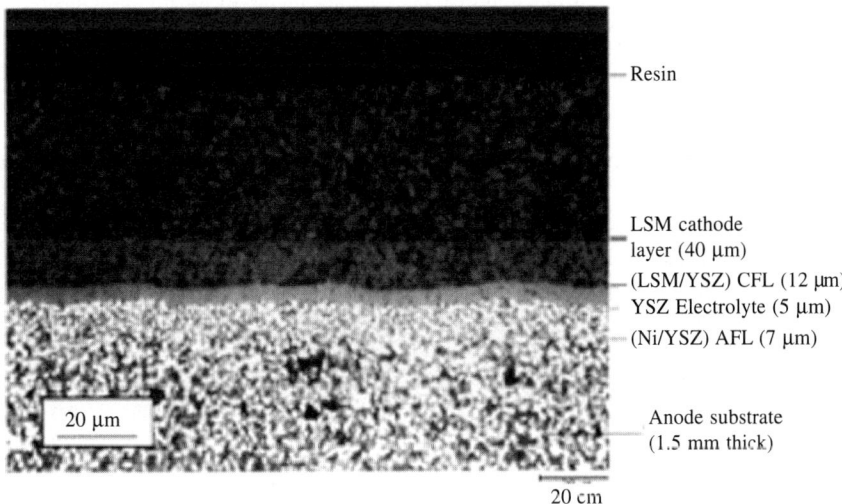

Fig. 10 Anode-supported thin film zirconia electrolyte support. Electrolyte film was co-fired along with anode functional layer (AFL) coated anode support and combination of the processing technique yielded a current density (~ 1 A cm^{-2} at 0.7) at 800°C (Basu et al. 2005).

Similar to vacuum slip casting, electrophoretic deposition (EPD) is also another colloidal approach. EPD is an extremely low cost, but potential technique to consolidate particulates on any dimension and shape (Randall et al. 2000). In EPD, charged particles from a stable suspension migrate towards a counter electrode under an applied field. The counter electrode is essentially the substrate for deposition. As particles migrate towards, and accumulate on the substrate electrode, a well-packed green film develops and grows. Several attempts have been made to utilize EPD to fabricate, particularly dense electrolyte film in the fuel cell structure (Ishihara et al. 1996, Negishi et al. 1999, Basu et al. 2001, Sarkar et al. 2005, Duquette et al. 2005). Typical green compaction of YSZ particulates and sintered film showing fully densed (gas-tight) YSZ film on porous LSM cathode tube obtained by EPD technique are shown in Fig. 12.

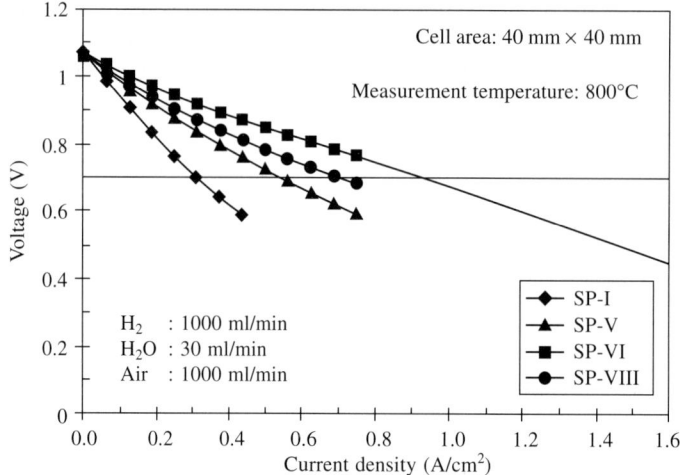

Fig. 11 Current voltage characteristics as a function of processing technique for operating temperature 800°C. The processing technique SP-VI shows the best cell performance (~ 1 A/cm²), referred to as the optimized combination of processing techniques and subsequent co-firing of zirconia electrolyte + anode functional layer + anode support (Basu et al. 2005).

Fig. 12 Processing of YSZ electrolyte film by EPD on porous LSM cathode tube (Siemens-Wesinghouse's tube): (a) Green deposition (unfired), (b) sintered surface and (c) sintered gas-tight film (Basu et al. 2001).

Fugitive graphite interlayer has been used in order to obtain such a high quality deposit. The graphite interlayer, however, burnt out completely while firing the ceramic coating. There are varieties of other techniques which are being employed to get such highly densed zirconia electrolyte film. In fact, using all these processing techniques it is possible to fabricate tapes or thin/thick films for other electrolyte materials such as doped CeO_2, doped $LaGaO_3$ etc.

3.2 Cathodes

The most commonly used and extensively studied cathode material in ceramic fuel cell is based on $LaMnO_3$ which is basically a perovskite oxide (ABO_3) with *p*-type conductivity. In order to achieve a high electronic conductivity it is doped heavily by acceptor which leads to an enhanced hole concentration and the resultant conductivity is due to hopping of an electron hole between Mn^{3+} and Mn^{4+}. It is doped at A-site or both at A- and B-sites with other cations. Normally cations with larger ionic radii (such as Ca^{2+}, Sr^{2+}) are preferred to substitute at A-site and cations with smaller ionic radii (Co, Fe, Ni, Mn, Cr) preferred to occupy B-site. The crystal structure is a function of the composition which is solely dependent on A- and B-site dopant and on oxygen nonstoichiometry which is influenced by temperature and oxygen partial pressure (Galasso 1969, Anderson 1992). The Sr-doped LaMnO$_3$ (commonly known as LSM) is rhombohedral at room temperature whereas pure LaMnO$_3$ (undoped) is orthorhombic. However, the transition from rhombohedral to tetragonal and even cubic structures can occur depending upon A-site substitution level and the temperature (Badwal et al. 1997).

Strontium enhances the electronic conductivity by increasing the Mn^{4+} content while substituting La^{3+} with Sr^{2+} and the process can be described by the following equation:

$$LaMnO_3 + x\,Sr^{2+} \rightarrow La^{3+}_{1-x}Sr^{2+}_x\,Mn^{3+}_{1-x}\,Mn^{4+}_x O_3 \qquad (10)$$

Fig. 13 shows the enhancement of electronic conductivity with the increase in Sr-level. The temperature dependence of the electrical conductivity gives straight line plots which suggests the small polaron hopping conduction and expressed by

$$\sigma T = (\sigma T)^0 \exp(-E_a/kT) = A(h\nu^0/k)c(1-c)\exp(-E_a/kT) \qquad (11)$$

where $(\sigma T)^0$ and E_a are the pre-exponential constant and the activation energy, respectively, A is a constant and c is the carrier occupancy on the sites and therefore, $c(1-c)$ indicates the probability of hopping from the carrier occupied site to the unoccupied sites (Yokokawa et al. 2003).

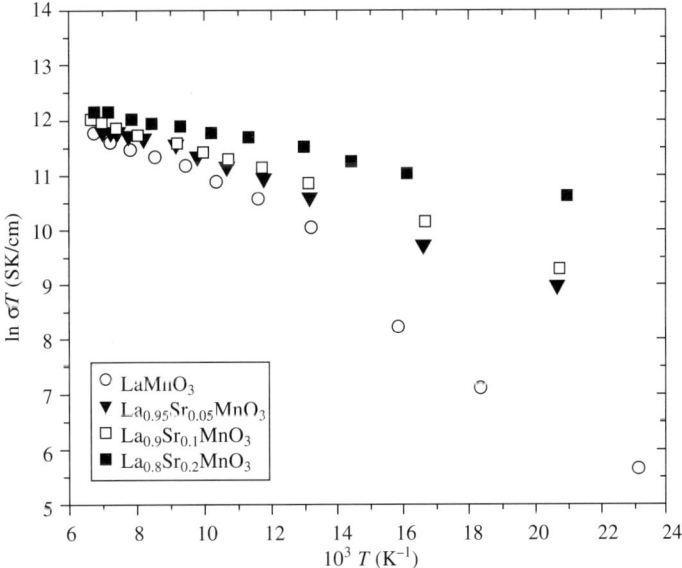

Fig. 13 Temperature dependence of conductivity for undoped and Sr-doped LaMnO$_3$ (Kuo et al. 1990).

It has been found that by increasing the Sr or Ca content in the A-site it is possible to increase the electronic conductivity due to change in the Mn^{3+}/Mn^{4+} ratio (Chakraborty, 1995). The resulting conductivity at 1000°C for La$_{0.8}$Sr$_{0.2}$MnO$_3$ may vary between 100 and 200 S/cm which shows maximum value for a Sr-content close to 55 mol%. However, their ionic conductivity is negligibly small, ~10^{-7} S/cm at 800°C in air. Sasaki et al. (1996) pointed out that this total conductivity (electronic and ionic) of LSM is not sufficient to completely neglect the in-plane resistance in the cathode—it is possible to minimize the corresponding losses by providing a relatively larger geometry and location of the current collector during cell operation.

LSM is stable in oxidizing atmosphere at operating temperature but starts decomposing beyond a certain oxygen partial pressure (10^{-13} atm) to La_2O_3 and MnO (Kuo et al. 1989). The same group later reported that although it shows stable electronic conductivity values, but it starts decreasing below pO_2 at around 10^{-10} atm at 1000°C (Kuo et al. 1990).

The thermal expansion coefficient for $La_{0.8}Sr_{0.2}MnO_3$ is 12.4×10^{-6}/K when measured between room temperature and 1000°C which is acceptable for SOFC application. As already mentioned it is possible to increase the electronic conductivity by increasing the A-site doping with strontium, but this leads to an increase in TEC. This trend could be controlled to some extent by doping with Ca. TEC also increases if B-site dopant increases and when it is completely replaced by Co, then the extraordinarily high TEC (19×10^{-6}/K) was obtained in case of composition $La_{0.8}Sr_{0.2}CoO_{3-y}$. Though, the conductivity of this composition is extremely high, but this high TEC value is not at all suitable as SOFC cathode.

Strontium doped $LaMnO_3$ is available commercially. It could also be produced in the laboratory following similar techniques as described in Section 3.1.3. It is well known that during processing (sintering) of LSM cathode in contact with YSZ electrolyte it starts forming an insulating phase from 1200°C. Due to this chemical interaction between these two cell component materials, an insulating phase, namely, $La_2Zr_2O_7$ is formed. Beside this phase another phase, $SrZrO_3$ is also formed when strontium content is high in LSM. The interaction is significant at high temperature and with longer duration. It has been found that a layer of $La_2Zr_2O_7$ up to 5 μm thick can be formed at the interface between $LaMnO_3$/YSZ when fired at 1450°C for 48 h. These are undesirable insulating phases and cause thermal stresses at the interface. The interaction between $LaMnO_3$/YSZ appears to proceed via the unidirectional diffusion of manganese, lanthanum or dopant cation like strontium, calcium into the YSZ (Minh et al. 1995). Badwal et al. (1997) attempted to summarize the overall reaction behavior and formation of additional phases at the $LaMnO_3$/YSZ interface (Fig. 14). In a recent study (Basu et al. 2004), the appearance of the $La_2Zr_2O_7$ phase was reported at the interface between $LaNi_{0.6}Fe_{0.4}O_3$ (LNF)/YSZ during co-firing

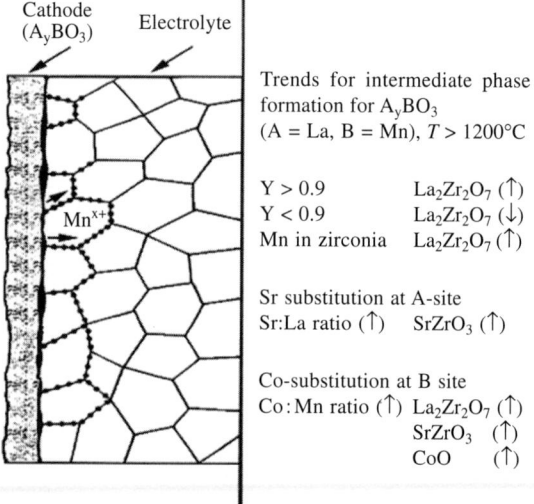

Cathode (A_yBO_3)   Electrolyte

Mn^{x+}

Trends for intermediate phase formation for A_yBO_3
(A = La, B = Mn), $T > 1200$°C

Y > 0.9 $La_2Zr_2O_7$ (↑)
Y < 0.9 $La_2Zr_2O_7$ (↓)
Mn in zirconia $La_2Zr_2O_7$ (↑)

Sr substitution at A-site
Sr:La ratio (↑) $SrZrO_3$ (↑)

Co-substitution at B site
Co:Mn ratio (↑) $La_2Zr_2O_7$ (↑)
            $SrZrO_3$ (↑)
            CoO (↑)

Fig. 14 Possible interfacial reactions between ABO_3 perovskite cathode materials and zirconia electrolyte (Badwal et al. 1997).

of the anode-supported structure at 1400°C (Fig. 15, Table 9) while investigating the compatibility of LNF (cathode) and YSZ (electrolyte). Nevertheless, this interaction could be minimized to a limited extent by a careful control of the composition and the preparation parameters, for example using a La-deficient $LaMnO_3$. Other possibility to control this interaction is to substitute La by a rare earth cation with smaller ionic radius such as Pr or other alternate Ce-based cathode material because the reaction products, $Pr_2Zr_2O_7$ and $Ce_2Zr_2O_7$, are thermodynamically unstable (Steele 1994).

Unfortunately, there is no single composition in LSM family which can fulfil the requirements of high electronic and ionic conductivities, more close thermal expansion (to that of electrolyte), optimum electrochemical performance and enhanced chemical stability. Therefore, it is desirable to have a new porous cathode material containing mixed oxides for fast transport of ionic and electronic defects through solid phase, rapid flow of gases through the pores and for efficient electrochemical reaction at the cathode/ electrolyte interface (Mogensen et al. 1996, Tanner et al. 1997, Dokiya 2002). The junction (cathode-

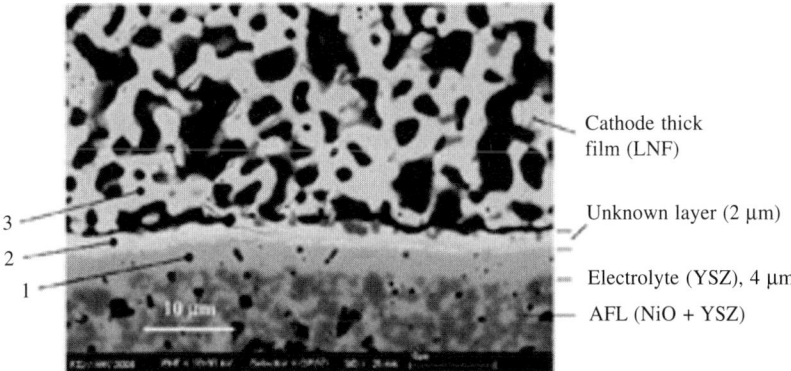

Cathode thick
film (LNF)

Unknown layer (2 μm)

Electrolyte (YSZ), 4 μm

AFL (NiO + YSZ)

Fig. 15 SEM image of a sintered (1400°C/5h) sample where only one cathode layer (LNF) was used. The EDX analysis was carried out at three different locations: electrolyte film (point 1), unknown layer (point 2) and cathode film (point 3) (Basu et al. 2004).

Table 9. **Results of chemical analysis carried out at three different points as indicated in SEM image (Fig. 15) (Basu et al. 2004)**

Name of the element	Point 1		Point 2		Point 3	
	Element (Wt%)	Possible composition	Element (Wt%)	Possible composition	Element (Wt%)	Possible composition
La	1.37	$Zr_{0.79}Y_{0.21}O_{2-x}$	39.26	$La_{1.65}Y_{0.39}Zr_{1.96}O_{7-x}$	58.13	$La_{1.10}Fe_{0.73}Ni_{0.35}O_{3-x}$
Zr	55.91		30.65		–	
Y	14.43		5.98		–	
Ni	1.65		2.35		11.83	
Fe	–		–		7.79	

electrolyte-gas phase) where this electrochemical reaction takes place is known as the triple phase boundary (TPB). In conventional ZrO_2-based SOFC, the normal practice to extend this triple phase boundary is to mix the LSM and YSZ in 1:1 ratio, also known as cathode functional layer (CFL). It is believed that CFL enhances the catalytic activity (Mogensen et al. 1996). However, for low temperature application (500-700°C), a large number of new cathode materials mostly on $LaCoO_3$ and $LaFeO_3$-based have been developed (Anderson et al. 1995, Petric et al. 2000, Ralph et al. 2001). The interactions of these cathode materials with that of YSZ and CGO electrolytes have been studied in detail. It has been found that CeO_2-based electrolyte does not react with these alternate cathode materials (Table 10). Very recently, Liu et al. (2004) claimed to have some innovative nanostructured electrodes which successfully enhance the cell performance. In their study they used a mixture of 70 wt% of $Sm_{0.5}Sr_{0.5}CoO_{3-\delta}$ and 30 wt% of $Sm_{0.1}Ce_{0.9}O_{3-\delta}$ (SDC) as the cathode material.

3.3 Anodes

SOFC anode is exposed to an extreme reducing environment ($< 10^{-18}$ atm) at high operating temperature (500-1000°C) and its material requirements are determined by its function and fabricated structure. The main functions of the anode could be broadly classified as to provide (i) reaction sites for the electrochemical

Table 10. Alternate cathode materials for low temperature operation (500-700°C) and their interactions with YSZ and CGO (Ralph et al. 2001)

Composition	Reaction products	
	With YSZ	With CGO
$La_{0.8}Sr_{0.2}CoO_3$	$La_2Zr_2O_7$, $SrZrO_3$	None
$La_{0.8}Sr_{0.2}Co_{0.8}Ni_{0.2}O_3$	$La_2Zr_2O_7$, $SrZrO_2$, NiO	None
$La_{0.8}Sr_{0.2}Co_{0.8}Fe_{0.2}O_3$	$La_2Zr_2O_7$, $SrZrO_3$	None
$La_{0.8}Sr_{0.2}Co_{0.8}Ni_{0.15}Cu_{0.05}O_3$	$La_2Zr_2O_7$, $SrZrO_3$	None
$Sm_{0.8}Sr_{0.2}FeO_3$	$Sm_2Zr_2O_7$, $SrZrO_3$	None
$La_{0.8}Sr_{0.2}FeO_3$	none	None
$La_{0.8}Sr_{0.2}Fe_{0.8}Ni_{0.2}O_3$	$La_2Zr_2O_7$, $SrZrO_3$	None
$La_{0.8}Sr_{0.2}Fe_{0.8}Ni_{0.15}Cu_{0.05}O_3$	$La_2Zr_2O_7$, $Sr_2Fe_2O_5$	None
$LaNiO_3$	$La_2Zr_2O_7$, La_2NiO_4, NiO	La_2NiO_4, NiO
$Nd_{0.8}Sr_{0.2}CoO_3$,	Not measured	Not measured
$Gd_{0.8}Sr_{0.2}CoO_3$,		
$La_{0.8}Sr_{0.2}Fe_{0.8}Co_{0.2}O_3$,		
$YBa_2Cu_3O_7$, $Bi_2Sr_2CaCu_2O_8$		

oxidation of the fuel and (ii) a path for electrons to be transported from the reaction sites to the interconnect. Fuel oxidation at the anode-electrolyte interface produces the electrons which flow through an external circuit besides water and/or carbon dioxide being the end products. SOFC anode has to have extreme fuel flexibility. It should accept hydrogen, natural gas, carbon monoxide, and various light hydrocarbons. Therefore, anode must be a good electronic conductor with high surface area and it should have high catalytic activity towards H_2 and CO oxidation reaction. Sufficient amount of porosities are necessary to allow the fuel and byproducts to be delivered and removed from the reaction sites without significant diffusion limitation. The performance of an SOFC depends strongly on the anode structure and because some aspects of electrochemical reactions are quite different from normal heterogeneous reactions. It is an established fact that the electrochemical reaction can only occur at the three-phase boundary (TPB), which is defined as the collection of sites where electrolyte, the electron-conducting metal phase, and the gas phase all come together. If there is a breakdown in connectivity in any one of the three phases, the reaction cannot occur. For example, if ions from the electrolyte cannot reach the reaction site, if gas-phase fuel molecules cannot reach the site, or if electrons cannot be removed from the site, then the site cannot contribute to the performance of the cell. However, the structure and composition greatly affects the size of the TPB also, which is suggested to be within 10 μm (Brown et al. 2000, Gorte et al. 2003). The charge transfer at the TPB has certain limitations as only a part of the anode-electrolyte takes active part. Several propositions have been reported to explain the appropriate role of anode TPB and its enhancement in order to have high performance cell (Mogensen et al. 2000, McEvoy, 2003). Fig. 16 illustrates three distinct possibilities of forming anodes and their TPBs (in presence of fuel) as shown by thick black portion. The options are: metal/electrolyte (Fig. 16(a)), cermet/electrolyte (Fig. 16(b)) and mixed ionic and electronic (MIEC) oxide/electrolyte (Fig. 16(c)). An extension of TPB, or in other words, the enhancement of catalytic activity can be obtained only in case of MEIC-oxide/electrolyte (Fig. 16(c)).

As already mentioned, so far, three types of materials were used for SOFC anode, viz. (a) pure metal, (b) cermet and (c) MIEC-oxide. A host of metals such as Ni, Pt, Co and Ru have so far been considered as

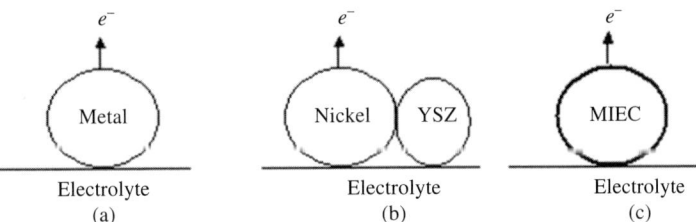

Fig. 16 Schematic of the anode reaction zone for the three distinct cases: (a) Metal/electrolyte, (b) Cermet/electrolyte and (c) MIEC/electrolyte. Thick black portion represents the reaction zone and MICE represents mixed ionic and electronic conductor. In case of MIEC the enhancement of reaction zone is shown.

SOFC anode (Minh et al., 1995). However, less expensive nickel has been finally emerged and established as the candidate anode material because of its good catalytic activity to steam reforming of natural gas despite certain less significant disadvantages like low resistance to sintering and grain growth. The reason for non-acceptance of other metals may be due to individual or combined effect of toxicity, availability and cost. Nevertheless, some of them have the potential to check the carbon deposition process at the anode and certain level of sulphur tolerance. The widely used SOFC anode is Ni-YSZ cermet. Pure metallic nickel has a very high thermal expansion coefficient (TEC) (16.9×10^{-6}/K), compared to the YSZ (10.5×10^{-6}/K). But when both are mixed, they form a Ni-YSZ cermet which has reduced TEC (12.7×10^{-6}/K, for 40vol% of Ni + 60vol% YSZ). Matching of thermal expansion is an essential criterion for this type of cermet composition when they are used as SOFC anode. Similarly the cermets must have good mechanical properties. The reported mechanical strength for Ni-YSZ cermet is less than 100 MPa (Badwal 1997).

Overpotential of Ni-YSZ cermet anode is significantly less compared to other ceramic fuel cell components. Normally, planar SOFCs having standard cell components is capable of producing more than 500 mW/cm^2 of power. However, a careful control of the anode substrate fabrication could even deliver very high power densities, even up to 1.8 W/cm^2 (~3.5 A/cm^2 at 0.5 V) at 800°C (Kim et al. 1999). The detailed information and fundamental studies on this aspect is available in numerous articles—calculations on the optimal morphology, microstructure, porosity and thickness of such cermet anodes have been reported by Minh et al. (1995), Costamagna et. al. (1998), Divisek et al. (1999), Sunde (2000) and McEvoy (2003).

The electrical conductivity of Ni-YSZ cermet is between 500 and 1000 S/cm at 1000°C (Badwal et al. 1997). Dees et al. (1987) has demonstrated through their S-shaped curve (conductivity vs. amount of nickel) that these cermets develop higher values of electronic conduction at or above about 30vol% Ni and below this nickel content, the conductivity is dominated by that of YSZ with little connectivity for nickel particles. Fig. 17 is a typical S-shaped curve obtained by Pratihar et al. (1999). This percolation behavior is explained in terms of the

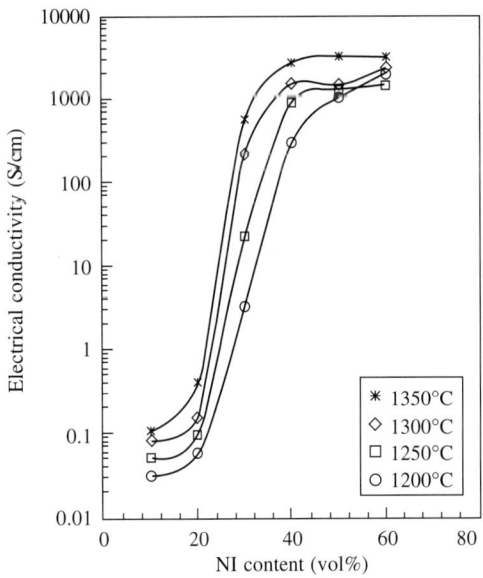

Fig. 17 Variation of electrical conductivity (measured at 1000°C) as a function of nickel concentration of the cermets at the indicated temperature (Pratihar et al. 1999).

existence of two conduction mechanisms e.g., the electronic conduction in nickel and the ionic conduction in YSZ. A schematic of a Ni YSZ cermet anode mechanisms which strongly depends on the synthesis procedure is shown in Fig. 18. There are various ways to prepare such cermets (Zhu et al. 2003). But none of the methods could provide a cermet structure where metallic nickel can cover around the zirconia particles and form a uniformly Ni-coated cermet. Schematic of such concept of uniform coating of nickel on YSZ powders is shown in Fig. 18(c) where a novel electroless technique was employed. In presence of palladium catalyst, individual YSZ particles are possible to coat uniformly by metallic nickel using a unique electroless-nickel-bath developed at author's group (Pratihar et al. 2004). The cermet thus obtained could be able to shift the percolation to much lower value than in the case of conventional cermet anode (30 vol% Ni +YSZ). Fig. 19 is typical photo-micrographs of 15 vol% and 20 vol% nickel-coated cermet samples prepared by the electroless technique. By improving the process parameters, it has been able to achieve conductivity value as high as 1440 S/cm at 1000°C in cermet having composition 20 vol% of Ni +80% YSZ.

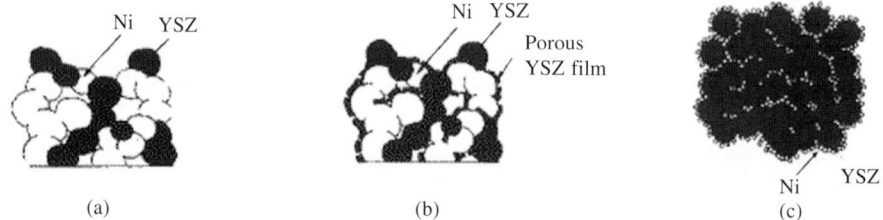

Fig. 18 Schematic of anode cermet structure showing interpenetrating networks of pores and conductors—nickel for electrons, yttria-stabilized zirconia for oxygen ions. Reactive sites are contact zones of the two conducting phases, also accessible to fuel through the porosity (McEvoy, 2003). Preparation methods are by (a) and (b) conventional slurry and vapor-phase, respectively (Ogumi et al. 1995); (c) electroless technique (author's lab).

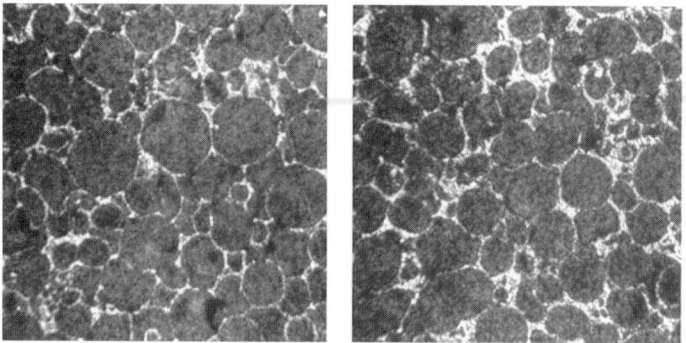

Fig. 19 Optical micrographs of polished Ni-YSZ cermets prepared by electroless technique having nickel content 15 vol% (left) and 20 vol% (right). Magnification used is 10×. The white patches surrounding the individual YSZ (black) is nickel. A clear indication of enhancement of the quantity of the nickel is seen. No pore former was used in these cermet samples.

Like Ni-YSZ, when the YSZ matrix is replaced with doped-CeO_2, the cermet of Ni-CeO_2 is formed which is also used as a SOFC anode. Attempts have also been made to develop alternate anode materials such as Ni-Al_2O_3, and Ni-TiO_2 in order to eliminate the thermal (TEC) expansion mismatching between YSZ electrolyte and traditional Ni-YSZ anode (Stöver et al. 1999). The TEC of YSZ and Ni/YSZ

(40/60 vol%) are 10.5×10^{-6}/K and 12.5×10^{-6}/K, respectively. The TECs for Al_2O_3 and TiO_2 are much smaller than YSZ. Therefore, it is possible to tailor made to a somewhat closer thermal expansion matching compared to other cell components including ferritic steel. In case of $Ni-Al_2O_3$, because of the low sinterability the sintered structure becomes more porous and thus missing electrical pathways in the cermet. Considering the other characteristics like reduction behavior, gas permeability and electrical conductivity, $Ni-TiO_2$ (40 vol% $Ni + TiO_2$) showed better performance and could be the replacement anode of the existing anode, Ni-YSZ.

Although, the potentiality of Ni-YSZ cermets anode is beyond doubt, but, still it suffers from some disadvantages such as agglomeration of nickel (nickel grain coarsening), sulfur poisoning and carbon deposition when natural gas is used as the fuel (in dry or less moist conditions). Such anode cermet also experienced large dimensional change upon redox cycling which may even cause cracking of the anode (cermet) and/or thin electrolyte coating. This may lead to gradual degradation of voltage/power and even stop generating power. Therefore, it is necessary to have new redox-stable anode material, which is basically a ceramic oxide having mixed ionic and electronic conduction (MIEC) (Fig. 16(c)) and addition of such MIEC oxides may help increasing the TPB (Fleig 2002).

Strontium titanate has also been proposed as an alternative anode material. It is chemically stable perovskite oxide and exhibits n-type electronic conduction upon reduction due to the presence of Ti^{3+}. Like other perovskites its electrical conductivity could be enhanced by donor doping with tri or pentavalent oxides such as La^{3+}, Y^{3+} or Nb^{5+} (Moos et al. 1997, Marina et al. 2002, Hui et al. 2002, Fu et al. 2005). Hui et al. (2002) reported that yttria-doped $SrTiO_3$ has a relatively high electronic conductivity at higher temperature (> 60 S/cm at 800°C) in highly reducing atmosphere (~ 10^{-19}atm) compared to other rare earth dopants. They have also studied the effect of acceptor doping (Fe, Co, Ni, Cr) at Ti-site and of all these, cobalt-doped sample, $Sr_{0.85}Y_{0.10}Ti_{0.95}Co_{0.05}O_3$, showed the highest conductivity and also highest resistance to re-oxidation than $Sr_{0.85}Y_{0.10}TiO_3$. However, according to Fu et al. (2005) further studies are required to establish the actual mechanism of redox irreversibility of this family of doped perovskites.

3.3.1 Anode Materials for Direct Oxidation SOFC

The development of SOFCs which can operate directly with dry hydrocarbon has great potential for applications such as transportation and distributed power generation (Gorte et al. 2000, 2003). But, avoiding hydrocarbon cracking is the key problem in operating SOFCs on hydrocarbon fuels. For reduced temperature operation of SOFCs (< 700°C), the direct hydrocarbon operation is attractive, because rates of hydrocarbon cracking reactions decrease with decreasing temperature, limiting problems with anode coking (Barnett 2003). But, the fuel utilization and power output is expected to be less because the kinetics of the anode reaction becomes considerably slower with methane than hydrogen. Traditional Ni-YSZ is modified in such cells. Ni-YSZ with an underlying Y-doped ceria (YDC) layer is used as an alternate anode. The maximum power density is reported to be 0.3 W/cm^2 at 650°C; the value is about five times less than for the same cells operated on hydrogen (Murray et al. 1999). This has been verified by impedance spectroscopy measurements carried out in two fuels (hydrogen and methane) at 600°C. Bigger arc for methane is observed, indicating a six times higher polarization resistance for methane than hydrogen. The effect of introduction of additional ceria layer in conventional Ni-YSZ cermet anode has also been demonstrated by impedance spectroscopy. The smaller arc is an indication of enhanced anode reaction. A factor of about six times decrease in interfacial resistance is observed when YDC layer is used. In both the cases humidified methane is the fuel. Fig. 20 represents the impedance spectra obtained for two different combinations of anode materials.

Traditional Ni-YSZ is not the choice of anode for direct oxidation fuel cell. Because at high temperature operation of SOFC, the catalytic combustion rate is high; however, carbon formation in the presence of dry hydrocarbons is thermodynamically favored as an intermediate product prior to complete combustion. Therefore, direct oxidation SOFCs are only possible if the anodes are inert for carbon formation. Copper based anodes, Cu-YSZ, has been demonstrated for this type of cells. Copper has been selected because it is cheap and an excellent electronic conductor. But it is a poor catalyst for C—C bond formation, a reaction that is likely related to coke formation. To overcome the preparation of Cu-YSZ (due to low melting temperature of copper) an innovative in-situ metal (Cu) impregnation technique has been introduce by Gorte et al. (2000). Their process involves a dual tape casting technique having both porous and dense YSZ tapes that finally gives a monolith structure. Copper nitrate is then impregnated on the porous YSZ matrix which upon firing at low temperature (450°C) forms the Cu-YSZ cermet. An enhancement in cell performance is observed when ceria is incorporated in the Cu-YSZ matrix.

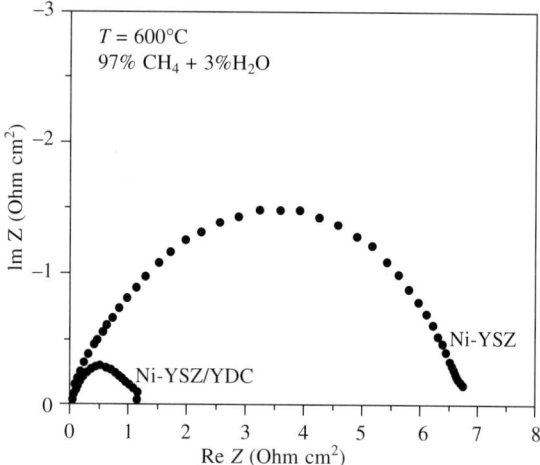

Fig. 20 Impedance spectra taken at 600°C in humidified methane for Ni-YSZ anodes with and without YDC layer showing a dramatic reduction in the polarization resistance due to the YDC (Murray et al. 1999).

3.4 Interconnects

Given that a single SOFC produces only 1 volt or less in open circuit condition, individual cells need to connect together – anode of one cell with the cathode of another cell. In other words, to obtain practical voltage output a number of individual cells consisting of a porous anode, a dense electrolyte, and a porous cathode are connected in series to form a stack. The connector is known as the interconnect. Essentially, interconnect must be fully densed and physically separates two individual cells, yet connects them electronically. The main functions of this cell component are to ensure: (i) the supply of fuel to the anode and oxidant to the cathode through gas channels and (ii) remove the reaction products in order to facilitate smooth electrochemical reaction (Anderson et al. 2003, Zhu et al. 2002). Fully densed structure of interconnect also prevents direct mixing of fuel and oxidant gases. It must withstand the operating temperature of the SOFC and exerts two extreme environments, approximately 10^{-18} atm to normal atmosphere.

SOFC interconnect could be either ceramic or metals/alloys. Ceramic interconnects are normally used between 800°C and 1000°C, whereas metallic interconnects are preferred for 750°C and below. While doped LaCrO$_3$ is used as ceramic interconnects, the Cr-based alloys and ferritic steels are the choice for metallic interconnects. In a recent review article, Sakai et al. (2004) has summarized the trend of research activities in the area of SOFC interconnects (Fig. 21). Very recently possibility of using stainless steel has also been examined by Ishihara et al. (2003) and Bance (2004). For the Siemens-Westinghouse tubular SOFC, the interconnection is deposited in the form of a ~ 85 μm thick, 9 mm wide strip along the air electrode tube length by plasma spraying. On the other hand, bipolar plates having channels on both sides are used for planar geometry. Several materials have been tested for SOFC interconnects, details of which are now available in the literature (Quadakkers 2003).

3.4.1 Ceramic Interconnects

Among the two perovskites, namely alkaline earth metal doped LaCrO$_3$ and YCrO$_3$ perovskite oxides, doped LaCrO$_3$-based perovskite is widely used as the ceramic interconnect material (Armstrong et al.

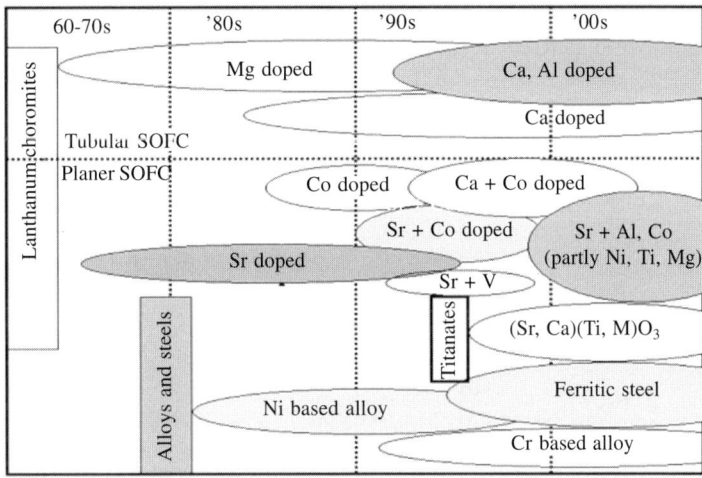

Fig. 21 Trend of R&D in SOFC interconnects material (Sakai et al. 2004).

1999). The present section will deal only with this variety of oxide. Pure LaCrO₃ is a p-type semiconductor and exhibits electrical conductivity only about 0.6-1 S/cm at 1000°C in air (Yasuda et al. 1993). These chromites are useful candidates as interconnects because at 1000°C (operating temperature) and highly reducing environment they remain as single phase and do not dissociate. However, conductivity starts decreasing with decreasing oxygen partial pressure as it becomes oxygen deficient $LaCrO_{3-\delta}$, and electron-holes are consumed through the formation of oxygen vacancy (Fergus, 2004). A-site of this perovskite can be partially substituted by Ca^{2+} ion (0.114 Å) or Sr^{2+} ion (0.132 Å) and B-site with smaller ion, Mg^{2+} (0.086 Å). The doping results in charge compensation by oxidizing Cr^{3+} to Cr^{4+} ions in an oxidizing atmosphere below 1400°C and the bulk conductivity becomes higher by two orders of magnitude than the pure $LaCrO_3$. The conduction mechanism of doped $LaCrO_3$ in air is based on an exchange reaction of electron between Cr^{3+} and Cr^{4+} ions, therefore the conductivity should be thermally activated hopping of small polarons. According to Mori et al. (1994), the electrical conductivity (σ) of such doped perovskites can be expressed by the following equation:

$$\sigma = (1/T)(e^2 a^2/6k_B\tau_0)[Cr^{3+}][Cr^{4+}] \exp (E/k_BT) \tag{12}$$

where T is the absolute temperature, e the elementary electrical charge, a the distance of jump, k_B the Boltzmann factor, τ_0 the average free time, $[Cr^{3+}]$ and $[Cr^{4+}]$ are the mole fractions and E the activation energy for conduction.

Fig. 22 represents examples of three abovementioned dopants where both oxidizing and reducing atmosphere were maintained. The electrical conductivity in air is well represented by Eq. (12). B-site substituted $LaCr_{1-x}Mg_xO_3$ samples showed lower conductivity compared to A-site substituted ($La_{1-x}Ca_xCrO_3$ and $La_{1-x}Sr_xCrO_3$) which is because of the decrease in chromium ion concentration. For the A-site substitution, the conductivity increased with increasing dopant concentration. However, marginally higher conductivity values were observed in case of Ca-doped samples compared to strontium. Mori et al. (1997) explained this conductivity differences in terms of lattice parameters. The lattice parameters of the Ca-doped perovskites were smaller than those of the Sr-doped perovskites. The reason for lowering in conductivities under reducing atmosphere is explained by the charge compensation from Cr^{4+} to Cr^{3+} ions by the formation of oxygen defects and the non-linear temperature dependence behavior of oxygen defects was due to the fact that the concentration of oxygen defects depends on pO_2 and temperature. The lowest conductivity in H_2 was obtained in case of Mg-doped $LaCrO_3$ (Fig. 22).

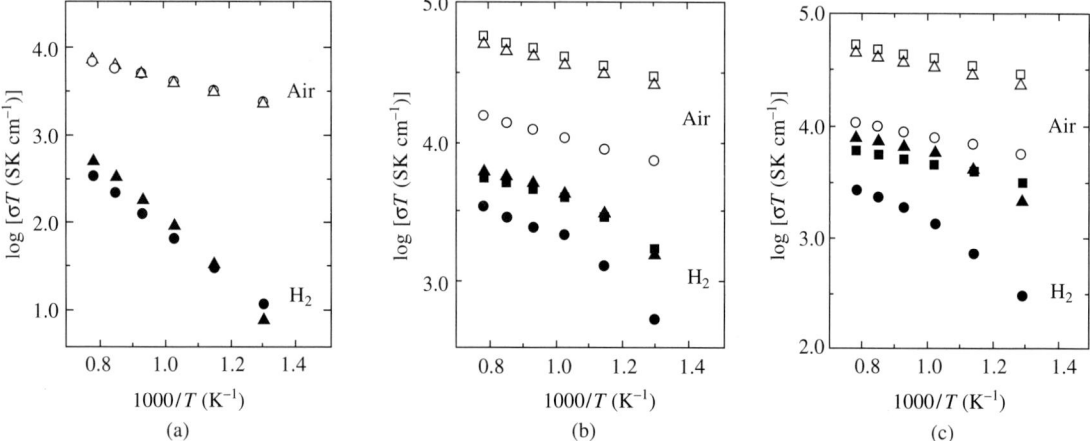

Fig. 22 Temperature dependence of electrical conductivity in air and H_2 atmosphere for the alkaline earth doped lanthanum chromite: (a) $LaCr_{1-x}Mg_xO_3$, (b) $La_{1-x}Ca_xCrO_3$ and (c) $La_{1-x}Sr_xCr_xCrO_3$. ($\bigcirc$), ($\bullet$), $x = 0.1$; ($\triangle$), ($\blacktriangle$), $x = 0.2$; ($\square$), ($\blacksquare$), $x = 0.3$ (Mori et al. 1997).

Table 11 gives a comprehensive list of ceramic interconnects showing electrical conductivities and thermal expansion coefficeints. Mori et al. (2001) has examined the nonlinear thermal expansion behavior during the phase transformation from orthorhombic to rhombohedral structure. For these studies they used the system $La_{0.9}Sr_{0.1}Cr_{1-x}M_xO_3$, where M = Mg, Al, Ti, Mn, Fe, Co, Ni; $0 \leq x \leq 0.1$ in order to prevent the phase transformation as well as to control the thermal expansion behavior. They finally suggested that the composition, $La_{0.9}Sr_{0.1}Cr_{0.96}Co_{0.02}Al_{0.02}O_3$, could be a candidate which can solve both the problems.

Table 11. **Electrical conductivity and thermal expansion coefficient of pure and doped (both A- and B-sites) $LaCrO_3$**

Composition	Electrical conductivity at 1000°C (S/cm)	Activation energy (kJ/mol)	Thermal expansion coefficient ($\times 10^{-6}$/K) at 1000°C
$LaCrO_3$	0.6-1.0	18	9.5
$LaCr_{0.9}Mg_{0.1}O_3$	3.0	19	9.5
$La_{0.8}Ca_{0.2}CrO_3$	35.0	13	10.0
$La_{0.7}Ca_{0.3}CrO_3$	43.0	30	10.2
$La_{0.9}Sr_{0.1}CrO_3$	14.0	12	10.7

Mori et al. (1997) earlier reported the mechanical properties of the A- and B-site doped compositions, viz., $LaCr_{0.9}Mg_{0.1}O_3$, $La_{0.9}Ca_{0.1}CrO_3$, $La_{0.9}Sr_{0.1}CrO_3$ and $La_{0.9}Ca_{0.12}CrO_3$ (Mori et al.) (Table 11). The maximum mechanical strength of 418 MPa was reported to be for Mg-doped chromite. Normally, the mechanical strength of ceramics decreases with increasing temperature; similar behavior is also observed in case of doped chromites (Table 12). The mechanical properties are largely dependent on the microstructure which is strongly related with the sintering mechanism. It has been reported that sintering mechanism for $LaCr_{0.9}Mg_{0.1}O_3$ under a low pO_2 atmosphere is solid-state sintering and liquid-phase in case of Ca- or Sr-doped $LaCrO_3$ (Table 12). While the microstructure of Mg-doped samples shows smaller grain size (3-6 μm) considerable grain growth is observed for liquid-phase sintering in case of chromites having A-site dopants. Normally these doped chromites need sintering temperature as high as 1600°C in oxygen atmosphere in order to get reasonably high densities. However, Chakraborty et al. (2000) reported that ultrafine powder of Ca-doped composition, $La_{0.7}Ca_{0.3}CrO_3$, could produce fully densed (> 99% of theoretical density) samples when sintered at a remarkably low sintering temperature of 1250°C (Fig. 23). The reason for such densification is expected to be due to the presence of minor impurity phase, $CaCrO_4$, which goes into the solid solution at that temperature and thus enhances the sintering via liquid phase sintering. The electrical conductivity and TEC values obtained at 1000°C are 43 S/cm and 10.2×10^{-6}/K, respectively, indicating both values are acceptable for SOFC application.

Table 12. Maximum bending strengths of the alkaline earth doped lanthanum chromites (Mori et al. 1997)

Temperature (°C)	Maximum bending strength (MPa)			
	$LaCr_{0.9}Mg_{0.1}O_3$	$La_{0.9}Ca_{0.1}CrO_3$	$La_{0.9}Sr_{0.1}CrO_3$	$La_{0.9}Ca_{0.12}CrO_3$
Room temperature	418	166	269	150
500	243	77	106	106
1000	186	36	77	33
Sintering mechanism	Solid	Liquid	Liquid	Liquid

As already mentioned $LaCrO_3$-based interconnect materials need very high temperature for sintering (~1600°C) when powders are synthesized in conventional solid state route. However, nanocrystalline/ultrafine, single-phase and homogeneous powder of doped-$LaCrO_3$ is possible to synthesize following similar auto-combustion technique which has been described in Section 3.1.3 and could possibly be sintered even at 1250°C to obtain full densification as shown in Fig. 23. Thus, the chromites are eventually the only choice as ceramic interconnects.

One of the major disadvantages of ceramic interconnects in SOFC is its poor thermal conductivity (< 5 W/mK). This causes non-uniform heat distribution in the fuel cell stack. As a result, significant thermal stresses are generated which ultimately leads to stack failure. Another problem which is often encountered

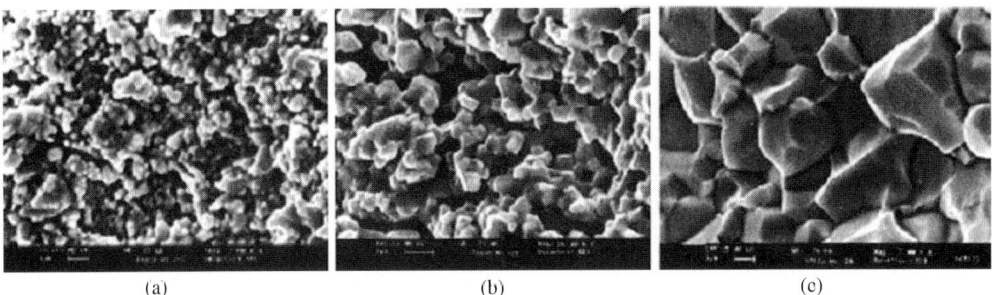

(a)   (b)   (c)

Fig. 23 Scanning electron micrographs of $La_{0.7}Ca_{0.3}CrO_3$ sintered for 6 h: (a) 1000°C, (b) 1150°C and (c) 1250°C (Chakraborty et al. 2000).

with this kind of interconnect material is the significant internal stress due to lattice expansion on the reducing side (anode side). This causes warpage of interconnect plates (for planar design) and catastrophic stack failure results after few hours of operation at 900-1000°C. To overcome this problem, small amount of dopants such as Al, Ti and Co are normally introduced into the $LaCrO_3$ system (Badwal et al. 1997, Mori et al. 2001).

3.4.2 Metallic Interconnects

As already mentioned in the previous section that the traditional $LaCrO_3$-based interconnect has difficulties such as machining, failure due to mechanical properties during cell operation, availability of matching sealants and the cost of chromium. Because of these limitations of ceramic interconnects, metal-based interconnects have been introduced in the anode supported thin film electrolyte structure which operates in the temperature range 700-800°C. The necessity to operate the ceramic fuel cell much below 1000°C (preferably below 750°C) has pushed the researchers to search for suitable metals or alloys. The metals have a number of advantages over ceramic interconnects.

Few advantages are, extremely high electrical conductivity; high thermal conductivity that provides uniform temperature distribution throughout the stack by eliminating the presence of thermal gradient both in plane and perpendicular; enhancement of the ability to accommodate thermal stresses which of course, is dependent on matching thermal expansion coefficient; reasonable mechanical strength and high creep resistance at elevated temperature; fabrication of any complex structure/shape and design is possible; joining with current collector/gas manifold is easier; robust structure and larger dimension than ceramic interconnect and cost-effective.

In spite of having many favorable characteristics, the metallic interconnects also suffer from certain drawbacks. Some of the pertinent issues are electrical contacts between metallic interconnect and ceramic electrodes (cathode and anode), matching of thermal expansion between the metallic interconnect and adjacent components, oxide scale formation on the metallic surface as well as cathode poisoning. All these issues need drastic improvement. For more than a decade, a number of alloys have been attempted, but the major interest for development of such metallic interconnect started only when SOFC developers started using metallic interconnects for SOFC operation, preferably $\leq$ 750°C.

High temperature oxidant resistant alloys containing aluminium, silicon and/or chromium have been tested. But the first two form insulating surface layers of alumina or silica are not suitable as interconnects. Chromia is a reasonably good conductor, and efforts are largely concentrated in chromia forming alloys. Oxide dispersion strengthen (ODS) chromium based alloy, Ducrolloy having composition, 94% Cr, 5% Fe and 1% Y_2O_3 ($Cr_5Fe_1Y_2O_3$) and developed by Plansee AG Austria, shows closely matching thermal expansion with YSZ, excellent oxidation resistance and satisfy other properties necessary for interconnect. In their first anode-supported (10×10 cm^2) ten cells stack, Forschungszentrum Jülich (Germany) has used Ducrolloy and operated the stack above 800°C (Buchkremer et al. 1996). This high chromia alloy was found to be not suitable option because of the fact that during cell operation the material leads to a rapid degradation of the electrical properties of SOFC due to chemical interactions at the cathode side. In addition, this alloy is very brittle and hence fabrication and joining becomes difficult and costly (Badwal 1997).

Most of the initial initiatives to replace the ceramic interconnect with metallic interconnects including other alloys such as, Haynes 230, Inconel 600 have been found unsuccessful, particularly, in electrolyte-supported stacks where operation temperature is around 1000°C. In contrast, ferritic steel or some combinations of Cr-based alloys are being used by most of the developers who are testing their anode-supported structures at $\leq$750°C (Zhu et al. 2003, Anderson et al. 2003, Hilpert et al. 2003, Fergus 2005). Detailed list of commercially available ferritic steels which are found to be suitable for SOFC interconnect are given by Horita et al. (2002) Hilpert et al. (2003), Fergus (2003). This variety of steel is preferred to be used within

750°C. For high temperature application, Cr content varies between 7 and 28%. However, formation of a protective, single-phase chromia layer and matching thermal expansion requires a Cr-content approximately within 17-20% depending upon the temperature, surface treatment, minor alloying additions, and impurities (e.g. C, S, P). With respect to oxidation resistance, the most important minor alloying additions (in ppm level) are Mn, Ti, Si and Al.

The presence of chromium in the feritic steel forms a very thin, electronically poor conductive Cr_2O_3 scale on the surface of the interconnect (in both oxidant and humidified fuel atmospheres), which in principle, should prevent further oxidation of the metal interconnect while the fuel cell is operating at high temperature ($> 650°C$) in an oxidizing atmosphere. But it causes degradation of the cell which is commonly known as chromium poisoning. According to Hilpert et al. (2003) it is based on the vaporization of Cr_2O_3 on the interconnect surface as CrO_3 (g) or $CrO_2(OH)_2$(g) which are the major gaseous species with chromium in the 6+ oxidation state. The vapor species are reduced at the triple phase boundary (electrolyte/cathode/oxidant) by an electrochemical reaction forming Cr_2O_3(s), which can react with the LSM. The Cr_2O_3 (s) or the reaction products thus formed inhibit the oxygen reduction (catalytic activity at cathode) necessary for the operation of SOFC and lead to high overpotential. As a result of which ohmic contact resistance starts increasing and the cell degrades.

To overcome the problem of the cathode poisoning or the suppression of Cr_2O_3 scale formation as described above, two approaches are more or less widely advocated: (i) development of new and more oxidant resistant ferritic steel alloys and (ii) use of a protective coating on the surface of ferritic steel. Several varieties of protective coatings have been developed and tested (Armstrong et al. 2003, Anderson et al. 2003, Basu et al. 2003). Teller et al. (2001) and later on Basu et al. (2003) used optimized ferritic steel compositions such as FeCr(Mn), FeCr(Mn, La, Ti), FeCr(Mn, La, Ti) and FeCrAl(Si, Mn) and LaCoO$_3$- and LaFeO$_3$ -based perovskites (e.g. $LaNi_{0.6}Fe_{0.4}O_3$) were applied on the top of these steel samples used as a protective coating. The composite (steel + coating) undergoes long-term heat treatment at 800°C. A typical time dependent electrical resistance behavior and corresponding microstructure are shown in Figs. 24 and 25, respectively. Among all the steel compositions, the two steel compositions containing Al and Si, formed an additional Al_2O_3 or SiO_2 layer in-between the steel and the reaction zone with the ceramic contact layer and may be the cause for showing higher resistance (Fig. 24). The corrosion scale is thinner in case of FeCr(Mn) compared to relatively thicker scale in case of FeCr(Mn, La, Ti). During long time exposure at 800°C both the elements La and Ti form their oxides and are also incorporated in the corrosion scales as dopants, leading to defects in the crystal lattice and therefore lower resistances (in case of metal alloy/LNF combination) (Basu et al. 2001).

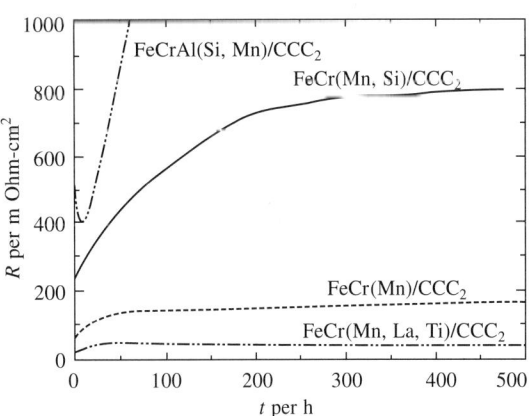

Fig. 24 Electrical resistances of material combinations composed of CCC$_2$ (cobaltite) together with FeCr(MnSi), FeCr(Mn), FeCr(Mn, La, Ti) and FeCrAl(Si, Mn). Experimental conditions: $T = 800°C$, $p(O_2) = 0.21$ atm, $i = 150$ mA/cm^2. Peroxidation of the steels: $T = 800°C$, $p(O_2) = $ atm, $t = 100$ h (Teller et al., 2001).

3.5 Sealing Materials

As already mentioned, there are two major designs of the SOFC, tubular and planar, which have been investigated extensively. Tubular design, also known as *sealless design*, does not require any sealant. On the other hand, planar configuration, depending on its design concept of stacking (externally or internally manifolding) must require suitable high temperature sealants compatible with the adjoining cell components

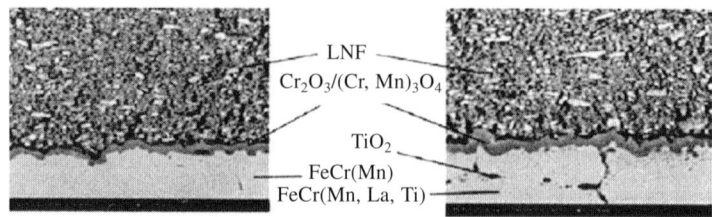

LNF
$Cr_2O_3/(Cr, Mn)_3O_4$
TiO_2
FeCr(Mn)
FeCr(Mn, La, Ti)

Fig. 25 SEM images of the material combinations FeCr(Mn)/LNF (left) and FeCr(Mn, La, Ti)/LNF (right) after exposure. Experimental conditions: $T = 800°C$, $p(O_2) = 0.21$ atm, $t = 600$ h and $i = 150$ mA/cm^2 (Basu et al. 2001).

like electrolyte, interconnect, electrodes and gas manifold. As anode supported thin film electrolyte is the most widely used planar SOFCs, a schematic illustration is shown in Fig. 26. This will give readers an idea about the location of the sealing. Irrespective of cell design, the sealants must fulfill the general requirements, such as complete gas tightness; matching thermal expansion; capable in withstanding thermal cycles with thermal stress relaxation ability during the heating and cooling operations; avoidance of unwanted reactions with adjoining cell components—chemical stability and low vapor pressure in both oxidizing and reducing (down to 10^{-18} atm) environment; avoidance of electrical short-circuiting (must be an insulator); dimensional stability; cost-effective fabrication and reliability and long term stability.

To avoid the above stringent requirements, many innovative sealing methods have been adopted in order to minimize the sealing area and sealing surfaces. Compressive sealing (with or without gasket) and high temperature glass or glass-ceramic sealants are the two methods of sealing which are now considered (Kendall et al. 2003). In case of compressive sealing, the amount of loading, perfect flatness of the cell components are the primary requirements and once the conditions are optimized, they are the ideal sealants. Between the above referred two sealing methods, glass/glass-ceramic based high temperature sealants are widely used and has drawn tremendous interests among the researchers (Larsen 1999, Barford et al. 2003, Fergus 2005). Since glass is a non-oxidizing, insulating material which can be tailored to adjustable parameters, essential for sealing. Two important criteria for selection of a suitable glass sealant are: (i) the glass transition temperature T_g and (ii) thermal expansion coefficient.

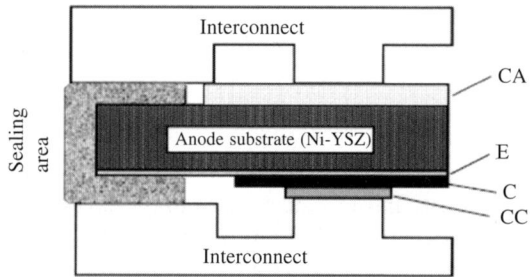

Fig. 26 Schematic drawing of sealing and contact layers within the stack: CA = contact layer anode (Ni-mesh); E = electrolyte; C = cathode; CE = contact layer cathode (Buchkremer et al. 1997).

Glass transition temperature T_g is important because glass must flow sufficiently to provide an adequate seal, while maintaining enough rigidity for mechanical integrity. Another important parameter is softening temperature T_s which is defined by the viscosity, and is thus a more direct measurement of flow characteristics of the glass. As the trends in T_s typically follow those for T_g. Thus, T_g data is available for a wider range of glass compositions. For this application, glass or glass-ceramic compositions is prepared in the traditional way of making glass. Large number of glass compositions are now available in the literature (Ley et al. 1996, Sohn et al. 2002, 2004, Zheng et al. 2004, Eichler et al. 1998; Larsen et al. 1998, 1999, Menzler et al. 2003, Bahadur et al. 2004). However, more updated information is available in a very recent review article by Fergus (2005). Some working glass compositions used by several groups are listed in Table 13. Once the glass is formed, they are ground and sieved to fine mesh sized powders. Using these powders,

Table 13. **Glass and glass-ceramic sealants with values of T_g and CTE in target range (Fergus 2005)**

Composition (mole%)						Properties	
			Alkaline-earth		Other	T_g (°C)	TEC ($\times 10^6$/K)
SiO_2	B_2O_3	Al_2O_3	BaO/SrO	CaO/MgO			
Boroaluminosilicates							
30	15	10	40 BaO		$5La_2O_3$	667	11.2
38	13	10	35 BaO		$5La_2O_3$	739	10.6
33	17	10	35 BaO		$5La_2O_3$	670	10.8
29	21	10	35 BaO		$5La_2O_3$	652	11.1
33	17	10	35 BaO		$5La_2O_3$	656	11.1
30	22	10	36 BaO		$2ZrO_2$	614	10.6
34	17	10	36 BaO		2 NiO	617	11.5
Borosilicates							
32	2	0	40 BaO	10 CaO	16 unspecified	660	10.7
33	3	0	40 BaO	10 CaO	14 unspecified	662	10.5
31	8	0	38 BaO	15 CaO	8 unspecified	626	11.3
34	8	0	42 BaO	8 CaO	8 unspecified	623	10.8
30	7	0	37 BaO	16 CaO	10 unspecified	630	11.4
Aluminosilicate							
50	0	5	45 BaO			730	10.7
Silicates							
35	0	0	44 BaO	11 CaO	10 unspecified	721	10.6
50	0	0	40 BaO	10 MgO		686	12.0
50	0	0	40 BaO		10 ZnO	676	10.7
Borate							
8	40	7	25 SrO		$20 La_2O_3$	760	11.5

pastes or tapes/sheets are made. When raised to high temperature, melting and wetting of the glass on the surface occurs, this ultimately provides gas-tight sealing. Nevertheless, these glass-based sealants suffered from certain problems due to the formation of undesirable phases, such as cordierite and cristobalite. Another disadvantage is that once the stack is sealed, the bond is permanent, and therefore, exchangeability is no longer an option.

Since the end of 1990s, very few information on sealants were available in the literature, however, for last couple of years a host of interesting results have started publishing (Jiang et al. 2001, Simmer et al. 2001, Sohn et al. 2002, Haanappel et al., 2005, Fergus 2005). Some attempts have been made towards solving the key issue of high temperature sealants that minimize the thermal stresses, generated during high temperature operation and thermal cycling (Taniguchi et al. 2000, Chou et al. 2002, Bansal et al. 2005). Stevenson's group has extensively studied and published their work on novel hybrid compressive mica sealing which is capable in delivering a remarkable low leak rate at 800°C (Simmer et al. 2001, Chou et al. 2002, 2003).

Some alternative ways of tackling the sealing problem in stack have also been attempted using metallic gaskets, o-ring etc. (Bram et al. 2001, Duquette et al. 2004). Their studies concentrated on the influence of contact load and internal pressure on the deformation behavior of various metallic gaskets, different in shape and fabricated from different alloys. The testing was carried out at room temperature. However, more work is necessary to establish issues related to its applicability at 800°C. Versa Power Systems Ltd.

(formerly Global Thermoelectric) is the only SOFC company who uses an innovative sealing method for their working stacks (Ghosh et al. 1999). This seal is basically a flexible pre-compressed ceramic fiber matrix seal impregnated with a plurality of solid particles ($d_{50} \leq 0.5$ μm). The ceramic fibers are selected from the group of materials comprising alumina, zirconia, titania, magnesia or silica. The solid particles may be ceramic, glass or other inert materials which are able to resist degradation and sintering at the operating temperature of the SOFC stack. The idea is not to allow the fibers and particles to sinter prior to stack testing and as a result of which the seal remains semi-flexible or can tolerate the thermal expansion (expansion and contraction) without cracking. However, the innovation lies in the way particles have filled the voids in the fiber matrix which ultimately provides the reliable gastight sealing. The required shape of the gaskets are first prepared out of these materials and are then fixed on the two metal interconnects to be sealed. The same seals are also used for sealing the gas-manifold through which fuel and air (oxidant) passes. When the components are fully assembled, compression is applied by ordinary bolts located mostly at the corner of the stack.

4. Future Directions

The most intense materials research is expected to be concentrated in the area of development of extremely high performance low temperature SOFC (lower than 650°C) where thinness of electrolyte (even below 1 μm) could play a crucial role. However, slow kinetics of the oxygen reduction at the cathode/electrolyte could be a major problem. As already mentioned, MIEC cathodes are suitable for low temperature application (Mogensen 2000). A host of research papers on low temperature compatible cathodes and anodes have been presented in the recently held international symposium on SOFCs at Quebec City in May 2005 which is an indicative of necessity of such composite materials. Low temperature operation has great technological advantages during stacking where stainless steel (SS) could be an interconnect. Interconnect research trends will probably be seen on the use of more SS interconnect in stacks and some suitable cathode contact materials may also require to be developed. Another area where concentrated research interest may possibly be noticed is the area of single chamber fuel cells. As a natural course, new materials development for these two activities will also get priority. Tailor made nanocrystallized cathode materials for single chamber SOFC has been very recently reported by Magnone et al. (2005). In parallel some activities may be seen on development of suitable sealing materials and methods for low temperature SOFC.

4.1 Low Temperature SOFCs

The operating temperature of SOFC strictly depends on the electrolyte thickness, thinner the electrolyte, lower is the operating temperature (but within certain limitation). SOFCs are best categorized into three operating temperature regimes: high temperature (HT) 900-1000°C; intermediate temperature (IT), 650-800°C and low temperature (LT), lower than 650°C. These temperature ranges are determined by the choice of electrolyte material as well as by the thickness of the electrolyte. There are two supported structures in ceramic fuel cells – electrolyte and electrode (anode or cathode). For electrolyte-supported cells (in the range of HT and IT) the minimum thickness is in between 150 and 250 μm for ZrO_2 and CeO_2-based electrolyte materials. In the case of a supported thin film electrolyte, a thickness of few microns is possible even for large cell areas. Weber et al. (2004) demonstrated the relationship between the operating temperature and the thinness of the electrolyte with two most popular electrolyte e.g., 8YSZ and LSGM. The voltage losses in the electrolyte at an acceptable current density of 300 mA/cm^2 as a function of operating temperature and electrolyte are shown in Fig. 27. Accepting electrolyte losses < 50 mV, the lowest operating temperature of electrolyte-supported single cells (LSGM or Sc-doped ZrO_2 electrolyte substrate, 150 μm thickness) is about 750°C, whereas a supported thin film electrolyte theoretically exhibits a minimum operating temperature of less than 500°C. Their prediction has recently been demonstrated by

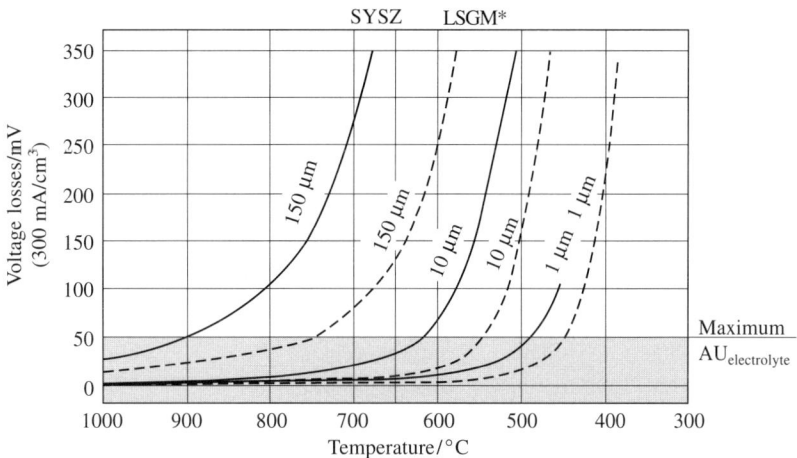

Fig. 27 Temperature dependence of voltage losses as a function of YSZ and LSGM electrolytes thickness (Weber et al., 2004). *Conductivity value, T. Ishihara et al., Proc. of 5th International Symposium on SOFC pp. 301-310, 1997, The Electrochemical Society, NJ.

Yan et al. 2005. A high performance SOFC has been constructed using 5 μm thick La$_{0.9}$Sr$_{0.1}$Ga$_{0.8}$Mg$_{0.2}$O$_{3-\delta}$ (LSGM) film as electrolyte, an extremely thin interlayer (~ 400 nm) made of Ce$_{0.8}$Sm$_{0.2}$O$_{2-\delta}$ (SDC) in between the anode substrate (NiO-Fe$_2$O$_3$-SDC) and the LSGM film to prevent the reaction between LSGM and NiO. Sm$_{0.6}$Sr$_{0.4}$CoO$_{3-\delta}$ (SSC) was used as cathode material. The electrolyte film was fabricated by pulsed laser deposition. The single cell thus constructed showed excellent power densities – 3, 2, 0.6 and 0.1 A/cm^2 at 700, 600, 500 and 400°C, respectively. Thin densed LSGM film and cell performance characteristics are shown in Fig. 28. Bance et al. (2004) also successfully deposited 10-30 μm CGO electrolyte film using porous stainless steel supported NiO-CGO anode. These metal supports happen to eliminate the scaling requirement as they can easily be welded with the interconnect. They fabricated very

Fig. 28 LT SOFC fabricated by pulsed laser deposition: (a) cross-section of anode supported LSGM thin film and (b) output performance of the single cell (after Yan et al. 2005).

large area samples (4 cm × 4 cm) and these cells produced power density of around 310 mW/cm². Table 14 summarizes the general trend of the materials used for reduced temperature operating SOFCs (650°C and below).

Table 14. Low temperature SOFCs—materials and performance at 650°C or below on H$_2$/air input gases (Haile 2003) (Modified)

Electrolyte	Thickness (μm)	Anode	Cathode	Peak power density (mW/cm²)	References
LSGMC	205	Ni-SDC	$Sm_{0.5}Sr_{0.5}CoO_{3-\delta}$	240-410	Haile 2003
LSGM+Ce$_{0.8}$Sm$_{0.2}$O$_{2-\delta}$	~ 5	Ni-Fe-SDC	$Sm_{0.6}Sr_{0.4}CoO_{3-\delta}$	1,951	Yan et al. 2005
$Ce_{0.9}Gd_{0.1}O_{1.95}$	~ 40	Ni-Ru-CGO	$Sm_{0.5}Sr_{0.5}CoO_{3-\delta}$	770	Haile 2003
$Ce_{0.85}Sm_{0.15}O_{2-\delta}$	~ 20	NI-SDC	$Ba_{0.5}Sr_{0.5}Co_{0.8}Fe_{0.3-\delta}$	1,010	Shao et al. 2004
YSZ	~ 10	Ni-YSZ	$La_{0.8}Sr_{0.2}FeO_{3-\delta}$	400	Haile 2003

LSGMC = $La_{0.8}Sr_{0.2}Ga_{0.8}Mg_{0.115}Co_{0.085}O_3$, SDC = $Ce_{0.85}Sm_{0.15}O_{2-\delta}$, CGO = $Ce_{0.9}Gd_{0.1}O_{1.95}$.

Reduction of area specific resistance (resistance per unit area) is beneficial in getting high performance SOFC. Very recently, Prinz's group of Stanford University demonstrated their breakthrough work where 100 nm YSZ film has been prepared by MEMS fabrication technology (Huang et al. 2005). Such a reduction in film thickness from micrometer to nano-scale lowers the SOFC operating temperature within the range between 250 and 400°C. The peak power density is claimed to be 100 times higher compared to SOFCs containing 10 μm thick YSZ electrolyte. The reason for such an extraordinary cell performance at such a low temperature (400°C), is not only due to drastic decrease in electrolyte thickness, but also due to an increase of charge transfer reaction rate at the cathode.

At lower temperature (and reduced thickness), the main contributor of resistance is not the electrolyte but the contributions of overpotential from the electrodes. Improved cell performance can be achieved by optimizing the processing conditions as well as microstructure of the electrochemically active layers. The benefits of such improved electrode performance can be used in three ways: (i) directly as enhanced power output, (ii) by lowering operating temperature in terms of longer component life and (iii) as a buffer for peak electricity demands during operation (Buchkremer et al. 1997, van Doorn et al. 1998, Jørgensen et al. 2000, Tietz et al., 2002).

4.2 Single Chamber Fuel Cells

Dyer (1990) is credited for examining the single chamber fuel cell concept. The concept might lead a compact, miniaturized high performance cell for low power applications. Particularly, SOFC developers could enjoy the benefits as sealing requirement in such cells has completely been eliminated. Simplicity of the design could even offer enhanced thermal and mechanical shock resistance which ultimately leads to quick start up and cool down. These advantages of single chamber fuel cell make it as great potential for applications whether it is mobile, stationary or transport. A brief and simple description of such a new concept has been given by Haile (2003) who finally correlates the concept with biofuel cells operating on aqueous glucose. However, long back it was realized by van Gool et al. (1965) that: (i) many fuels do not directly react with oxygen at typical fuel cell temperature and (ii) the catalytic activity of anode and cathode materials are quite selective to certain types of reactions. This is the basis of single chamber fuel cells which simply means a fuel cell without having separate chambers for cathode and anode. This is quite different from conventional fuel cell design where electrolyte not only mediates the electrochemical reactions taking place at anode and cathode, but also separates the fuel from the oxidant to prevent direct combustion (Fleig et al. 2003). Fig. 29 illustrates a single chamber fuel cell where a hydrocarbon fuel is partially oxidized at the anode producing CO and H_2 and consuming O_2. The resulting oxygen partial pressure gradient drives the electrochemical reactions of the fuel cell. In such cell, the cathode and anode are selective by appropriate activity of catalysts. When a controlled mixture of fuel and oxygen is fed through a lone inlet of the cell, the half cell reactions are given as follows:

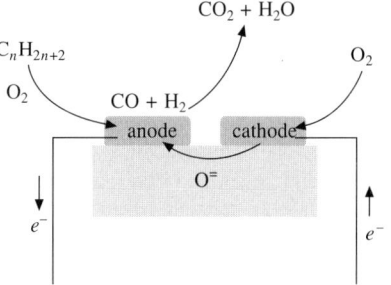

Fig. 29 Schematic of a single chamber fuel cell (Haile 2003).

Anode:
$$CH_4 + \frac{1}{2} O_2 \rightarrow CO + 2H_2 \text{ (Chemical)} \tag{13}$$

$$H_2 + O_2^= \rightarrow H_2O + 2e^- \text{ (Electrochemical)} \tag{14}$$

$$CO + O_2^= \rightarrow CO_2 + 2e^- \text{ (Electrochemical)} \tag{15}$$

Cathode:
$$\frac{1}{2} O_2 + 2e^- \rightarrow O_2^= \text{ (Electrochemical)} \tag{16}$$

In their first single chamber fuel cell Hibino et al. 1996 used Gd-doped $BaCeO_3$ electrolyte with gold and platinum electrodes for selective reduction of oxygen and methane respectively. However, the same group has also demonstrated SOFC using conventional cell components viz., YSZ (electrolyte) and LSM and NiO-YSZ (electrodes) at an operating temperature of 950°C. Later a successful attempt of operating 500°C (Ce-based electrolyte and electrodes) single chamber fuel cell having a moderate power density has also been reported (Hibino et al. 2000, 2002). The fuel and oxidant mixture used were ethane or propane and air. Very recently, Napporn et al. (2005) demonstrated single chamber SOFC. They used both dry and humidified fuels and concluded that dry fuel is beneficial as some water vapor is expected to be available inside the single chamber. However, the chemical reactions inside the single chamber are complex in nature and more research is necessary to establish such types of fuel cells in reality. Among few debated issues there are concerns about the explosion (during operation) and the internal (actual) temperature rise during fuel combustion.

References

Anderson, H.U. (1992), Review of p-type doped perovskite materials for SOFC and other applications, *Solid State Ionic*, **52**, 33-41.

Anderson, H.U., Tai, L.W., Chen, C.C., Nasrallah, M.M. and Huebner, W. (1995), Review of the structural and electrical properties of the (LaSr)(CoFe)O_3 system, in: Proceedings of the Forth International Symposium on Solid Oxide Fuel Cells (SOFC-IV), Eds. M. Dokiya, O. Yamamoto, H. Tagawa and S.C. Singhal. The Electrochemical Society Proceedings Series, Pennington, NJ, p. 375-384.

Anderson, H.U. and Tietz, F. (2003), Interconnects, in: High Temperature Solid Oxide Fuel Cells: Fundamentals, design and Applications, Eds. S.C. Singhal and K. Kendall, Elsevier, UK, pp. 173-195.

Arachi, A., Sakai, H., Yamamoto, O., Takeda, Y. and Imanishi, N. (1999), Electrical conductivity of the ZrO$_2$–Ln$_2$O$_3$

(Ln=lanthanides) system, *Solid State Ionics*, **121**, 133-139.

Armstrong, T.R., Hardy, J.S., Simmer, S.P. and Stevenson, J.W. (1999), Optimizing lanthanum chromite interconnects for solid oxide fuel cells, Proceedings of the Sixth International Symposium on Solid Oxide Fuel Cells (SOFC-VI), Eds. S.C. Singhal and M. Dokiya. The Electrochemical Society Proceedings Series, Pennington, NJ, pp. 706-715.

Armstrong, T.J., Homel, M.A. and Virkar, A.V. (2003), Evaluation of metallic interconnects for use in intermediate temperature SOFC, Eigth International Symposium on Solid Oxide Fuel Cells, Eds. S.C. Singhal and M. Dokiya, pp. 841- 850. The Electrochemical Society, Pennington, NJ, USA.

Badwal, S.P.S. and Foger, K. (1997), Materials for Solid Oxide Fuel Cells, *Materials Forum*, **21**, 187-224.

Badwal, S.P.S. (1992), Zirconia-based solid electrolytes: microstructure, stability and ionic conductivity, *Solid State Ionics*, **52**, 23-32.

Badwal, S.P.S., Ciaachi, F.T. and Milosevic, D. (2000), Scandia-zirconia electrolytes for intermediate temperature solid oxide fuel cells, *Solid State Ionics*, **136-137**, 91-99.

Bahadur, D., Lahl, N., Singheiser, L. and Hilpert, K. (2004), Influence of Nucleating agents on chemical interaction of $MgO-Al_2O_3-SiO_2-B_2O_3$ glass sealant with components of SOFCs, *J. Electrochem. Soc.*, **151**, A558-A562.

Bance, P., Brandon, N.P., Grivan, B., Holbeche, P., O'Dea, S. and Steele, B.C.H. (2004), Spinning-out a fuel cell company from a UK University—2 years of progress at Ceres Power, *J. Power Sources*, **131**, 86-90.

Bansal, N.P. and Gamble, E.A. (2005), Crystallization kinetics of a SOFC seal glass by DTA, *J. Power Sources*, Available online.

Barford, R., Koch, S., Liu, Y.-L, Larsen, P.H. and Hendriksen, P.V. (2003), Long-term tests of DK-SOFC cells, Eighth International Symposium on Solid Oxide Fuel Cells, Eds. S.C. Singhal and M. Dokiya, pp. 1158-1166. The Electrochemical Society, Pennington, NJ, USA.

Barnett, S.A. (2003), Direct hydrocarbon SOFCs, Handbook of Fuel Cells – Fundamentals, Technology and Applications, Edited by W. Vielstich, H. A. Gasteiger and A. Lamm, Volume 4: Fuel Cell Technology and Applications, pp. 1098-1108, John Wiley & Sons, Ltd.

Basu, R.N., Randall, C.A. and Mayo, M.J. (2001), Fabrication of dense zirconia electrolyte films for tubular solid oxide fuel cells by electrophoretic deposition, *J. Am. Ceram. Soc.* **84**, 33-40.

Basu, R.N., Altin, O., Mayo, M.J., Randall, C.A. and Eser, S. (2001), Pyrolytic carbon deposition on porous cathode tubes and its use as an interlayer for solid oxide fuel cell zirconia electrolyte fabrication, *J. Electrochem. Soc.*, **148**, A506-A512.

Basu, R.N., Tietz, F., Teller, O., Wessel, E., Buchkremer, H.P. and Stöver, D. (2003), $LaNi_{0.6}Fe_{0.4}O_3$ as a cathode contact material for solid oxide fuel cells, *J. Solid State Electrochem.*, **7**, 416-420.

Basu, R.N., Tietz, F., Wessel, E. and Stöver, D. (2004), Interface reactions during co-firing of solid oxide fuel cell components, *J. Mater. Process. Technol.*, **147**, 85-89.

Basu, R.N., Blass, G., Buchkremer, H.P., Stöver, D., Tietz, F., Wessel, E. and Vinke, I.C. (2005), Simplified processing of anode-supported thin film planar solid oxide fuel cells, *J. Euro. Ceram. Soc.*, **25**, 463-471.

Battle, P.D., Catlow, C.R.A., Drennan, J. and Murray, A.D. (1983), *J. Phys. C: Solid State Phys.* **16**, L561, The structural properties of the oxygen conducting δ-phase of Bi_2O_3.

Baur, E. and Preis, H. (1937), Über Brennstoffketten mit Festkörpern (On fuel chains with solids), *Z Elektrochem.*, **43**, 727-732.

Beaudet-Savignat, S., Lima, A., Brathet, C. and Henry, A. (2003), Elaboration and ionic conduction of apatite-type rare-earth oxides, Proc. of the Eighth International Symposium on Solid Oxide Fuel Cells (SOFC-VIII), Eds. S.C. Singhal and M. Dokiya, pp. 372-378.

Boivin, J.C. and Mairesse, G. (1998), Recent material developments in fast oxide ion conductors, *Chem. Mater.* **10**, 2870-2888.

Bonanos, N., Knight, K.S. and Ellis, B. (1995), Perovskite solid electrolytes: Structure, transport properties and fuel cell applications, *Solid State Ionics*, **79**, 161-170.

Bram, M., Brünings, S.E., Meschke, F., Meulenberg, W.A., Buchkremer, H.P., Steinbrech, R.W. and Stöver, D. (2001), Application of metallic gaskets in SOFC stacks, Proc. Seventh International Symposium on Solid Oxide Fuel Cells (SOFC-VII), Eds. H.Yokokawa and S.C. Singhal, pp. 875-884, The Electrochemical Soc., Pennington, NJ, USA.

Bram, M., Reckers, S., Drinovac, P., Monch, J., Steinbrech, R.W., Buchkremer, H.P. and Stöver, D. (2004), Deformation behavior and leakage tests of alternating sealing materials for SOFCs stacks, *J. Power Sources*, **138**, 111-119.

Brisse, A., Barthet, C., Sauvet, A.L., Beaudet-Savignat, S. and Fouletier, J., Study of a new solid oxide fuel cell operated at intermediate temperature 650°-700°C, Ninth International Symposium on Solid Oxide Fuel Cells (SOFC-IX), Eds. S.C. Singhal and J. Mizusaki, pp. 363-370. The Electrochemical Soc., Inc., Pennington, NJ, USA.

Brown, M., Primdahl, S. Mogensen, M. (2000), Structure/performance relations for Ni/yttria-stabilized zirconia anodes for solid oxide fuel cells, *J. Electrochem. Soc.* **147**, 475-485.

Buchkremer, H.P., Diekmann, U. and Stöver, D. (1996), Components manufacturing and stack integration of an anode supported planar SOFC systems, Proceedings of the Second European Solid Oxide Fuel Cell Forum, Vol. 1, Ed. B. Thorstensen. Göttingen, Germany, pp. 221-228.

Buchkremer, H.P., Diekmann, U., de Haart, L.G., Kabs, H., Stimming, U. and Stöver, D. (1997), Advances in the anode-supported planar SOFC technology, Proc. of the Fifth International Symposium on Solid Oxide Fuel Cells (SOFC-V), Eds. U. Stimming, Singhal, S.C., Tagawa, H. and Lehnert, W. pp. 160-170, The Electrochemical Soc. Inc., Pennington, NJ, USA.

Chakraborty, A. (1995), Preparation and characterization in pure and substituted LaMnO₃ for use as cathode material for solid oxide fuel cells, PhD Thesis, Jadavpur University, Calcutta, India.

Chakraborty, A., Basu, R.N. and Maiti, H.S. (2000), Low temperature sintering of La(Ca)CrO₃ prepared by an autoignition process, *Mater. Letts.*, **45**, 162-166.

Chou, Y.S. and Stevenson, J.W. (2002), Thermal cycling and degradation mechanisms of compressive mica-based seals for SOFCs, *J. of Power Sources*, **112**, 376-83.

Chou, Y.S., Stevenson, J.W. and Chick, L.A. (2002), Ultra-low leak rate of hybrid compressive mica-seals for SOFCs, *J. Power Sources*, **112**, 130-136.

Chou, Y.S. and Stevenson, J.W. (2003), Phlogopite mica-based compressive seals for SOFCs: effect of mica thickness, *J. Power Sources*, **124**, 473-78

Costamagna, P., Costa, P. and Antonucci, V. (1998), Micro-modeling of solid oxide fuel cell electrodes, *Electrochem. Acta*, **43**, 375-394.

Choudhary, C.B., Maiti, H.S. and Subbarao, E.C. (1980), Solid Electrolytes and their Applications, Ed. E.C. Subbarao, Plenum Press, pp. 34-49.

Dees, D.W., Claar, T.D., Easler, T.E., Fee, D.C. and Mrazek, F.C. (1987), Conductivity of porous Ni/ZrO₂-Y₂O₃ cermets, *J. Electrochem. Soc.*, **134**, 2141-2146.

de Jonghe, L.C., Jacobson, C.P. and Visco, S.J. (2003), Supported electrolyte thin film synthesis of solid oxide fuel cells, *Annu. Rev. Mater. Res.*, **33**, 169-182.

de Souza, S., Visco, S.J. and De Jonghe, L.C. (1997), Thin-film solid oxide fuel cell with high performance at low temperature, *Solid State Ionics*, **98**, 57-61.

Deng, X. and Petric, A. (1999), A solution to anode-electrolyte reaction in lanthanum gallate fuel cells, Processing and characterization of electrochemical materials and devices, pp. 87-94, Eds. P.N. Kumta, R. Manthiram, S.K. Sundaram and Y.M. Chiang, American Ceramic Society, Westerville, OH, USA.

Divisek, J., Wilkenhöner, R. and Volfkovich, Y. (1999), Structure investigations of SOFC anode cermets – Part I: Porosity investigations. *J. Appl. Electrochem.* **29**, 153-163.

Dokiya (2002), SOFC system and technology, Solid State Ionics, **152-153**, 383-392.

Drennan, J., Zelizko, V., Hay, D., Ciacchi, F., Rajendran, T., Rajendran, S. and Badwal, S.P.S. (1997), Characterization, conductivity and mechanical properties of the oxygenion conductor, La₀.₉Sr₀.₁Ga₀.₈Mg₀.₂O₃₋ₓ, *J. Mater. Chem.* **7**, 79-83.

Duquette, J. and Petric, A. (2004), Silver wire seal design for planar solid oxide fuel cell stack, *J. Power Sources*, **137**, 71-75.

Duquette, J., Basu, R.N., Deng, X., Zhitomirsky, I. and Petric, A. (2005), Fabrication of cathode supported SOFC by colloidal processing, Ninth International Symposium on Solid Oxide Fuel Cells (SOFC-IX), Eds. S.C. Singhal and J. Mizusaki, pp. 482-488. The Electrochemical Soc., Inc., Pennington, NJ, USA.

Dyer, C.K. (1990), A novel thin film electrochemical device for energy conversion, *Nature*, **343**, 547-548.

Eichler, K., Solow, G., Otschik, P. and Schaffrath, W. (1999), BAS (BaO-Al₂O₃-SiO₂) glasses for high temperature applications, *J. Euro. Ceram. Society.*, **19**, 1101-1104.

Feng, M. and Goodenough, J.B. (1994), A superior oxide-ion electrolyte. *Euro. J. Solid State Inorg. Chem.*, **31**, 663-672.

Fergus, J.W. (2004), Lanthanum chromite-based materials for solid oxide fuel cell interconnects, *Solid State Ionics*, **171**, 1-15.

Fergus, J.W. (2005), Metallic interconnects for solid oxide fuel cells, *Mater. Sci. & Engr.*, **A397**, 271-283.

Fergus, J.W. (2005), Sealants for solid oxide fuel cells, *J. Power Sources*, **147**, 46-57.

Fleig, J., Kreuer, K.D. and Maier, J. (2003), Ceramic Fuel Cells, Handbook of Advanced Ceramics, Editor-in-Chief, S. Sømiya, Vol. II, pp. 59-105.

Fleig, J. (2002), On the width of the electrochemically active region in mixed conducting solid oxide fuel cell cathodes, *J. Power Sources*, **105**, 228-238.

Fu, Q., Tietz, F. and Stöver, D. (2005), Electrical conductivity and redox behaviour of yttrium-substituted SrTiO₃: Dependence on preparation and processing procedures, Ninth International Symposium on Solid Oxide Fuel Cells (SOFC-IX). Eds. S.C. Singhal and J. Mizusaki, pp. 1417-1428. The Electrochemical Soc. Inc., Pennington, NJ, USA.

Galasso, F.S. (1969), Structure, properties and preparation of perovskite-type compounds, Pergaman Press, Oxford.

Gao, W. and Sammes, N.M. (1999), An Introduction to Electronic and ionic Materials, World Scientific, Singapore.

Ghosh, D., Wang, G., Brule, R., Tang, E. and Huang, P. (1999), Performance of anode supported planar SOFC cells, Proc. of the Sixth International Symposium on Solid Oxide Fuel Cells (SOFC-VI), Eds., S.C. Singhal and M. Dokiya, pp. 822-829. The Electrochemical Society Inc., Pennington, NJ, USA.

Gödickemeier, M. and Gauckler, L.J. (1998), Engineering of solid oxide fuel cells with ceria-based electrolytes. *J. Electrochem. Soc.*, **145**, 414-421.

Goodenough, J.B. (2003), Oxide-ion electrolytes, *Annu. Rev. Mater. Res.*, **33**, 91-228.

Gorte, R.J., Park, S., Vohs, J.M. and Wang, C. (2000), Anodes for direct oxidation of dry hydrocarbons in a solid oxide fuel cell, *Adv. Mater.*, **12**, 1465-1469.

Gorte, R.J. and Vohs, J.M. (2003), Novel SOFC anodes for the direct electrochemical oxidation of hydrocarbons, *Journal of Catalysis*, **216**, 477-486.

Haile, S.M., Fuel cell materials and components, *Acta Materialia*, **51**, 5981-6000 (2003).

Hannappel, V.A.C., Shemet, V., Vinke, I.C. and Quadakkers, W.J. (2005), A novel method to evaluate the suitability of glass sealant-alloy combinations under SOFC stack conditions, *J. Power Sources.*, **141**, 102-107.

Hannappel, V.A.C., Shemet, V., Gross, S.M., Koppitz, TH., Zahid, M. and Quadakkers, W.J. (2005), Behaviour of various glass-ceramic sealants with ferritic steels under simulated SOFC stack conditions, *J. Power Sources.*, Available online on 5th April.

Hibino, T., Ushiki, K. and Kuwahara, Y. (1996), New concept of simplifying SOFC system, *Solid State Ionics*, **91**, 69-74.

Hibino, T., Hashimoto, A., Inoue, T., Tokuno, J., Yoshida, S. and Sano, M. (2000), A low-operating temperature solid oxide fuel cell in hydrogen-air mixtures, *Science*, **288**, 2031-2033.

Hibino, T., Hasahimoto, A., Yano, M., Sizuki, M., Yoshida, S. and Sano, S. (2002), High performance anodes for SOFCs operating in methane-air mixture at reduced tempertures, *J. Electrochem. Soc.* **149**, A133-A136.

Higgins, S., Sammes, N. and Smirnova, A. (2005), Proton-conductive electrolyte materials for protonic ceramic fuel cells (PCFCs), Proc. of the ninth international symposium on solid oxide fuel cells (SOFC-IX), Eds., S.C. Singhal and J. Mizusaki, Vol. 2, pp. 1149-1155. The Electrochemical Society Inc., Pennington, NJ, USA.

Hilpert, K., Quadakkers, W.J. and Singheiser, L. (2003), Interconnects, Handbook of Fuel Cells: Fundamentals, Technology and Applications, Eds. W. Vielstich, H.A. Gasteiger and A. Lamm. Vol. 4: Fuel Cell Technology and Applications, pp. 1037-54.

Herbstritt, D., Warga, C., Weber, A. and Ivers-Tiffee (2001), Long-term stability of SOFC with Sc-doped zirconia electrolyte, Proc. Seventh International Symposium on Solid Oxide Fuel Cells (SOFC-VII), Eds. H.Yokokawa and S.C. Singhal, pp. 349-357, The Electrochemical Soc., Pennington, NJ, USA.

Horita, T., Xiong, Y., Yamaji, K., Sakai, N. and Yokokawa, H. (2002), Chracterization of Fe-Cr alloys for reduced operation temperature SOFCs, *Fuel Cells*, **2**, 189-194.

Horvat, M., Samardzija, Z., Hole, J. and Bernik, S. (1999), Subsolids phase equilibria in the La₂O₃-Ga₂O₃-CeO₂ system, *J. Mater. Res.*, **14**, 4460-4462.

Huang, H., Nakamura, M., Su, P., Fasching, R., Saito, Y. and Prinz, F. (2005), MEMS fabrication and performances of nano-thin solid oxide fuel cell, Extended abstract, Electrochemical Society 208th Meeting, October 16-21, Los Angeles, California.

Huang, K., Feng, M. and Goodenough, J.B. and Milliken, C. (1997), Electrode performance test on single ceramic fuel cells using electrolyte Sr- and Mg-doped LaGaO₃, *J. Eletrochem. Soc.*, **144**, 3620-3624.

Huang, K., Feng, M. and Goodenough, J.B. (1998), Synthesis and electrical properties of dense $Ce_{0.9}Cd_{0.1}O_{1.95}$ ceramics. *J. Am. Ceram. Soc.*, **81**, 357-362.

Huang, K., Wan, J-W and Goodenough, J.B. (2001), Increasing power density of LSGM-based solid oxide fuel cells using new anode materials, *J. Electrochem. Soc.*, **148**, A788-A794.

Huang, P. and Petric, A. (1996), Superior oxygen ion conductivity of lanthanum gallate doped with strontium and magnesium, *J. Electrochem. Soc.*, **143**, 1644-1648.

Hui, S. and Petric, A. (2002), Electrical conductivity of yttria doped $SrTiO_3$: Influence of transition metal additives, *Mater. Res. Bull.*, **37**, 1215-1231.

Hui, S. and Petric, A. (2002), Electrical properties of yttrium-doped strontium titanate under reducing condition, *J. Electrochem. Soc.*, **149**, J1-J10.

Isenberg, A.O. (1981), Energy conversion via solid oxide electrolyte electrochemical cells at high temperatures, *Solid State Ionics*, **3/4**, 431-437.

Ishihara, T., Matsuda, H. and Takita, Y. (1994), Doped LaGaO₃ perovskite-type oxide as a new oxide ion conductor, *J. Am. Chem. Soc.* **116**, 3801-3803.

Ishihara, T., Furutani, H., Honda, M., Yamada, T., Shibayama, T., Akbay, T., Sakai, N., Yokokawa, H. and Takita, Y. (1999), Improved oxide ion conductivity in $La_{0.8}Sr_{0.2}Ga_{0.8}Mg_{0.2}O_3$ by doping Co, *Chem. Mater.*, **11**, 2081-2088.

Ishihara, T., Sato, K. and Takita, Y. (1996), Electrophoretic deposition of Y_2O_3-stabilized ZrO_2 electrolyte films in solid oxide fuel cells, *J. Am. Ceram. Soc.*, **79**, 913-919.

Ishihara, T., Novel electrolytes operating at 400-600°C (2003), Handbook of Fuel Cells – Fundamentals, Technology and Applications, Edited by W. Vielstich, H.A. Gasteiger and A. Lamm, Volume 4: Fuel Cell Technology and Applications, pp. 1109-1122, John Wiley & Sons Ltd.

Ishiahara, T., Sammes, N.M. and Yamamoto, O. (2003), Electrolytes, in: High Temperature Solid Oxide Fuel Cells: Fundamentals, design and Applications, Eds. S.C. Singhal and K. Kendall, Elsevier, UK, pp. 83-117.

Iwahara, H., Uchida, H. and Tanaka, S. (1983), High temperature type proton conductor based on SrCeO₃ and its application to solid electrolyte fuel cells, *Solid State Ionics*, **9-10**, 1021-1025.

Iwahara, H., Uchida, H., Ono, K. and Ogaki, K. (1988), Proton conduction in sintered oxides based on BaCeO₃, *J. Electrochem. Soc.*, **135**, 529-533.

Jiang, S.P., Christiansen, L., Hugan, B. and Foger, K. (2001), Effect of glass sealant materials on microstructure and performance of Sr-doped LaMnO₃, *J. Mater. Sci.*, **20**, 695-97.

Jiang, Y. and Virkar, A.V. (2001), A high performance, anode-supported solid oxide fuel cell operating on direct alcohol, *J. Electrochem. Soc.*, **148**, A706-A709.

Jiang, Y. and Virkar, A.V. (2003), Fuel composition and diluent Effect on gas transport and performance of anode-supported SOFCs, *J. Electrochem. Soc.*, **150**, A942-A951.

Jørgensen, M.J., Holtappels, P. and Appel, C.C. (2000), Durability test of SOFC cathodes, *J. Appl. Electrochem.*, **30**, 411-418.

Joshi, A.V., Steppan, J.J., Taylor, D.M. and Elangovan, S. (2004), Solid electrolyte materials, devices, and applications, *J. Electroceramics*, **13**, 619-625.

Kendall, K., Minh, N.Q. and Singhal, S.C. (2003), Cell and stack design, High Temperature Solid Oxide fuel Cells: Fundamentals, Design and Applications, Eds.: S.C. Singhal and K. Kendall, Elsevier, UK, p. 197-228.

Kilner, J.A. (2005), Brian Steele's contributions to solid state ionics, Ninth International Symposium on Solid Oxide Fuel Cells (SOFC-IX), Eds. S.C. Singhal and J. Mizusaki, pp. 13-19. The Electrochemical Soc. Inc., Pennington, NJ, USA.

Kim, J.-W, Virkar, A.V., Fung, K.-Z., Mehta, K. and Singhal, S.C. (1999), Polarization effects in intermediate temperature, anode-supported solid oxide fuel cells, *J. Electrochem. Soc.* **146**, 69-78.

Kramer, S. Spears, M. and Tuller, H.L. (1994), Conduction in titanate pyrochlores – role of dopants, *Solid State Ionics*, **72**: 59-66.

Kreuer, K.D. (2003), Proton conducting oxides, *Ann. Rev. Mater. Res.*, **33**, 333-360.

Kuo, J.H., Anderson, H.U. and Sparlin, D.M. (1989), Oxidation reduction behavior of undoped and Sr-doped LaMnO₃ nonstoichiometry and defect structure, *J. Solid State Chem.*, **83**, 52-60.

Kuo, J.H., Anderson, H.U. and Sparlin, D.M. (1990), Oxidation reduction behavior of undoped and Sr-doped LaMnO₃ defect structure, electrical conductivity, and thermoelectric power, *J. Solid State Chem.*, **87**, 55-63.

Larsen, F.H. and James, P.F. (1998), Chemical stability of MgO/CaO/Cr₂O₃-Al₂O₃-B₂O₃-phosphate glasses in SOFC environment, *J. Mater. Sci.*, **33**, 2499-2507.

Larsen, P.H. (1999), Sealing Materials for Solid Oxide Fuel Cells, Ph.D. Thesis (Sheffield University, UK), Risø National Laboratory, Roskilde, Denmark.

Ley, K.L., Krumplet, M., Kumar, R., Meiser, J.H. and Bloom, I. (1996), Glass-ceramic sealant for SOFC: part-I, physical properties, *J. Mater. Res.*, **11**, 1489-1493.

Lu, C., Worrell, W.L., Gorte, R.J. and Vohs, J.M. (2003), SOFCs for direct hydrocarbon fuels with samaria-doped ceria electrolyte, *J. Electrochem. Soc.*, **150**, A354-A358.

Liu, Y., Zhu, S. and Liu, M. (2004), Novel nanostructured electrodes for solid oxide fuel cells fabricated by combustion chemical vapor deposition, *Adv. Mater.*, **16**, 256-260.

Ma, G., Shimura, T. and Iwahara, H. (1999), Simultaneous doping with La³⁺ and Y³⁺ for Ba²⁺- and Ce⁴⁺ -sites in BaCeO₃ and the ionic conduction, *Solid State Ionics*, **120**, 51-60.

Majewski, P., Rozumek, M. and Aldinger, F. (2001), Phase diagram studies in the systems La₂O₃-SrO-MgO-Ga₂O₃ at 1350-1400°C in air with emphasis on Sr and Mg substituted LaGaO₃, *J. Alloys Compounds*, **329**, 253.

Magnone, E., Traversa, E. and Miyayama, M. (2005), Synthesis and characterization of strontium and iron-doped lanthanum cobaltite nanocrystaline powders for single chamber solid oxide fuel cells, Ninth International Symposium on Solid Oxide Fuel Cells (SOFC-IX), Eds. S.C. Singhal and J. Mizusaki, pp. 1617-1626. The Electrochemical Soc. Inc., Pennington, NJ, USA.

Marina, O.A., Canfield, N.L. and Stevenson, J.W. (2002), Thermal, electrical, and electrochemical properties of lanthanum-doped strontium titanate, *Solid State Ionics*, **149**, 21-28.

McEvoy, A. (2003), Anodes, in: High Temperature Solid Oxide Fuel Cells: Fundamentals, design and Applications, Eds. S.C. Singhal and K. Kendall, Elsevier, UK, pp. 149-171.

Menzler, N.H., Bram, M., Buchkremer, H.P. and Stöver, D. (2003), Development of a gastight sealing material for ceramic components, *J. Euro. Ceram. Soc.*, **23**, 445-454.

Menzler, N.H., Sebold, D., Zahid, M., Gross, S.M. and Koppitz, T. (2005), Interaction of metallic SOFC interconnect materials with glass-ceramic sealant in various atmospheres, *J. of Power Sources.*, Available online.

Meyer, M., Nicoloso, N. and Jaenisch, V. (1997), Percolation model for the anomalous conductivity of fluorite-related oxides, *Phys. Rev. B*, **56**, 5961-66, 1997.

Minh, N.Q. and Takahashi, T. (1995), Science and Technology of Ceramic Fuel Cells, Elsevier, Amsterdam.

Minh, N.Q. (2004), Solid Oxide fuel cell technology – features and applications, *Solid State Ionics*, **174**, 271-277.

Mogensen, M. and Skaarup, S. (1996), Kinetic and geometric aspects of solid oxide fuel cell electrodes, *Solid State Ionics*, **86-88**, 1151-1160.

Mogensen, M., Sammes, N.M. and Tompsett, G.A. (2000), Physical, chemical and electrochemical properties of pure and doped ceria, *Solid State Ionics*, **129**: 63-94.

Mogensen, M., Primdahl, S., Jørgensen, M.J. and Bagger, C. (2000), Composite Electrodes in Solid Oxide Fuel Cells and Similar Solid State Devices, *J. Electroceramics*, **5:2**, 141-152.

Moos, R. and Härdtl, K.H. (1997), Defect chemistry of donor-doped and undoped strontium titanate ceramics between 1000° to 1400°C, *J. Am. Ceram. Soc.*, **80**, 2549-2562.

Mori, M., Yamamoto, T., Itoh, H., Abe, T., Yamamoto, S., Takeda, Y. and Yamamoto, O. (1994), Electrical conductivity of alkaline earth metal (Mg, Ca, Sr) doped lanthanum chromites, First European Solid Oxide fuel Cells Forum, Ed. U. Bossel, pp. 465-473, Lucerne, Switzerland.

Mori, M., Yamamoto, T., Itoh, H. and Watanabe, T. (1997), Compatibility of alkaline earth metal (Mg, Ca, Sr)-doped lanthanum chromites as separators in planar-type high-temperature solid oxide fuel cells, *J. Mater. Sci.*, **32**, 2423-2431.

Mori, M., Hiei, Y. and Yamamoto, T. (2001), Control of the thermal expansion of strontium-doped lanthanum chromite perovskites by B-site doping for high temperature solid oxide fuel cell separators, *J. Am. Ceram. Soc.*, **84**, 781-86.

Murray, E.P., Tsai, T. and Barnett, S.A. (1999), A direct-methane fuel cell with a ceria-based anode, *Nature*, **400**, 649-651.

Nakayama, S. and Sakamoto, M. (2001), Ionic conductivities of apatite-type La$_x$(GeO₄)₆O$_{1.5x-12}$ (x = 8-9.33) polycrystals, *J. Mater. Sci. Letts.*, **20**, 1627-1629.

Nakayama, S. and Sakamoto, M. (1998), Electrical properties of new type high oxide ionic conductor $RE_{10}Si_6O_{27}$ (RE = La, Pr, Nd, Sm, Gd, Nd), *J. Euro. Ceram. Soc.*, **18**, 1413-1418.

Napporn, T.W., Savoie, S., Roberge, R., Jacques-Bedard, X. and Meunier, M. (2005), Single-chamber SOFC: comparing dry and humidified conditions, Proc. of the ninth international symposium on solid oxide fuel cells (SOFC-IX), Eds., S.C. Singhal and J. Mizusaki, Vol. 1, pp 371-377. The Electrochemical Society Inc., Pennington, NJ, USA.

Negishi, H., Sakai, N., Yamaji, K., Horita, T. and Yokokawa, H. (1999), Fabrication of small tubular SOFCs by electrophoretic deposition technique, Proc. of the Sixth International Symposium on Solid Oxide Fuel Cells (SOFC-VI), Eds. S.C. Singhal and M. Dokiya, pp. 885-892, The Electrochemical Soc. Inc., Pennington, NJ, USA.

Nguyen, T.L. and Dokiya (2000), Electrical conductivity, thermal expansion and reaction of $(La, Sr)(Ca, Mg)O_3$ and $(La, Sr)AlO_3$ system, Solid State Ionics, **132**, 217-226.

Ogumi, Z., Ioroi, T., Uchimo, Y. and Tekehara, Z. (1995), Novel method for preparing nickel/cermet by a vapor-phase process, *J. Am. Ceram. Soc.*, **78**, 593-598.

Pal, U. and Singhal, S.C. (1990), Electrochemical vapor deposition of YSZ, *J. Electrochem Soc.*, **137**, 2937-41, 1990.

Perednis, D. and Gauckler L.J. (2004), Solid oxide fuel cells with electrolytes prepared via spray pyrolysis, *Solid State Ionics*, **166**, 229-239.

Petric, A., Huang, P. and Tietz, F. (2000), Evaluation of La-Sr-Co-O perovskites for solid oxide fuel cells and gas separation membranes, *Solid State Ionics*, **135**, 719-725.

Pratihar, S.K., Das Sharma, A., Basu, R.N. and Maiti, H.S. (2004), Preparation of nickel coated YSZ powder for application as an anode for solid oxide fuel cells, *J. Power Sources*, **129**, 138-142.

Pratihar, S.K., Basu, R.N., Mazumder, S. and Maiti, H.S. (1999), Electrical conductivity and microstructure of Ni-YSZ anode prepared by liquid dispersion method, Eds. S.C. Singhal and M. Dokiya, Sixth International Symposium on Solid Oxide Fuel Cells (SOFC-VI). The Electrochemical Soc., Pennington, NJ, USA.

Quadakkers, W.J., Piron-Abellan, J., Shemet, V. and Singheiser, L. (2003), Metallic interconnectors for SOFCs—a review, *Materials at High Temperatures*, **20**, 115-127.

Ralph, J.M. and Kilner, J.A. (1997), Grainboundary conductivity enhancement in ceria-gadolina solid solution, Proc. of the Fifth International Symposium on Solid Oxide Fuel Cells (SOFC-V), Eds. U. Stimming, S.C. Singhal, H. Tawa and W. Lehnert, pp. 1021-1030, The Electrochemical Society, Pennington, NJ, USA.

Ralph, J.M., Kilner, J.A. and Steele, B.C.H. (2000), Improving Gd-doped ceria electrolytes for low temperature SOFC, in New Materials for Batteries and Fuel Cells, Eds. D.H. Doughty, L.F. Nazar, M. Arakawa, H.P. Brack and K. Naoi, MRS Symposium Proc., Vol. 575, pp. 309-314.

Ralph, J.M., Schoeler, A.C. and Krumpelt, M. (2001), Materials for lower temperature solid oxide fuel cells, *J. Mater. Sci.*, **36**, 1161-1172.

Randall, C.A., Van Tassel, J.V., Hitomi, A., Daga, A., Basu, R.N. and Lanagan, M. (2000), Electroceramic device opportunities with electrophoretic deposition, *J. Mater. Education*, **22**, 131-145.

Sakai, N., Yokokawa, H., Horita, T. and Yamaji, K. (2004), Lanthanum chromite-based interconnects as key materials for SOFC stack development, *Int. J. Appl. Ceram. Technol.*, **1**, 23-30.

Sammes, N.M., Tompsett, G.A., Näfe, H. and Aldinger, F. (1999), Bismuth based oxide electrolyte—structure and ionic conductivity, *J. Euro. Ceram. Soc.*, **19**, 1801-1826.

Sansom, J.E.H., Sermon, P.A. and Slater, P.R. (2005), Synthesis and conductivities of the apatite-type phases, $La_{9.33}Si_{6-x}Ge_xO_{26}$, $La_9BaSi_{6-x}Ge_xO_{26.5}$, and related titanium doped systems, Proc. of the ninth international symposium on solid oxide fuel cells (SOFC-IX), Eds., S.C. Singhal and J. Mizusaki, Vol. 2, pp. 1156-1164, The Electrochemical Society Inc., Pennington, NJ, USA.

Sarkar, P. Rho, H., Liu, M., Yamarte, L and Johanson, L. (2005), High power density tubular SOFC for portable applications, Proc. of the ninth international symposium on solid oxide fuel cells (SOFC-IX), Eds., S.C. Singhal and J. Mizusaki, Vol. 2, pp. 411-418. The Electrochemical Society Inc., Pennington, NJ, USA.

Sasaki, K., Wurth, J-P., Gschwend, R., Gödickemeier, M. and Gauckler, L.J. (1996), Microstructure-property relations of solid oxide fuel cell cathodes and current collectors—cathode polarization and ohmic resistance, *J. Electrochem. Soc.* **143**, 530-543.

Shao, Z. and Haile, S.M. (2004), A high-performance cathode for the next generation of solid-oxide fuel cells, *Nature*, **431**, 170-173.

Shima, D. and Haile, S.M. (1997), The influence of cation non-stoichiometry on the properties of undoped and gadolinia-doped barium cerate, *Solid State Ionics*, **97**, 443-455.

Sohn, S.B. and Choi, S.Y. (2004), Suitable glass-ceramic sealant for planar solid oxide fuel cells, *J. Am. Ceram. Soc.*, **87**, 254-60.

Sohn, S.B., Choi, S.Y., Kim, G.H., Song, H.S. and Kim, G.D. (2002), Stable sealing glasses for planar SOFC, *J. Non-cryst. Solid.*, **297**, 103-12.

Simmer, S.P. and Stevenson, J.W. (2001), Compressive mica seals for SOFC applications, *J. Power Sources.*, **102**, 310-316.

Singhal, S.C. and Kendall, K. (2003) (Editors), High Temperature Solid Oxide Fuel Cells: Fundamentals, design and Applications, Elsevier, UK.

Skowron, A., Huang, P. and Petric, A., Structural study of $La_{0.8}Sr_{0.2}Ga_{0.85}Mn_{0.15}O_{2.825}$ (1999), *J. Solid State Chem.*, **143**, 202-209.

Slater, P.R., Irvine, J.T.S. Irvine, Ishihara, T. and Takita, Y. (1998), The structure of the solid oxide ion conductor $La_{0.9}Sr_{0.1}Ga_{0.8}Mg_{0.2}O_{2.85}$ by powder neutron diffraction, *Solid State Ionics*, **107**, 319-323.

Steele, B.C.H. (1993), Materials for electrochemical energy conversion and storage systems, *Ceramic International*, **19**, 269-277.

Steele, B.C.H. (1994), State-of-the-Art SOFC Ceramic Materials, in Proceedings of the 1st European SOFC Forum, Ed. U. Bossel, Switzerland, 1994, pp. 375-397.

Steele, B.C.H. (2000), Appraisal of $Ce_{1-y}Gd_yO_{2-y/2}$ electrolytes for IT-SOFC operation at 500°C, *Solid State Ionics*, **129**: 95-110.

Stöver, D., Diekmann, U., Flesch, U., Kabs, H., Quadakkers, W.J., Tietz, F. and Vinke, I.C. (1999), Proc. of the sixth international symposium on solid oxide fuel cells (SOFC-VI), Eds., S.C. Singhal and M. Dokiya, pp. 812-821. The Electrochemical Society Inc., Pennington, NJ, USA.

Stöver, D., Buchkremer, H.P. and Huijsmans, J.P.P. (2003), MEA/cell preparation methods: Europe/USA, Handbook of Fuel Cells—Fundamentals, Technology and Applications, Ed. W. Vielstich, H. A. Gasteiger and A. Lamm, Volume 4: Fuel Cell Technology and Applications, pp. 1015-1031. John Wiley & Sons.

Strickler, D.W. and Carlson, W.G. (1964), Ionic conductivity of cubic solid solutions in the system $CaO-Y_2O_3-ZrO_2$, *J. Am. Ceram. Soc.*, **47**, 122-127.

Subbarao, E.C. and Maiti, H.S (1984), Solid electrolytes with oxygen ion conduction, *Solid State Ionics*, **11**, 317.

Sunde, S. (2000), Simulations of composite electrodes in fuel cells, *J. Electroceram.*, **5**, 153-182.

Takahashi, T., Esaka, T. and Iwahara, H. (1977), Conduction in bismuth (III) oxide-based oxide ion conductors under low oxygen pressure. I. Current blackening of the bismuth (III) oxide-yttrium oxide electrolyte, *J. Appl. Electrochem.*, **7**, 299-302.

Taniguchi, S., Kadowaki, M., Yasuo, T., Akiyama, Y., Miyake, Y. and Nishio, K. (2000), Improvement of thermal cycling characteristics of a planar-type solid oxide fuel cell by using ceramic fibre as sealing material, *J. Power Sources*, **90**, 163-69.

Tanner, C.W., Fung, K.Z. and Virkar, A.V. (1997), The effect of porous composite electrode structure on solid oxide fuel cell performance, *J. Electrochem. Soc.*, **144**, 21-30.

Tao, S.W., Poulsen, F.W., Meng, G.Y. and Sorensen, O.T. (2000), High temperature stability study of the oxygen-ion conductor $La_{0.9}Sr_{0.1}Ga_{0.8}Mg_{0.2}O_{3-x}$, *J. Mater. Chem.*, **10**, 1829-1833.

Teller, O., Meulenberg, W.A., Tietz, F., Wessel, E. and Quadakkers, W.J. (2001), Improved material combinations for stacking of solid oxide fuel cells, Proc. of the Seventh International Symposium on Solid Oxide Fuel Cells (SOFC-VII), Eds. H. Yokokawa and S.C. Singhal, pp. 895-903, The Electrochemical Soc., Pennington, NJ, USA.

Tietz, F., Buchkremer, H.P. and Stöver, D. (2002), Components manufacturing for solid oxide fuel cell, *Solid State Ionics*, **152-153**, 373-381.

Tsai, T. and Barnett, S.A. (1997), Increased solid-oxide fuel cell power density using interfacial ceria layers, *Solid State Ionics*, **98**, 191-196.

Tuller, H.L. (1997), Semiconduction and mixed ionic-electronic conduction in nonstoichiometric oxides: impact and control, *Solid State Ionics*, **94**, 63-74.

van Doorn, R.H.E., Bouwmeester, H.J.M., Burggraaf, A.J. (1998), Kinetic decomposition of $La_{0.3}Sr_{0.7}CoO_{3-\delta}$ perovskite membranes during oxygen permeation, *Solid State Ionics*, **111**, 263-272.

van Gool, W. (1965), The possible use of surface migration in fuel cells and heterogeneous catalysis, *Philips Research Report*, **20**, 81-93.

Veith, M., Mathur, S., Lecerf, N., Huch, V. and Decker, T. (2000), Sol-gel synthesis of nano-scaled $BaTiO_3$, $BaZrO_3$ and $BaTi_{0.5}Zr_{0.5}O_3$ oxides via single-source alkoxide precursors and semi-alkoxide routes, *J. Sol-Gel Sci. Technol.*, **15**, 145-158.

Verkerk, M.J. and Burggraff, A.J. (1981), High oxygen ion conduction in sintered oxides of the Bi_2O_3-DY_2O_3 system, *J. Electrochem. Soc.*, **128**: 75-82.

Wan J.-H., Yan, J.-Q. and Goodenough, J.B. (2005), LSGM-based solid oxide fuel cell with 1.4 W/cm^2 power density and 30 day long-term stability, *J. Electrochem. Soc.*, **152**, A1511-A1515.

Waschsman, E.D., Ball, G.R., Jiang, N. and Stevenson, D.A. (1992), Structural and defect studies in solid oxide electrolytes, *Solid State Ionics*, **52**, 213-218.

Weber, A. and Ivers-Tiffée, E. (2004), Materials and concepts for solid oxide fuel cells (SOFCs) in stationary and mobile applications, *J. Power Sources*, **127**, 273-283.

Williams, M.C., Strakey, J.P. and Surdoval, W.A. (2005), U.S. DOE Solid Oxide Fuel Cells: Technical Advances, Proc. of the Ninth International Symposium on Solid Oxide Fuel Cells (SOFC-IX), Eds., S.C. Singhal and J. Mizusaki, Vol. 1, pp 20-31, The Electrochemical Society Inc., Pennington, NJ, USA.

Wolfenstine, J., Huang, P. and Petric, A. (1999), Creep behavior of doped lanthanum gallate versus cubic zirconia, *Solid State Ionics*, **118**, 257.

Yahiro, H., Ohuchi, T., Eguchi, K. and Arai, H. (1988), Electrical properties and microstructure in the system ceria-alkaline earth oxide, *J. Mater. Sci.*, **23**, 1036-1041.

Yan, J., Matsumoto, H., Enoki, M. and Ishihara, T. (2005), High-power SOFC using $La_{0.9}Sr_{0.1}Ga_{0.8}Mg_{0.2}O_{3-\delta}$/$Ce_{0.8}Sm_{0.2}O_{2-\delta}$ composite film, *Electrochem. and Solid State Letts.*, **8**, A389-A391.

Yamaji, K., Negishi, H., Horita, T., Sakai, N. and Yokokawa, H. (2000), Vaporization process of Ga from doped $LaGaO_3$ electrolytes in reducing atmospheres. *Solid State Ionics*, **135**, 389-396.

Yamamoto, O. (2000), Solid oxide fuel cells: fundamental aspects and prospects, *Electrochemica Acta*, **45**, 2423-2435.

Yamamuru, Y., Kawasaki, S. and H. Sakai (1999), Molecular dynamics analysis of ionic conduction mechanism in yttria-stabilized zirconia, *Solid State Ionics*, **126**, 181-89.

Yasuda, I., Matsuzaki, Y., Yamakawa, T. and Koyama, T. (2000), Electrical conductivity and mechanical properties of alumina-dispersed doped lanthanum gallates, *Solid State Ionics*, **135**, 381.

Yasuda, I. and Hikita, T. (1993), Oxygen potential profile and ionic leak current in $LaCrO_3$-based interconnect materials, Proc. of the third international symposium of solid oxide fuel cells (SOFC-III). Eds. S.C. Singhal and H. Iwahara, pp. 354-363.

Yokokawa, H. and Horita, T. (2003), in High Temperature Solid Oxide Fuel Cells: Fundamentals, Design and Applications, Ed. S.C. Singhal and K. Kendall, Elsevier, UK, pp. 119-147.

Zacate, M.O., Minervini, L., Bradfield, D.J., Grimes, R.W. and Sickafus, K.E. (2000), Defect cluster formation in M_2O_3-doped cubic zirconia, *Solid State Ionics*, **128**, 243-54.

Zheng, R., Wang, S.R., Nie, H.W. and Wen, T.L. (2004), SiO_2-CaO-B_2O_3-Al_2O_3 ceramic glaze as sealant for planar IT-SOFC, *J. Power Sources*, **128**, 165-172

Zhu, W.Z. and Deevi, S.C. (2002), Development of interconnect materials for solid oxide fuel cells, *Mater. Sci. & Eng.*, **A348**, 227-243.

Zhu, W.Z. and Deevi, S.C. (2003), A review on the status of anode materials for solid oxide fuel cells, *Mater. Sci. & Eng.*, **A362**, 228-239.

Zhu, W.Z. and Deevi S.C. (2003), Opportunity of metallic interconnects for solid oxide fuel cells: a status on contact resistance, *Materials Research Bulletin*, **38**, 957-972.

Recent Trends in Fuel Cell Science and Technology
Edited by S. Basu
Anamaya Publishers, New Delhi, India

13. Fuel Cell Power-Conditioning Systems

Sudip K. Mazumder

Department of Electrical and Computer Engineering
Laboratory for Energy and Switching-Electronics Systems (LESES), University of Illinois, Chicago

1. Need for Alternative Energy Systems

Currently, USA is the largest consumer of energy in the world with projected consumption of 5207 billion kWhs in 2025 (http://www.eia.doe.gov/), with the actual and forecast demand for electrical energy is growing in all the end-use sectors. The highest annual growth rates are projected for the commercial, industrial, and residential sectors at about 2.2, 1.6 and 1.4%, respectively, as shown in Fig. 1(a). From 2000 to 2003, 69 GWs of peaking capacity was added and 112 GWs of combined-cycle capacity, which is efficient in both baseload and cycling applications, was installed. To meet increasing base and peak power demands, the demand for energy resources is rapidly increasing and unfortunately, as Fig. 1(b) shows, progressively depleting resources, such as natural gas and coal, which cause environmental pollution, still account for more than 50% of the total energy-generation resources (http://www.eia.doe.gov/). Furthermore, as the demand for natural gas continues to rise, domestic production dwindles and world-wide demand and supply scenario for oil and gas approach a delicate balance, we are forced with greater reliance on importation (which incidentally, is still insufficient to meet the demand) and economic vulnerability. Clearly, there is an urgent need for alternative sources of energy.

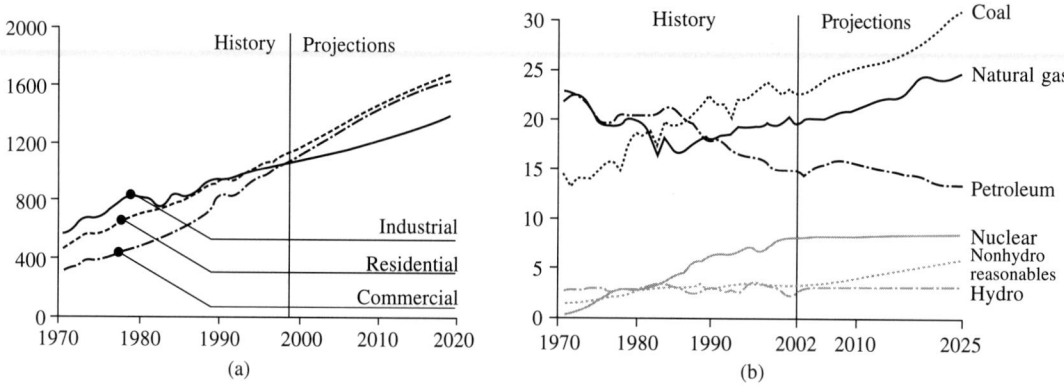

Fig. 1 (a) Current and projected (a) electricity sales (TWh) by sector and (b) energy production (quadrillion BTU) by fuel in USA (http://www.eia.doe.gov/).

2. Overall Requirements for Stationary Fuel Cell Power Systems

Hydrogen based fuel cell energy is one of the two front-runner alternative-energy solutions to address and alleviate the imminent and critical problems of existing fossil-fuel-energy systems: environmental pollution due to high emission level and rapid depletion of fossil fuel. The framework for integrating these "zero-emission" alternative-energy sources to the existing energy infrastructure has been provided by the concept of microgrid or distributed generation (DG), which provide an additional advantage: reduced reliance on existing and new centralized power generation, thereby saving significant capital cost. However, to achieve the projected world-wide target of $50 billion by 2015, the fuel cell energy systems have to address *cost, durability and reliability, and energy efficiency.*

Currently, the higher cost for this energy system (as compared to conventional energy systems) is primarily due to the energy source (fuel cell) and the power system. Recent U.S. Department of Energy (DOE) studies have predicted that for fuel-cell energy systems to be economically viable, the cost of the power systems have to be ≤ $400/kW while the cost of the power electronics needs to be ≤ $40/kW. The issues of durability and life of the fuel cells are of key importance.

Recent comprehensive studies by National Energy Technology Laboratory (NETL) have shown that power-electronics, which interfaces directly to the fuel cell stacks, has a significant impact on the long-term durability (life) and reliable energy efficiency of the fuel cells. Energy-conversion efficiency (of the power-conditioning system) for fuel cell is of significant importance primarily in light of the lower kW/dollar of some of the conventional energy systems, a selective few of which (e.g. combined cycle), incidentally, have achieved near-comparable efficiencies in recent years. Today, the efficiencies of power–electronics conversion technology have exceeded 90%; however, under severe cost constraints, most of these technologies are not economically viable. As such, achieving high power-conversion efficiency at significantly low cost for the viability of fuel cell power systems is a daunting challenge.

3. Overview of Power-Conditioning Topologies for Stationary Fuel Cell Energy Systems

Fig. 2 shows a typical fuel-cell power-conditioning system. Because the DC voltage generated by a fuel cell stack varies widely and is usually low in magnitude (< 100 V for a 5-10 kW system, < 350 V for a 300 kW system), a step-up DC/DC converter stage is essential to generate a regulated higher voltage DC

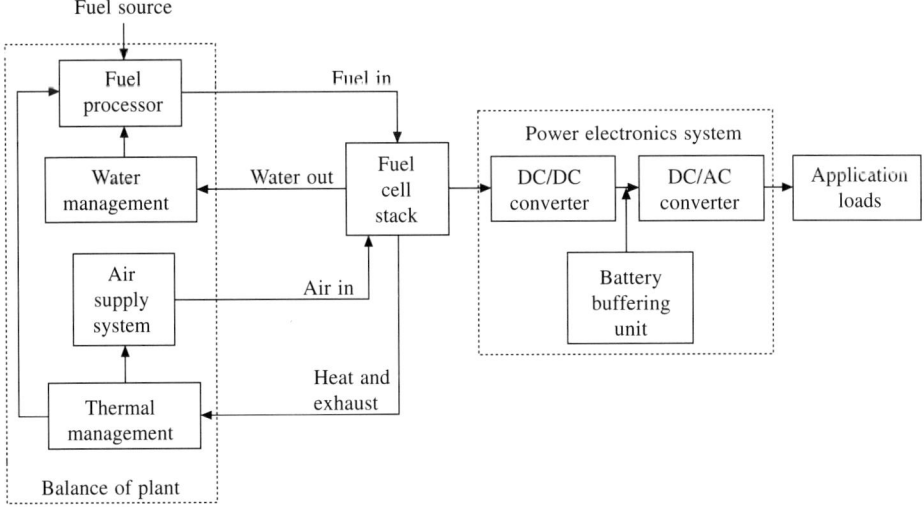

Fig. 2 A block diagram of a typical stationary fuel cell power conditioner supplying loads.

(400 V typical for 120/240 V AC output). The DC/DC conversion stage is responsible for drawing power from the fuel cell and, therefore, should be designed to match the fuel cell ripple current specifications. Further, the DC/DC converter should not introduce any negative current into the fuel cell. A DC/AC conversion stage is essential for converting the DC/AC power at 60/50 Hz frequency. An output filter stage connected to the inverter filters the switching frequency harmonics and generates a high quality sinusoidal AC waveform suitable for the load.

While Fig. 2 shows the general requirements of a fuel cell power-conditioning system, Fig. 3 (a-c) illustrates three mechanisms for achieving galvanic isolation between the fuel cell stack and the output load. The conventional scheme shown in Fig. 3 (a) achieves isolation by placing 50/60 Hz line-frequency transformers at the output of the DC/AC inverter. Such a transformer is bulky because it has to handle line frequency. The other two options in Fig. 3 (b, c) achieve isolation by placing the isolation transformer within the power-conditioning system. The high-frequency isolation can be included in the DC/AC converter stage, as shown in Fig. 3 (b), or in the DC/DC converter stage, as illustrated in Fig. 3 (c). For either case, because the isolation transformer operates at a high frequency, its size is significantly smaller than that of the line transformer in Fig. 3 (a).

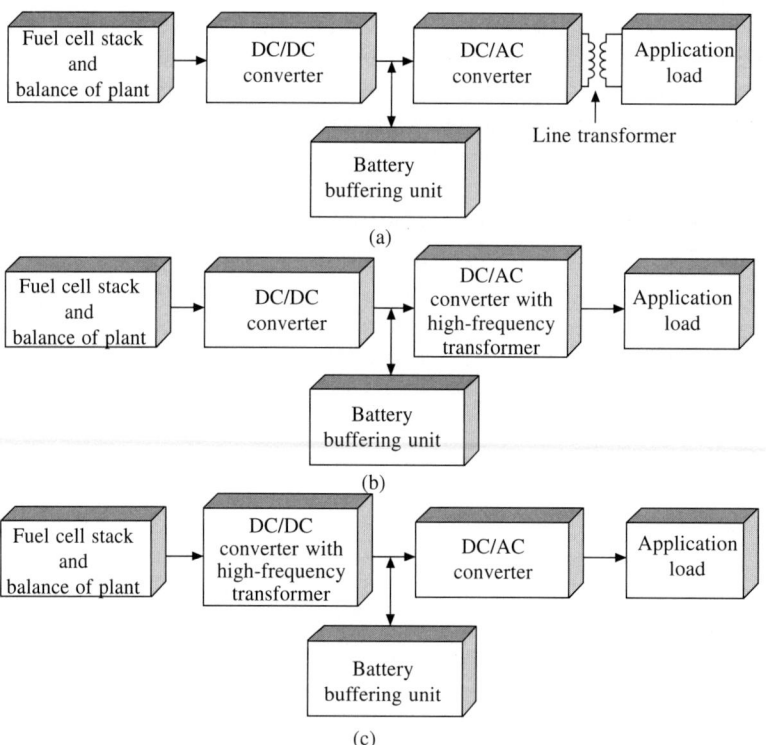

Fig. 3 Power-conditioning scheme with (a) line-frequency transformer, (b) high-frequency transformer in the DC/AC stage and (c) high-frequency transformer in the DC/DC stage.

3.1 DC/DC Converter Topologies

In this section, several DC/DC converter topologies are examined, which have been employed in fuel cell power systems. We only investigate the strengths and weaknesses of the topologies and leave the reader to review (Erickson and Maksimovic, 2001; Mohan et al, 2002; Krein, 1997; Kassakian et al, 1991) for details on operating modes.

3.1.1 Boost Converter

Fig. 4 is a simplified schematic of a boost converter. The circuit consists of a control block, a switch Q, a diode D, inductor L and output capacitor C. The key problems with the boost converter are two-fold: because D carries high current due to lower fuel cell stack voltage, its reverse-recovery when Q turns on could be a problem; and non-minimum phase and nonlinear behavior of the converter limiting the system bandwidth.

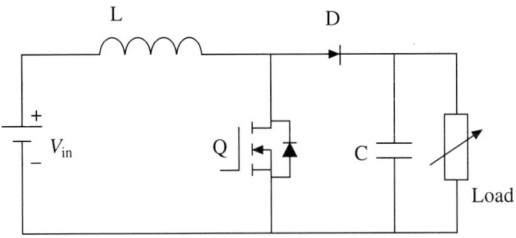

Fig. 4 Circuit schematic of a single-switch boost converter.

3.1.2 Multiphase Isolated Boost Converter

Fig. 5 is a circuit schematic of a multi-phase isolated boost converter (Lai, 2005). The circuit consists of twelve controlled switches (S_1-S_{12}) and six diodes (D_1-D_6) to generate a DC voltage output. Because the current is shared by three-phases, the current stress on each device reduces. As a result, the problems of diode recovery and higher switch and diode losses are substantially alleviated. Also, because the topology is a buck-derived type of converter, the controller implementation is simplified.

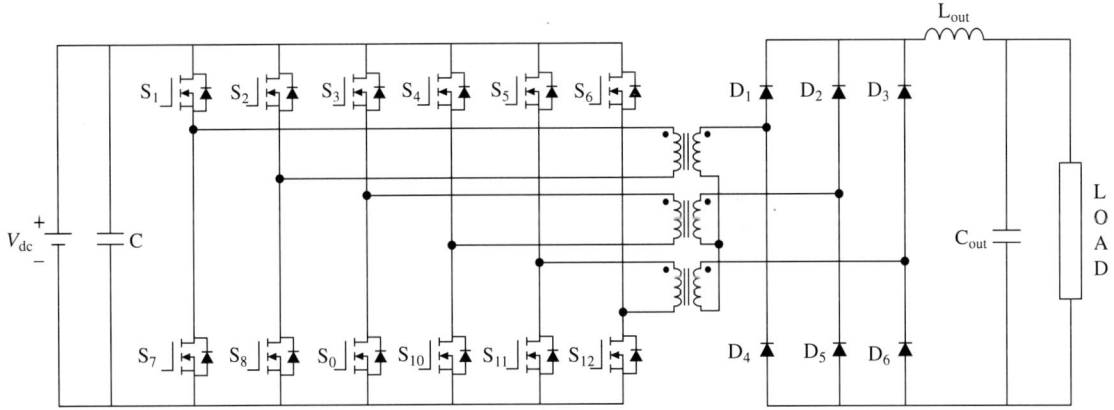

Fig. 5 Circuit schematic of a multi-phase boost converter.

3.1.3 Isolated Ćuk Converter

Fig. 6 shows the schematic of the isolated Ćuk converter (Middlebrook and Ćuk, 1983), which steps-up the input voltage to a high intermediate DC voltage and provides galvanic isolation as well. The isolated Ćuk converter attains fewer components, enables integrated magnetics leading to low input and output current ripples and lower electromagnetic interference (EMI) as compared to other conventional isolated step-up converters. The magnetic integration yields significant savings in weight and volume as well. However, because the current through transformer sees an almost instantaneous change due to the switching of S_1, the leakage inductance Ćuk converter has to be really low; otherwise, the voltage spike across S_1 (and S_2) could potentially destroy the device. A higher voltage rating of S_1 (S_2) leads to higher losses. Non-dissipative

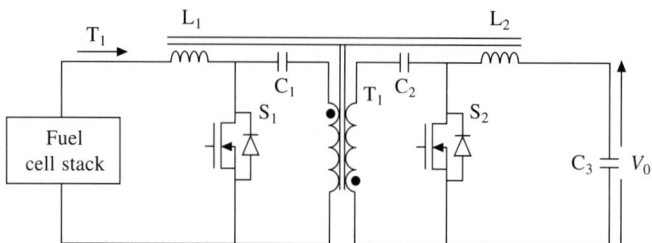

Fig. 6 Isolated Ćuk converter.

(resonant) or dissipative snubbers (RC and RCD) can be used to overcome the voltage spike problem. Also, resonant clamp circuit across the transformer primary can be used to clamp the voltage overshoot to a pre-decided value. However, all of these options have a price tag from the standpoint of efficiency.

3.1.4 Push-Pull Converter

The push-pull topology, as shown in Fig. 7, utilizes a step-up high-frequency transformer with a center-tapped winding and boosts the input voltage to a higher output. The push-pull topology can handle higher power and has a simple design. However, if the two halves of the center-tapped winding are not equal or symmetrically wound or if the switching times of the power devices as well as their on-state drops are not equal, the transformer core can saturate possibly leading to converter failure. Possible solutions to this problem include (i) gapping the core, whereby an air gap is introduced in the core to avoid saturation but, increases the size of the converter or (ii) using current-mode control (ensuring equal current in each switch) to alleviate the problem of flux-imbalance. The leakage inductance of the high-frequency transformer also needs to be addressed to reduce voltage spikes.

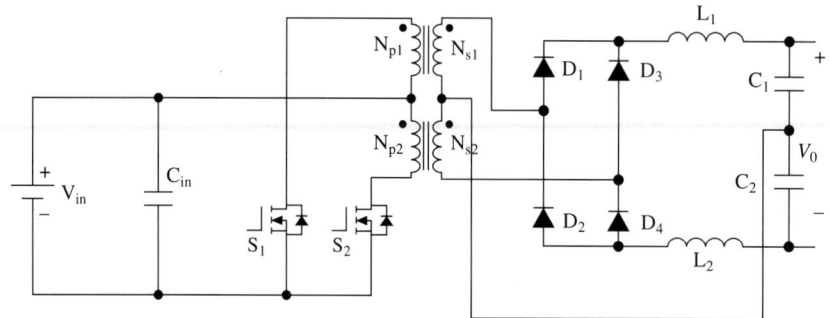

Fig. 7 Push-pull DC/DC converter with voltage doubling center-tapped DC output.

3.1.5 Weinberg Converter

By using an input and output magnetic coupling, the Weinberg converter (Weinberg and Schreuders, 1985), as shown in Fig. 8, overcomes the strict switching requirements for a voltage-fed push-pull converter. Further, no output inductor is employed. Nonetheless, the Weinberg converter is also susceptible to voltage spikes due to the existence of leakage inductance in not only the center-tapped transformer but also the input magnetics.

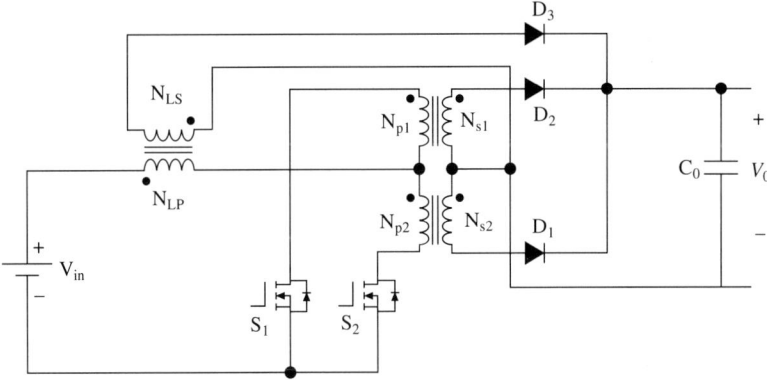

Fig. 8 Topology of the Weinberg converter.

3.1.6 Full-Bridge DC/DC Converter

Fig. 9 shows a voltage-fed full-bridge DC/DC converter topology, which is a strong contender for fuel cell applications because of symmetrical transformer flux and ease of modulation. This topology also offers the possibility of soft-switching using the transformer parasitics and without the need for auxiliary circuits. However, for input voltage levels greater than 500 V, the voltage stress on the semiconductors increases and reliability of the power electronics is an issue. Multilevel arrangement of switches (Canales et al., 2002; Pinheiro and Barbi, 1993) reduces the voltage stress on individual switches and improves the reliability.

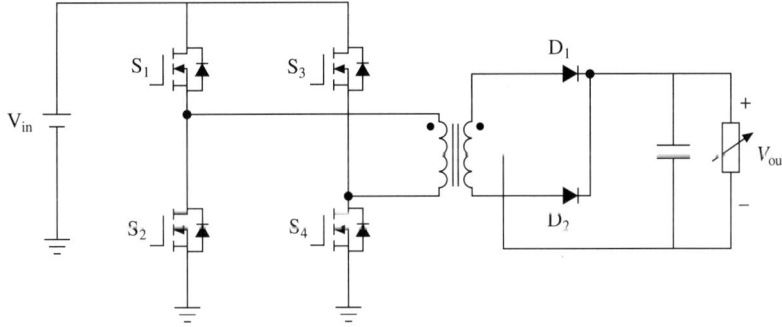

Fig. 9 Voltage-fed full-bridge DC/DC converter.

3.2 DC/AC Converter Topologies

The DC/AC inverter is an essential part of a fuel cell power conditioner for generating single- and three-phase AC outputs. Usually, three varieties of DC/AC inverter outputs are required to power most loads: (i) single-phase AC output for domestic loads (120 V and 60 Hz); (ii) single-phase AC dual output for powering homes (e.g. 120 V/240 V 60 Hz for domestic/industrial loads) and (iii) three phase fixed voltage, fixed frequency AC output for utility interface and backup power systems (208 V/230 V/480 V and 60 Hz). In this section, DC/AC inverters are examined in the context of fuel cell power conversion.

3.2.1 Half-Bridge Non-isolated DC/AC Inverter

Fig. 10 shows a single phase half-bridge inverter. The switches S_1 and S_2 are modulated sinusoidally to generate a line frequency output. Though this topology is very simple to implement, it suffers from increased voltage stress on the devices. Moreover, if there is any mismatch in the values of the input capacitors, there would be an offset in the output voltage and have a degrading effect on the performance of the power conditioning system.

3.2.2 Full-Bridge Non-isolated DC/AC Inverter

Fig. 11 shows the topology of a full-bridge DC/AC inverter, which consists of four switches. For the control of the inverter switches, two reference sinusoidal signals 180 degrees out of phase from each other are used to intersect the same triangular carrier signal. Though this topology is widely used for inverter applications, the voltage stresses across the switches are high. So, the utility of the topology decreases significantly for very high-voltage fuel cell applications.

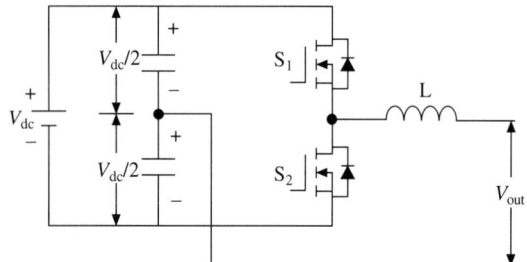 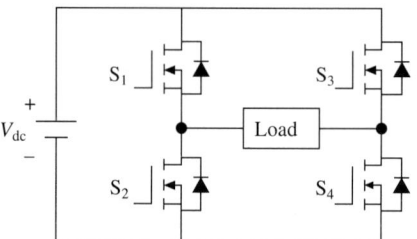

Fig. 10 Circuit schematic of a half-bridge single phase inverter.

Fig. 11 Schematic of a single phase full-bridge DC/AC inverter.

3.2.3 Push-Pull Isolated DC/AC Converter

Fig. 12 shows a push-pull based DC/AC converter (Jin and Enjeti, 2004), which consists of a front-end boost converter to regulate the DC voltage. A battery is connected to the intermediate DC bus to provide fast dynamic response during load-transients. The push-pull converter operates with a 50% duty ratio and provides rectangular voltage pulses at the secondary of the high-frequency transformer. The cycloconverter is sinusoidally modulated to produce a 60 Hz line-frequency voltage at the output. The push-pull converter

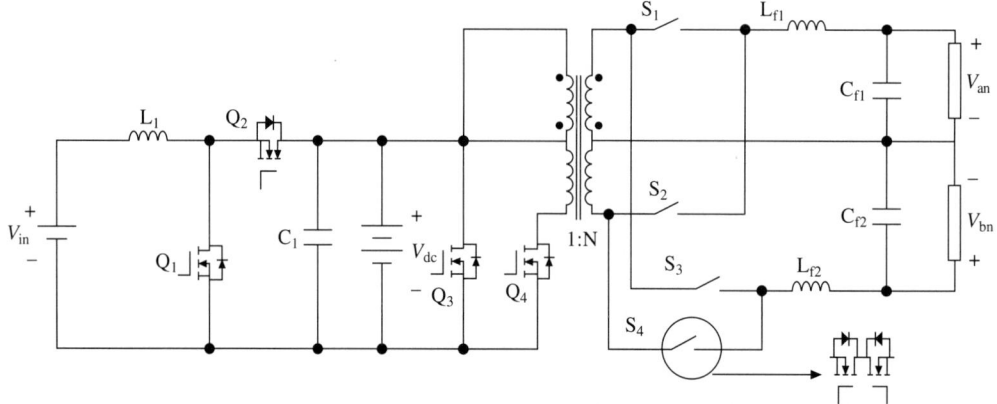

Fig. 12 Push-pull converter based DC/AC converter followed by a full-bridge cycloconverter.

used in this topology is simple and can handle high-power applications. However, the transformer core could saturate if the switching times of the switches Q_3 and Q_4 are not equal or if there is any parametric difference in these two switches. Also, with this approach, we do not have the option of line-frequency switching of the cycloconverter. Because all of the switches operate at high-frequencies, careful design is required to alleviate higher switching losses.

3.2.4 Full-Bridge Isolated DC/AC Converter

Figure 18 shows a full-bridge based DC/AC converter (Kawabata et al, 1990; Mazumdar et al, 2002; Krein et al, 2004). The switches of the full-bridge converter are sinusoidally modulated, while the switches of the cycloconverter operate at line frequency to generate a 60 Hz voltage at the output. The turns ratio of the transformer is adjusted to generate the desired amplitude at the output of the converter. Because the voltage across the transformer is symmetric, the full-bridge converter based topology does not suffer from problems due to transformer core saturation. Further, because the cycloconverter switches are operated at line-frequency, the switching losses of the converter are lower, which leads to increased efficiency of the inverter. However, because of the higher turns-ratio required with this topology, the transformer parasitic leakage could be significant, leading to increased losses and increased voltage stresses across the switches due to spikes. To alleviate this problem, the high-frequency transformer must be designed optimally and the switches for the cycloconverter need to be selected carefully with particular attention to the body-diode and output-capacitance characteristics.

3.2.5 Multilevel Single- and Two-Phase Inverters

For higher power fuel cell hybrid systems, the fundamental AC output voltage from an inverter must be increased to medium voltage levels (>1000 V) to keep the current levels low. To accomplish this, the corresponding DC-link voltage (V_{DC}) must be increased and the power semiconductor devices (Canales et al, 2002; Pinheiro and Barbi, 1993; Mazumder and Burra, 2005) must be connected in series. This necessitates the use of matched pair of semiconductor devices and complicated dynamic voltage sharing schemes. Multilevel inverter topologies have the advantage of generating higher voltage AC output with simultaneous voltage sharing of series connected devices with clamped diodes.

4. Towards Low-Cost, Energy-Efficient and Reliable Residential Power Electronics

In this section, we describe three fuel-cell power conversion schemes for standalone residential loads. Our first power-electronics system (PES) comprises a *two-stage* power conditioner comprising an isolated parallel DC/DC Ćuk converter at the front end followed by a full-bridge DC/AC inverter at the back end. This converter can be used for a power rating of < 5 kW. The second PES comprises a front end phase-shifted high-frequency (HF) full-bridge inverter followed by a step-up high-frequency transformer and a forced cycloconverter. This *single-stage* DC/AC converter approach meets the higher energy efficiency requirement of DOE without exceeding the DOE cost constraint. The third PES is also a *two-stage* low-cost power conditioner. The front end of the PES is a zero-ripple boost converter (ZRBC), followed by a two-stage DC/AC converter comprising a soft-switched phase shifted sinewave-pulse-width-modulation (SPWM) multilevel high-frequency inverter and a line-frequency switched cycloconverter.

4.1 Isolated Parallel Ćuk Converter followed by a Voltage-Source Inverter

In this approach, the isolation is provided in the isolated DC/DC Ćuk converter, which is the front end, followed by a voltage-source inverter (VSI), which feeds AC power to the load. The overall design leads to lesser number of components and lower electromagnetic interference (EMI) as compared to other conventional isolated step-up converters. Besides, the similarity of the voltage waveforms across the magnetic components of the Ćuk converter facilitates the possibility of integration of the magnetic components that can potentially yield significant savings in the weight and volume of the power electronics system. Further, the characteristic of reduced or even better, zero-ripple current at the input side enhances the durability of the fuel cell stack (Mazumder et al, 2004; Gemmen, 2001). Due to the power requirements and the high voltage step-up ratio, we implement two Ćuk converter modules with their inputs connected in parallel and outputs connected in series. Using this approach, we ensure that each module handles half of the required power, which potentially reduces the stresses on the individual switches. The overall schematic for the initial approach is shown in Fig. 13. Module 1 of the Ćuk converter acts as a dedicated master and generates the PWM signals for itself and the slave converter (Module 2). Fig. 13 (b) shows the controllers for the Ćuk converter and the VSI. A current-mode controller is used for the Ćuk converter to regulate the output voltage and control one of the inductor currents. The latter is important because the Ćuk converter is a non-minimum phase, nonlinear system and the current control provides better robustness during transient conditions (i.e. against feed-forward and feedback disturbances). The voltage-loop controller

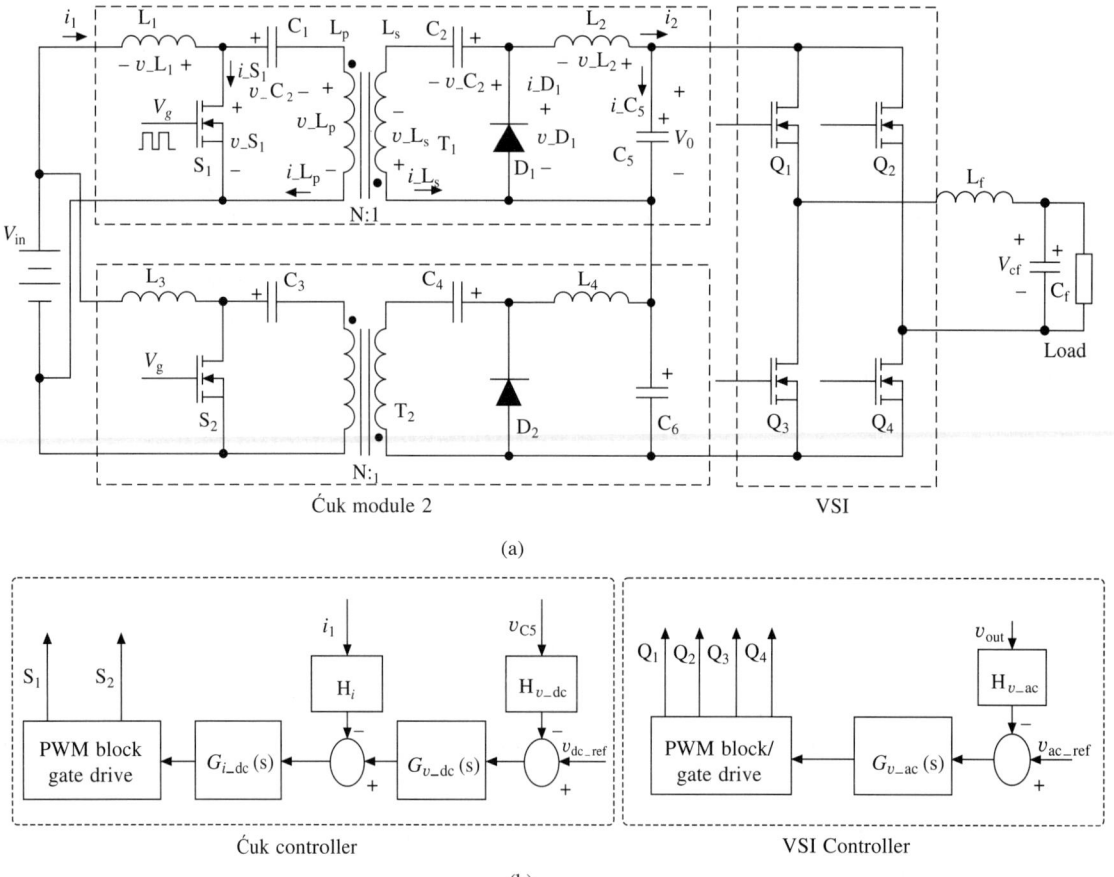

(a)

(b)

Fig. 13 (a) Isolated parallel Ćuk converter, which is followed by a VSI inverter and (b) Control schematic of the Ćuk converter and the VSI.

comprises of a PI compensator. The inner current-loop currently consists of a proportional compensator and provides robustness against transient and parametric variations. A voltage-mode controller is used for the VSI using a PI compensator. The output voltage and current obtained with this converter and the corresponding efficiencies are shown in Fig. 14. However, for the Ćuk converter, the transformer needs to be carefully designed. This is because when the primary switch of the Ćuk converter is turned off, the current in the primary winding of the transformer changes direction. This sudden change in the current causes a voltage spike across the primary switch of the Ćuk converter. As a result, we observe a significant voltage spike across the primary switches of the Ćuk converter, as shown in Fig. 15. To ensure that the switches do not fail due to this spike, we have to either use high-voltage devices to accommodate for the additional voltage stress across the device or use additional clamp circuit (Pomilio and Spiazzi, 1994).

4.1.1 Modes of Operation of the Overall Converter
The schematic waveforms for the Ćuk converter and the switching sequence of the VSI and the operating modes of the converter are shown in Figs. 16 and 17, respectively.

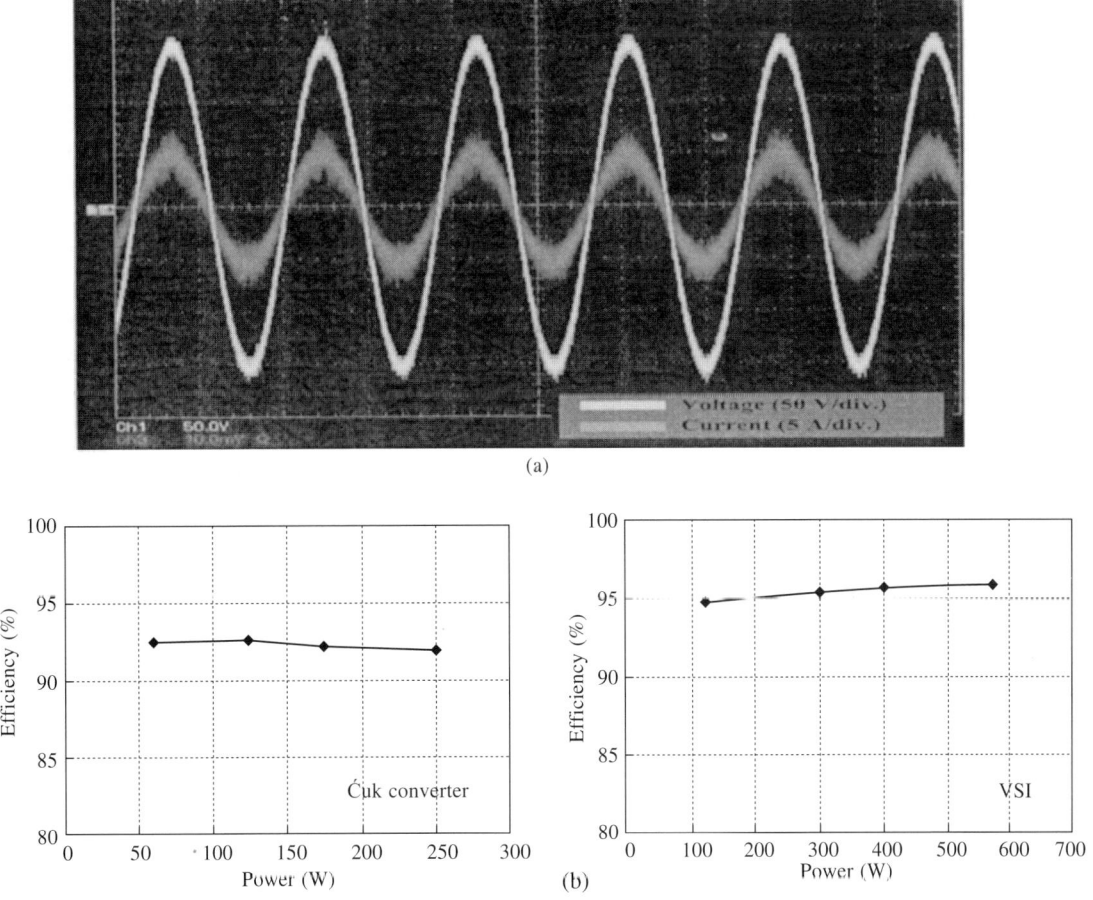

Fig. 14 (a) VSI Output Voltage, 50 V/div and output current 5 A/div for an input voltage of 200 V and a switching frequency of 20 kHz and (b) efficiency for the Ćuk converter and the VSI.

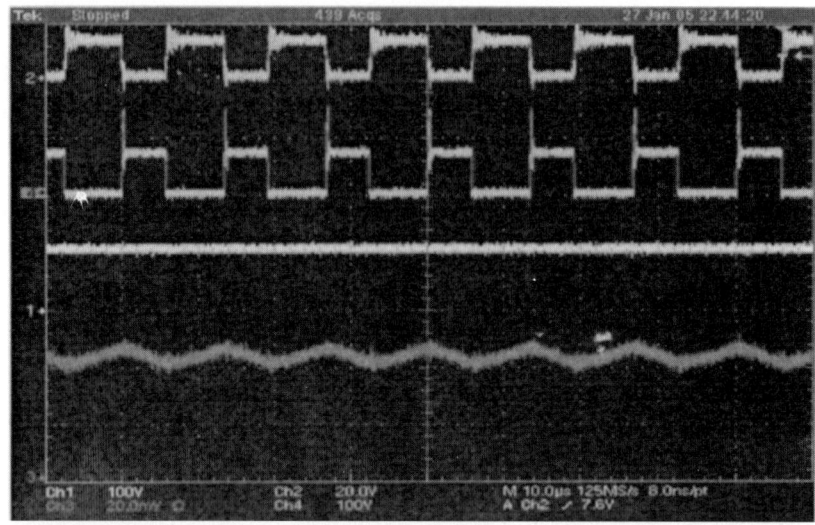

Fig. 15 Experimental results for the Ćuk converter showing gate drive signal 20 V/div, voltage stress on the switch 100 V/div, output voltage 100 V/div, input current 4 A/div for 30 V input.

Mode 1 (Fig. 17(a)): During this period, the switch S_1 is turned on, the current flowing through the input inductor, L_1, increases and the inductor stores energy. At the same time, the capacitor C_1 discharges through S_1, and there is, thus, energy transfer through the transformer T_1. The capacitor C_2 is discharged to the circuit formed by L_2, C_5, and the load. For the VSI, switches Q_1 and Q_4 are turned on, so energy is transferred from the bus to the output. This state corresponds to the active state for the VSI.

Mode 2 (Fig. 17(b)): For the Ćuk converter, this mode is the same as Mode 1. For the VSI, switches Q_3 and Q_4 are turned. During this state there is no connection between the output and the bus, so this state corresponds to the zero state for the VSI.

Mode 3 (Fig. 17(c)): The switch S_1 is off, the current in L_1 charges C_1 and C_2 using the energy, which was stored while S_1 was on. During this time, the load is supplied by L_2 and C_2. The operation of the VSI in this mode is same as that of Mode 1.

Mode 4 (Fig. 17(c)): In this mode, the Ćuk converter operates just like in Mode 3 and the VSI operates like in Mode 2.

4.2 Isolated Phase-Shifted High-Frequency Inverter followed by a Forced Cycloconverter

The topology described in this section achieves a direct power conversion and does not require any intermediate energy storage components. For high power (>3 kW) fuel cell power conditioning applications where high energy efficiency and power density is a requirement, this topology could be a choice. As shown in Fig. 18, the topology consists of a high-frequency (HF) inverter followed by a HF transformer and a forced cycloconverter. The switches (Q_1-Q_4) of the HF inverter are sine-wave modulated to create a HF three-level AC voltage (V_{p1} and V_{p2}), as shown in Fig. 19. The three-level AC at the output at the secondary of the HF transformer is converted to line frequency (60 Hz) AC by the forced cycloconverter and the LC filter. The turns ratio of the HF transformer are adjusted so that the desired AC output voltage (120 V, 60 Hz) can be obtained for the least input voltage of the fuel cell. For an input of 30 V, the turns ratio of the HF transformer is $N = 7$. Because of the significant turns ratio, care must be taken in fabricating the transformer. Higher turns ratio yields enhanced secondary leakage inductance and secondary winding-

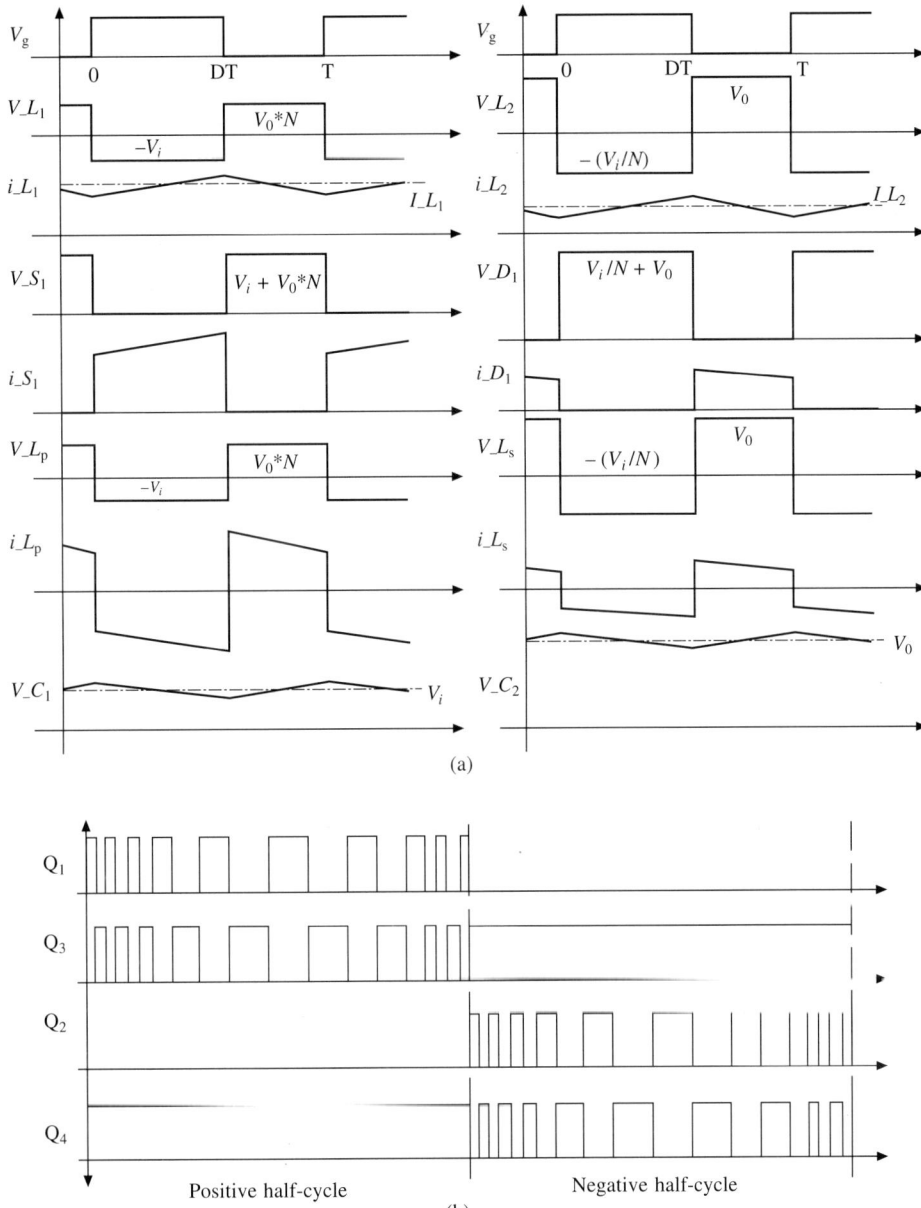

Fig. 16 (a) Idealized voltage and current waveforms of the primary and secondary side components of the Ćuk converter and (b) switching sequence for the VSI.

resistance, which results in a loss of duty cycle and higher secondary copper losses. Higher leakage also leads to higher voltage spike (Fig. 20), which added to the high nominal voltage of the secondary necessitate the use of high-voltage power devices. Such devices have higher on resistance and slower switching speeds. To minimize the conduction losses and increase the overall efficiency of the converter (Fig. 21), we use two parallel cycloconverter modules as shown in Fig. 18.

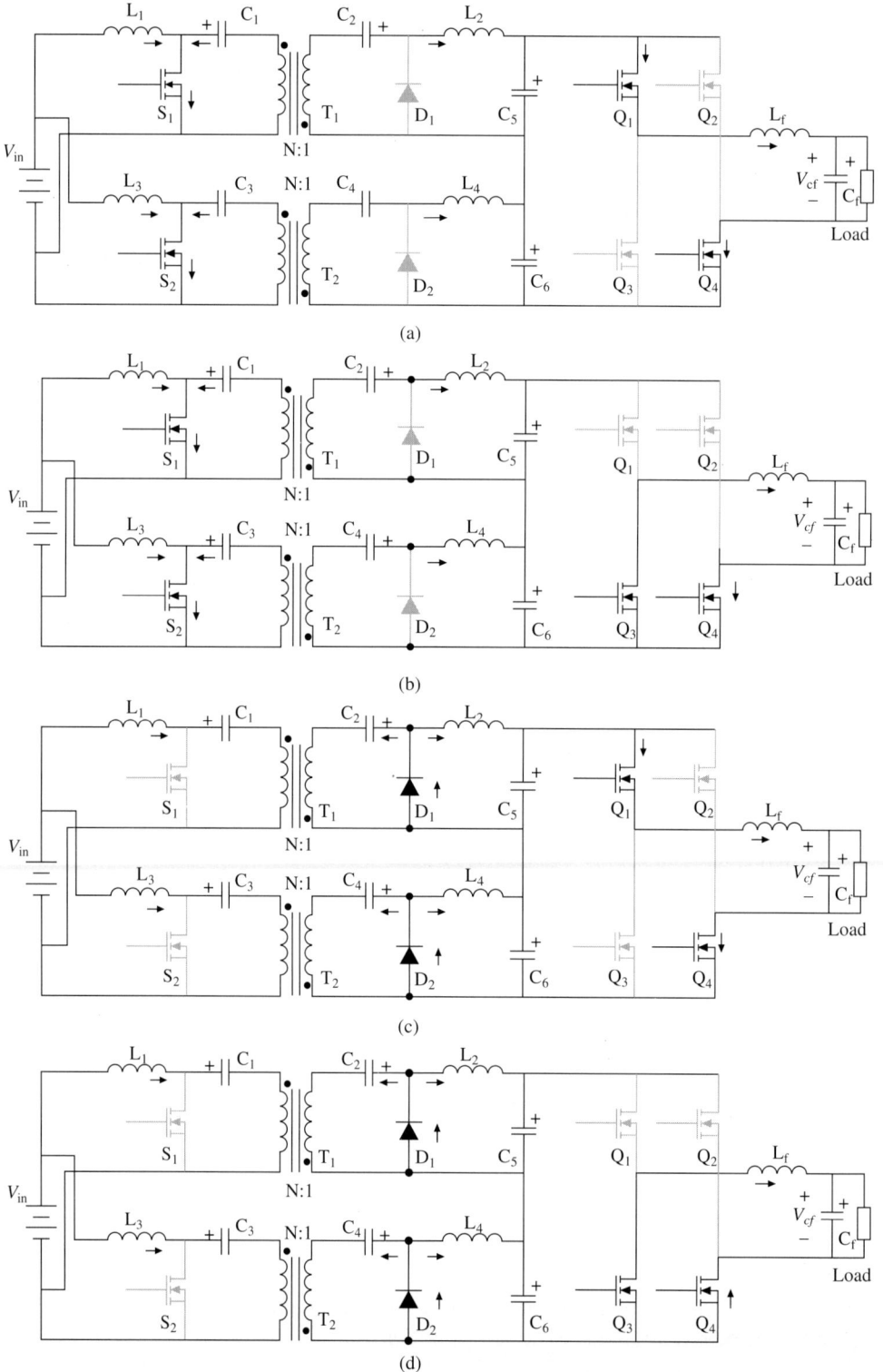

Fig. 17 Operating modes of the isolated parallel Ćuk converter followed by a VSI.

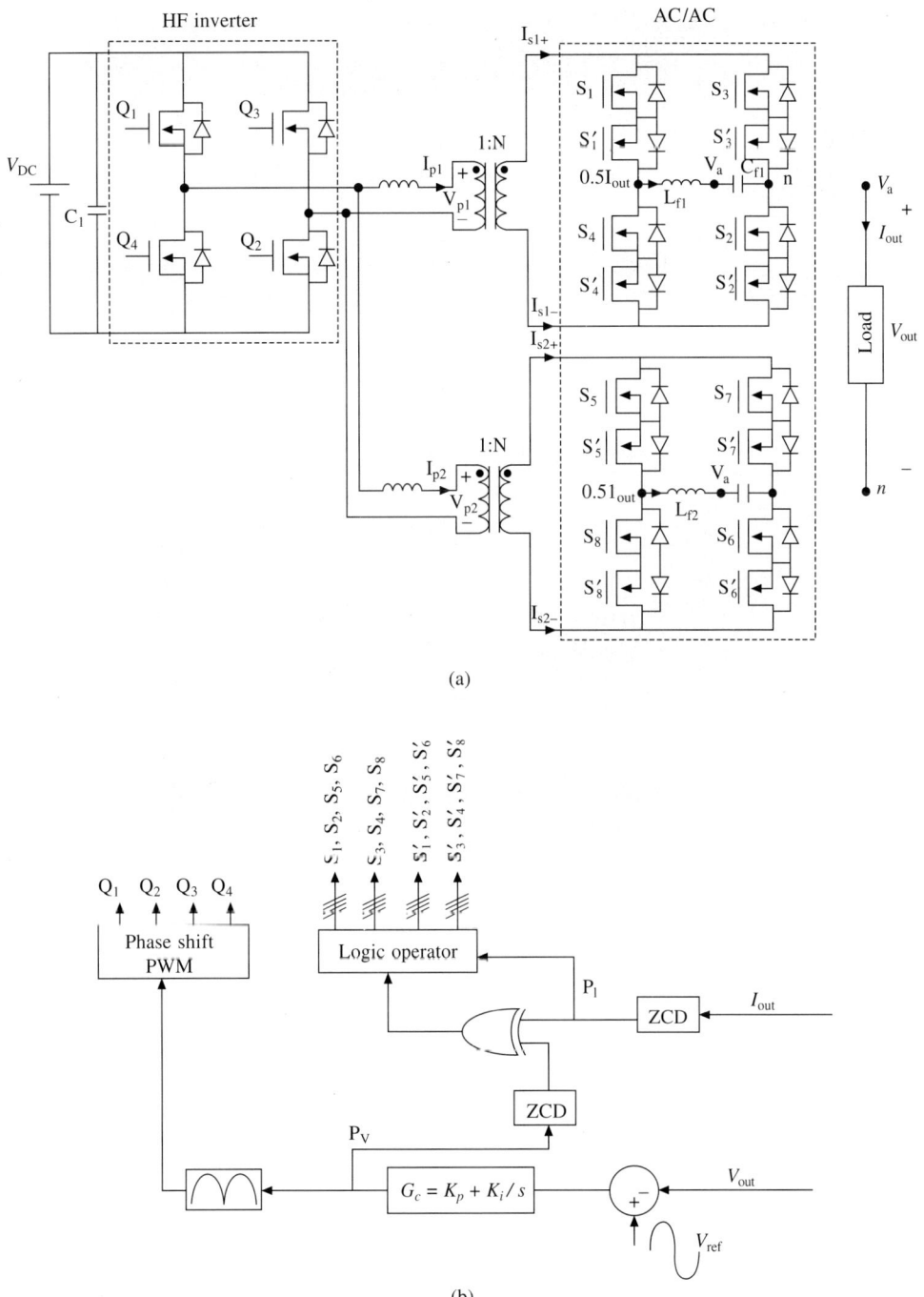

(a)

(b)

Fig. 18 Schematics of the (a) isolated phase-shifted high frequency inverter followed by a forced cycloconverter topology and (b) control scheme of the PES.

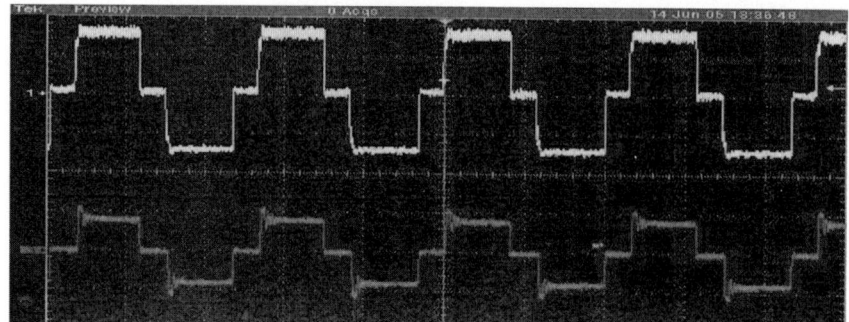

Fig. 19 HF transformer primary voltage (Top, 200 V/div) and HF transformer secondary voltage (Bottom, 500 V/div) for a transformer turns ratio of 1:2.4.

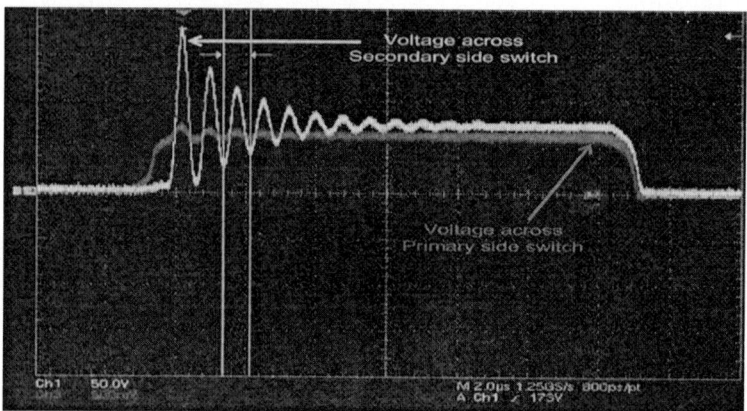

Fig. 20 Experimental results showing voltage spike across the switches of the primary and the secondary sides of the isolated phase-shifted high frequency inverter followed by a forced cycloconverter topology.

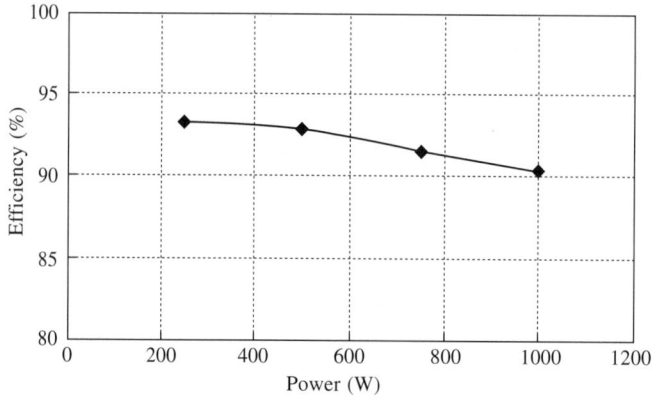

Fig. 21 Converter measured efficiency and THD final approach (possible variation: ± 1%).

4.2.1 Modes of Operation

Figure 22 shows the waveforms of the five operating modes of the phase-shifted HF inverter and a positive filter-inductor current. Modes 2 and 4 show the zero-voltage switching (ZVS) turn-on mechanism for switches Q_3 and Q_4, respectively. Unlike conventional control scheme for cycloconverter (Kawabata et al., 1990; Krein et al., 2004) which modulates the switches at high-frequency, the switches of the proposed cycloconverter are operated at the line frequency. The switches are commutated at high-frequency only when the polarities of the output current and voltage are different (Mazumder and Burra, 2005; Deng and Mao, 2003). For unity-power-factor operation, this duration is negligibly small and therefore, the switching loss of the AC/AC cycloconverter is considerably reduced compared to the conventional control method. A similar set of 5 modes exists for a negative primary current as well.

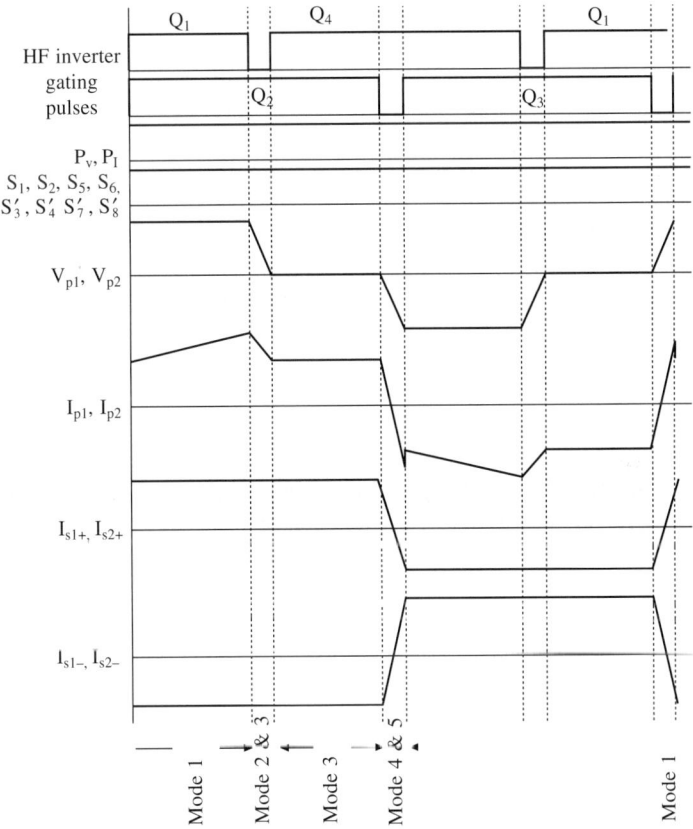

Fig. 22 Waveforms corresponding to the operating modes of the HF inverter when primary currents are positive.

Mode 1 (Fig. 23(a)): During this mode, switches Q_1 and Q_2 of the HF inverter are on and the transformer primary current I_{p1} and I_{p2} is positive. The load current splits equally between the two cycloconverter modules. For the top cycloconverter module, the filter-inductor current ($0.5 \times I_{out}$) is positive and flows through the switches pair S_1 and S'_1, the output filter L_{f1} and C_{f1}, switches S_2 and S'_2, and the transformer secondary. Similarly, for the bottom cycloconverter module, the filter inductor current ($0.5\, I_{out}$) is positive and flows through the switches pair S_5 and S'_5, the output filter L_{f2} and C_{f2}, switches S_6 and S'_6, and the transformer secondary.

Mode 2 (Fig. 23(b)): At the beginning of this interval the gate voltage of the switch Q_1 undergoes a high-to-low transition. As a result, the output capacitance of Q_1 begins to accumulate charge and at the same time the output capacitance of switch Q_4 begins to discharge. Once the voltage across Q_4 goes to zero, it can be turned on under ZVS. The transformer primary current I_{p1} and I_{p2} and the load current I_{out} continue to flow in the same direction. This mode ends when the switch Q_1 is completely turned off and its output capacitance is charged to V_{DC}.

Mode 3 (Fig. 23(c)): This mode initiates when Q_1 turns off. The transformer primary current I_{p1} and I_{p2} is still positive, and free wheels through Q_4 as shown in Fig. 23(c). Also the load current continues to flow in the same direction as in Mode 2. Mode 3 ends at the commencement of turn off Q_2.

Mode 4 (Fig. 23(d)): At the beginning of this interval, the gate voltage of Q_2 undergoes a high to low transition. As a result of this, the output capacitance of Q_2 begins to accumulate charge and at the same time, the output capacitance of switch Q_3 begins to discharge as shown in Fig. 23(d). The charging current of Q_2 and the discharging current of Q_3 together add up to the primary current I_{p1} and I_{p2}. The transformer current makes a transition from positive to negative. Once the voltage across Q_3 goes to zero, it is turned on under ZVS. The load current flows in the same direction as in Mode 3, but makes a rapid transition from the bidirectional switches S_1 and S_1' and S_2 and S_2' to S_3 and S_3' and S_4 and S_4' and during this process the filter-inductor current splits between the two legs of the cycloconverter modules as shown in Fig. 23(d). Mode 4 ends when the switch Q_2 is completely turned off and its output capacitance is charged to V_{DC}. At this point, it is necessary to note that since S_1 and S_2 are off simultaneously, each of them support a voltage of V_{DC}.

Mode 5 (Fig. 23(e)): This mode starts when Q_2 is completely turned off. The primary current I_{p1} and I_{p2} is negative, while the load current is positive as shown in Fig. 23(e).

4.3 Zero-Input-Ripple Boost Converter followed by an Isolated Multilevel Inverter and a Forced Cycloconverter

The topology described in this section is a slight variation of the PES described in the previous section. Such a topology is suitable for high power applications where the intermediate high voltage DC bus. Fig. 24 shows the power-conditioning system (PCS), which comprises a fuel-cell powered DC/DC zero-ripple boost converter (ZRBC), which generates a high voltage DC at its output, followed by a soft-switched, isolated DC/AC inverter, which generates a 110 V AC. The high-frequency inverter switches are arranged in a multilevel fashion and are modulated using a fully-rectified sine wave to create a high-frequency, three-level AC voltage as shown in Fig. 24(b). Multilevel arrangement of the switches is particularly useful when the intermediate DC voltage is >500 V. The high-frequency inverter is followed by the AC/AC cycloconverter which converts the three-level AC to a voltage that carries the line frequency sinusoidal information (as illustrated in Fig. 24(c)).

4.3.1 Zero-Ripple Boost Converter (ZRBC)

The ZRBC is a standard non-isolated boost converter, with the conventional inductor replaced by a zero-ripple inductor (ZRI). The secondary winding of the ZRI is connected to an external trimming inductor and a filter capacitor as shown in Fig. 24. Unlike the boost converter, the input current in a ZRBC is split between the two windings of the ZRI, as shown in Fig. 25. The primary winding conducts the DC current while the secondary winding conducts the AC current, thereby, making the output current of the fuel cell ripple free.

4.3.2 Multilevel Inverter Followed by the Forced Cycloconverter

Figure 26 shows the waveforms of the five operating modes of the HF converter for a positive primary current. Modes 2 and 4 show the ZVS turn-on mechanism for switches S_4 and S_3, respectively. The AC/AC stage comprises a single-phase cycloconverter and an output LC filter. The cycloconverter has two bi-

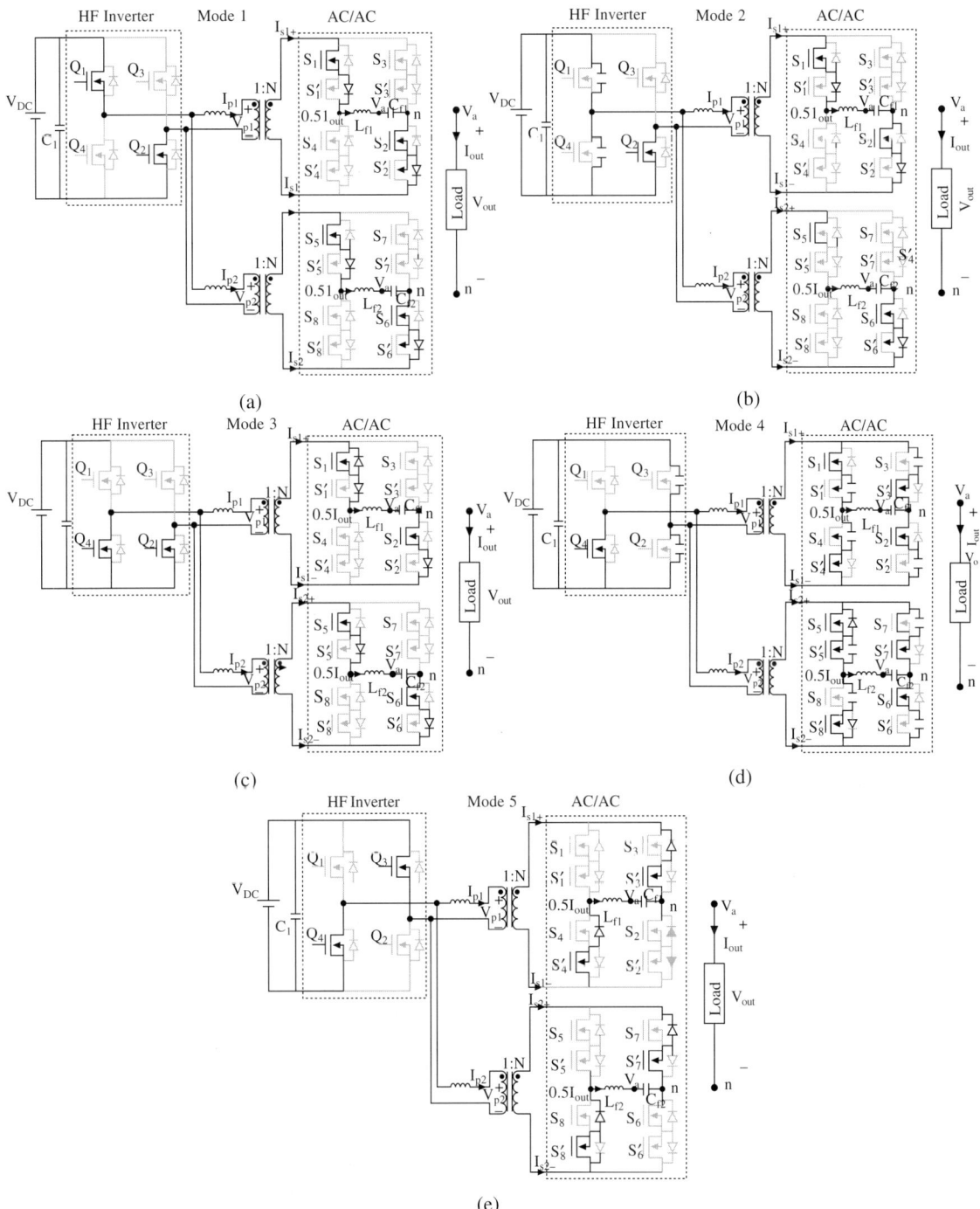

Fig. 23 Topologies describing the five operating modes of the overall DC/AC converter for positive primary current and for power flow from the input to the load; the dark lines indicate the direction of current flow.

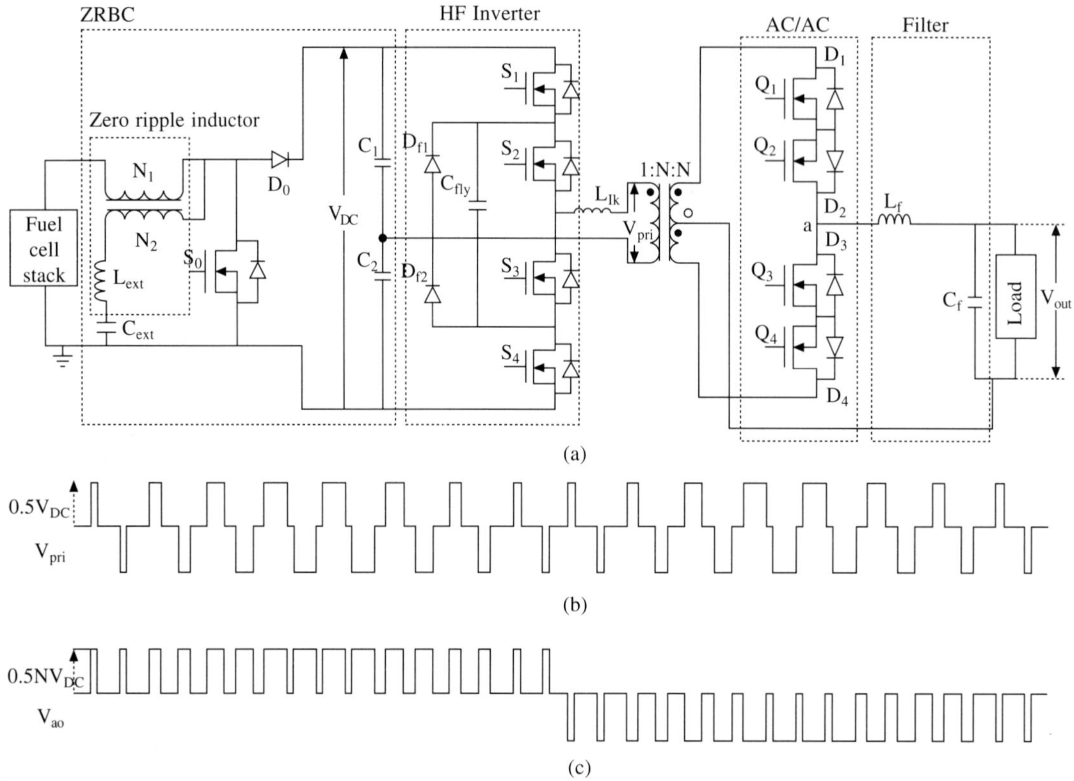

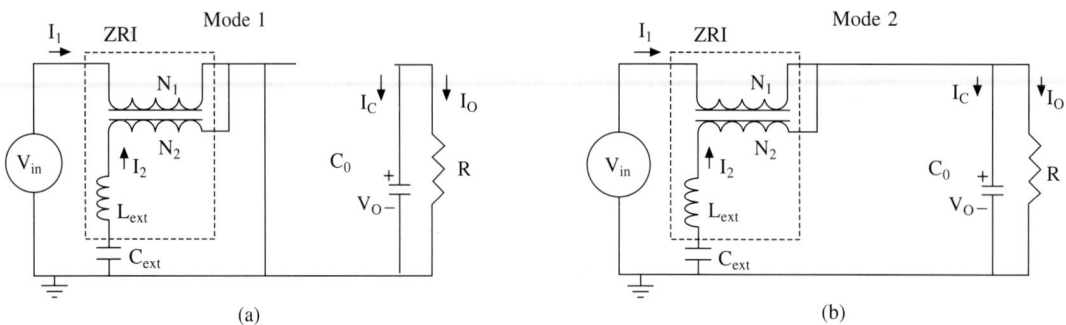

Fig. 24 Schematic of the proposed PCS (Mazumder and Burra, 2005; Mazumder et al., 2003).

Fig. 25 Topological modes of the ZRBC when (a) S_0 is turned on (b) S_0 is turned off.

directional switch pairs Q_1 and Q_2 and Q_3 and Q_4 for single-phase outputs. Unlike conventional control scheme (Kawabata et al, 1990; Krein et al, 2004), which modulates the switches at HF, the proposed cycloconverter operates at line frequency (Mazumder and Burra, 2005). The switches are commutated at high frequency only when the polarities of output current and voltage are different as shown in Fig. 27. Usually this duration is very small and therefore, the switching loss of the AC/AC cycloconverter is considerably reduced compared to the conventional control method.

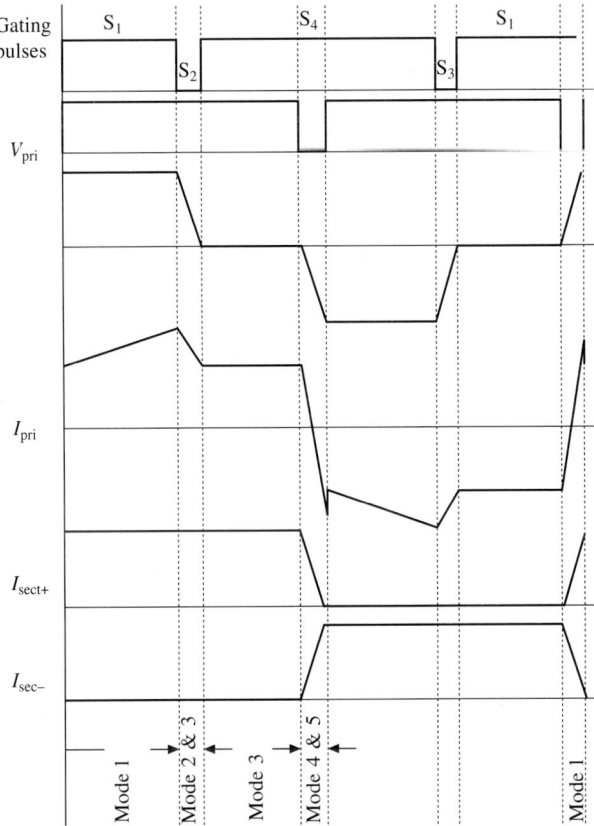

Fig. 26 Schematic waveforms show the operating modes of the
HF inverter operation when primary current is positive.

Modes of Operation

Fig. 28 shows 5 modes of the DC/AC converter operation. A set of five modes exists for a negative primary
current as well.

ZVS Range of the HF Inverter

The SPWM results in the loss of ZVS for each switch, twice in every line cycle. The extent of loss of ZVS
is a function of the output current I_{out} (Eq. (1)) and the load power factor (cos ϕ). Eq. (2) is the expression
of the available ZVS range (shown in Fig. 29) as a percentage of the line cycle.

$$I_{out} = \frac{\sqrt{2}\, P_{in}}{V_0 \cos \phi} \tag{1}$$

$$\frac{t_{ZVS}}{t_{Line\,cycle}} = \frac{2}{\pi} \sin^{-1}\left(\left(\frac{1}{4}\,\frac{V_{in}^2\left(\frac{4}{3}\,C_{oss} + \frac{1}{2}\,C_T\right)}{I_{out}^2\, L_{lk}}\right)^{1/2}\right) \tag{2}$$

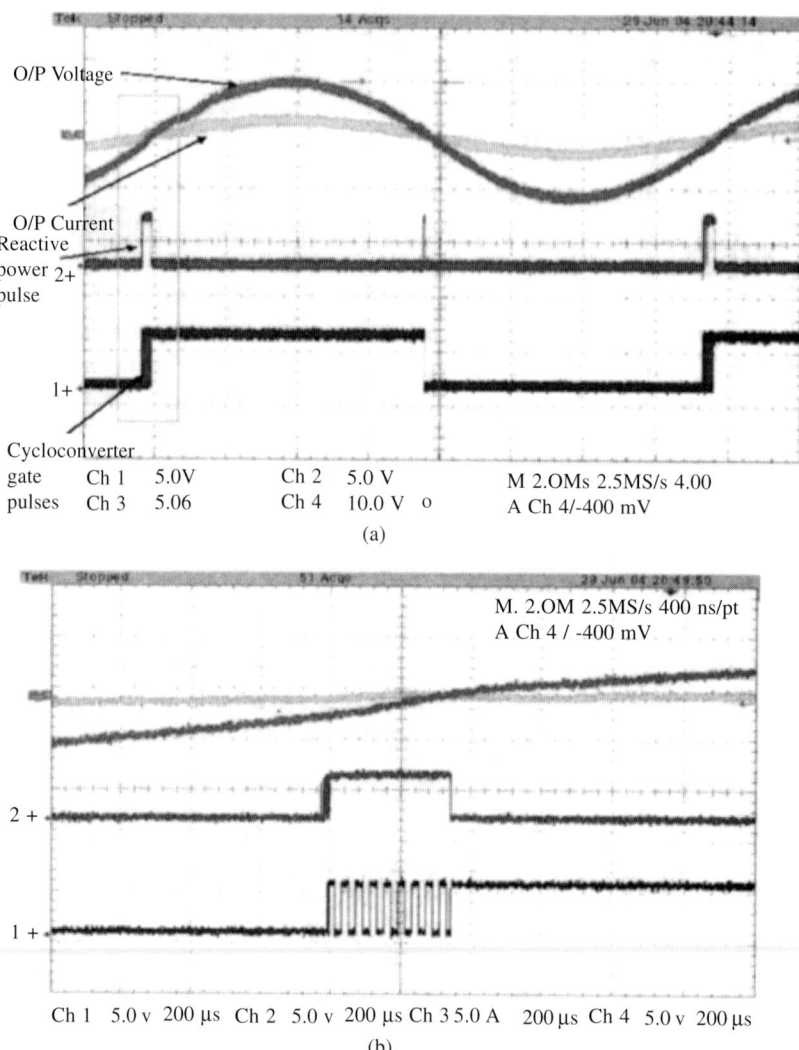

Fig. 27 Experimental illustration of the AC/AC converter operation. (a) PES output voltage and current (top), reactive power information (center), gating pulse for switch Q₁ (bottom) and (b) an expanded view of the rectangular portion of plot (a).

where P_{in} is the input power, V_0 is output voltage, V_{in} the input voltage, L_{lk} is leakage inductance of the primary transformer, C_{oss} the effective output capacitance of the MOSFETs of the HF inverter, C_T is equivalent parasitic capacitance of the HF transformer and $t_{Linecycle}$ the line-cycle time period (e.g. for a 60 Hz line frequency it is 16.67 ms).

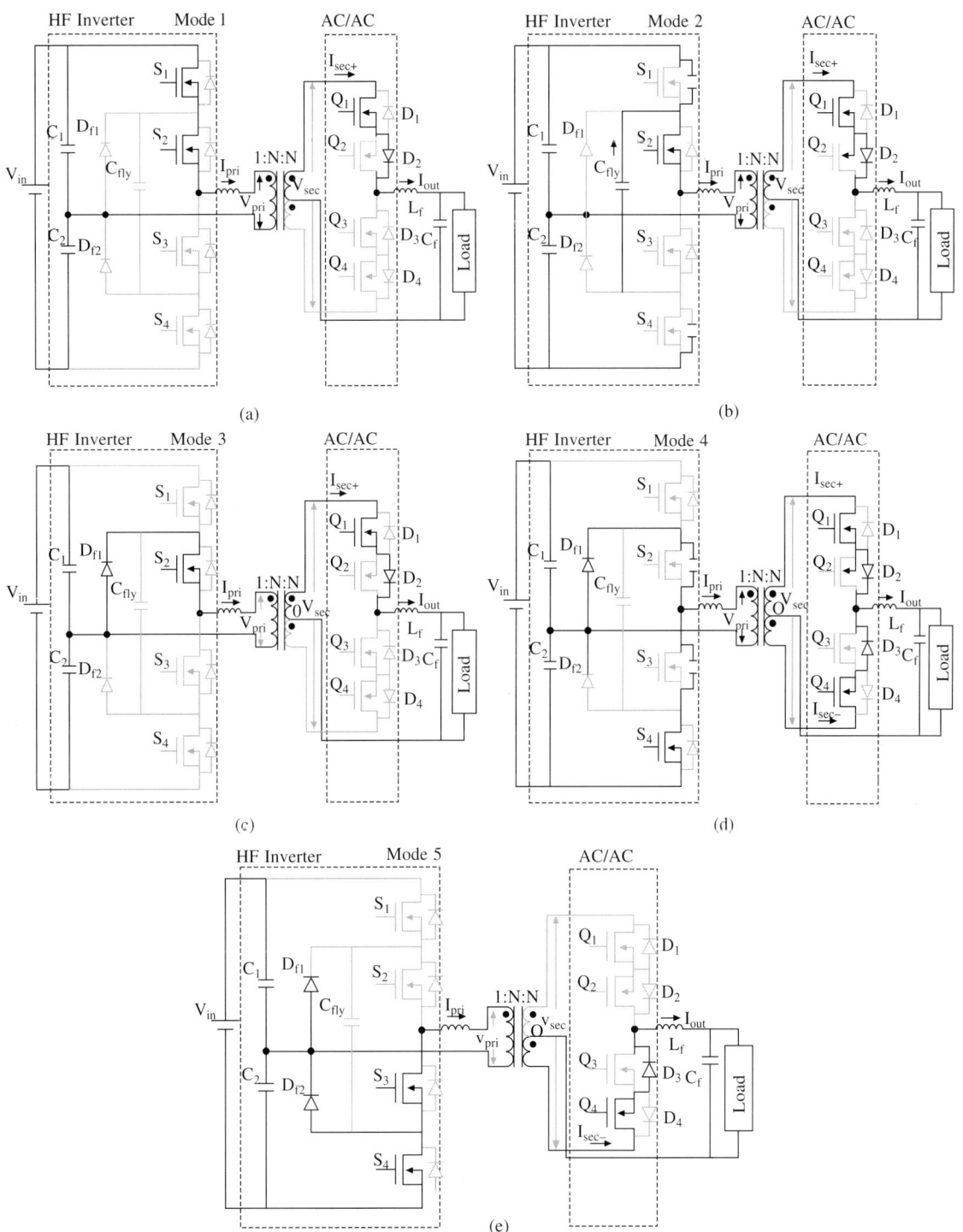

Fig. 28 Topologies corresponding to the five operating modes of the DC/AC converter for positive primary current and for power flow from the input to the load.

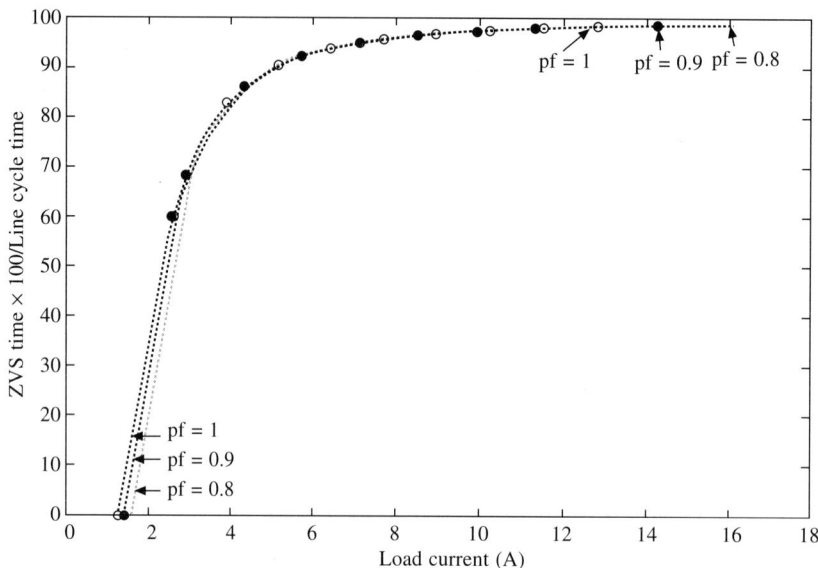

Fig. 29 Available ZVS range for the HF inverter switches (expressed as a percentage of the line cycle time period) with variation in load current. Also the effect of variation of the power factor for same average output power.

Acknowledgements

The work described in this chapter is supported in part by the California Energy Commission (CEC) under Award No. 53422A/03-02. A non-provisional U.S. patent has been filed by the University of Illinois at Chicago on the topology described in Section IV.C of this chapter. The presented work is also supported in part by the U.S. Department of Energy (DOE), under Award No. DE-FC2602NT41574. However, any opinions, findings, conclusions, or recommendations expressed herein are those of the authors and do not necessarily reflect the views of the CEC and DOE.

References

http://www.eia.doe.gov/

R.W. Erickson and D. Maksimovic, *Fundamentals of Power Electronics*, Springer, 2nd edition, January 2001.

N. Mohan, T.M. Undeland and W.P. Robbins, *Power Electronics: Converters, Applications, and Design*, Wiley, 3rd edition, October, 2002.

P.T. Krein, *Elements of Power Electronics*, Oxford University Press, September, 1997.

J.G. Kassakian, M.F. Schlecht and G.C. Verghese, *Principles of Power Electronics*, Prentice Hall, July 1991.

J. Lai, "Power electronic technologies for fuel cell power systems", *Proceedings of the Sixth Annual SECA Workshop*, April 18-21, 2005, also available at http://www.netl.doe.gov/publications/proceedings/05/SECA_Workshop/ SECAWorkshop05.html.

R.D. Middlebrook and S. Ćuk, *Advances in switched-mode power conversion*, vols. I and II, TESLACO, 1983.

A. Weinberg and J. Schreuders, "A high power high voltage DC-DC converter for space applications", *IEEE Power Electronics Specialists Conference*, pp. 317-329, 1985.

F. Canales, P. Barbosa, and F.C. Lee, "A zero-voltage and zero-current switching three-level DC/DC converter", *IEEE Transactions of Power Electronics*, vol. 17, no. 6, pp. 898-904, November 2002.

J.R. Pinheiro and I. Barbi, "The three-level ZVS-PWM DC-to-DC converter", *IEEE Transactions on Power Electronics*, vol. 8, no. 4, pp. 486-492, October 1993.

Y. Jin and P. Enjeti, "A high frequency link direct dc-ac converter for residential fuel cell power systems", *IEEE Power Electronics Specialists Conference*, pp. 4755-4761, June 2004.

T. Kawabata, H. Komji, K. Sashida et al., "High frequency link DC/AC converter with PWM cycloconverter", *IEEE Industrial Application Society Conference*, pp: 1119-1124, 1990.

J. Mazumdar, I. Batarseh and N. Kutkut et al., High frequency low cost DC-AC inverter design with fuel cell source for home applications, *IEEE Industry Applications Conference*, 2002, pp. 789-794.

P.T. Krein, R.S. Balog and X. Geng, "High-frequency link inverter for fuel cells based on multiple carrier PWM", *IEEE Transaction of Power Electronics*, vol. 19, no. 5, pp. 1279-1288, September 2004.

S.K. Mazumder and R.K. Burra, "Fuel cell power conditioner for stationary power system: towards optimal design from reliability, efficiency and cost standpoint", Keynote Lecture, *ASME Third International Conference on Fuel Cell Science, Engineering and Technology*, FUELCELL2005-74178, May 23-25, 2005.

S.K. Mazumder, K. Acharya, C. Haynes, R. Williams, M.R. von Spakovsky, D. Nelson, D. Rancruel, J. Hartvigsen and R. Gemmen, "Solid oxide fuel cell performance and durability: resolution of the effects of power-conditioning systems and application loads", *IEEE Transactions on Power Electronics*, vol. 19, no. 5 , pp. 1263-1278, 2004.

R. Gemmen, "Analysis for the effect of inverter ripple current on fuel cell operating condition", *Proceedings of IMECE: ASME International Mechanical Engineering Congress and Exposition,* 2001.

J.A. Pomilio and G. Spiazzi, "Soft-commutated Ćuk and SEPIC converters as power factor preregulators", *20th International Conference on Industrial Electronics, Control and Instrumentation*, pp. 256-261, 1994.

S. Deng and H. Mao, "A new control scheme for high-frequency link inverter design", *IEEE Applied Power Electronics Conference and Exposition*, pp. 512-517, 2003.

S.K. Mazumder, R.K. Burra and K. Acharya, "Novel efficient and reliable dc/ac converter for fuel cell power conditioning", *Non-provisional patent application No. 60/501,955,* September 2003.

Recent Trends in Fuel Cell Science and Technology
Edited by S. Basu
Anamaya Publishers, New Delhi, India

14. Future Directions of Fuel Cell Science and Technology

Suddhasatwa Basu

Department of Chemical Engineering, Indian Institute of Technology Delhi,
New Delhi-110016, India

1. Case for Hydrogen Energy!

Future development and implementation of fuel cell technology would depend on upward trend in global oil price, depletion of oil wells, fall in oil well discovery and the improvement of hydrogen energy infrastructure. The concern for environmental pollution and damage from the emission of automobile, thermal power plant, petroleum-crude refinery would catalyse the process of development unless financial benefits in terms of lowering of pollution damage cost are perceived by users and the manufacturers. Infrastructure development of the hydrogen energy encompasses production, distribution, dispensing and safety regulations of fuels (e.g., hydrogen, alcohol, esters and natural gases, naphtha and synthesis gases), which is directly fed to the fuel cells or to the fuel processor. Out of these hydrogen and alcohols can be generated from renewable sources (wind, solar power in water electrolysis, biomass gasification and fermentation) and others including hydrogen and alcohol can also be generated from fossil fuel. In the former case, green house gas emission is much lower and almost negligible. One can dream of zero emission of air pollutant and green house gases from automobiles and stationary power plants except for the case of biomass gasification. In the latter case, the air pollutant will be generated in a centralized location and cities will be free of pollution, which is otherwise generated from automobile using internal combustion engine. It should be noted that the hydrogen fuel cell vehicle (H_2FCV) and H_2FCV-hybrid electric vehicle offers least environmental damage among all the advanced options. When fuelled with hydrogen derived from natural gas, pollution damage costs are 1/8 as large as for today's gasoline internal combustion engine vehicles without CO_2 sequestration and 1/15 as large with CO_2 sequestration (Ogden et al. 2004). Although economics does not work out at present for PEMFC (Proton Exchange Membrane Fuel Cell) based automobile or SOFC (Solid Oxide Fuel cell) based stationary power plant with present inadequate hydrogen infrastructure but it is hoped that with the increase in crude price, no new crude or gas reserve findings, increase in fuel cell stack efficiency and decrease in cost of the fuel cell and improvement of hydrogen energy infrastructure facility, the Fuel Cell Vehicle (FCV) and distributed power generation from fuel cell will become more profitable leaving aside the cost benefits due to less environmental pollution. Optimists are looking at 20% use of FCV in world wide, 10% share of the domestic power generation from fuel cell source and 50% share of portable electronic equipment powered by fuel cell by 2020.

Let us check the reality of such a situation in terms of past and present data of fossil fuel discoveries, crude oil demand and price fluctuations and model predictions for the future (Sørensen, 2005). Table 1 shows that there is hardly any new crude well discoveries in recent times whereas its consumption is increasing exponentially on the face of rather disturbing oil price fluctuation coupled with political instabilities in most of the oil producing nations of the world. In such a situation, the options for hydrogen energy are beneficial to humanity.

Table 1. Trend in crude discoveries, consumption and price with time

Crude oil discoveries	Year	G bbl/Year
	1960	60
	1980	30
	2000	10
	2020	5
Crude oil consumtion	Year	$\times 10^{19}$ J/year
	1960	25
	1980	120
	2000	160
	2020	200
Crude oil price	Year	US$/bbl (5.74 GJ)
Note: crude price is highly	1960	20
fluctuating in last four years	1980	55
(2001-2005); average being	2000	30
US$ 30/bbl. In 2005, it was	2020	65-70
US$ 55/bbl		

bbl: barrels of oil equivalent

2. Question on Well-to-Wheel Efficiency?

Some investigators criticize that the well-to-wheel efficiency of hydrogen energy is much lower than well-to-wheel efficiency of conventional crude (Wald, 2004). Recent publications of National Research Council of USA (2004) on hydrogen economy and Wang (2005) on well-to-wheel efficiency suggest that CNG and diesel based internal combustion engine (ICE) and hybrid electric vehicle (HEV), hydrogen fuel cell vehicle (H_2FCV), H_2FCV-HEV give more well-to-wheel energy than any other types of vehicle. The details of the energy consumption in different type of passenger cars are given in Table 2.

Table 2 shows that diesel ICE-HEV, CNG-ICE-HEV, H_2FCV and H_2FC HEV have similar range of energy consumption but from the point of view of the green house gas emission H_2FCV and H_2FC HEV is much less polluting than diesel ICE-HEV and CNG-ICE-HEV. Typical carbon dioxide emission from

Table 2. Energy consumption in different types of passenger cars

Type	Energy consumption, kJ/km		
	Well-to-tank	Tank-to-wheel	Well-to-wheel
Gasoline ICE	694	2777	3471
Diesel ICE	377	2314	2691
CNG ICE	423	2834	3257
H_2ICE	1424	2136	3560
Gasoline ICE-HEV	479	1915	2394
Diesel ICE-HEV	251	1543	1794
CNG-ICE-HEV	241	1615	1856
H_2ICE-HEV	1102	1653	2755
H_2FCV	771	1157	1928
H_2FC-HEV	740	1111	1851

different types of passenger car is given in Fig. 1. Carbon dioxide emission is highest in gasoline ICE and next to it are CNG-ICE, H$_2$-ICE and diesel ICE and gasoline ICE-HEV. Thus the case for future implementation of H$_2$FCV and H$_2$FC HEV is quite favourable. The details of relative emission of SO$_x$, NO$_x$, CO, VOCs and particulate matter from different types of vehicles are summarized in Wald (2004). In the context of stationary power generation, it is well known that combined heat and power from PEMFC and SOFC stationary power plant or hybrid coal derived carbon fuel cell and gas based turbine generators is more efficient than common thermal power plant. If we can use natural gas or naphtha as fuel (which is possible in the case of SOFC) the efficiency of fuel cell technology will be way higher than the ICE technology or coal based thermal power plant (Larminie and Dick, 2003).

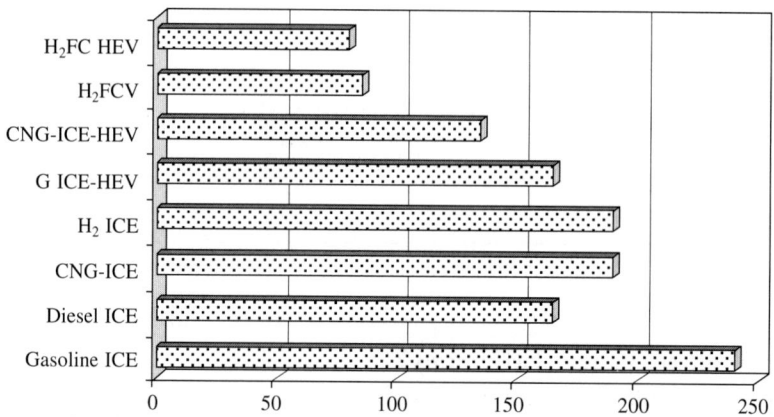

Fig. 1 Carbon dioxide emission from different types of passenger cars.

It is now evident from the above discussion that the development of fuel cell science and technology would bring about a sea change in energy conversion technology with higher conversion efficiency and low cost and lower detrimental environment impact. Before going to the requirement for future development and strategies to be followed in fuel cell technology, let us discuss the viable fuel cell technologies and their potential applications.

3. Choice of Fuel Cell, Fuel and Its Application

3.1 Proton Exchange Membrane Fuel Cell (PEMFC)
PEMFC is an ideal case for vehicle application, low load power generation for portable application as well as for building complex and blocks (utility market in the form of combined heat and power). Hydrogen is the energy carrier (or fuel) for PEMFC, which can be produced through different pathways, e.g., utilizing solar/wind power in electrolysis of water to hydrogen production, biomass to synthesis gas to hydrogen, coal to hydrogen with carbon dioxide sequestration, natural gas to hydrogen with separation of unwanted gases, nuclear/thermo-chemical reaction route to hydrogen. In the initial implementation stage, fixed route vehicles: buses, trucks, delivery vans, should be tried. ICE/FCV or FC-HEV should be the starting point of implementation to have smooth transition from ICV to FCV in terms of economy, infrastructure, technology development and safety.

3.2 Direct Alcohol Fuel Cell (DAFC) and Direct Alcohol Alkaline Fuel Cell (DAAFC)
DAFC is constructed based on proton exchange membrane (PEM) or anion exchange membrane (AEM). DAFC and DAAFC are suitable for power provider to portable electronics equipments e.g. laptops, mobile

phones, video pods, camcorder replacing the exiting use of Li-ion batteries with possible commercialisation in near future. Note that the market for portable application is increasing by leaps and bounds with the advancement of micro-electronics, micro-fluidics, micro-chemical plants and a lab-on-a-chip. Thus, demand for small portable power pack will increase and portable fuel cell is a good option as the duration of power supply is much higher than Li-ion battery. As for example, Li-ion battery lasts 4 to 6 h in laptop application whereas alcohol-based fuel cell would be able to provide uninterrupted power supply of 12 to 15 h. The primary candidates as fuel in DAFC and DAAFC are methanol, ethanol, formic acid, esters and sugar with the improved catalysts and membrane technology. Note that ethanol can provide energy density of 7.44 kWh/kg (5.9 kWh/l), whereas that for methanol and hydrogen is 6 kWh/kg (5 kWh/l) and 33 kWh/kg (2.77×10^{-3} kWh/l at atmospheric condition).

3.3 Molten Carbonate Fuel Cell (MCFC) and Phosphoric Acid Fuel Cell (PAFC)

MCFC and PAFC would be used as stationary power supply where grid electricity is not available due to geographical constraints, defence application and mobile power pack. PAFC was commercialised (200 kW) by UTC Fuel Cells with approximately 250 installations at subsidized rates. PAFC production was stopped recently because it is not economically viable (\$ 4500/kW). MCFC is in advanced stage of commercialisation. Fuel is for MCFC synthesis gas and natural gas. On the similar line investigators are working on direct carbon fuel cell which can take coal derived fuel and operates at lower temperature than SOFC. MCFC and PAFC may be a temporary phenomenon in the evolution of fuel cell technology development.

3.4 Solid Oxide Fuel Cell (SOFC)

SOFC technology should be used in stationary power generation with high load and combined heat and power for higher efficiency. Eventually SOFC might be used in powering submarines, aeroplanes, ships and automobiles; but it is far from reality in terms of cost and scaling up issues. Auxiliary power unit (APU) for heavy vehicles, fork lift is an immediate future application. Primarily hydrogen has been looked as carrier of energy but as SOFC operates at high temperature, internal reforming is possible. Thus, natural gas or syngas from fossil fuel or biofuels can be used as fuel in SOFC.

The targets in terms of technology and cost in fuel cell development are itemized below.

4. Technology and Cost Targets

4.1 PEMFC

Fuel cell vehicle (FCV) with PEMFC technology should run without any major maintenance for a minimum of 5000 hours (currently achieved 1500 hours). Only change in MEA in the fuel cell stack may be required during servicing of the vehicle. It is like changing lube oil in internal combustion engine of a vehicle. The complete change of fuel cell stack may be required after 15,000 hours of operation. Higher reliability of PEMFC stack is very important for automobile application. In stationary application it is possible to have back-up arrangement but not in automobile. Twin stack arrangement may be possible if weight of the stack is considerably decreased from its present value 3.77 kg/m^2 of electrode area. Cost targets for automobile application are: (a) stack system cost: US\$ 70-100/kW (currently, 900 US\$/kW), (b) fuel cost: US\$ 1-1.5/kg (current: US\$ 2.5-3.5/kg), (c) energy cost (operations and maintenance): less than US\$ 0.03/kWh. The target in terms of well-to-wheel efficiency is 40%-50% (currently around 23% for FCV and 17% for ICV). There is a good scope for improvement in PEMFC based FCV performance. The target and improvement in PEMFC stack performance year-wise is shown in Fig. 2 (UTC). The learning curve from Fig. 2 gives the possible trend in future development of PEMFC. The cost target of PEMFC system used

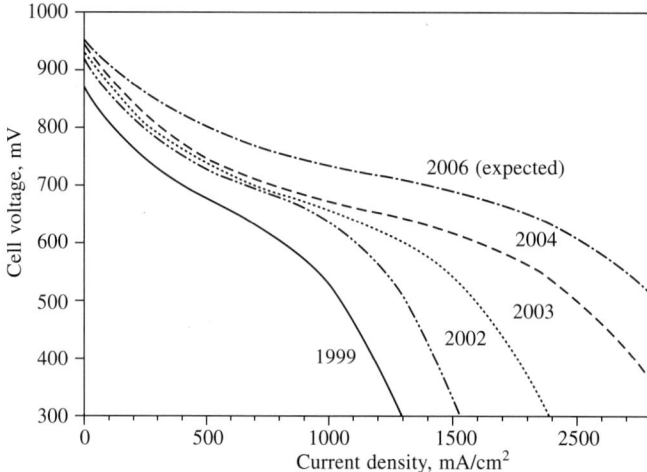

Fig. 2 Improvement in cell stack performance year-wise.

for stationary application is US$ 200-700/kW and design life of 40,000 hours with less than 10% degradation. The cost targets and projections for fuel cell systems are highly variable of capacity of the fuel cell and number of units produced. These targets are well acknowledged by most of the organization in the world working on fuel cell stack development. However, the target set by Ballard is quite stringent and optimistic in nature (Ballard News Release 2005). The target for current density in Ballard PEMFC stack at 50% no load is 3000 mA/cm^2 and the currently achieved target is 1000-1600 mA/cm^2. Similarly, power density target of DOE is 2000 watt net/kg whereas that for Ballard is 2500 watts net/kg (current status is 777-1205 watts net per kg). The target for hydrogen purity to be used in PEMFC is 40 ppm CO; currently it is 15 ppm. The freeze start targets are: in 30 seconds 90% power at –20°C (DOE), 30 seconds 50% power at –30°C (Ballard) (current status in 100 seconds to 50% power at –20°C). The year wise target achieved in start stop endurance test of FCV is presented later. The detailed data on the target of endurance test is available in Ballard New Release 2005.

4.2 DAFC/DAAFC

The Portable power supply to electronic equipment from DAFC (direct alcohol fuel cell) and DAAFC (direct alcohol alkaline fuel cell) should have minimum recharging interval of 10-12 hours. Recharging means changing of methanol and ethanol fuel cartridges. Cost targets are: (a) stack cost: US$ 150-200/kW, (b) fuel cost US$ 1-1.5/kg. Fuel Efficiency is 30-40%.

4.3 SOFC/MCFC

The life span target of power generation from SOFC or MCFC or DCFC (direct carbon fuel cell) without interruption of power supply is 10,000 hours (minor maintenance work allowable with no major shut down). Major shut down may be required after 40,000 to 50,000 hours of operation. Cost targets are: (a) stack cost: US$ 400/kW (Currently US$ 900/kW). Note, it is a utility service for community and industry, which should be sustainable and have higher reliability than automobile application, high load control and longer life. (b) fuel cost: US$ 1-1.5/kg. The electricity cost of the SOFC expected to be US$ 5-6/kWh. Fuel Efficiency is 50%-70% (LHV) with combined heat and power (CHP). The capacity of SOFC could be from watts to kilowatt to megawatts with combine turbine generator and the cost factor would change accordingly.

5. Further Scope and Improvement Required in Fuel Cell Technology

5.1 Proton Exchange Membrane Fuel Cell

1. Increase in operating temperature of PEMFC to 150°C from currently operating maximum temperature of 80°C would lead to multiple benefits like faster reaction kinetics, greater efficiency and increase in CO tolerance level in hydrogen fuel such that electrodes are not poisoned. The modification of proton exchange membrane is required such that it allows transfer of proton at the elevated temperature. It is necessary to prepare polymer structure with higher glass transition temperatures and which can solvate the mobile cations in a polar phase that contain the anions and the solvation groups. The solvation groups being researched world-wide include imidazoles, which is tethered to the polymer matrix (Schuster et al. 2005). It is required to carry out research to provide a total solid-state membrane with no leachable components. PolyFuel offers new membrane which works at a higher temperature without much deterioration to reactants and by-products forms (Ashley 2005). Other new membranes such as sulfonated poly (arylene ether sulfonate) (BPSH), sulfonated poly (arylene thioether sulfone) (PATS), sulfonated poly(imide) (sPI) copolymers were found to be as good as Nafion® if not better than that in many respect (Hickner and Pivovar, 2005). Recently, high temperature proton exchange polyimide electrolyte membranes having fluorenyl and sulfopropoxy groups with higher proton conductivity are tested by Zhou et al. (2005).

2. A completely new proposition and methodology in the membrane technology is welcome. PTFE may not have to be used as the backbone of the membrane as most of the manufacturer do. Can some ceramic membrane be developed as a substitute for PEM and used in fuel cell? Ceramic material and metal oxides are used in high temperature SOFC. Why not ceramic membrane be developed for PEM like fuel cell which may be operated at a lower temperature (200°C)? Impregnation of nano-metal oxide in membrane may help in proton transport. Membrane should have characteristics e.g., greater chemical and mechanical stability for higher durability, control over undesired side reactions, greater tolerance to contamination caused by fuel impurity or by-products, decreased electro-osmotic drag, increased conductivity of protons and no electron conductivity.

3. Water and thermal management issues should be handled with great care in new types of gas diffusion layer (GDL) e.g., carbon cloth or metallic-carbon cloth for better electron transfer and mass transport of the reactants and products. A well defined fluid flow path in GDL and electron transfer through solid metallic structure would reduce over potential losses.

4. GDL is key to the success of MEA (membrane electrode assembly) with higher efficiency. It should be such that activation over-potential at anode and mass transport over potential at cathode are minimal. GDL should keep a balance between water flooding condition (increase of mass transport resistance hence loss in voltage) and water drying condition (decrease in proton conductivity through membrane).

5. Designing of bipolar plate of high electrical (> 10 S/cm) and thermal (20 W/mK) conductivity material for excellent stack performance is the key to success of minimization of cost and higher efficiency as mentioned before. Traditionally investigators have been giving maximum importance to membrane and electrode-catalyst development. But following Fig. 3 on cost components of a typical PEMFC stack, it becomes obvious that commercialisation of PEMFC highly depends on development of cheaper and efficient bipolar plate and GDL. Towards this end, the

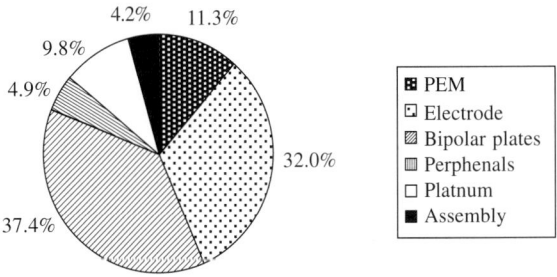

Fig. 3 Cost components of a typical PEMFC stack.

investigators are working on carbon fibre doped with metals, new carbon material and carbon nano-tubes.

6. Anode catalyst with less amount of noble metal in bi- or tri- metallic form and non-noble metal catalysts like Ru, Cr, Co, Ni and Sn should be tried keeping efficiency of fuel oxidation at same level. Recent development of different Pt-bimetallic or Pt-trimetallic anode catalyst with newer catalyst spreading technology with high surface area has cut down the cost drastically. However, our present goal should be to introduce CO tolerant anode catalyst, which will drastically reduce the hydrogen production cost. This is also possible by elevating the temperature of the cell (see item 1). Use of carbon nano tubes and nano-particles to increase the surface area/unit mass of catalyst (Rojas et al. 2005) should be taken up seriously for the better performance and lowering of cost of PEMFC.

7. Cathode catalyst should not produce hydrogen peroxide which may lead to increase in over potential at cathode.

The PEMFC is in pre-commercial stage of development and needs further improvement as mentioned earlier to become more competitive in terms of cost, reliability and efficiency. PEMFC applications can be considered in the niche market, particularly in the under 25 kW size, because in this size range PEMFC must compete with existing diesel genset (DG) technologies that have heating and cooling system applications and are reliable, durable, and low cost. If there is a sizeable market, DG could provide PEMFC manufacturing experience, enhancing the learning curve for PEMFC and hastening its automotive applications, which has much more stringent volume and cost requirements (NRC and NAE of National Academy, USA, 2004). The success of FCV hinges on hydrogen infrastructure development and light weight, compact, safe, inexpensive, quick to refuelling on board hydrogen storage system.

5.2 Direct Alcohol Fuel Cell

Most of issues related to future development of PEM technology as discussed in the previous section is applicable to direct alcohol fuel cell (DAFC). Few pertinent points peculiar to DAFC are mentioned below.

1. The most of the portable electronic equipments work at room temperature and in some cases at sub-zero temperature where electronic equipments are used for outdoor activities. The start up time for the stack should be few seconds (5-10 s) even in the sub-zero temperature. Although DAFC works at a relatively lower range of temperatures compared to other types of fuel cell. Higher operating temperature would give better reaction kinetics during fuel oxidation at anode, the operating temperature cannot be much higher than ambient temperature because of its application in portable electronic equipment.

2. In DAFC, main hurdle to be over come is the development of membrane such that it does not allow crossover of fuel (methanol, ethanol) through the membrane restricting the fuel oxidation at cathode and minimizing the over voltage losses.

3. Although most of the development on DAFC is taking place with methanol as fuel, ethanol is preferred as a fuel as it is renewable in nature and can be produced in large quantity from agro products through fermentation route. Development of new catalyst, which breaks C-C bond of ethanol at a lower temperature, would provide tremendous impetus towards the growth of direct ethanol fuel cell based on PEM technology or direct ethanol alkaline fuel cell. Ru-Ni catalyst shows promise towards this end (Tarasevich et al. 2005). Gupta et al. (2004) claimed breakage of C-C bond of ethanol in the presence of CuNiPt and CuNiPtRu alloys during electro-oxidation of ethanol in alkaline medium.

4. Development of anion exchange membrane some what in the similar line of proton exchange membrane for direct alcohol alkaline fuel cell, would facilitate commercialisation of fuel cell use in portable electronic equipments (Varcoe and Slade, 2005).

5.3 Solid Oxide Fuel Cell

1. Electrolyte should have excellent ionic conductivity with negligible electronic conductivity and zero porosity. SOFC developers and researchers are working on material that can provide high ionic conductivity at low temperature like 700°C or below. Lowering of temperature of SOFC from 1000 to 650°C and even to a lower value would be a big step towards the commercialisation of SOFC as stainless steel can be used in manufacturing of SOFC stack. This will significantly reduce the cost and simplify its operation at high temperature. While Yitria stabilized Zerconia (YSZ) remains as the electrolyte of choice, other material such as doped Lanthanum Gallate, doped Ceria, Sc-doped Zirconia are promising towards this end (Singhal and Mizusaki, 2005).

2. Researchers and product developers are geared towards development of ceramic/perovskite coatings for metallic interconnects for high temperature operation at 1000°C. Improved alloy composition may be looked at as metallic interconnect.

3. Gas seal in planar stack is a major issue to be addressed. High temperature operation and thermal cycling leads to the failure of gas sealing. Glass-metal composites may work as good sealing material.

4. Electrodes must have excellent electron conductivity, ionic conductivity, zero porosity and minimum chemical interaction with electrolyte material. Also, electrodes and electrolyte material should be compatible in terms of coefficient of thermal expansion. Major challenge is to optimise cathode microstructure, obtaining higher electronic conductivity and lowering of cathode overpotentials. Perovskites, e.g., Lanthanum Strontium Manganites (LSM), Lanthanum Strontium Cobaltite (LSCo), Lanthanum Strontium Ferrites (LSF), are good candidate for cathode material (Singhal and Mizusaki, 2005). Ni-YSZ, having good catalytic activity and electronic conductivity, are used as anode with most people fixated on the idea of fuel reforming to hydrogen as a first stage either internally or externally. But, redox and thermal cyclability of stacks raise major problems. Ni-YSZ is a poor anode when it comes to redox cycling and thus new anode materials need to be developed. Since direct use of hydrocarbon in SOFC is key to its success, the use of other catalysts like, Cu/CeO_2, La chromite, Sr-titanates and various other mixed metal oxides are emerging as anode material (Park et al. 2000; McInosh et al. 2004). Poisoning of SOFC electrodes by sulphur (hydrogen sulfide, mercaptans and thiophenes), nitrogen compounds and halides should be studied and anode which tolerates above poisoning compounds without affecting the electro-oxidation process should be developed.

5. Both tubular and planar design has its advantages and disadvantages. Tubular design gives better current density than planar design but its cost is higher. On the other hand, the planar design has the ability of utilizing traditional and commercially accepted ceramic processing techniques like tape casting, screen printing. However, they have been tested only for about 2-3 years. It is expected that the prices for SOFCs will reduce significantly for planar design, compared to other available designs. The main hurdle which planar design needs to overcome is the high temperature sealant, which separates both fuel and oxidant component. A right kind of glass-sealant development is one of the main focus of researchers. For anode-supported planar SOFC, availability of good ferritic steel interconnect is important.

6. Life span of SOFC having high efficiency need to be increased and capital costs need to be reduced. These can be improved only if pre-reforming is not necessary and the electrodes give considerably less over-potential. Basic performance requirements indicate that the stack should be stable under high thermal gradients within a cell, as reforming and oxidising is taking place in the same stack. Most of the SOFC producers are focusing on optimising the above parameters for 5-100 kW range.

In addition to SOFC as stand alone DG or in CHP system, SOFCs are being developed in an SOFC/gas turbine hybrid configuration. In the hybrid configuration, the fuel cell converts fuels eg., either direct hydrogen, syngas from fossil fuels, or biofuels, into electricity and water along with by-product heat. The residual fuel from the fuel cell is then burned by a gas turbine for additional electricity production. The product could have range from 1 to 10 MW with more than 65% efficiency and it could be cost effective compared to conventional coal or gas based thermal power station.

6. Engineering Issues

1. Stack engineering and control is the next biggest hurdle once the above things are taken care of. The biggest challenge is in the computer controlling of fuel cell stack for load variation with minimum response time, better stack design (improve in heat integration, less fuel flow resistance, by-products removal), lowering of stack mass per unit volume and cyclic endurance. The improvement of cyclic endurance year-wise is shown in Fig. 4 and it is to be perfected further.

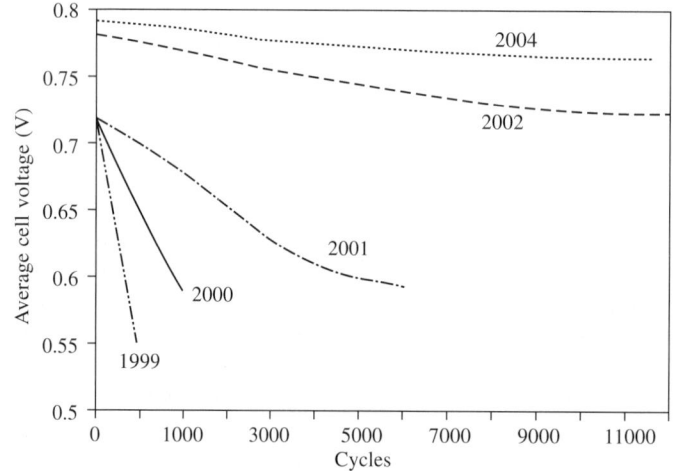

Fig. 4 Improvement of cyclic endurance year-wise.

2. Other challenges related to successful implementation of fuel cell technology are: DC to DC conversion through electronic transformer which increases current density by compensating through drop in voltage and power conditioner i.e. DC to AC conversion for utility services and domestic appliances.

Finally, the fuel cell vehicle or fuel cell for stationary power have to be mass produced and the development of automated mass production technology is very important in terms of bulk production, cost reduction and reproducibility of fuel cell performance. Most of the development work on MEA preparation is through hands-on experience and hand held technology and it is very difficult to reproduce MEA with similar performance as some amount of manual error always creeps in. Thus automation and development of mass production technology of MEA preparation and assembly line for stack production should be looked into very seriously especially for FCVs.

References

Ashley, S., On the road to fuel cell cars, Scientific American, March 2005, 50-57, (2005).

Ballard Power Systems Inc. Commercially Viable Fuel Cell Stack Technology Ready by 2010, News Release March 20, (2005).

Gupta, S.S., Mahapatra, S.S. and Datta, J., A potential anode material for the direct alcohol fuel cell, *J. Power Sources*, **131**, 169-174 (2004).

Hickner, M.A. and Pivovar, B.S., The chemical structural nature of proton exchange membrane fuel cell properties, Fuel Cells, **5**(2) 213-229, (2005).

Larminie and A. Dick, Fuel Cell Systems Explained, 2nd Ed John Wiley, p 371 (2003).

McIntosh, S., He, H., Lee S-I., Costa Nunes, O., Krishnan, V.V., Vohs, J.M. and Gorte, R.J., An examination of carbonaceous deposits in direct utilization SOFC anodes, *J. Electrochem. Soc.*, **151**(4) A604-A 608 (2004).

National Research Council (NRC) and National Academy of Engineering (NAE) of the National Academy USA, The Hydrogen Economy opportunities, costs, and barriers and R&D needs, The National Academies Press, Washington, (2004).

Ogden, J.M., Williams, R.H. and Larson, E.D. Societal lifecycle costs of cars with alternative fuels/engines, *J. Power Sources*, **31**(1), 7-27 (2004).

Park, S., Vohs, J.M. and Gorte, R.J., Direct oxidation of hydrocarbons in a solid oxide fuel cell, *Nature*, **404** (6775), 265, (16th March 2000).

Rojas, S., Garcia-Garcia, F.J., Martinez-Huerta, M.V., Fierro, J.L.G., Boutonnet, M., Preparation of carbon supported Pt and Pt-Ru nano-particles from micro-emulsion electro-catalysts for fuel cell application, Appl. Catal. A: Gen (in press).

Schuster, M., Roger, T., Noda, A., Kreuer, K.D., Maier, J., 'About the choice of the protogenic group in PEM separator materials for intermediate temperature, low humidity operation: A critical comparison of sulfonic acid, phosphonic acid and imidazole functionalised model compounds' Fuel Cells, in press (2005).

Singhal, S.C. and Mizusaki, J. (Eds.) Solid Oxide Fuel Cells, IX, Proc. International Symposium, Vol. 1, Cells, Stacks and Systems, (2005-07), ECS Publication.

Sørensen, B., Hydrogen and Fuel cells, Elsevier Academic Press, London p 387 (2005).

Tarasevich, M.R., Karichev, Z.R. and Bogdanovskaya, V.A., Kinetics of ethanol electrooxidation at RuNi catalysts, Electrochem. Commun, **7**, 141-146 (2005).

Varcoe, J.R., Slade, C.T., Prospect for alkaline anion exchange membrane in low temperature fuel cells, Fuel Cells **5**(2) 187-200, (2005).

Wald, M. L., Question about hydrogen economy, Scientific American, **290**(5), 66, (2004).

Wang, M.J. Power Sources, Fuel choices for fuel cell vehicles: well-to-wheels energy and emission impacts, **112**, 307-322 (2002).

Zhou, H., Miyatake, K. and Watanabe, M., Polyimide electrolyte membranes having fluoronyl and sulphopropoxy groups for high temperature PEFCs, Fuel Cells **5**(2) 296-301 (2005).

Index